D0146664

Bioorganic Chemistry:
Nucleic Acids

Topics in Bioorganic and Biological Chemistry
A Series of Books in Support of Teaching and Research

Series Editors:
Sidney M. Hecht, University of Virginia
Richard L. Schowen, University of Kansas

Bioorganic Chemistry:
Nucleic Acids

Edited by

Sidney M. Hecht

University of Virginia

New York   Oxford
OXFORD UNIVERSITY PRESS
1996

Oxford University Press

Oxford   New York
Athens   Auckland   Bangkok   Bombay
Calcutta   Cape Town   Dar es Salaam   Delhi
Florence   Hong Kong   Istanbul   Karachi
Kuala Lumpur   Madras   Madrid   Melbourne
Mexico City   Nairobi   Paris   Singapore
Taipei   Tokyo   Toronto

and associated companies in
Berlin   Ibadan

Library of Congress Cataloging-in-Publication Data
Bioorganic chemistry : nucleic acids / edited by Sidney M. Hecht.
        p. cm.—(Topics in bioorganic and biological chemistry)
Includes bibliographical references and index.
ISBN 0-19-508467-5
1. Nucleic acids.   I. Hecht, Sidney M.   II. Series.
QP620.B58 1996
574.87'328—dc20   95-21401

9 8 7 6 5 4 3 2 1

Printed in the United States of America
on acid-free paper

Preface

The explosive growth of bioorganic chemistry in the past two decades has extended the reach of organic chemistry into areas that were previously within the domain of biology. The participation of organic chemists in studies at the interface between chemistry and biology has enriched both disciplines and provided important insights into the workings of biological systems at a molecular level.

However, the breadth of studies now undertaken by organic chemists at this interface has created a problem in the education of future generations of practitioners, for the range of topics that might be thought to constitute a reasonable "core" of expertise has become quite large. As part of a series of books intended to support teaching and research in bioorganic and biological chemistry, I have edited a set of three books for organic chemists interested in biology. This first volume deals with nucleic acids; two more will appear shortly in the areas of peptides and proteins and of carbohydrates.

Although these volumes may prove to be of more general utility, they have been written specifically to support the teaching of graduate students in bioorganic chemistry. Each book is written as a set of chapters whose numbers approximate the number of weeks in a semester and whose subject has been identified as critical to an appreciation of ongoing research activity within the broader field summarized in that book. The chapters are organized in much the same fashion as lectures in special topics courses. Each chapter begins with an introduction that includes basic principles, a summary of key findings that underpin current research in the field, and an overview of current research activity. The remainder of each chapter deals in greater detail with a small number of recent studies that illustrate the nature of ongoing activity in the field. A complete set of figures for all chapters is available as overheads to facilitate classroom presentation.

Thanks are due to the authors for their time and effort in writing chapters tailored to the needs of a new type of volume and to my colleague Bob Ireland for numerous discussions that helped to shape the volumes. I thank Oxford University Press for the willingness to participate in an educational experiment and Bob Rogers, Senior Editor, for his help and advice. This volume was typed by Vickie Thomas, whose hard work and patience are greatly appreciated. Finally, I wish to acknowledge the enthusiastic participation of dozens of graduate and postdoctoral students at the University of Virginia, to whom I have already taught special topics courses in the areas of nucleic acids and peptides and proteins based on the first two volumes.

Charlottesville   S. M. H.
January 1996

Contents

Contributors

Kenneth J. Addess
University of California, Los Angeles

Serge L. Beaucage
Food and Drug Administration

Marvin H. Caruthers
University of Colorado at Boulder

Juli Feigon
University of California, Los Angeles

Laurence H. Hurley
University of Texas, Austin

Sergei A. Kazakov
SRI International

Christopher J. Larson
Harvard University

David M. J. Lilley
The University, Dundee

Eric C. Long
Indiana University-Purdue University,
Indianapolis

Larry W. McLaughlin
Boston College

Paul S. Miller
Johns Hopkins University

John A. Mountzouris
University of Texas, Austin

Maryanne J. O'Donnell
Boston College

Sam J. Rose
Behring Diagnostics

Francis J. Schmidt
University of Missouri

M. Sriram
University of Illinois

Thomas D. Tullius
Johns Hopkins University

Reuel B. Van Atta
Naxcor

Mark Van Cleve
Becton Dickinson Research Center

Gregory L. Verdine
Harvard University

Andrew H.-J. Wang
University of Illinois

Bioorganic Chemistry:
Nucleic Acids

Fundamentals of Nucleic Acids

Eric C. Long

Nucleic acids occupy a position of central importance in biological systems. Remarkably, even though based on relatively simple nucleotide monomers, these biopolymers partici- pate in an impressive array of complex cellular functions. For example, from the de- oxyribonucleic acid (DNA) duplex structure, genetic information is stored, accessed, and replicated as a linear nucleotide code. In partnership with DNA, ribonucleic acid (RNA) is an essential biopolymer which, among other functions, transports genetic information from DNA to the site of protein manufacturing, the ribosome (Watson et al., 1987).

The flow of genetic information alluded to above, DNA *transcribed* into RNA which is ultimately *translated* into a protein, constitutes the so-called central dogma of mo- lecular biology originally set forth by Francis Crick (Fig. 1-1) (Crick, 1970). This scheme implies the presence of a hierarchy among biomolecules in which nucleic acids alone may store information or specify the sequence of protein products. In contrast, proteins are never able to specify a particular nucleic acid or protein sequence. While the basic tenets of the central dogma have withstood the test of time and experimen- tation, certain additions have been necessary. These include the occurrence of RNA- directed RNA synthesis and RNA-directed DNA synthesis (reverse transcription), as found in some viruses and plant species (Weiss et al., 1985).

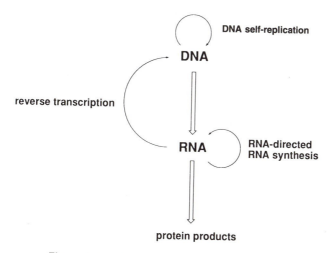

Figure 1-1. The central dogma of molecular biology.

The foregoing comments underscore the importance of nucleic acids in the processes that permit life as we know it and, perhaps, in the origin and evolution of life itself. Recent findings that indicate increasingly complex biological roles for RNA (Altman et al., 1986; Cech & Bass, 1986; Sharp, 1987) have fueled speculation that RNA may have been the primary self-replicating molecule from which life originated (Benner et al., 1989; Gesteland & Atkins, 1993). In this primordial "RNA world," RNA polymers not only served as the storehouse of genetic information, but also participated in the catalysis of their own replication and chemical manipulation. Eventually, this RNA-based system evolved such that the chemically more stable DNA molecule was used for information storage, and the diverse chemical properties of amino acid side chains were employed to mediate an expanded repertoire of chemical reactions, resulting in the development of proteinaceous enzyme catalysis.

Given their importance, it should not be surprising that nucleic acids constitute a primary target for binding or chemical modification by several classes of molecules. These agents can take the form of gene regulatory proteins which are necessary to repress or stimulate the natural flow of genetic information through DNA or RNA (Ptashne, 1987; Steitz, 1990). Alternatively, low molecular weight species from extracellular sources may also artificially alter or inhibit the activities of RNA or DNA. These exogenous agents can be based on organic (Bailly & Henichart, 1991; Dervan, 1986; Nielsen, 1991) or inorganic species (Pyle & Barton, 1990) and may alkylate, associate noncovalently, or induce the strand scission of nucleic acids. Such molecules, accessible from either natural sources or by synthesis, have played a major role in the development of chemotherapeutic regimens (Abrams & Murrer, 1993; Gale et al., 1972) and have also contributed to our understanding of the molecular recognition of nucleic acids.

As one might expect, the study of nucleic acids has become a broad area spanning many diverse fields of science. Primary research aims include not only an understanding of the fundamental nature of nucleic acids, but also more focused endeavors such as the discovery of chemotherapeutic agents or the development of molecular tools for use in biotechnology. While the subsequent chapters of this book will discuss several important research areas and aspects of nucleic acids in depth, the present chapter seeks to acquaint the reader with the fundamentals of nucleic acid chemistry including structure, chemical reactivity, and basic experimental techniques.

DNA and RNA Structures

The primary structures of DNA and RNA consist of phosphodiester-linked nucleotide units that contain a 2'-deoxy-D-ribose or D-ribose sugar ring (in DNA and RNA, respectively) and an aromatic nucleobase (Fig. 1-2). The nucleobases found in DNA include the purines adenine (A) and guanine (G) and the pyrimidines cytosine (C) and thymine (T) while RNA incorporates the pyrimidine uracil (U) in place of T. The resulting polynucleotide chain has a consistent $5' \rightarrow 3'$ polarity with both a negatively charged sugar-phosphate backbone and an array of relatively hydrophobic nucleobases, amphiphilic features which ultimately drive the assembly and maintenance of secondary and tertiary nucleic acid structures.

While at first glance the primary structures of RNA and DNA appear rather similar, the C2'—OH substituents of RNA cause this nucleic acid to adopt conformations that differ from those of DNA (Saenger, 1984; Wang et al., 1982). Along with this

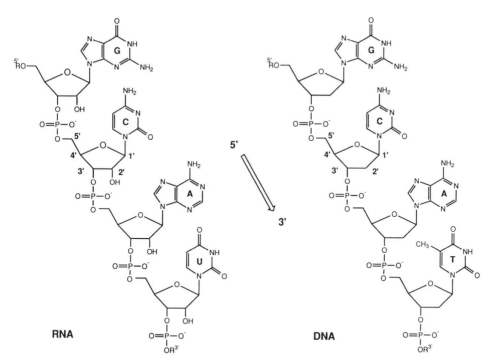

Figure 1-2. The primary structures of DNA and RNA.

fundamental difference, the primary sequences of biologically active RNAs and DNAs can be differentiated also by *posttranscriptional* nucleobase modifications (Hall, 1971), examples of which are found in Figure 1-3. RNA molecules, particularly transfer RNAs (tRNAs), often contain modifications to a large percentage of the nucleotides present in their primary structures. These modified nucleobases, which are thought to "tune" the stability and three-dimensional shape of the nucleic acid to suit particular biological roles, are almost always derived from one of the parent purine or pyrimidine heterocycles and can vary considerably from organism to organism. Likewise, ribosomal RNAs (rRNAs) and messenger RNAs (mRNAs) can also contain modified nucleobases. Unlike the wide variety of modified nucleobases found in RNAs, the DNAs of eukaryotes contain relatively simple modifications involving the methylation of the 5-position of cytosine or the exocyclic amino group of adenine (Fig. 1-3). These modifications support one mechanism by which eukaryotic systems regulate, and in some instances inactivate, the expression of individual genes at the DNA level (Cedar, 1988; Holliday, 1989).

B-Form DNA. The predominant DNA structure found under physiological conditions is referred to as the *B-form*. This conformation of DNA contains two *antiparallel* strands of nucleic acid connected by Watson–Crick A·T and G·C base pairs that spiral around a central polymer axis (Fig. 1-4). The specific nature of Watson–Crick base pairing results in a duplex composed of single strands that are *self-complementary*; thus knowledge of the nucleobase order in one strand is sufficient to define the primary sequence of the other, a feature that facilitates the replication and repair of DNA.

B-form DNA adopts a right-handed helical structure (Fig. 1-5) containing a hydrophobic interior of Watson–Crick base pairs stacked nearly perpendicular to the central axis at 3.4-Å intervals (Dickerson et al., 1982; Wing et al., 1980). The π–π stacking interactions that occur between these aromatic planes provide a substantial stabilizing force that helps to maintain the duplex nature of DNA (Saenger, 1984). Each base-pair plane of B-form DNA is rotated approximately 36° relative to the one preceding it, resulting in a complete right-handed helical turn for every 10 contiguous base pairs and thus a helical pitch of ~34 Å (3.4 Å/repeat unit × 10 base pairs/helical turn). With the Watson–Crick base pairs inside, the anionic sugar-phosphate backbone spirals around the outside of the helix, creating a hydrophilic exterior with a net charge of −2 for each repeat unit. The high density of negative charge associated with this biopolymer also dictates to some extent the character of agents which associate with it; agents that bind most efficiently to DNA often possess a net overall positive charge. Two other important structural details of the sugar-phosphate backbone include the deoxyribose ring conformation and the N-glycosidic bond angles; in B-form DNA the deoxyribose ring adopts a C2'-endo conformation, while the N-glycosidic bond angle is in an *anti*-configuration (Figs. 1-6 and 1-7).

The overall structure of B-form DNA creates two distinct helical grooves, the minor and the major, which spiral around the surface of the double strand (Fig. 1-8). In B-form DNA, the minor groove is narrow, while the major groove is wide, with both grooves possessing a moderate, nearly equivalent depth. Importantly, these two grooves create unique microenvironments for the binding and recognition of ligands (i.e., proteins or small molecules). The floors of the major and minor grooves are defined by the opposite sides of the stacked Watson–Crick base-pair planes (see Fig. 1-4) which create patterns of hydrogen-bond donor and acceptor sites within the plane of the base

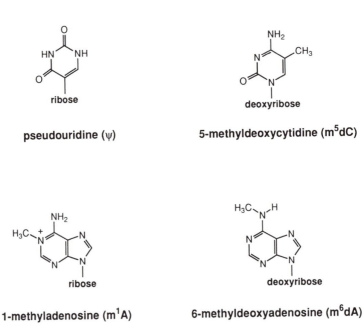

pseudouridine (ψ)

5-methyldeoxycytidine (m^{5}dC)

1-methyladenosine (m^{1}A)

6-methyldeoxyadenosine (m^{6}dA)

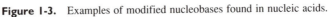

Figure 1-3. Examples of modified nucleobases found in nucleic acids.

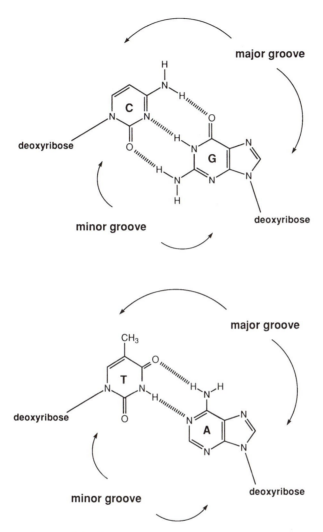

Figure I-4. The structure of Watson–Crick base pairs.

pair (Seeman et al., 1976). Thus, the floors of the grooves of the DNA helix would differ for individual base-pair sequences with respect to their patterns of H-bond donors, H-bond acceptors, and sites available for hydrophobic interactions (e.g., the C5 methyl group of thymidine). These differences form the basis, in part, for the sequence-selective binding of certain ligands (Steitz, 1990).

While the description of B-form DNA given above constitutes an idealized view, it is becoming increasingly evident that each base pair step along the DNA helix can be considered a unique structure (Drew & Travers, 1984). Thus, a 5′-GT-3′ step differs from a 5′-AT-3′ step not only in base composition, but also in the overall fashion in which each base pair is positioned in space relative to its neighboring (flanking) base pairs. This *microheterogeneity* of the DNA results from localized, sequence-dependent changes in the shape of the helix which occur to limit steric interactions between the

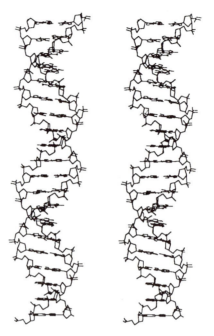

Figure I-5. Stereoview of B-form DNA.

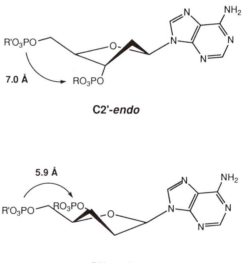

purines of adjacent base pairs (Calladine, 1982). Adjustments in the structural relationship between adjacent base pairs can be achieved through several operations including (1) changes in the local helical twist, (2) translation along the long base-pair axis (slide), (3) an overall tilting of the mean base-pair planes (roll), and (4) the opposite rotation of paired bases along their long axis (propeller twist) [see EMBO Work-

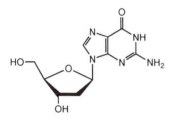

anti deoxyguanosine

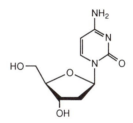

anti deoxycytidine

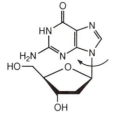

syn deoxyguanosine

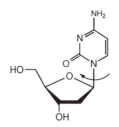

syn deoxycytidine

Figure 1-7. The *N*-glycosidic bond angles found in DNA structures.

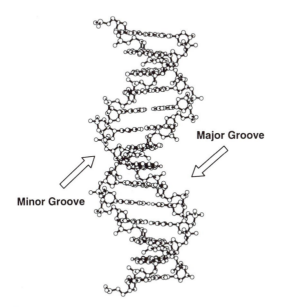

Major Groove

Minor Groove

Figure 1-8. The location of the major and minor grooves of B-form DNA.

9

shop (1989) for a good description of these structure parameters]. Importantly, these subtle differences in structure may be another factor that contributes to the selective recognition of the DNA helix by proteins or small molecules (Drew et al., 1990; Hunter et al., 1989).

In addition to differences at the base-pair level, the overall helical structure of DNA is also polymorphic, adopting several distinct conformations that are influenced by ionic strength, extent of hydration, temperature, solvent, and nucleotide composition (Saenger, 1984). Along with the B-form, two other relatively well-studied structures of DNA include the A-form and the left-handed Z-form (Fig. 1-9). A comparison of the helix structural parameters for crystallized examples of these three forms of DNA is shown in Table 1-1 (Dickerson et al., 1982).

A-Form DNA. B-form DNA transforms into an A-form helix as the relative humidity of its environment decreases to 75% and the NaCl concentration drops below 10%. The A-form of DNA is a stubby structure in comparison to the B-form and results in a right-handed helix with 11 base pairs per complete turn and a helical pitch of ~28 Å (Dickerson et al., 1982; Frederick et al., 1989). The most pronounced feature of this structure is the 20° tilting of the base-pair planes and their net displacement away from the central axis. Indeed, looking down the central axis of A-form DNA would reveal a hollow core distinctly different from that of B-form DNA. In addition to these features, the A-form helix also adopts a C3'-endo sugar pucker conformation (Fig. 1-6) as opposed to the C2'-endo conformation present in B-form DNA. These structural

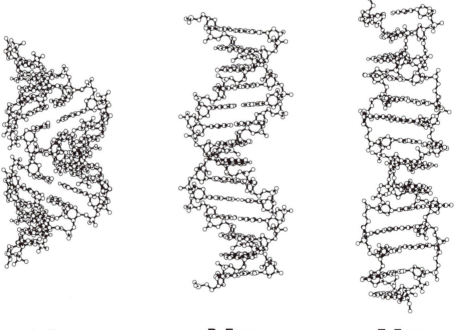

A–Form   **B–Form**   **Z–Form**

Figure 1-9. A comparison of the overall structures of A-, B-, and Z-form DNAs.

Table 1-1 Summary and Comparison of Double-Helical DNA Structural Parameters

Structural Parameter	DNA Conformation		
	A-Form	B-Form	Z-Form
Helical sense	Right-handed	Right-handed	Left-handed
Base pairs per repeat unit	1	1	2
Average base pairs per helical turn	10.7	10	12
Degrees of inclination of base pair from normal to the helical axis	19	−1	−9
Rise per base pair (Å)	2.3	3.3–3.4	3.8
Pitch per helical turn	24.6	33.2	45.6
Glycosyl angle conformation	*Anti*	*Anti*	*Anti* at cytosine *Syn* at guanine
Deoxyribose ring conformation	C3′-endo	O1′-endo to C2′-endo	C2′-endo at cytosine, C2′-exo to C1′-exo at guanine (C3′-endo)
Helical diameter (Å)	~26	~20	~18
Major groove	Narrow and deep	Wide and deep	Flattened
Minor groove	Wide and shallow	Narrow and deep	Narrow and deep

features result in a deep and narrow major groove and a very shallow and wide minor groove. (While the terms "major" and "minor" are only directly applicable to B-form DNA, the designation here is used to indicate a particular location within the DNA helix.) In addition to being a helical conformation adopted by double-stranded DNA, the A-form helix is also adopted by RNA–DNA hybrids (Milman et al., 1967; Wang et al., 1982) and double-stranded RNAs (Arnott et al., 1973) as a result of the C2′—OH substituent which, due to steric interactions, forces the sugar to assume a C3′-endo conformation.

Z-Form DNA. While the most prominent feature of Z-DNA is its ability to adopt a *left-handed* helical structure, this form of DNA is not simply the mirror image of the B-form or A-form helices (Rich et al., 1984; Wang et al., 1979). The Z-form helix is elongated and slender with 12 base pairs per complete turn and a helical pitch of ~45 Å. Z-form DNA contains a wide and shallow major groove with a narrow and extremely deep minor groove. Z-form DNA also adopts sugar puckering and *N*-glycosidic torsion angles that *alternate* between C2′-endo and C3′-endo and between *anti* and *syn*, respectively, making the actual repeating unit for the Z-form helix two base pairs in contrast to the single-base-pair repeat unit of A- and B-form DNAs. This alternating pattern, in combination with the 180° rotation of the bases about the glycosidic bond, results in a left-handed helix with a phosphate backbone that appears to zigzag around the helical structure, hence the name Z-DNA.

The transition of DNA into the Z-form generally requires high salt (3–4 M NaCl) and an alternating pyrimidine–purine sequence. Additionally, methylation of the C5 position of cytosine and negative torsional stress in supercoiled DNAs also appear to facilitate Z-form helix formation. While at first it may seem unreasonable to expect Z-form DNA to appear within living cells, ongoing research efforts are being directed toward identifying Z-form DNA in vivo and understanding its possible role in the regulation of gene expression (Jaworski et al., 1987; Zacharias et al., 1988).

Unusual DNA Structures. Along with the linear, repeating polymers described above, current research also indicates that DNA can adopt a host of so-called unusual structures (Wells et al., 1988; Wells & Harvey, 1987). These structures are often highly localized to a particular base sequence and require a precise set of conditions to extrude from a native B-form DNA. Examples of these alternative structures include (1) cruciforms that appear within palindromic DNA sequences (Lilley, 1989), (2) triple-stranded (Rajagopol & Feigon, 1989) or H-form structures (Htun & Dahlberg, 1988) that involve a duplex DNA hydrogen bonded to a third single strand of nucleic acid, and (3) "quartet" structures consisting of four DNA strands (Guschlbauer et al., 1990; Sundquist & Klug, 1989). Some noncanonical DNA structures can also be inherent to the base-pair composition and may not require an abrupt transformation of the native B-form. An example of this type of structure includes "bent" DNA which exhibits sequence-dependent curvature of the helix (Hagerman, 1986; Hsieh & Griffith, 1988; Koo & Crothers, 1988). Current research is attempting both to elucidate the precise structural details of these forms of DNA and to determine what, if any, biological significance they may possess. A thorough discussion of this topic is included in Chapter 7.

RNA Structures. Unlike the linear, repeating nature of most DNAs, RNA structures are strikingly diverse, reflecting the many biological functions of this polymer. The emergence of an RNA world likely depended on the ability of this nucleic acid to adopt many different forms, resulting in RNA molecules tailor-made to suit particular activities. Indeed, current research supports the idea that the precise function of an RNA molecule relies on its ability to maintain a proper three-dimensional structure (Altman, 1984; Cech, 1987; Rould et al., 1989). In this regard, the folded domains of RNA molecules (Murphy & Cech, 1993) can be likened to those of globular proteins and do not easily fit into categories like the DNA conformations previously discussed. The following sections will, therefore, illustrate RNA secondary and tertiary structure through a description of the three main functional classes of RNAs found within a cell.

Transfer RNA. Our most detailed view of RNA structure is derived from transfer RNA (tRNA) molecules that have been examined at high resolution (Brown et al., 1985; Rich, 1977; Westhof et al., 1985). Transfer RNAs are relatively compact single strands of nucleic acid that contain 60–95 nucleotides. Individual tRNAs transport a covalently attached amino acid to the ribosome and facilitate its proper incorporation into a protein sequence, the latter of which is specified by the sequence of nucleotides, read as a triplet code, in the mRNA. Thus, a tRNA molecule contains two key functional domains: (1) the site of covalent attachment for a particular amino acid and (2) a triplet anticodon base sequence that is complementary to a codon in the mRNA. Each tRNA is specific for only one particular amino acid; for example, the tRNA used to transport the amino acid phenylalanine would be designated tRNA^{Phe}. The triplet code displays degeneracy, however (Watson et al., 1987), and consequently there are often several different tRNA molecules [isoacceptor tRNAs (Sprinzl et al., 1991)] for one particular amino acid, each with a different three-base anticodon. Surprisingly, even though all tRNA molecules are believed to adopt very similar tertiary structures, the enzymatic attachment of the correct amino acid onto its cognate tRNA by an aminoacyl-tRNA synthetase is carried out with profound fidelity (Schimmel, 1987). Of particular current interest therefore is the determination of those features of a tRNA molecule which lead to the attachment of the proper amino acid. One strategy employed in this regard is the dissection of a tRNA into its component parts thereby creating "minihelices" or "microhelices" (Hou et al., 1989). These "reconstructed" tRNAs seek to identify the key features necessary for discrimination by the synthetase enzyme.

The primary and secondary structures of tRNA molecules are often depicted as a cloverleaf pattern containing double-stranded stems and stems connected to single-stranded loops, two recurring elements of RNA secondary structure (Fig. 1-10). Each A-form stem consists of hydrogen-bonded base pairs (G · C and A · U) that occur through self-complementary regions of the tRNA primary sequence. While many tRNAs exist in nature, they all contain several constant structural regions that can

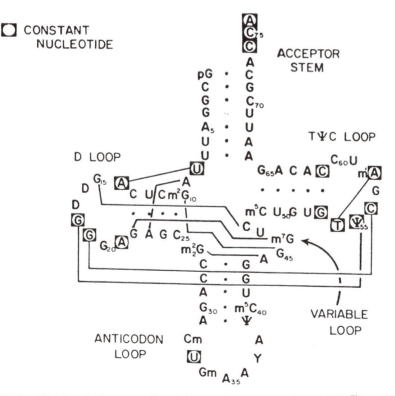

Figure 1-10. The cloverleaf representation of the secondary structure of yeast tRNA^{Phe}. Solid lines indicate nucleotides involved in tertiary hydrogen bonding interactions. Reprinted from Rich, 1977, with permission.

be described using the cloverleaf motif. At the 5'-termini, each tRNA is phosphorylated, and virtually all have a seven-base-pair structure referred to as the *acceptor stem*; it is so named because it provides the point of connection for the appropriate amino acid. The 3'-end of the tRNA, which is the position of amino acid attachment, always terminates in the sequence -CCA-3' and provides a free 3'-OH group for attachment of the amino acid as an activated ester by the cognate aminoacyl-tRNA synthetase. The acceptor stem is also unique in that it often contains non-Watson–Crick base pairs such as G · U. In addition to the acceptor stem, the *TψC loop* contains a five-base-pair stem with a highly conserved seven-base loop region that incorporates the modified nucleobase pseudouracil (ψ). All tRNAs also contain (1) a *D-loop* that frequently contains the modified nucleobase dihydrouracil (D) and (2) a *variable loop* which differs in nucleotide length between individual tRNAs. The remaining key structural feature of a tRNA is the *anticodon stem*, a stem-loop structure that contains the anticodon triplet responsible for hydrogen bonding to the complementary sequence of mRNA and consequent delivery of the proper amino acid to the ribosomal machinery.

The tertiary structure of tRNA resulting from these secondary interactions is distinctly globular in appearance, resembling an L shape (Fig. 1-11). In this structure, one arm of the tRNA, consisting of the acceptor stem and the TψC loop, is folded into an A-form double helix, while the D-loop and the anticodon loop similarly form the other arm of the L. Each arm of this structure is approximately 60 Å in length, with the anticodon and the acceptor stems at opposite ends of the molecule. The folded structure of the tRNA is maintained by non-Watson–Crick hydrogen bonding interactions (Fig. 1-10) which serve to crosslink distant regions of the tRNA, locking it into the desired tertiary structure. In addition to hydrogen bonding, stacking interactions within the interior of the molecule are also extremely important in maintaining this structure; over 90% of the bases present are involved in this form of interaction. While this extensive stacking renders most of the tRNA interior inaccessible to solvent, those regions necessary for intermolecular interactions (i.e., the anticodon loop and the acceptor stem) are placed at readily accessible locations.

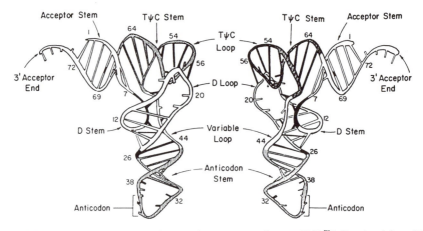

Figure 1-11. Views of both sides of the tertiary structure of yeast tRNA[Phe]. Reprinted from Rich, 1977, with permission.

Ribosomal RNA. In common with tRNA, ribosomal RNA (rRNA) is also a highly structured nucleic acid. Ribosomal RNA is the major component of the ribosome, encompassing approximately two-thirds of the gross weight and forming the functional portion involved in protein synthesis (Dahlberg, 1989). The other component of the ribosome consists of several proteins that are believed to assist in the maintenance of the required three-dimensional structure of the active rRNA, and which presumably have other functions as well.

The *E. coli* ribosome contains three well-studied rRNAs. This ribosome, which has a sedimentation coefficient of 70S, can be separated into two subunits designated 50S and 30S. The 30S subunit contains one large rRNA, 16S rRNA, and 21 individual proteins, while the 50S subunit contains two rRNAs, 5S and 23S rRNAs, and 32 different proteins. These rRNAs vary considerably in length; the 23S, 16S, and 5S rRNAs are 2904, 1542, and 120 nucleotides long, respectively.

Examination of the primary sequence of rRNAs by methods that predict higher-order structure (Zucker, 1989) suggest that they incorporate many of the elements of secondary structure employed by tRNA molecules. As shown in Fig. 1-12 for the *E. coli* 16S rRNA, rRNA molecules contain an array of double-stranded stems and single-stranded loops that resemble the structural elements in the cloverleaf structure of tRNA. In addition, there are interior loops, bulge loops, and multibranched loop structures. While prediction methods can generate many plausible forms, analysis of the 16S rRNA from several species consistently gave rise to the four-domain structure shown in Fig. 1-12 (Stern et al., 1989). This pattern is also consistent with nuclease digestion studies—that is, in experiments useful in determining enzyme-accessible portions of the RNA molecule and regions involved in secondary or tertiary structures (Brimacombe, 1988).

Along with the secondary structure elements discussed above, there is increasing evidence to suggest that rRNAs (and other forms of RNA) often contain a structural unit termed the *pseudoknot* (Dam et al., 1992; Schimmel, 1989). This unique structural element is formed from stem-loop structures that hydrogen-bond through the loop to an *additional* strand of RNA, thereby creating a new stem structure. Examples of pseudoknots that are believed to occur in 16S rRNA are illustrated in Fig. 1-13. While the exact function(s) of RNA pseudoknots are still a topic of active investigation, it is speculated that they create unique tertiary structural elements for recognition by other RNAs or proteins and may also provide a conformational switching mechanism between two structural forms of RNA (Dam et al., 1992).

Although the exact three-dimensional structure of the ribosome is not known at high resolution, models have been constructed based on electron micrograph data and *footprinting* studies (Brimacombe, 1988). These footprinting techniques (discussed later) have assisted in elucidating positions of protein–RNA binding within the ribosomal complex, and they have also helped to elucidate regions likely to be involved in tertiary folding. Eventually, elucidation of the complete tertiary structure of the ribosome will assist in understanding the process of translation and further our knowledge of structure–function relationships in RNA.

Messenger RNA. In stark contrast to tRNA and rRNAs, messenger RNAs (mRNAs) are thought to be single-stranded nucleic acids in their biologically active form although local secondary structures specific for individual mRNAs do form, as discussed below. Given its role as a carrier of the genetic code from DNA to the ribosome, the primary

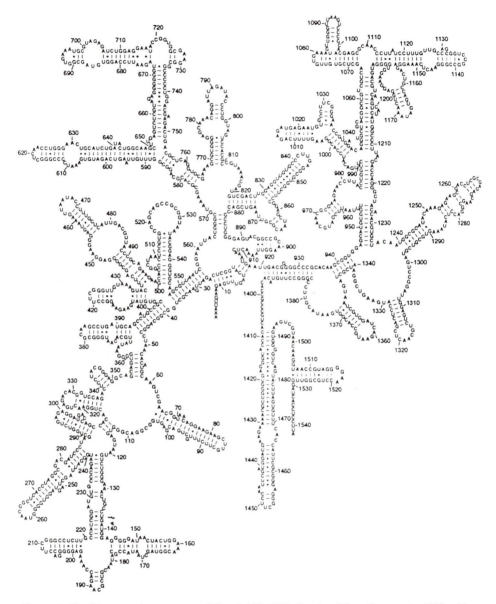

Figure 1-12. The secondary structure of *E. coli* 16S rRNA. Reprinted from Stern et al., 1989, with permission.

activity of mRNA is to permit this code to be "read" through the formation of complementary hydrogen bonds with the triplet anticodon of a tRNA. To facilitate access to this information, mRNA is likely to remain relatively less structured than other RNAs in the biological milieu, although eukaryotic mRNAs, at least, are associated with cellular proteins (Amero et al., 1992).

Although the single-stranded form of mRNA predominates, hairpin or stem-loop structures can occur and provide a means of controlling the rate of expression of a particular gene by providing stopping and pausing points for the ribosomal machinery. These regions, called *terminators* or *antiterminators*, have been studied within the *trp* operon and attest to the ability of RNA to actively participate in controlling its own rate of transcription and translation (Stroynowski et al., 1983; Yanofsky, 1988). In addition to the stem-loop features found in the structures mentioned above, an increasing body of evidence suggests that some mRNAs also incorporate pseudoknots within their ribosomal binding sites (Schimmel, 1989).

While not as structurally complex as the previous examples of RNA, knowledge of their linearity is being exploited in the development of oligonucleotides capable of binding to a complementary sequence of mRNA to form double-stranded hybrids. This exciting new area of research, known as the *antisense oligonucleotide strategy*, could allow a synthetic oligonucleotide to bind to a portion of mRNA as it passes from the nucleus to the ribosome (Cohen, 1989), leading, in principle, to the modification or destruction of a particular mRNA, or simply rendering it dysfunctional for the expres-

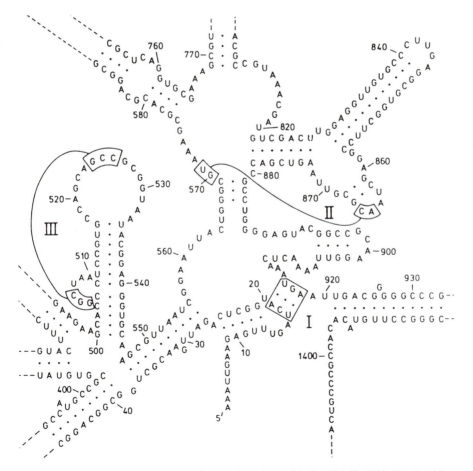

Figure 1-13. The proposed structures of pseudoknots located in *E. coli* 16S rRNA. Reprinted from Dam et al., 1992, with permission.

sion of its protein product. This approach, the topic of Chapter 12 of this volume, may eventually provide a means to prevent the synthesis of a deleterious protein or control the activity of a cell.

Overall, RNA structures are quite distinct and differ substantially from the predominantly linear, repeating polymers formed by DNA. While their global tertiary structures appear diverse, several fundamental units of secondary structure recur including A-form double-stranded stems, stem-loops, and pseudoknots. It seems likely that as more examples of RNA are examined by high-resolution methodologies, additional underlying structural motifs will be revealed.

The Chemistry of Nucleic Acids

The wide variety of structural features and chemical functional groups found in nucleic acids enable them to interact with many different molecules, often promoting the noncovalent association of a binding ligand or leading to their structural modification through alkylation or polymer strand scission. Such interactions, while varying substantially in mechanism, can also be classified with regard to their sequence preferences; nucleic acid binding or modification can be either *sequence-selective*, discriminating between individual nucleotides or groups of nucleotides, or *sequence-neutral*, leading to interactions at every position along the polymer.

Noncovalent Binding Interactions. Noncovalent interactions are extremely important in promoting the molecular recognition of nucleic acids. The nucleobase heteroatoms and the anionic phosphates participate in the formation of hydrogen bonds, van der Waals interactions, hydrophobic contacts, and salt bridges. In addition, the hydrophobic surfaces of the helical grooves or tertiary folds of a nucleic acid also assist in the binding or orientation of a ligand. While not a directed contact like a hydrogen bond or a salt bridge, these surface features can complement the structure of a binding agent in a fashion analogous to ligand–protein receptor association. Clearly, the extent of individual types of interactions will vary as a function of the conformation or tertiary structure adopted by the particular nucleic acid.

One of the key participants in noncovalent interactions is the negatively charged ribose-phosphate backbone. Importantly, the anionic phosphates of the backbone interact electrostatically with cations such as Na^+ or Mg^{2+}, increasing the structural stability of the polymer by screening the phosphate–phosphate repulsive interactions that would otherwise occur within a DNA double helix or folded RNA (Saenger, 1984). The polyanionic nature of nucleic acids is also essential during the initial stages of the binding of positively charged proteins or small molecules; in bulk solution, the overall negative charge of DNA and RNA serves to increase the relative local concentration of an oppositely charged ligand.

Once the nucleic acid and potential binding ligand are in close proximity, short-range, weak molecular forces begin to participate in their interaction. These weak interactions are usually a combination of hydrogen bonds with the functional groups of the nucleobases, hydrophobic interactions, and salt bridges between a positively charged moiety of the binding ligand and an anionic phosphate. With proteins that bind to RNA or DNA (Fig. 1-14) these interactions usually involve the side chains of amino acids which can act as H-bond donors or acceptors, provide sites of hydrophobicity, or participate in

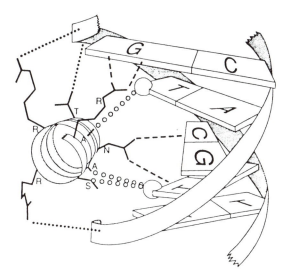

Figure 1-14. An illustration of the noncovalent forces involved in the binding of a helical protein region to the major groove of B-form DNA. Dashed lines, dotted lines, and open circles represent hydrogen bonds, salt bridges, and hydrophobic interactions, respectively. Small capital letters represent the single-letter amino acid code. Reprinted from Ellenberger et al., 1992, with permission.

ionic interactions (Ollis & White, 1987; Steitz, 1990). Specifically, in the case of tRNA, these features permit the correct recognition and attachment of an amino acid by the aminoacyl-tRNA synthetase, while for B-form DNA they are exploited for the binding of gene regulatory proteins, restriction enzymes, and other important DNA binding factors that require a high degree of sequence discrimination. In B-DNA, as mentioned previously, this recognition process is aided also by the functional groups found along the floors of the grooves (Seeman et al., 1976); while it is appealing to think that a precise set of hydrogen bonding interactions could lead to the sequence-selective binding of proteins to nucleic acids, in reality this process is quite complex and can involve many other factors including the inherent shape (Otwinowski et al., 1988), flexibility (Wu & Crothers, 1984), and microheterogeneity of the nucleic acid structure.

In addition to promoting the association of proteins, noncovalent forces also influence the binding of relatively small molecules to nucleic acids, the topic of Chapter 10 of this volume. In B-form DNA, the minor groove is the preferred binding site of low molecular weight species such as the antitumor agents distamycin and netropsin (Coll et al., 1989; Kopka et al., 1985; Tabernero et al., 1993; Zimmer & Wahnert, 1986). The size and chemical features of these particular drugs complement the narrow minor groove of AT-rich DNA regions resulting in a tight, sequence-selective binding interaction (Fig. 1-15). Agents such as distamycin are also capable of displacing a series of organized water molecules bound along the minor groove of B-form DNA referred to as the *spine of hydration*, a feature which provides an entropic driving force favoring complex formation.

Intercalation. Another important noncovalent mechanism of ligand binding involves the *intercalation* of ligands into nucleic acids (Lerman, 1961; Waring, 1970). Intercalation occurs when a planar, aromatic molecule slides between the stacked base pairs

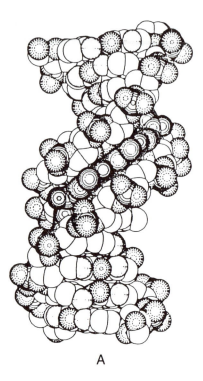

A

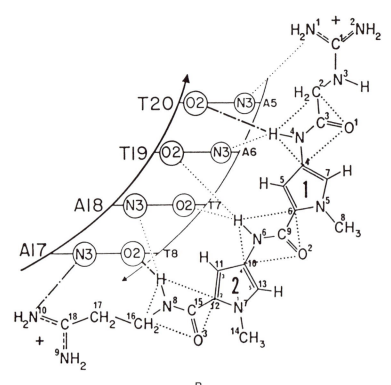

B

Figure 1-15. Two representations of the binding of netropsin to B-form DNA: (**A**) Van der Waals diagram of netropsin bound within the minor groove of the helix (from Coll et al., 1989). (**B**) Schematic figure of the network of hydrogen bonds formed between netropsin and the floor of the DNA minor groove. Reprinted from Kopka et al., 1985, with permission.

of the hydrophobic interior of a helical DNA (Fig. 1-16). While most of our knowledge concerning intercalative binding is derived from studies using DNA substrates, the tertiary structures of RNA also appear to enable ligands to bind through this mechanism (Chow & Barton, 1990). This unique mode of binding is utilized by agents ranging from simple organic molecules, such as ethidium bromide, to natural products that exhibit antitumor activity, like daunomycin (Fig. 1-17) (Pindur et al., 1993).

In DNA, intercalation is a two-stage process involving (1) an initial diffusion-controlled association of the usually positively charged intercalator with the exterior of the helix followed by (2) a slower insertion of the planar moiety between the stacked base pairs (Berman & Young, 1981; Long & Barton, 1990; Wilson & Jones, 1982). Intercalation results in the formation of a rigid, oriented complex between the DNA polymer and the bound ligand. Upon binding, an intercalator causes a pronounced structural perturbation in the helical polymer in order to accommodate the new ligand between the π stacked base pairs. These structural distortions cause the DNA host to unwind and lengthen, leading to distinct changes in the hydrodynamic behavior of the polymer (i.e., an increase in viscosity or altered gel electrophoretic mobility). Intercalation of the helical stack also leads to thermal stabilization of the duplex structure and is accompanied by electronic perturbations (hypochromicity and bathochromic shifts) in the intercalating chromophore due to its new local electronic environment.

Covalent Binding Interactions. Nucleic acids also participate in chemical reactions that result in the covalent modification of their structures through alkylation or metallation reactions. Agents that lead to these DNA or RNA modifications can often func-

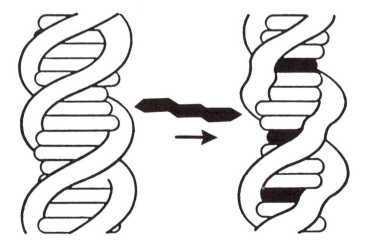

Figure 1-16. Schematic representation of the binding of an intercalator (dark bars) with the B-form DNA helix. Reprinted from Pindur et al., 1993, with permission.

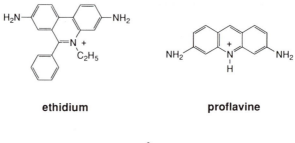

ethidium   proflavine

2-hydroxyethanethiolato-2,2',2"-terpyridine platinum(II)

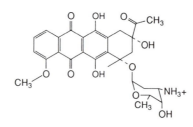

daunomycin

Figure 1-17.   The structures of several intercalating molecules.

tion in nucleobase-selective modification reactions. Such agents sometimes also possess antitumor properties, or, ironically, act as carcinogenic or mutagenic substances.

The interaction of mitomycin C with DNA constitutes a much studied example of the alkylation of a nucleic acid by a natural product. This agent is used in anticancer chemotherapy and is believed to function through the formation of monofunctional *and* bifunctional DNA adducts resulting from crosslinked inter- (Tomasz et al., 1988) and intrastrand (Bizanek et al., 1992) guanine residues. Initially, a reductively activated mitomycin associates with DNA in proximity to the exocyclic N2 amino group of guanine, a position accessible from the minor groove of B-form DNA. The resulting orientation promotes the attack of the N2 amino group of guanine on the C1 aziridine moiety of the drug leading to a monofunctional adduct (Fig. 1-18). At this point, the reaction can terminate or continue on through interaction of the C10 carbamate moiety of the drug with the N2 of an *adjacent* guanine, yielding a crosslinked DNA adduct.

In addition to mitomycin, simple organic reagents like dimethyl sulfate (DMS) that do not preassociate with the nucleic acid strand also react with the nucleobases. One of the major lesions that is formed in the reaction with DMS, an N7-methylated guanine (Lawley & Brookes, 1963), imparts a positive charge to the heterocycle and promotes the hydrolysis of the *N*-glycosidic bond, affording an apurinic site (i.e., a nucleotide that has lost its purine nucleobase but still maintains the ribose-phosphate backbone of the polymer) (Fig. 1-19) (Kalnik et al., 1989). Apurinic sites, or abasic sites in general, also form spontaneously due to hydrolysis of the *N*-glycosidic bond and can be recognized and cleaved (Fkyerat et al., 1993) or utilized in the attachment of spectroscopic probes to nucleic acids (Ide et al., 1993).

Along with the alkylation of nucleic acids, the coordinate binding of metal ions can also result in modification of a nucleic acid (see Chapter 9). Besides the association of alkali metal ions with the anionic phosphate oxygens, the greatest opportunity for direct metal ion ligation to a nucleic acid occurs through the coordination of functional

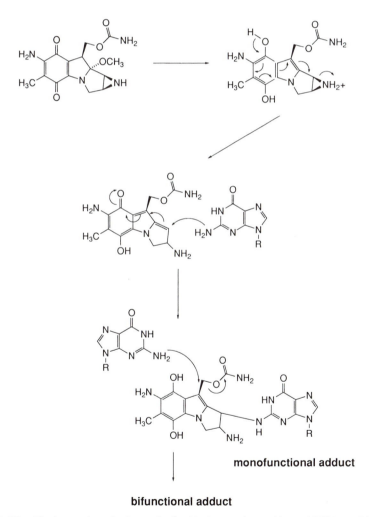

monofunctional adduct

bifunctional adduct

Figure 1-18. The interaction of mitomycin C with the guanine residues of DNA resulting in the formation of monofunctional and bifunctional covalent adducts.

groups found on the nucleobases; the endocyclic nitrogens and other functional groups provide ligands with differing propensities for transition metal ligation (Martin, 1985). While some nucleobase positions are not accessible due to their involvement in Watson–Crick base pairing, those functional groups that are available are exposed within the grooves of the nucleic acid. Like the alkylation reactions discussed previously, such interactions can either be monofunctional, as observed for Cu^{2+} bound to the N7 position of guanines in Z-form DNA (Geierstanger et al., 1991; Kagawa et al., 1991), or bifunctional as in the binding of the antitumor agent cisplatin [*cis*-dichloro-diammine platinum(II)] to the N7 positions of adjacent intrastrand guanine residues (Fig. 1-20) (Sherman & Lippard, 1987). In the case of cisplatin, the DNA lesion formed induces a localized unwinding of the helix (~13°) and a pronounced bending of the DNA backbone (~34°) toward the major groove, features which may contribute to the anticancer properties of this drug (Pil & Lippard, 1992).

Nucleic Acid Strand Scission. Nucleic acids are also chemically altered through strand scission of their polymeric backbones. This form of nucleic acid modification can occur through several mechanisms including (1) oxidation of the ribose or deoxyribose sugar ring, (2) alkylation or oxidation of the aromatic nucleobase, or (3) hydrolysis of the phosphodiester backbone. Agents that mediate these changes vary from organic

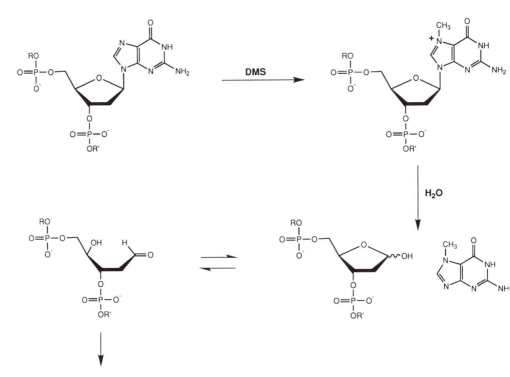

further reaction with aldehyde

Figure 1-19. Modification of the N7 position of a guanine residue by dimethylsulfate (DMS) leading to the formation of an apurinic site.

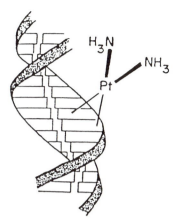

Figure 1-20. Schematic representation of the binding of cisplatin to adjacent intrastrand guanine residues via the N7 atoms located in the major groove of the helix. Reprinted from Sherman & Lippard, 1987, with permission.

natural products and transition metal complexes to highly reactive, reduced forms of oxygen.

Oxidative degradation of the sugar moiety often results from the homolytic or heterolytic breakage of a C—H bond on the ring, resulting in the formation of a carbon-centered radical or cation (Stubbe & Kozarich, 1987). The formation of such reactive intermediates ultimately results in fragmentation of the polymer through a variety of mechanisms which depend upon, for example, the position of the initial H abstraction (i.e., C4'—H versus C1'—H), the oxidative agent employed, and the concentration of dioxygen present in the medium. One possible mechanism of C4'—H abstraction in the presence of O_2 is shown in Fig. 1-21.

A variety of agents are capable of inducing this form of nucleic acid strand scission, including several which are highly selective for one C—H position versus another. The observed differences between such agents in terms of cleavage site selectivities presumably reflect their location of binding (e.g., the minor or major groove) and precise orientation. For example, the metal complex $Cu(phen)_2^{2+}$ has been shown to promote deoxyribose oxidation through the breakage of the C1'—H bond located in the minor groove of the DNA double helix (Goyne & Sigman, 1987; Sigman, 1986). Thus, the orientation of the proposed Cu—O intermediate active in strand scission is such that ready access to this position is provided. Similarly, the activated Fe complex of the antitumor agent bleomycin interacts exclusively with the C4'—H bond of the deoxyribose moiety of DNA (Hecht, 1986; Stubbe & Kozarich, 1987).

Along with agents that employ a metal ion, wholly organic structures as represented by the enediyne class of antitumor agents [e.g., neocarzinostatin (Goldberg, 1991), calicheamicins (Lee et al., 1991), and dynemicin (Sugiura et al., 1990)] can also induce strand scission through C—H bond breakage. This class of cleavage agents generates a drug-centered carbon radical which again is capable of inducing the breakage of C—H bonds located in the minor groove of the DNA helix (i.e., C1'—H, C4'—H, and C5'—H).

In addition to their selectivity for one C—H position versus another, the agents men-

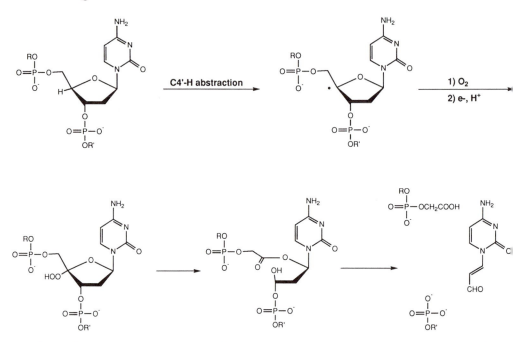

Figure 1-21. A possible mechanism of DNA strand scission resulting from abstraction of C4'—H from DNA.

tioned above often exhibit high degrees of sequence-selectivity due to their noncovalent association with the nucleic acid prior to strand scission. This effect is also magnified as the nucleic acid substrate becomes more complex; in the case of Fe·bleomycin, strand scission of DNA occurs with selectivity for 5'-GC-3' and 5'-GT-3' sites, while for RNA substrates ribose oxidation results in a cleavage pattern that is dependent upon the secondary and tertiary structures of the substrate (Carter et al., 1990; Holmes et al., 1993).

An additional mode of nucleic acid strand scission involves oxidation of the aromatic nucleobases. This mechanism, which often results in the selective modification of guanines due to their low oxidation potential, has been used to explain DNA cleavage induced by photosensitizers of $^1O_2$ (Fleisher, 1988) and some Ni^{2+} complexes (Chen et al., 1991). While the oxidation of the guanine heterocycle appears to be a common mechanism of DNA modification, it does not necessarily result in the *direct* strand scission of the polymer backbone; often, complete strand fragmentation requires alkaline hydrolysis of the lesions that are formed initially (e.g., abasic sites).

Along with compounds that catalyze the oxidative strand scission of nucleic acids, there is much current interest in the development of agents that can induce the *hydrolysis* of a nucleic acid polymer. Among such agents, the most efficient by far are enzymatic systems. As discussed in Chapter 3 of this volume, enzymes specific for RNA or DNA can have *exonuclease* or *endonuclease* activities, targeting either the nucleic acid termini or interior sites, respectively. Indeed, the discovery and application of *restriction endonucleases*, which induce the double-strand hydrolysis of DNA at *specific* 4 to 8-base-pair sites, has provided researchers with the means to cleave DNA at precise locations (Nathans & Smith, 1975; Yuan, 1981). The ability to manipulate DNA with restriction enzymes is fundamental to genetic engineering and biotechnology; unlike oxidative degradation, hydrolytic strand scission yields nucleic acid termini which are amenable to further enzymatic manipulation, allowing DNA to be specifically cut *and* re-ligated.

At the present time, unfortunately, the specific hydrolysis of nucleic acids is limited to those sites targeted by naturally occurring enzymes. Consequently, the development of artificial hydrolases with unique polymer recognition sequences could add to the available repertoire of naturally occurring restriction enzymes (Basile & Barton, 1989). While several inorganic (Morrow et al., 1992; Stern et al., 1990) and organic systems (Barbier & Brack, 1988; Yoshinari et al., 1991) that mediate the hydrolysis of RNA or simple phosphate monoesters have been developed, the hydrolysis of the DNA backbone has been much more difficult to achieve. This is due to the absence of a 2′-OH group which, when deprotonated, can assist in the cleavage of phosphodiester bonds between ribonucleotides (Fig. 1-22).

While the design of artificial hydrolases is a topic of much current interest, the discovery of naturally occurring agents such as bleomycin or the enediyne antibiotics has also promoted the development of agents that oxidize nucleic acids (Barton, 1986; Basile & Barton, 1989; Dervan, 1986; Sigman, 1986). Among those that have received considerable attention is methidiumpropyl-EDTA-Fe(II) (MPE-Fe), which contains an intercalating moiety (methidiumpropyl) tethered to an Fe-chelating EDTA ligand (Hertzberg & Dervan, 1984). This novel hybrid molecule was designed to carry the destructive oxygen radicals generated by $Fe^{2+} + O_2$ to the nucleic acid backbone via the agency of the intercalating moiety, resulting in nucleic acid degradation. Similarly, the simple metal complex Fe(II)·EDTA is also capable of mediating cleavage, even

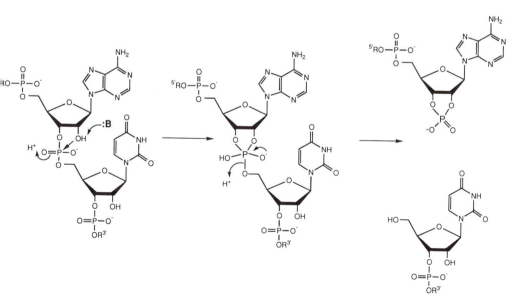

Figure 1-22.   Mechanism of base-promoted RNA hydrolysis involving the deprotonated C2′—OH functionality.

though intimate association of this negatively charged complex with a nucleic acid is not likely (Tullius & Dombrowski, 1986; Tullius et al., 1987). As described below and elaborated in Chapter 5, these "artificial nucleases" have been central to the development of chemical probes of nucleic acid structure.

Nucleic Acid Self-Processing Reactions. Along with strand scission by exogenous reagents, some RNAs are also capable of effecting their own modification through *self-processing* reactions. While several examples of self-processing RNAs have been reported (see Chapter 13), perhaps the best known are the intervening sequence of the *Tetrahymena thermophilia* pre-rRNA (Cech, 1987) and the M1 RNA component of RNase P (Altman et al., 1986). These two RNAs catalyze, respectively, the excision of a 413-nucleotide intron from the pre-rRNA to yield a mature rRNA and the removal of "extra" nucleotides present at the 5'-terminus of tRNA precursors. In both cases, the catalytic activity of these so-called ribozymes is *independent* of any protein-based enzymatic component.

In the *Tetrahymena* intron, self-processing results from a series of transesterification reactions that are initiated by the free 3'-OH of a guanosine nucleotide cofactor (Fig. 1-23). While guanosine is certainly a key element in this reaction, the RNA molecule also participates indirectly by providing a three-dimensional structure that properly orients the reactive groups involved in the reaction. Current research also indi-

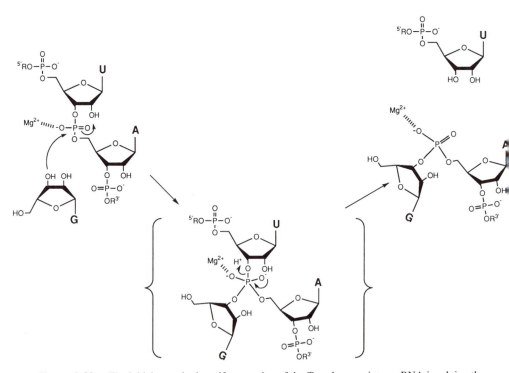

Figure I-23. The initial steps in the self-processing of the *Tetrahymena* intron rRNA involving the attack of the guanosine C3'—OH group on the UpA phosphate bond.

cates that the structural stability and catalytic efficiency of ribozymes is increased by the presence of metal ion cofactors (Pyle, 1993). Metal ions, as described previously, serve to decrease phosphate–phosphate repulsive interactions within the RNA tertiary structure and may also assist directly in the self-processing reactions of some ribozymes through Lewis acid catalysis. These self-processing systems illustrate the intimate connection between structure and function in RNA and may eventually lead to the development of artificial nucleic acid hydrolases.

Basic Techniques in the Study of Nucleic Acids

The study of nucleic acids has promoted the development of a number of fundamental techniques essential for the examination of their structure and chemistry. In a very general sense, the key elements of concern in this area include the ability to (1) obtain nucleic acids of a defined sequence for detailed chemical, biochemical, and biological studies, (2) determine the exact sequences of bases present in nucleic acid fragments, (3) assess the binding of ligands to nucleic acids in solution, and (4) understand the structure of nucleic acids and their complexes at high resolution.

Synthesis of Nucleic Acids. The ability to obtain reasonable quantities of synthetic RNA or DNA of defined sequence has been of immeasurable value to nucleic acid research (Itakura et al., 1984). Synthetic oligonucleotides have aided structural investigations and have also facilitated the development of a detailed understanding of the interactions that occur between nucleic acids and binding/modifying ligands. However, the overwhelming majority of applications of synthetic oligodeoxynucleotides exist in the area of biotechnology where they are applied as primers in the enzymatic synthesis or amplification of DNA (see Chapters 3 and 14 for details), used in site-directed mutagenesis studies, or used as diagnostic hybridization probes (Maniatis et al., 1982). As mentioned previously, synthetic oligonucleotides are also finding use in the developing area of antisense probes of cellular mRNA (Cohen, 1989).

The basic strategy in oligonucleotide synthesis is similar to the stepwise synthesis of oligopeptides; a chemically protected nucleotide is connected to a growing oligonucleotide chain, deprotected, then allowed to couple to the next nucleotide monomer of the desired sequence (Gait, 1984). Initially, the coupling of nucleotide monomers was carried out in solution using the *phosphodiester method* (Khorana, 1979). This method set the stage for the development of the automated methodologies utilized today which involve the *phosphoramidite method* (Caruthers et al., 1987; Gait, 1984) and the sequential coupling of nucleotide monomers to a 3'-terminal nucleotide attached to a solid support.

The key to the synthesis of RNA and DNA is the correct formation of internucleotide phosphodiester bonds. This is complicated by the presence of reactive functional groups within the nucleotide monomers, necessitating the protection of these groups prior to their incorporation into the DNA or RNA oligonucleotide. In the case of DNA and RNA nucleotides, two and three hydroxyl functionalites are present, respectively, which require proper protection during the coupling to the next nucleotide of the chain. In addition to the hydroxyl groups, the functional groups of the nucleobases also require protection during the internucleotide coupling reaction. The detailed steps involved in the synthesis of RNA and DNA oligonucleotides is presented in Chapter 2 of this volume.

Determining the Sequences of Nucleic Acids. In addition to the ability to synthesize nucleic acids of defined sequence, there is also a routine need to determine the sequences of *unknown* fragments of DNA and RNA. Presently, there are two main methods for determining the sequences of DNA: (1) the base-specific chemical cleavage of nucleic acids and (2) chain termination methods which rely on the ability to truncate the enzymatic synthesis of a complementary nucleic acid chain. As discussed below, RNA sequencing also employs similar strategies.

In DNA sequencing by base-specific cleavage, a radioactive label is incorporated enzymatically at the 5'- or 3'-terminus of the DNA sequence of interest to permit the detection of cleavage fragments through autoradiography. Samples of this labeled DNA of uniform length are then modified through nucleobase-specific chemical reactions developed by Maxam and Gilbert (1977, 1980); specific chemical sequencing reactions have been developed to modify guanines exclusively (G reaction), adenines and guanines (A + G reaction) but not pyrimidines, cytosine alone (C reaction), or cytosine and thymine (C + T reaction) but not purines. These base-specific reactions alter the nucleic acid strand such that upon subsequent treatment with piperidine, polymer strand scission occurs *only* at the nucleotide modified in the base-specific reaction. It is important to note that this procedure relies on the ability to specifically cleave a *population* of end-labeled DNA fragments of uniform length so that an individual nucleic acid strand is chemically modified, on average, only once. This treatment leads to a family of cleaved nucleic acid fragments that can be separated at single nucleotide resolution by polyacrylamide gel electrophoresis (PAGE) and identified by autoradiography (see Fig. 1-24). Application of each of the base-specific chemical reactions (G, G + A, C, C + T) to samples of the same DNA fragment identifies the position of each type of nucleotide in the polymer, permitting identification of the complete sequence.

In contrast to sequencing via base-specific chemical cleavage, the chain termination method relies not on the degradation of a nucleic acid strand, but on the controlled *synthesis of a complementary strand* of DNA. In this method, the unknown nucleic acid fragment provides a *template* for the enzymatic synthesis of its Watson–Crick DNA complement through the use of the Klenow fragment of DNA polymerase I. This enzyme elongates a DNA strand in the 5' → 3' direction in the presence of the template, a *primer* (which provides the necessary double-stranded starting point for the synthesis of the complementary strand), and the four deoxynucleotide triphosphate monomers, dTTP, dCTP, dGTP, and dATP (Smith, 1980).

When applied as a method of DNA sequencing, this system incorporates the DNA fragment of unknown sequence as the template, a suitable primer, and the four deoxynucleotide triphosphates, one of which also includes a radioactive phosphorus which provides a label for autoradiography. In addition to these components, a 2',3'-*dideoxy*nucleotide triphosphate of adenine, guanine, cytosine, or thymine is included in a sequencing reaction specific for one of these nucleotides, respectively (see Fig. 1-25). The dideoxynucleotide is enzymatically incorporated into the growing nucleotide chain at the appropriate position dictated by the template sequence; however, due to its lack of a 3'-OH group, incorporation of this nucleotide *terminates* further synthesis. This reaction results in the enzymatic synthesis of the complementary strand which is truncated at controlled locations dictated by the type of dideoxynucleotide employed

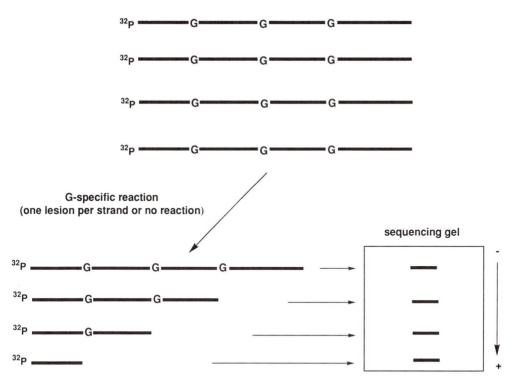

Figure 1-24. Schematic illustration of the Maxam–Gilbert G-specific DNA sequencing reaction.

in a given reaction. Thus, carrying out four parallel reactions with either the A, G, T, or C derivative of the dideoxynucleotide results in a population of truncated fragments which can be separated at single nucleotide resolution by gel electrophoresis, detected by autoradiography, and used to determine the sequence of nucleotides in the chain. This method thus yields information about the strand *complementary* to the template fragment of unknown sequence.

Like the chain termination method used for DNA, RNA sequencing also employs an enzyme-based protocol (Gosh et al., 1980). In this case, the enzyme *reverse transcriptase* is used to synthesize a complementary strand of DNA (cDNA) from an RNA template containing the unknown sequence. Once a cDNA is synthesized, either of the above protocols can be applied to determine the sequence of the cDNA, and hence that of the original complementary strand of RNA. Alternatively, a direct nucleobase-specific cleavage protocol similar to that employed for DNA can be used to determine the sequence of nucleotides present in a radioactively end-labeled RNA strand.

Chemical Probes of Nucleic Acids in Solution. While one would ideally like to study the structure of a nucleic acid or a nucleic acid–ligand complex through high-resolution methods, these experiments often can be problematic and time-consuming as discussed below. To circumvent this problem, chemical and enzymatic methods have been developed to study nucleic acids and nucleic acid–ligand complexes in solution. While these methods do not provide high-resolution information, they can often be uti-

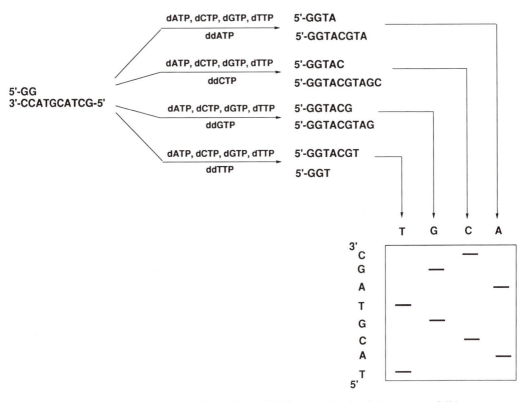

Figure 1-25. The chain termination method of DNA sequencing involving the use of dideoxynucleotide triphosphates (ddNTPs).

lized in the development of working models of the structure of nucleic acids and their ligand-bound complexes.

Presently, several solution methods are employed routinely in the determination of the binding sites of ligand–nucleic acid complexes. Among these, two basic strategies have emerged including (1) *footprinting*, where a bound ligand protects a portion of a nucleic acid from the action of a sequence-neutral nucleic acid cleavage agent and (2) *affinity cleavage*, where the binding ligand of interest is synthetically transformed into an agent which is also capable of inducing the strand scission of a nucleic acid at the site of binding (Dervan, 1986, 1991). These methods have proven applicable to a wide variety of nucleic acid substrates, including structured RNAs, and to binding agents ranging from relatively small drugs to much larger proteins.

The footprinting procedure employs a radiolabeled nucleic acid substrate which is bound by the ligand of interest and subsequently digested in the presence of a sequence-neutral strand scission reagent (Fig. 1-26). Commonly used footprinting reagents include the enzyme DNase I and the ·OH radical-producing agents methidiumpropyl-EDTA-Fe(II) or EDTA-Fe(II) (Tullius et al., 1986, 1987). These agents induce strand scission of the nucleic acid substrates with a high frequency at all positions except for those "protected" by the bound ligand. After digestion with the sequence-neutral reagent, the fragmented nucleic acid is examined at single nucleotide resolution by PAGE; identification of the nucleotide positions protected from cleavage reveals the site(s) of ligand binding. By utilizing several end-labeled nucleic acid fragments, numerous sequences may be tested, allowing the determination of relative binding site preferences.

In a fashion that is complementary to the footprinting methodology, affinity cleavage also reveals information about the site of interaction of a nucleic acid-bound ligand (Dervan, 1991). In this procedure, an end-labeled nucleic acid is also employed but instead of inducing cleavage through the use of a sequence-neutral agent, the binding ligand itself is transformed into an agent capable of inducing the strand scission of nucleic acids. This transformation can be carried out through several means including the synthetic attachment of the Fe(II)-chelating moiety EDTA or, in the case of proteins, the Cu(II)- or Ni(II)-chelating domain NH_2-Gly-Gly-His.

Once the ligand of interest has been altered to permit binding *and* strand scission of nucleic acids, the hybrid molecule is allowed to react with an end-labeled nucleic acid substrate resulting in strand scission at the site of binding. After the cleavage reaction, the fragmented nucleic acid substrate is analyzed by PAGE, revealing the location of the site of binding through strand scission (Fig. 1-27). Information regarding the location *and* relative orientation of a bound ligand can often be obtained from these procedures; when hybrid molecules which produce diffusible oxidants are employed in the reaction, a ligand bound in the minor groove of B-form DNA creates a 3'-asymmetric pattern of cleavage, while radicals generated through a major groove binding

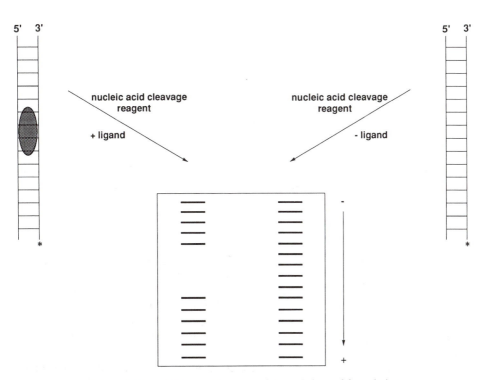

Figure 1-26. Schematic illustration of the technique of footprinting.

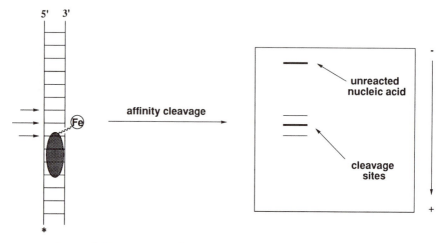

Figure 1-27. Schematic illustration of the technique of affinity cleavage.

interaction reveal two 3'-asymmetric cleavage patterns that are 5'-shifted in relation to one another (Dervan, 1991).

In addition to finding utility in the characterization of ligand–nucleic acid interactions, reagents such as Fe(II)·EDTA and inorganic species (Pyle & Barton, 1990), as well as other chemical and enzymatic systems (Wells et al., 1988), have been employed successfully as conformational probes of nucleic acids in solution. In most instances, the techniques employed are similar to the affinity cleavage methodology described previously; an end labeled nucleic acid molecule is allowed to interact with the probe molecule, and the position(s) of interaction is determined following strand scission or covalent modification. Tullius and his coworkers have also used Fe(II)·EDTA directly as a probe of conformation. As discussed in detail in Chapters 5 and 8 of this volume, such probes have been useful in studying the conformations of unusual DNAs such as "bent" DNA (Burkhoff & Tullius, 1987) and in understanding the solvent-accessible locations of structured RNA molecules (Chen et al., 1993; Chow et al., 1992; Latham & Cech, 1989). In each case the utility of the probe relies on the chemical reactivity, accessibility, or subtle structural features of the nucleic acid and its ability to react or complement the structure of the probe molecule. Information derived from these studies can be used to map conformational changes along a DNA helix (Pyle & Barton, 1990) or to develop models of tertiary structure in the absence of high-resolution structure data.

Crystallography and Nuclear Magnetic Resonance Spectroscopy of Nucleic Acids.
While the preceding methods permit the study of nucleic acids and their complexes with relative ease, the resolution of these techniques do not produce a detailed molecular structure. Among other techniques (Lilley & Dahlberg, 1992), the two main methods currently employed for the high-resolution characterization of nucleic acids involve single-crystal X-ray diffraction and nuclear magnetic resonance (NMR) methodologies.

Single-crystal X-ray diffraction studies depend on the ability to grow suitable crystals of the nucleic acid or nucleic acid complex of interest (Kennard & Hunter, 1989,

1991; Steitz, 1990). In the case of nucleic acids or their complexes with proteins or small molecules, well-ordered crystalline materials can be challenging to obtain. In those cases where a crystalline material is successfully obtained, the packing between molecules can result in crystals with intermolecular channels that contain large amounts of disordered solvent molecules. This does not facilitate structural analysis of the crystalline material, but does lend support to the concept that the crystal structure of such biomolecules accurately reflects their conformation in solution. Once suitable crystals are in hand, an X-ray diffraction pattern is obtained which is deciphered as an electron density map and ultimately as an atomic model. The data obtained from the diffraction pattern, however, do not necessarily always yield a high-resolution image; the quality or packing of the crystal can limit the interpretation to a low-resolution structure where only the rough contour of the system of interest is revealed. These comments notwithstanding, X-ray crystallography can be an extraordinarily powerful tool for the definition of nucleic acid structure; a detailed analysis is provided in Chapter 4.

Along with X-ray crystallography, two-dimensional NMR studies of nucleic acids and their complexes, as discussed in Chapter 6, hold great promise in understanding their structures in solution (Clore & Gronenborn, 1989; Gao & Patel, 1989; van de Ven & Hilbers, 1988; Wüthrich, 1986). In the procedures employed, intra- or intermolecular nuclear Overhauser effects between protons within the nucleic acid or between the nucleic acid and those of a bound ligand yield distance information that can be interpreted in terms of a molecular model. While this technique does not require the growth of a crystalline material, it is restricted in terms of the size of the molecules employed for study; molecules that do not tumble rapidly in solution do not yield high-resolution data.

While both techniques have their disadvantages with regard to the amount of sample required and difficulty of data interpretation, recent advances in molecular biology and computer science have greatly assisted this field. The ability to clone particular genes and overproduce proteins, in addition to the ability to synthesize large quantitites of oligonucleotides, have provided researchers with adequate amounts of material for study by both NMR and crystallographic techniques. In addition, increased computational hardware and software capabilities have also facilitated data collection, interpretation, and modeling.

Summary

The study of nucleic acids has grown tremendously since the initial report of the structure of B-form DNA by Watson and Crick more than 40 years ago. Surprisingly, however, intriguing new details concerning the structure of DNA and the complexes that it forms both in vitro and in vivo continue to be revealed. While DNA continues to be a major focus, the tertiary structures and complexes adopted by RNA are in the spotlight with increasing frequency as we grow to appreciate the many roles that this nucleic acid plays within the cell. As illustrated in subsequent chapters of this volume, the quest for knowledge of the structure and chemistry of nucleic acids has drawn from many diverse areas of chemistry and biology, a fact which again underscores the central role of nucleic acids in life processes.

2

The Chemical Synthesis of DNA/RNA

Serge L. Beaucage and Marvin H. Caruthers

Over the past 40 years, several procedures have been developed for the synthesis of oligodeoxyribonucleotides and oligoribonucleotides. These include the phosphate diester, phosphate triester, *H*-phosphonate, and phosphoramidite methods. Although one, the phosphate diester approach, is now only of historical interest, the remainder are being used currently for different applications in nucleic acids research. This chapter has several objectives. One is to describe, especially from a mechanistic view, the chemistries currently used to synthesize oligodeoxyribonucleotides. Another is to extend this analysis to the synthesis of oligoribonucleotides. Perhaps the major difference in the synthesis of these two oligomers is the selection of the 2′-hydroxyl protecting group in oligoribonucleotides and its impact on the other parameters involved in preparing these macromolecules. These criteria will be addressed as well. The last part of the chapter is devoted to a discussion of DNA and RNA analogues that have potential therapeutic and diagnostic applications. Other important aspects of polynucleotide synthesis, such as the detailed description of nucleobase and phosphorus protecting groups, are not discussed, because these topics have been reviewed elsewhere in detail (Beaucage & Iyer, 1992; Kössel & Seliger, 1975; Sonveaux, 1986). Because of space limitations, this chapter is not a comprehensive review of the scientific literature. As a consequence, a number of excellent contributions have regrettably been omitted.

Chemical Synthesis of Oligodeoxyribonucleotides

The Phosphotriester Method. The first chemical synthesis of a dinucleotide containing a (3′ → 5′)-internucleotide linkage identical to natural DNA was reported in the mid-1950s (Michelson & Todd, 1955). Specifically, the activation of 5′-*O*-acetylthymidine-3′-*O*-benzyl hydrogen phosphonate **1** to the corresponding phosphorochloridate **2** was effected by *N*-chlorosuccinimide. The reaction of **2** with 3′-*O*-acetylthymidine produced the dinucleoside phosphate triester **3**, which upon removal of protecting groups afforded the natural dinucleoside monophosphate **4**. This pioneering work paved the way for the future development of the phosphotriester approach.

More than a decade later, it was shown that the condensation of the 5′-*O*-protected deoxyribonucleoside **5** with phosphate monoester **6**, activated with 2,4,6-trimethylbenzenesulfonyl chloride (MS-Cl), afforded the nucleoside phosphodiester **7**. Addition of a deoxyribonucleoside having a free 5′-hydroxyl function (**8a** or **8b**) and 2,4,6-tri-

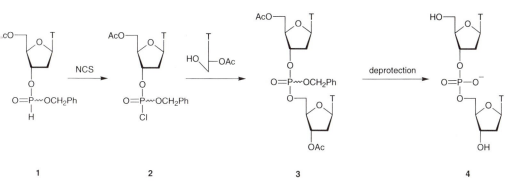

Ac = acetyl; NCS = N-chlorosuccinimide; T = thymin-1-yl

isopropylbenzenesulfonyl chloride (TPS-Cl) to a reaction mixture containing **7** produced the dinucleoside phosphate triester **9a** or **9b**, respectively, which were easily separated from the ionic starting material by silica gel chromatography (Letsinger & Ogilvie, 1967, 1969). Further treatment of **9b** with hydrazine hydrate in pyridine-acetic acid removed the benzoylpropionyl protecting group without unblocking the phosphotriester function (Letsinger et al., 1969; Letsinger & Miller, 1969). These developments led to a strategy for the stepwise synthesis of oligodeoxyribonucleotides having mixed sequences and also to the preparation of polydeoxyribonucleotides via the use of short oligodeoxyribonucleotide blocks (Letsinger et al., 1969). However, it became apparent that the 2-cyanoethyl phosphate protecting group was too labile to sustain the multiple work-up and purification steps involved in the synthesis of large oligodeoxyribonucleotides.

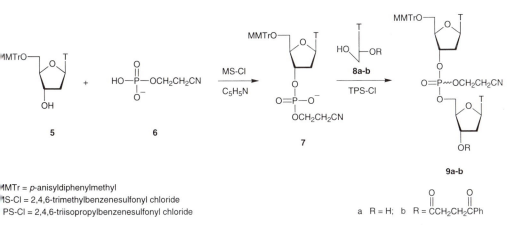

MMTr = p-anisyldiphenylmethyl
MS-Cl = 2,4,6-trimethylbenzenesulfonyl chloride
TPS-Cl = 2,4,6-triisopropylbenzenesulfonyl chloride

a R = H; b R = CCH₂CH₂CPh

The search for stable phosphate protecting groups, as well as for methods leading to their removal, was pursued intensively (Reese, 1978; Sonveaux, 1986). As a result of many investigations, aryl groups were shown to be preferred for blocking phosphate groups (Reese & Pei-Zhuo, 1993) even though their removal by alkaline hydrolysis led to significant cleavage of internucleotide linkages (Cusak et al., 1973). The extent of this cleavage was, however, controlled by incorporation of electron-withdrawing substituents in the phenyl group (Reese, 1978) (**10a** and **10b**) and replacement of the alkaline hydrolysis step by treatment with the conjugate base of either (*E*)-2-nitroben-zaldehyde oxime (**11**) or *syn*-pyridine-2-carboxaldehyde oxime (**12**) (Reese et al., 1978; Reese & Zard, 1981). Typically, reaction of dinucleoside phosphotriester **10a** with the 1,1,3,3-tetramethylguanidinium salt of **11** resulted in the complete removal of the aryl phosphate protecting group within 30 min. Cleavage of the internucleotide linkage occurred to less than 0.1% (Reese & Zard, 1981). These observations prompted studies to determine the mechanism for this reaction. The most definitive work demonstrated that the conjugate base of *syn*-4-nitrobenzaldoxime reacted with diphenyl-(4-nitro-phenyl) phosphate (Reese & Yau, 1978). The product of this reaction, the oxime ester **13**, was detected by ^{31}P nuclear magnetic resonance (NMR) spectroscopy and shown to undergo base-catalyzed decomposition to produce diphenyl phosphate (**14**) and 4-nitrobenzonitrile (**15**). As a consequence of these elegant investigations, the 2-chlorophenyl group was shown to be the preferred aryl protecting group for oligodeoxyribonucleotide synthesis by the phosphotriester approach. It is completely stable during chain assembly but readily removed by selected oximate ions with little cleavage of internucleotide linkages.

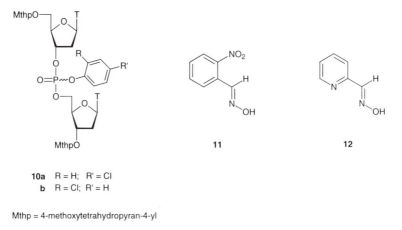

10a R = H; R' = Cl
b R = Cl; R' = H

Mthp = 4-methoxytetrahydropyran-4-yl

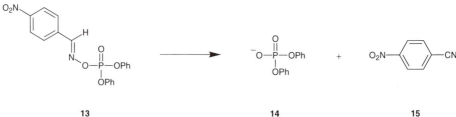

13   14   15

Through the years, the synthesis of building blocks analogous to **7** has evolved so as to eliminate the formation of symmetrical (3′ → 3′)- and (5′ → 5′)-phosphotriesters (Reese, 1978). When bifunctional phosphorylating agents, such as phenyl phospho-rodichloridate (**16**) or reagents similar to **6**, are used in the phosphorylation of stoi-chiometric amounts of protected nucleosides similar to **5**, some symmetrical (3′ → 3′)-phosphotriester **18** is formed along with the activated intermediate **17**. Both **17** and unreacted **16** can then condense with the incoming nucleoside (analogous to **8b**) to produce the desired asymmetrical (3′ → 5′)- and the respective symmetrical (5′ → 5′)-phosphotriesters **19** and **20**. These symmetrical by-products (**18** and **20**) are undesirable not only because they lower the yield of **19**, but also because they create purification problems, especially with high molecular weight oligodeoxyribonucleotides (Arentzen & Reese, 1977). Consequently, monofunctional phosphorylating agents are the reagents of choice for preparing building blocks analogous to **7**. Among the various reagents reported in the literature, the phosphorochloridates **21** and **22** are the most frequently used (Sonveaux, 1986). For example, reaction of 5′-*O*-di-*p*-methoxytrityldeoxynucleosides (**23**) with phosphorochloridate **21** in the presence of 1-methylimidazole produces the deoxyribonucleoside phosphate triester **24**. Further treatment of **24** with triethylamine in pyridine affords the corresponding phosphodiester **25** (Adamiak et al., 1976; Crea et al., 1978). Alternatively, the reaction of the phosphotriester **26** with activated zinc and catalytic amounts of 2,4,6-triisopropylbenzenesulfonic acid in pyridine rapidly (2 min) removes the 2,2,2-trichloroethyl group to give the phosphodiester **27** in quantitative yield (de Rooij et al., 1979). Another highly successful synthesis of **24** involves the reaction of **23** with the bis-triazolide of *p*-chlorophenyl phosphate (**28**) followed by the addition of 3-hydroxypropionitrile. The fully protected triester can then be isolated in near quantitative yield by silica gel column chromatography (Itakura et al., 1974). Unlike phosphorodichloridates, the bis-triazolide derivatives do not lead to the formation of symmetrical dimers.

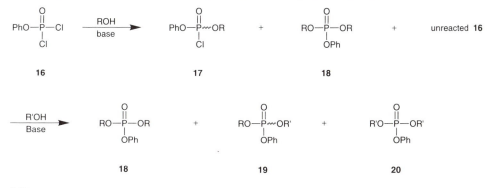

ROH = 5′-protected deoxyribonucleoside; R'OH = 3′-protected deoxyribonucleoside

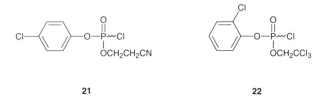

A critical step in the phosphotriester methodology is the activation of phosphodiester building blocks, such as **25**, prior to chain extension. Arylsulfonyl chlorides (phenyl-, tolyl-, 2,4,6-trimethylphenyl-, 2,4,6-triisopropylphenyl-, and 8-quinolylsulfonyl chlorides) have been used extensively for this purpose (Lohrmann & Khorana, 1966; Takaku et al., 1979). To minimize sulfonylation of the free 5′-hydroxyl function of incoming nucleosides or nucleotides, sterically hindered arylsulfonyl chlorides are preferred for activation of phosphodiesters. In this context, sterically hindered aryl-

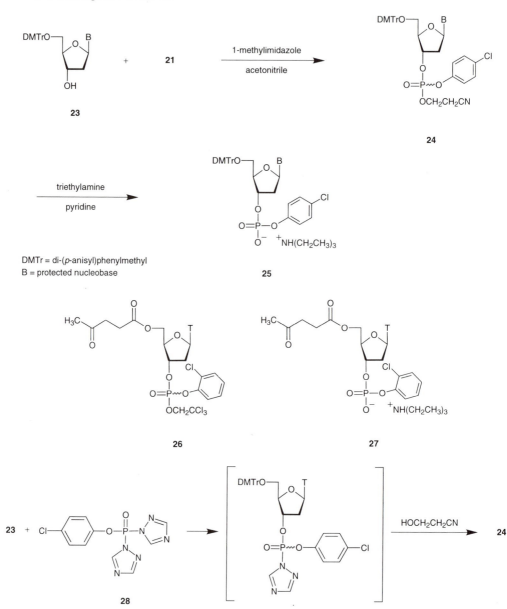

DMTr = di-(*p*-anisyl)phenylmethyl
B = protected nucleobase

sulfonyl azolides, because of their low reactivity toward alcoholic functions, have been used even more successfully than the chlorides. While various imidazolides (Berlin et al., 1973), triazolides (Katagiri et al., 1974), and 4-nitroimidazolides (van Boom et al., 1977) proved to be only marginally useful as coupling agents, tetrazolide **29** (Stawinski et al., 1976) and 3-nitrotriazolide **30** (Chattopadhyaya & Reese, 1979; Reese et al., 1978) had satisfactory condensation kinetics in pyridine. It must be noted that the arylsulfonyl 3-nitrotriazolide **30** is considerably more stable than **29**, which must be prepared immediately prior to use.

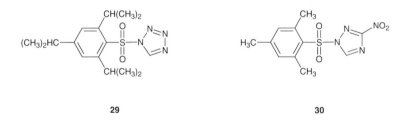

29   30

The mechanism for coupling a phosphodiester salt (e.g., **25**) with the 5'-hydroxyl of 3'-O-acetylthymidine in the presence of an arylsulfonyl chloride or azolide has been studied by ^{31}P NMR spectroscopy (Chandrasegaran et al., 1984; Zarytova & Knorre, 1984). It would appear that the reaction of **25** with 2,4,6-trimethylbenzenesulfonyl chloride (MS-Cl) or **30** in pyridine produces the mesitylenesulfonic acid-phosphate mixed anhydride **31** and the symmetrical pyrophosphate tetraester **32**. This step is catalyzed by pyridine or 1-methylimidazole. The reaction of **31** and **32** with 3'-O-acetylthymidine to produce the dinucleoside phosphate triester **34** proceeds slowly in the absence of any catalyst, but rapidly (10 min at 22°C) in the presence of 1H-tetrazole, 3-nitro-1,2,4-triazole, or 1-methylimidazole. It has been proposed that the interaction of 1-methylimidazole or the other azoles with **31** and **32** generates a very reactive intermediate, such as **33**, which rapidly reacts with 3'-O-acetylthymidine to afford **34** (Efimov et al., 1982; Zarytova & Knorre, 1984).

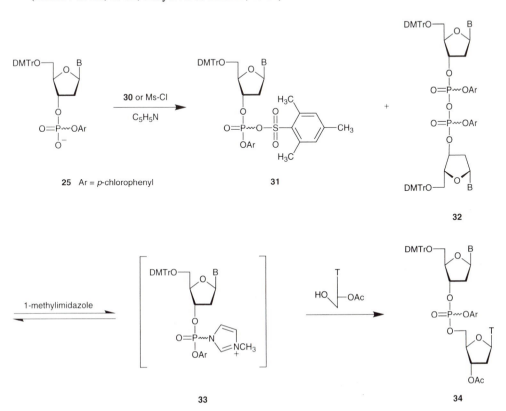

25   Ar = p-chlorophenyl   31   32

33   34

In addition to 1-methylimidazole, certain 4-substituted pyridine 1-oxides have been recommended as powerful nucleophilic catalysts in phosphotriester condensations that involve mesitylenesulfonyl chloride or triisopropylbenzenesulfonyl chloride as the activating agents (Efimov et al., 1985). A complete coupling reaction takes place within 1–2 min when either **35** or **36** is used as the nucleophilic catalyst. It must be noted that condensation reactions in the presence of **35** proceed at a much slower rate in pyridine than in methylene chloride. It should also be noted that 1-hydroxy-4,6-dinitrobenzotriazole (**37**) is a very efficient nucleophilic catalyst. For example, a fully protected dinucleoside phosphotriester can be synthesized within 4 min when mesitylenesulfonyl chloride and **37** are used as condensing agent and nucleophilic catalyst, respectively (Reese & Pei-Zhuo, 1993).

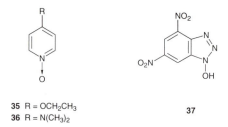

35 R = OCH$_2$CH$_3$
36 R = N(CH$_3$)$_2$

37

One important side reaction during oligodeoxyribonucleotide synthesis by the phosphotriester approach is the modification of N^2-acylguanine and thymine. These nucleobases are sensitive to sulfonylation by condensing agents and produce substitution products in the presence of nucleophiles (Reese & Ubasawa, 1980; Reese & Skone, 1984). In this context (Efimov et al., 1985), the combination of condensing agents (mesitylenesulfonyl chloride, 2,4,6-triisopropylbenzenesulfonyl chloride, and mesitylenesulfonyl-3-nitrotriazole) with the nucleophilic catalyst **35** or **36** produces considerably less modification of N^2-acylguanine residues (0.3% to 0.5% per min) than that observed with N-methylimidazole (1% to 3% per min).

Another important contribution to phosphotriester methodology was the development of 2-chlorophenyl-O,O-bis-[1-benzotriazolyl] phosphate (**38**) as a phosphorylating agent (Marugg et al., 1984). This method consists of reacting a suitably protected nucleoside such as **23** with **38** to give the hydroxybenzotriazole phosphotriester **39**. Addition of excess deoxyribonucleoside produced **40** and a dimer having a (3′ → 3′)-phosphotriester linkage. Depending upon the nature of **23**, the yield of **40** varied between 50% and 77%. The coupling reaction time reported from this study was about 1 hr, but others have now shown that the rate of phosphotriester bond formation can be accelerated severalfold in the presence of 1-methylimidazole or 4-substituted pyridine 1-oxides (Efimov et al., 1985). The advantage of this method, relative to other phosphotriester approaches, is the absence of an activating agent which may otherwise lead to modification of guanine and thymine nucleobases (Marugg et al., 1984).

The Phosphotriester Method on Polymer Supports. The pioneering strategy that led to the development of the phosphotriester chemistry (Letsinger & Ogilvie, 1967) was used originally in the stepwise synthesis of di-, tri-, and tetranucleotides on an insoluble polymeric support (Letsinger & Mahadevan, 1965, 1966). The support, a copolymer composed of styrene, p-vinylbenzoic acid, and p-divinylbenzene, was insoluble in water and the organic solvents used in oligodeoxyribonucleotide synthesis. The acid

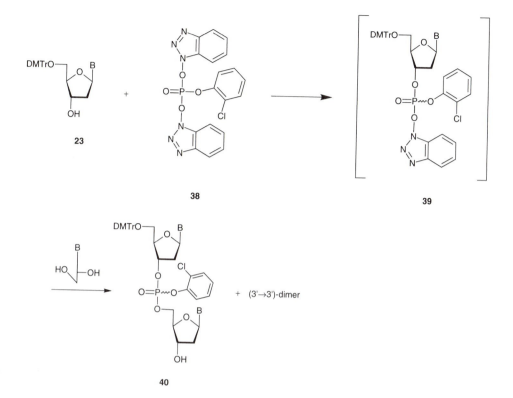

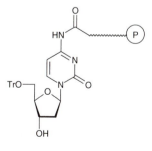

chloride form of the polymer was reacted with the exocyclic amino (Letsinger & Mahadevan, 1965, 1966), 3'-hydroxyl (Letsinger et al., 1967a), or 5'-hydroxyl groups (Letsinger et al., 1967b; Shimidzu & Letsinger, 1968) of appropriately protected nucleosides and afforded, for example, the polymer-bound nucleoside **41**. Chain extension from **41** was achieved using the phosphodiester or phosphotriester chemistries, and the resulting oligomer was cleaved from the support under basic conditions. These first attempts at polymer-supported oligodeoxyribonucleotide synthesis had many of the advantages that were first experienced in peptide synthesis (Erickson & Merrifield, 1976; Letsinger & Kornet, 1963; Merrifield, 1962). These attractive features included separation of the growing oligomer product from solvents, excess reagents, and solu-

41   Tr = triphenylmethyl
        P = polystyrene-based polymer

ble by-products by simple filtration, which avoided numerous time-consuming isolation steps.

This innovative work on polymer-supported oligodeoxyribonucleotide synthesis promoted a tremendous amount of research directed toward devising new solid supports, methods for linking the leader nucleoside to the support, and the development of new strategies for synthesizing polynucleotides on insoluble supports. These studies were important because synthesis on supports must be accompanied by essentially quantitative yields at each step with very few, if any, support-linked by-products. If these criteria are not met, then the final product is severely contaminated with intermediates and by-products with almost no possibility of purifying the oligodeoxyribonucleotide to homogeneity.

Over the years, a large number of different support matrices have been tested for the synthesis of oligodeoxyribonucleotides. Early work focused almost exclusively on organic supports such as polystyrenes, polyamides, and polysaccharides (Amarnath & Broom, 1977; Gait, 1980; Kaplan & Itakura, 1987). However, the swelling of these polymers in organic solvents or the presence of complex, highly porous structures led to limited and unpredictable diffusion of reagents and solvents through the supports. These problems compromised their suitability for the automation of oligodeoxyribonucleotide synthesis. Another important problem that plagued early solid-phase oligodeoxyribonucleotide synthesis strategies was that the chemistries available for the condensation of nucleotide units were inefficient and generated many side products. The combination of these factors painted a very bleak picture for the future of oligodeoxyribonucleotide synthesis on insoluble carriers.

However, rapid progress became possible when high-performance liquid chromatography (HPLC) grade silica was found to give reproducibly high, stepwise yields of oligodeoxyribonucleotides (Matteucci & Caruthers, 1980, 1981). Presumably, good results were obtained because these supports had been designed so that solvents and solutes rapidly diffuse both into and out of the matrix. After these initial observations on silica supports, controlled-pore glass (CPG), which has the same design characteristics and utilizes the same nucleotide chemistries, was found to give somewhat higher oligodeoxyribonucleotide yields (Gough et al., 1981). During the past decade, CPG has been used extensively in the automated synthesis of oligodeoxyribonucleotides. At the present time, the leader nucleoside is attached covalently to the commercially available "long-chain alkylamine" controlled-pore glass (LCAA-CPG) **42** through a succinate linkage. Specifically, the support **42** is first activated under acidic conditions, treated with triethylamine, and then reacted with succinic anhydride to afford **43**. Treatment of the succinylated support **43** with properly protected nucleosides in the presence of 1,3-dicyclohexycarbodiimide (DCC) or 1-(3-dimethylaminopropyl)-3-ethylcarbodiimide (DEC) produced the derivatized matrix **44** (Damha et al., 1990). It is to be noted that amino and carboxy functions from unreacted **42** and **43**, respectively, are chemically inactivated to prevent the formation of oligodeoxyribonucleotidic side products during the following synthesis steps.

The solid-phase synthesis of oligodeoxyribonucleotides by the phosphotriester approach begins with the treatment of the support **44** with a protic acid to remove the 5'-O-dimethoxytrityl group of the anchored nucleoside. Condensation of **45** with the monomeric nucleoside phosphodiester **46**, using **30** as an activating agent and 1-methylimidazole as catalyst, leads to the solid-phase linked dinucleoside phosphotriester **47**. The coupling reaction proceeds for 15 min (Christodoulou, 1993). Conden-

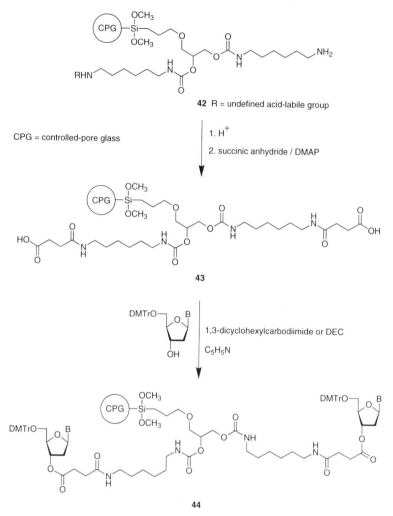

42 R = undefined acid-labile group

CPG = controlled-pore glass

1. H⁺

2. succinic anhydride / DMAP

43

1,3-dicyclohexylcarbodiimide or DEC

C₅H₅N

44

DMAP = 4-dimethylaminopyridine; DEC = 1-(3-dimethylaminopropyl)-3-ethylcarbodiimide

sation reaction time can be reduced to 2–5 min when 1-methylimidazole is replaced by a 4-substituted pyridine 1-oxide such as **35** or **36** (Efimov et al., 1985). The 5'-hydroxyl groups of oligomers that do not undergo condensation (failure sequences) are then treated with acetic anhydride and 4-dimethylaminopyridine or, preferably, 1-methylimidazole in an inert solvent. This capping step prevents failure sequences from reacting during the next condensation step which, in turn, limits the length of failure sequences and facilitates separation of the final product from these side products. In summary, three steps are required for the addition of one nucleotide: deblocking, condensation, and capping. Repetition leads to the full-length oligodeoxyribonucleotide. Following synthesis of the oligodeoxyribonucleotide, the solid support is treated with *syn*-2-nitrobenzaldoxime and 1,1,3,3-tetramethylguanidine in aqueous dioxane to remove the 2-chlorophenyl protecting groups and release the partially deprotected oligodeoxyribonucleotide from the solid support. Complete removal of blocking groups

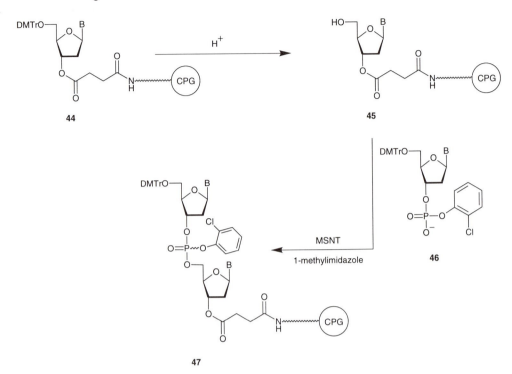

MSNT = 1-(2-mesitylenesulfonyl)-3-nitro-1,2,4-triazole

then follows standard procedures (Christodoulou, 1993). Of interest, the solid-phase synthesis of oligodeoxyribonucleotides using hydroxybenzotriazole-activated phosphotriester intermediates has also been reported (Marugg et al., 1983).

The phosphotriester approach is, however, most attractive for the synthesis of oligodeoxyribonucleotides in solution. Nucleotide building blocks (**25**) and phosphotriester intermediates (**34**) are very stable and easy to handle. Furthermore, only a slight excess of monomer, dimer, and larger building blocks are needed, because small quantities of moisture present in reaction mixtures are eliminated by the excess of condensing agent (Reese & Pei-Zhuo, 1993). In addition to having been applied extensively to the synthesis of linear oligodeoxyribonucleotides, the phosphotriester method has also been useful in the synthesis of cyclic (Rao & Reese, 1989) and branched cyclic oligodeoxyribonucleotides (Sund et al., 1993).

The Phosphoramidite Method. In the mid-1970s, when the phosphotriester approach for the synthesis of oligodeoxyribonucleotides was still sluggish and inefficient, the phosphite triester methodology was developed (Letsinger et al., 1975; Letsinger & Lunsford, 1976). This approach is exemplified by the reaction of a 5′-*O*-protected thymidine (**48**) with the bifunctional phosphitylating agent (2,2,2-trichloroethyl) phosphorodichloridite (**49**) to generate the deoxyribonucleoside-3′-*O*-phosphorochloridite intermediate **50** within 5 min at −78°C. The addition of 3′-*O*-protected thymidine followed by aqueous iodine oxidation resulted in the rapid formation of the dinucleoside phosphate triester **52** in addition to the symmetrical (3′ → 3′)- and (5′ → 5′)-dinucleoside phosphate triesters **53** and **54**, respectively. The preparation of **52** was complete within 1 hr. At the time, the formation of internucleotide linkages with such kinetics was unprecedented.

In the early 1980s, synthesis of the deoxyribonucleoside chlorophosphite **55** and its reaction with a deoxyribonucleoside covalently attached to a silica support (**45**) through a 3'-O-succinate linkage was achieved (Matteucci & Caruthers, 1980, 1981). The dinucleoside phosphate triester **56** was produced in a yield greater than 90%. By adding 1*H*-tetrazole to the deoxyribonucleoside chlorophosphites, condensation rates and coupling yields could be increased significantly relative to the parent chlorophosphites. Such improvements led to the solid-phase preparation of a dodecanucleotide in an overall yield of 55%. These results were very exciting because they demonstrated for the first time that simple mononucleotide synthons could be used to produce biochemically useful oligodeoxyribonucleotides rapidly in high yields. It must, however, be noted that the routine application of nucleoside chlorophosphites or the corresponding tetrazolides to the solid-phase synthesis of oligodeoxyribonucleotides was accompanied by several serious problems. The preparation of these reagents from reactive bifunctional phosphitylating agents had to be performed at low temperature in the absence of moisture and under an inert atmosphere. Because of sensitivity to air oxidation and hydrolysis, they also had to be prepared immediately prior to use and stored under an inert atmosphere at −70°C. An especially troublesome problem was the presence of variable amounts of (3' → 3')-dinucleoside phosphite triesters which were generated during the synthesis of **55**. As a consequence, each preparation of **55** had to be titrated before use in order to ensure that these synthons were present in amounts ad-

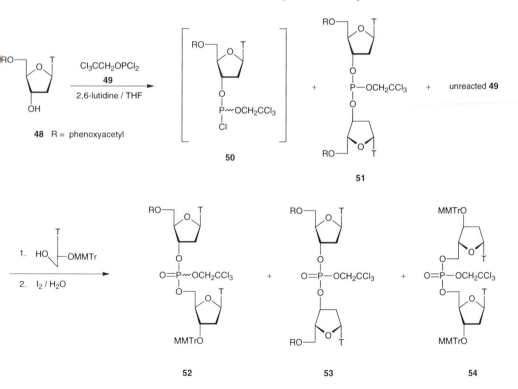

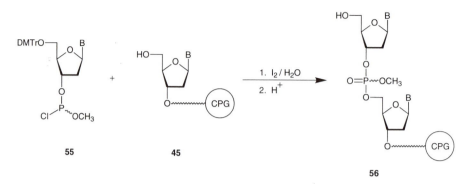

55   **45**

56

equate to afford high yields of the desired product. These various undesirable features prevented synthons such as **55** from being useful in the routine, automated synthesis of oligodeoxyribonucleotides.

These problems were overcome by the development of a new, very important class of reagents called *deoxyribonucleoside phosphoramidites*. Although discovered some time ago (Beaucage & Caruthers, 1981), these reagents are still the synthons of choice for oligodeoxyribonucleotide synthesis and have been absolutely essential in the development of methods for the automated synthesis of oligodeoxyribonucleotides on insoluble polymer supports. There are many reasons why these synthons have proven to be so valuable in oligodeoxyribonucleotide synthesis. For example, deoxyribonucleoside phosphoramidites are stable to hydrolysis and air oxidation under normal laboratory conditions, but readily activated to form an internucleotide linkage in essentially quantitative yield with very few side products. Thus, two of the most important criteria for the solid-phase synthesis of oligodeoxyribonucleotides—high yields and few side products—are met by deoxyribonucleoside phosphoramidites. The synthesis of these reagents is also straightforward. The approach consists of reacting the protected deoxyribonucleosides **23** with chloro-(*N*,*N*-dimethylaminomethoxyphosphine and *N*,*N*-diisopropylethylamine. The products (**57**) are isolated by conventional extraction procedures, precipitated, and stored as dry powders.

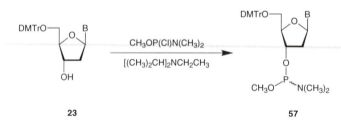

23   **57**

The elegance of the phosphoramidite approach emerges from the conversion of the relatively stable deoxyribonucleoside phosphoramidite derivatives into highly reactive intermediates suitable for oligodeoxyribonucleotide synthesis. Specifically, the interaction of *N*,*N*-dimethylaniline hydrochloride with **57** affords the corresponding deoxyribonucleoside chlorophosphite **55** which, upon reaction with 3'-*O*-levulinylthymidine, affords the (3' → 5')-dinucleoside phosphite triesters **58** in almost quantitative yield. Tertiary amine hydrochlorides are, however, hygroscopic; this creates a problem for reliable oligodeoxyribonucleotide synthesis, because anhydrous conditions are required for optimum coupling reactions. The search for nonhygroscopic,

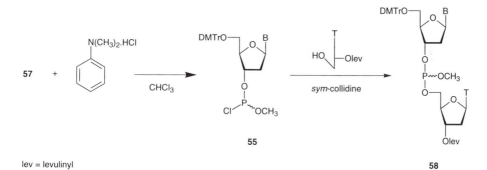

55

58

lev = levulinyl

weak acids capable of activating deoxyribonucleoside phosphoramidites led to the commercially available 1*H*-tetrazole which can be purified and dried by sublimation. The addition of 1*H*-tetrazole to **57** and 3'-*O*-levulinylthymidine in dry acetonitrile affords **58** in quantitative yields within a few minutes as estimated by ^{31}P NMR spectroscopy (Beaucage & Caruthers, 1981). This strategy—that is, the use of deoxyribonucleoside phosphoramidites with 1*H*-tetrazole—has been successfully applied to the solid-phase synthesis of oligodeoxyribonucleotides of varying chain lengths (Caruthers, 1981; Caruthers et al., 1982; Josephson et al., 1984).

The use of deoxyribonucleoside phosphoramidites **57** in automated systems proved to be unreliable because these monomers were unstable in acetonitrile, the solvent used during condensations. This instability, which varied from hours to weeks, appeared to be due to the presence of variable amounts of acidic contaminants in the deoxyribonucleoside phosphoramidites. As a result of these observations, several *N*,*N*-dialkylamino substituents were tested as potential derivatives for the automated solid-phase synthesis of oligodeoxyribonucleotides (Adams et al., 1983; McBride & Caruthers, 1983). For example, the deoxyribonucleoside phosphoramidites **59**, unlike **57**, could be purified by silica gel chromatography and were stable in acetonitrile for at least a month without significant decomposition. These phosphoramidites were used to synthesize several oligodeoxyribonucleotides on silica gel (Dörper & Winnacker, 1983), CPG (Sinha et al., 1984a), or cellulose filter discs (Ott & Eckstein, 1984). Similarly, the deoxyribonucleoside phosphoramidites **60** did not show significant decomposition in acetonitrile. These derivatives were used to prepare 51-mers on CPG which, at the time, were the largest DNA segments ever synthesized chemically (Adams et al., 1983).

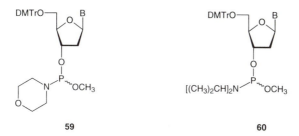

59

60

One disadvantage in using deoxyribonucleoside phosphoramidites **60** in the automated synthesis of oligodeoxyribonucleotides was the thiolate treatment which was required for the removal of the methyl phosphate protecting groups (Andrus & Beaucage, 1988; Daub & van Tamelen, 1977). In order to eliminate this postsynthesis deprotection protocol, the phosphoramidites **61** and **62** were prepared from protected deoxyribonucleosides and the monofunctional phosphitylating reagents **63** and **64**, respectively, under conditions similar to those reported by Beaucage and Caruthers (Sinha et al., 1984b). Although phosphoramidites **61** were more stable than phosphoramidites **60** in wet acetonitrile, as measured by ^{31}P NMR spectroscopy (Zon et al., 1985), cleavage of the β-cyanoethyl phosphate protecting group from oligodeoxyribonucleotides could be completed under the basic conditions required for the deprotection of nucleobase protecting groups. This property, which eliminated thiophenol from deprotection protocols, was useful for the automated solid-phase synthesis of oligodeoxyribonucleotides.

An alternate strategy for synthesizing deoxyribonucleoside phosphoramidites involves the reaction of properly protected deoxyribonucleosides with bis-(pyrrolidino)methoxyphosphine activated by 4,5-dichloroimidazole (Beaucage, 1984; Moore & Beaucage, 1985). ^{31}P NMR spectroscopy indicated that the corresponding deoxyribonucleoside phosphoramidites **65** were produced in yields exceeding 86% within 10 min. Without further purification, these monomers could be activated with 1*H*-tetrazole and used to synthesize oligodeoxyribonucleotides. This approach eliminated problems associated with the isolation and purification of deoxyribonucleoside phosphoramidites as well as those pertaining to the stability of the phosphoramidites in acetonitrile. In spite of these useful features, the methodology does not produce deoxyribonucleoside phosphoramidites *in situ* with complete chemoselectivity; (3′ → 3′)-dinucleoside methyl phosphite triester contaminants (less than 10%) were also detected by ^{31}P NMR spectroscopy. The selective activation of bis-(*N*,*N*-dialkylamino) alkoxyphosphines was investigated by others subsequently. It was shown that the reaction of deoxyribonucleosides **23** with the phosphorodiamidite **66** and limited amounts of 1H-tetrazole or its *N*,*N*-diisopropylammonium salt afforded, within about 1 hr, the corresponding deoxyribonucleoside phosphoramidites **60** in isolated yields varying between 82% and 92% (Barone et al., 1984; Lee & Moon, 1984). Less than 1% of (3′ → 3′)-dimers were observed. This approach has become the method of choice for preparing deoxyribonucleoside phosphoramidite monomers because of the relative stability of phosphorodiamidites, the high selectivity of activation, and the mildness of the phosphitylation conditions.

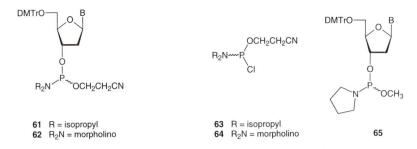

61 R = isopropyl
62 R₂N = morpholino

63 R = isopropyl
64 R₂N = morpholino

65

The preparation of deoxyribonucleoside phosphoramidites from suitably protected deoxyribonucleosides and monochlorophosphoramidite or phosphorodiamidite derivatives is facile and proceeds in high isolated yields. Furthermore, deoxyribonucleoside phosphoramidites structurally related to **61**, upon activation with 1*H*-tetrazole, can be used to synthesize relatively large oligomers (150-mer) on nonporous silica microbeads

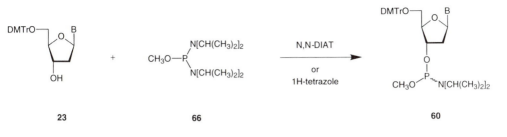

23   **66**   **60**

N,N-DIAT = N,N-diisopropylammonium salt of 1H-tetrazole

(Seliger et al.,1989) or rigid nonswelling polystyrene beads (McCollum & Andrus, 1991; Vu et al., 1990). Stepwise yields average 98% to 99%. The dinucleotide phosphoramidites **67** and **68** can also be used for solid-phase synthesis of oligodeoxyribonucleotides (Kumar & Poonian, 1984; Wolter et al., 1986). Activation of **67** with 1H-tetrazole led to coupling efficiencies (~99%) similar to those observed with monomeric phosphoramidites (Kumar & Poonian, 1984). Similarly, activation of **68** with 5-(p-nitrophenyl)-1H-tetrazole was used to prepare a large oligomer (101-mer; Wolter et al., 1986).

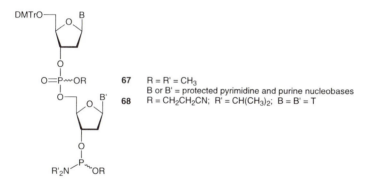

67   R = R' = CH$_3$
B or B' = protected pyrimidine and purine nucleobases
68   R = CH$_2$CH$_2$CN; R' = CH(CH$_3$)$_2$; B = B' = T

Recently, the mechanism of activation of deoxyribonucleoside phosphoramidites by 1H-tetrazole has attracted considerable attention. In this context, diethoxy-(N,N-diisopropylamino)phosphine (**69**) and diethoxy-(N-tetrazolyl)phosphine (**70**) were used in attempts to assess whether activation of **60** by 1H-tetrazole generates **71** as an intermediate (Berner et al., 1989). Analysis of **70** by ^{31}P NMR spectroscopy indicated a resonance at 126 ppm. This resonance was also observed upon treatment of **69** or **60** with 1H-tetrazole in acetonitrile. Furthermore, the Rp or Sp diastereoisomers of **60** underwent rapid epimerization at phosphorus under similar conditions and generated

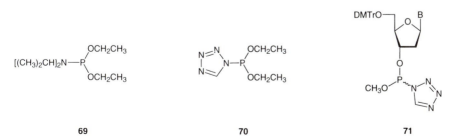

69   **70**   **71**

racemic deoxyribonucleoside phosphite triesters upon reaction with ethanol (Stec & Zon, 1984). These experiments were consistent with the proposal that activation of deoxyribonucleoside phosphoramidites by 1*H*-tetrazole occurred through a rapid protonation followed by the reversible and relatively slow formation of the phosphorotetrazolide intermediates **71**. In addition to 1*H*-tetrazole, several other reagents have been used to activate deoxyribonucleoside phosphoramidites. These include *N*-methylanilinium trifluoroacetate (Fourrey & Varenne, 1984), *N*-methylanilinium trichloroacetate (Fourrey et al., 1987), *N*-methylimidazole hydrochloride (Hering et al., 1985), *N*-methylimidazolium trifluoromethanesulfonate (Hostomsky et al., 1987), 5-trifluoromethyl-1*H*-tetrazole (Hering et al., 1985), 5-(*p*-nitrophenyl)-1*H*-tetrazole (Froehler & Matteucci, 1983), and 1-hydroxybenzotriazole (Claesen et al., 1984).

Dahl and his colleagues investigated the factors that influence the coupling rates of activated deoxyribonucleoside phosphoramidites. They observed that these rates varied with the *O*-alkyl/aryl and *N*,*N*-dialkylamino functions of the phosphoramidites. Decreased coupling rates were observed in the order *O*-methyl > *O*-(2-cyanoethyl) > *O*-(1-methyl-2-cyanoethyl) > *O*-(1,1-dimethyl-2-cyanoethyl) >> *O*-(*o*-chlorophenyl) and *N*,*N*-diethylamino > *N*,*N*-diisopropylamino > *N*-morpholino > *N*-methylanilino (Dahl et al., 1987). The steric bulk of substituents at the guanine exocyclic amino function also significantly altered coupling rates and yields. Specifically, the activation of **72** with 1*H*-tetrazole and a condensation time of 10 min resulted in coupling yields averaging 45% on a CPG support (Sekine et al., 1986). Such coupling yields were far below those obtained with the commonly used deoxyribonucleoside phosphoramidites (98% to 99%), thus highlighting the importance of steric hindrance when designing nucleobase protecting groups or modified bases for incorporation into oligodeoxyribonucleotides.

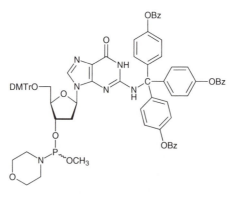

72 Bz = benzoyl

Despite the high coupling efficiency of the phosphoramidite approach, chain extension does not occur quantitatively even under optimum conditions. Consequently, the final, unpurified oligodeoxyribonucleotide product also contains a mixture of shorter oligomers. For the longer products, isolating the desired *n*-mer oligodeoxyribonucleotide from the large number of (*n* − 1)-mers can be very difficult. In order to obviate this problem, unphosphitylated oligomer intermediates are acetylated (capping step) in each synthesis cycle. As a consequence, these oligomers are unreactive during subsequent condensation steps and, therefore, do not produce a significant (*n* − 1)-

mer fraction of side products. Instead, these failure sequences produce a mixture of variable-length side products that are easily separated from the full-length oligodeoxyribonucleotide. The most effective capping formulation is a mixture of acetic anhydride, 2,6-lutidine, and 1-methylimidazole in tetrahydrofuran (Farrance et al., 1989). In addition to preventing the elongation of unphosphitylated oligomers, the capping reagent also effectively reduces the concentration of O6-phosphitylated guanine residues that are generated during the coupling step (Eadie & Davidson, 1987; Pon et al., 1986). These latter adducts, if not eliminated as phosphites by capping or a basic aqueous wash (Beaucage & Caruthers, 1981; Drahos et al., 1982), serve as secondary sites for further oligomer synthesis which leads to branched oligodeoxyribonucleotides. The presence of these ring adducts and the need to remove them has been recognized for some time. All the early synthesis protocols contained an aqueous wash or a capping step prior to oxidation (Caruthers, 1981, 1985; Caruthers et al., 1982, 1987a; deHaseth et al., 1983; Drahos et al., 1982).

Oxidation of the internucleotide phosphite triester to the phosphate triester represents an important step in the automated synthesis of oligodeoxyribonucleotides by the phosphoramidite approach. This is because phosphite triesters are very labile to acid. Thus, if the pentavalent derivative is not formed prior to each acidic detritylation step, degradation of the trivalent phosphite internucleotide linkage occurs (Matteucci and Caruthers, 1981). An aqueous solution of iodine in 2,6-lutidine (Letsinger et al., 1975) or pyridine (Usman et al., 1985) is generally used for this task primarily because these reagents are stable and rapidly generate the oxidized product without any evidence of side reactions. The oxidation occurs with an overall retention of configuration (Cullis, 1984). For certain applications, nonaqueous oxidizing reagents may advantageously offer an alternative to aqueous iodine for the oxidation of oligodeoxyribonucleoside phosphite triesters. Particularly, dinitrogen tetroxide (Bentrude et al., 1989), tert-butyl hydroperoxide (Hayakawa et al., 1986), di-tert-butyl hydroperoxide (Hayakawa et al., 1986), cumene hydroperoxide (Hayakawa et al., 1986), hydrogen peroxide (Hayakawa et al., 1986), bis-trimethylsilyl peroxide in the presence of catalytic amounts of trimethylsilyl triflate (Hayakawa et al., 1986), m-chloroperbenzoic acid (Tanaka & Letsinger, 1982), iodobenzene diacetate (Fourrey & Varenne, 1985), tetra-n-butylammonium periodate (Fourrey & Varenne, 1985), and molecular oxygen in the presence of 2,2'-azobis(2-methylpropionitrile) (AIBN) under thermal or photochemical conditions (Bentrude et al., 1989) have been effective.

Interestingly, aqueous iodine oxidation of the internucleotide o-methylbenzyl phosphite ester, which is generated by coupling **73** to the growing DNA segment, generates a phosphate diester (Caruthers et al., 1987b). Thus, the o-methylbenzyl protecting group is lost during oxidation. It was further demonstrated that the absence of phosphate protecting groups did not inhibit the subsequent elongation of the DNA chain. Indeed, the deoxyribonucleoside phosphoramidite **73** was used to prepare an oligothymidylic acid (20-mer) on a solid support with an average stepwise yield of 96%. It was speculated that phosphate–phosphite mixed anhydrides, which presumably formed from the interaction of phosphate diesters with activated deoxyribonucleoside phosphoramidites, were cleaved by excess $1H$-tetrazole to regenerate deoxyribonucleoside phosphorotetrazolide intermediates. An extension of this work provides new insights on the mechanism for removing the phosphorus protecting group (Nielsen and Caruthers, 1988). It was shown that alkyl deoxyribonucleoside phosphites having alkyl groups which stabilize the S_N1 character of the second stage of the Arbuzov reaction

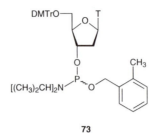

73

react selectively with most electrophiles and eliminate only the appropriate tertiary or benzylium carbocation protecting group. For example, when the protecting group is 2-cyano-1,1-dimethylethyl (**74**), selective elimination occurs under iodine oxidation conditions and produces a ^{31}P NMR signal corresponding to the deoxydinucleoside phosphoryl iodide **75**. Based upon this concept, it was further shown that iodine oxidation in the presence of water, alcohols, and amines produces the corresponding dinucleoside phosphate **76**, phosphate triester **77**, and phosphoramidate **78**, respectively. It was also demonstrated that the reaction of **74** with azides led to phosphoramidates (**78**) and phosphoroazoamidates, depending on the azide used. The reaction was, therefore, postulated to proceed by first selectively eliminating the appropriate alkyl protecting group as the carbocation under oxidative, Arbuzov-type conditions and produce a dinucleoside phosphoryl iodide (**75**). The next step would be attack of the appropriate nucleophile on the phosphoryl iodide to generate the pentavalent derivative. Another possi-

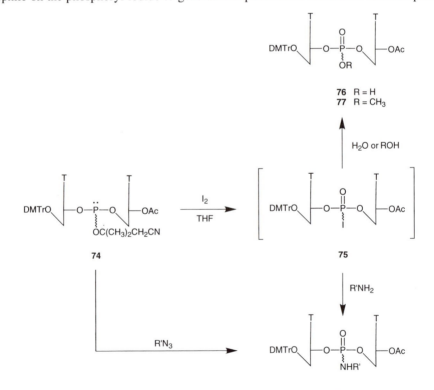

ble reaction pathway could involve the iodophosphonium ion intermediate. However, irrespective of the mechanism, the use of this class of protecting groups could provide a versatile pathway for generating a large variety of oligodeoxyribonucleotide analogs from a common synthon.

The versatility of deoxyribonucleoside phosphoramidites has been further illustrated by the solid-phase synthesis of oligodeoxyribonucleotides from 5′-*O*-DMTr-2′-deoxyribonucleoside-3′-*O*-(*N*,*N*-diisopropylamino)methoxyphosphines having unprotected nucleobases. The success of the approach depends upon a modified protocol involving treatment with an equimolar solution (0.1 M) of pyridine hydrochloride and aniline in acetonitrile immediately after each coupling reaction and prior to the oxidation step. This procedure cleaves nucleobase adducts originating from the oligodeoxyribonucleotidic chain (Gryaznov & Letsinger, 1991). This strategy should facilitate the solid-phase synthesis of oligodeoxyribonucleotides bearing base-sensitive functional groups.

The H-Phosphonate Method. The application of nucleoside *H*-phosphonates to the synthesis of oligodeoxyribonucleotides was reported in the 1950s. Typically, the interaction of 2′,3′-*O*-isopropylideneuridine-5′-hydrogenphosphonate (**79**) with diphenylphosphorochloridate produced a reactive intermediate, presumably a mixed anhydride, which reacted with 2′,3′-*O*-isopropylideneadenosine to afford the corresponding dinucleoside hydrogen phosphonate **80** in high yields (Hall et al., 1957). The phosphonate linkage was easily oxidized by *N*-chlorosuccinimide to the phosphodiester **81**.

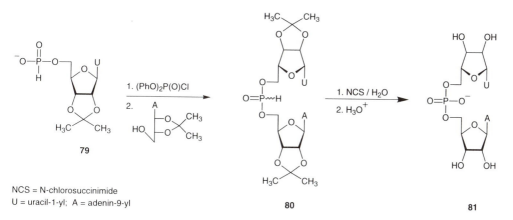

NCS = N-chlorosuccinimide
U = uracil-1-yl; A = adenin-9-yl

Polymer-supported synthesis of short oligodeoxyribonucleotides using nucleoside *H*-phosphonates anchored to a polymeric support was described in the early 1970s. Specifically, the nucleosidic polymer **82** was converted to **83** upon reaction with phosphorus trichloride and subsequent hydrolysis. The nucleoside hydrogen phosphonate **83** was then activated with mercuric chloride in pyridine, in the presence of unprotected deoxyribonucleosides, to produce the dinucleotides **84** in yields varying between 46% and 80% (Kabachnik et al., 1971). Tri- and tetranucleotides were synthesized by this approach in considerably lower yields. Oligodeoxyribonucleotides were released from the solid support by treatment with dilute trifluoroacetic acid.

The formation of internucleotide linkages via *H*-phosphonate intermediates was rein-

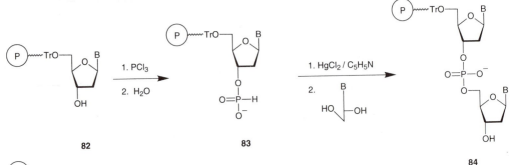

P = polystyrene-based polymer

vestigated in the mid-1980s. It was found that the nucleoside *H*-phosphonate **85** reacted rapidly with 3'-*O*-benzoylthymidine, in the presence of chlorodiphenylphosphate and 1-methylimidazole, to give the dinucleoside hydrogen phosphonate **87** in near quantitative yields (Garegg et al., 1985). It was postulated that the reactive intermediate was the mixed anhydride **86**. One serious limitation of this approach is the activation of **85** which requires very careful control of the reaction conditions. For example, when activated **85** was kept in pyridine for more than 5 min before adding 3'-*O*-benzoylthymidine, lower coupling yields were observed. This was due to the formation of trinucleoside metaphosphite **88** (Garegg et al., 1987b), which, depending on the amount of diphenylchlorophosphate used, was slowly converted to the nucleoside phosphorodichloridite **89** according to ^{31}P NMR spectroscopy (Stawinski et al., 1988a). Acyl chlorides, such as pivaloyl chloride (Froehler et al., 1986) and adamantoyl chloride (Andrus et al., 1988), have also been successfully used for activation of nucleoside *H*-phosphonates. Similar to activation with diphenylchlorophosphate, treatment of the nucleoside *H*-phosphonate **85** with pivaloyl chloride rapidly generated the mixed anhydride **90** (de Vroom et al., 1987; Garegg et al., 1987a). This intermediate reacted with 3'-*O*-benzoylthymidine to produce the dinucleoside H-phosphonate **87**. In the absence of the reactive nucleosides, excess pivaloyl chloride reacted with **90** to generate the bis-acylphosphite **91** and, at a much slower rate, the side product **92** from an Arbuzov-type reaction (de Vroom et al., 1987). It is, therefore, important to limit the coupling reaction time to about 1 min in order to minimize the production of several undesirable by-products. In contrast, it has recently been shown that activation of deoxyribonucleoside *H*-phosphonates with dipentafluorophenyl carbonate reduced the formation of side products considerably during preactivation of nucleoside *H*-phos-

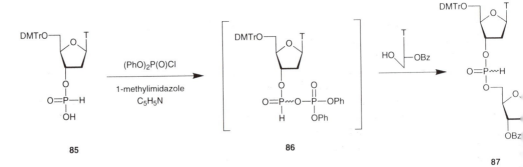

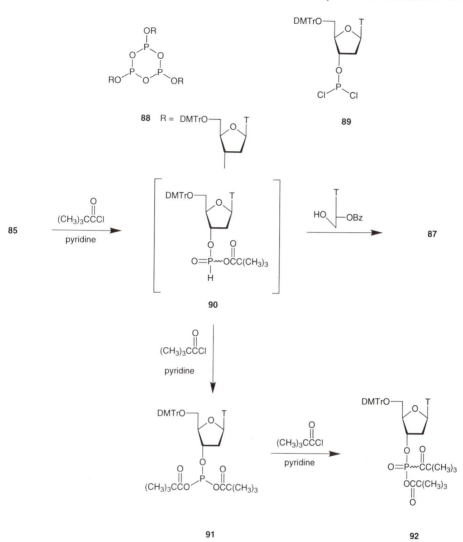

phonates (Efimov et al., 1993). This carbonate may become the activator of choice for nucleoside *H*-phosphonate monomers. Activation of nucleoside *H*-phosphonates followed by coupling to a nucleoside linked to a solid support resulted in the synthesis of oligodeoxyribonucleotide *H*-phosphonates (**93**) which could be oxidized to natural oligomers (**94**) by treatment with aqueous iodine (Froehler et al., 1986). Alternatively, oxidation with elemental sulfur in carbon disulfide and triethylamine afforded oligodeoxyribonucleotide phosphorothioates (**95**) (Froehler, 1986). By appropriately choosing the oxidants, other analogs are also possible. For example, the use of a primary or secondary amine produced the corresponding phosphoramidates **96**, whereas oxidation in the presence of methanol or *n*-butanol generated the phosphotriesters **97** (Froehler, 1986). Although the *H*-phosphonate procedure is not as efficient as the phosphoramidite methodology for the synthesis of oligodeoxyribonucleotides, it would appear that this approach is well-suited for the synthesis of selected analogs.

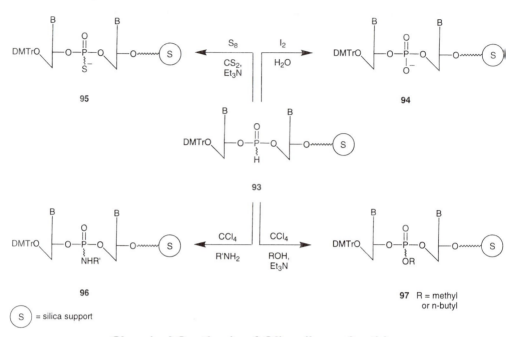

95   94

93

96   97   R = methyl or n-butyl

(S) = silica support

Chemical Synthesis of Oligoribonucleotides

Because an additional protecting group is required for the 2'-hydroxyl, RNA synthesis methods are more complex and generally afford lower yields than do comparable DNA procedures. This protecting group must remain stable during chain assembly and, for most synthetic strategies, be the last blocking group cleaved during the deprotection cycle. It must also be removed quantitatively under conditions that will not lead to cleavage or migration of the internucleotide linkages. Ultimately, the choice for 2'-protection will restrict and define the type of blocking groups that can be used on the nucleobases, phosphorus atom, and the 5'-hydroxyl group. Another important consideration is the steric bulk of the 2'-protecting group which may affect the rate of chemical processes occurring at the 3'-position. Regioselective incorporation of a 2'-hydroxyl protecting group in the cis-2',3'-diol system of ribonucleosides (**98**) can most easily be initiated by the use of 1,3-dichloro-1,1,3,3-tetraisopropyldisiloxane (**99**) to block the 5'- and 3'-hydroxyl functions simultaneously (**100**) (Markiewicz and Wiewiórowski, 1978; Markiewicz, 1985). This protecting group (TIPDS) is relatively stable to acidic and basic conditions but can be cleaved quantitatively by 1 M tetra-n-butylammonium fluoride in tetrahydrofuran within 10 min. Because of its facile introduction, stability, and compatibility with a variety of protecting groups, the TIPDS group is generally useful for the incorporation of most ribonucleoside 2'-hydroxyl protecting groups.

The question of whether 2'-protection should be with an acid- or base-labile group has received considerable attention. A number of years ago (Brown et al., 1956), it was shown that partial hydrolysis of cytidine 3'-O-benzylphosphate (**101**) under basic conditions afforded a mixture of 2'- and 3'-monophosphates (**103** and **104**). The latter products were postulated to form through the intermediacy of an unstable cyclic phosphate (**102**). Alternatively, the partial hydrolysis of **101** under acidic conditions pro-

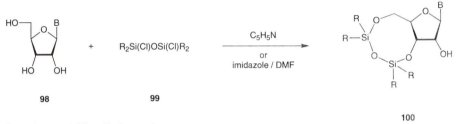

98   **99**   **100**

B = purine or pyrimidine; R = isopropyl

duced the aralkylphosphate ester **106** via the hydrolytically labile cyclic phosphate **105**, in addition to the monophosphates **103** and **104**. This model experiment indicated that the conversion of a $(3' \rightarrow 5')$ to a $(2' \rightarrow 5')$-internucleotide linkage can occur with phosphodiester and protected phosphotriester RNA oligomers under either basic or acidic conditions (Reese, 1978). This is a serious problem, because the separation of oligoribonucleotides containing one or multiple $(2' \rightarrow 5')$-internucleotide linkages from natural oligoribonucleotides is impossible via current technologies. It is, therefore, very important to avoid phosphoryl migration or chain cleavage during removal of the 2'-hydroxyl protecting group.

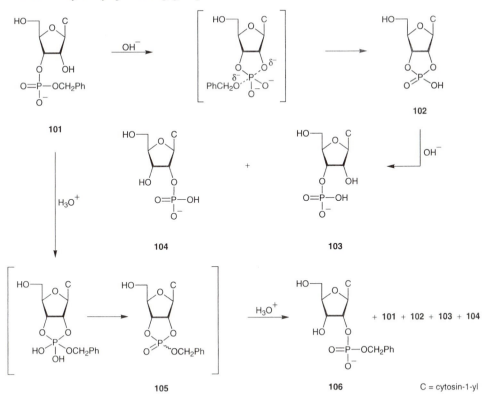

101   **104**   **103**   **102**

105   **106**   C = cytosin-1-yl

Given the above considerations, the acid-labile tetrahydropyran-2-yl group (Thp) has been tested as a 2'-OH protecting group (**107**) in oligoribonucleotide synthesis (Griffin & Reese, 1964). The conditions necessary for its complete removal were relatively mild (0.01 N HCl, ~3–4 hr at 20°C) and did not induce significant phosphoryl migration. This group has been used extensively in the synthesis of oligoribonucleotides by the phosphotriester approach, but its application to solid-phase synthesis is limited to relatively short segments (20-mers or less) (Kierzek et al., 1986; Tanimura et al., 1987). This is because current strategies for oligoribonucleotide synthesis involve 5′-*O*-pixyl or 5′-*O*-dimethoxytrityl protection of ribonucleoside synthons. These groups are removed with acid under conditions that also partially hydrolyze the 2′-Thp group (Christodoulou et al., 1986). Following partial deprotection, acid-catalyzed internucleotide bond cleavage and scrambling to mixtures of (2′ → 5′)- and (3′ → 5′)-linkages lead to the rapid accumulation of various side products and limit the usefulness of the approach. One additional disadvantage inherent to the Thp group is its chirality, which leads to the formation of diastereoisomers and complicates the purification of ribonucleoside building blocks. The achiral 4-methoxytetrahydropyran-4-yl (Mthp) has therefore replaced the Thp group for 2′-protection (**108**) and is used widely in the synthesis of oligoribonucleotides (Reese et al., 1970). However, due to its acid lability, the 2′-*O*-Mthp is also incompatible with the 5′-*O*-pixyl or 5′-*O*-dimethoxytrityl groups for solid-phase oligoribonucleotide synthesis. To circumvent this problem, the base-labile 9-fluorenylmethoxycarbonyl (Fmoc) blocking group has been tested at the 5′-position. Using 5′-*O*-Fmoc and 2′-*O*-Mthp as complementary base- and acid-labile protecting groups, respectively, ribonucleoside phosphoramidites **109** were applied to a solid support oligoribonucleotide synthesis strategy (Lehmann et al., 1989). The activation of **109** with 5-(*p*-nitrophenyl)-1*H*-tetrazole produced a coupling efficiency of about 96% in 2–10 min. The 5′-*O*-Fmoc group was cleaved, in a stepwise manner, with DBU; the release of Fmoc was monitored spectrophotometrically at 305 nm.

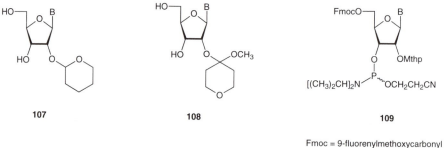

107   108   109

Fmoc = 9-fluorenylmethoxycarbonyl
Mthp = 4-methoxytetrahydropyran-4-yl

The search for a 2′-*O*-protecting group that was stable under the acidic conditions required for complete removal of the 5′-*O*-pixyl or 5′-*O*-dimethoxytrityl groups led to the development of the 1-[(2-chloro-4-methyl)phenyl]-4-methoxypiperidin-4-yl (Ctmp) group for the 2′-protection of ribonucleosides (Reese et al., 1986). This protecting group was designed in such a way that the tertiary anilino function of the acetal **110** would be mostly unprotonated at pH 2–2.5. Hydrolysis would therefore occur at a rate similar to that observed with the Mthp group. At lower pH, however, the protonation of the anilino function (**111**) would generate a strong inductive effect and enhanced stability under the conditions used to remove 5′-dimethoxytrityl or pixyl blocking groups. For example, the rate of hydrolysis of 2′-*O*-Mthp uridine was faster by a factor of 140 at pH 1 when compared to that observed at pH 3. In contrast, the rate of hydrolysis of

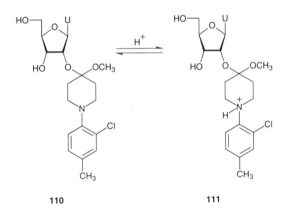

110   111

2'-O-Ctmp uridine was faster at pH 1 than at pH 3 by only a factor of about 2, and both Ctmp and Mthp groups hydrolyzed at similar rates at pH 2. Because of its enhanced stability at low pH, the compatibility of the Ctmp group with 5'-O-pixyl was evaluated by solid-phase synthesis of a nonadecaribonucleotide. Activation of the ribonucleoside phosphoramidites **112** with 5-(*p*-nitrophenyl)-1*H*-tetrazole led to an average stepwise yield of 93% with a coupling time of 15 min (Rao et al., 1987). The *H*-phosphonate approach was also used to test the combination of 5'-O-DMTr and 2'-O-Ctmp protecting groups during the solid-phase synthesis of an octadecaribonucleotide (Sakatsume et al., 1989). Although Ctmp was satisfactory for protecting the ribonucleoside 2'-hydroxyl group, the more accessible 1-(2-fluorophenyl)-4-methoxypiperidin-4-yl (Fpmp) group was proposed as an alternative because acidic hydrolysis of Fpmp proceeds in a manner and rate analogous to Ctmp (Reese & Thompson, 1988). The synthesis of ribonucleoside phosphoramidites protected with Fpmp (**113**) and the application of these monomers to the solid-phase synthesis of oligoribonucleotides has been reported recently (Beijer et al., 1990).

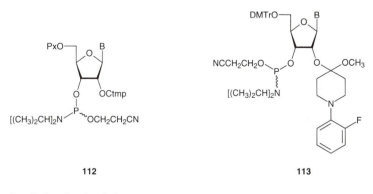

112   113

Px = (9-phenyl)xanthen-9-yl

The use of esters such as acetyl and benzoyl to protect the 2'-hydroxyl group of ribonucleosides has also been investigated (Fromageot et al., 1968). The main disadvantage of such protection is the difficulty in obtaining 2'-O-acyl ribonucleoside derivatives free from contaminating 3'-O-acyl derivatives, because acyl migration between O2' and O3' occurs under mild base-catalyzed conditions. Consequently, acyl groups are not recommended for protection of the 2′-hydroxyl function of ribonucle-osides.

A major contribution to oligoribonucleotide synthesis was the development of the *tert*-butyldimethylsilyl (TBDMS) as a 2′-hydroxyl protecting group (Ogilvie, 1973). This group (**114**) is reasonably stable to either acidic or basic conditions and is cleaved rapidly by fluoride ion within 30 min at 20°C (Ogilvie et al., 1978). The usefulness of TBDMS as a 2′-hydroxyl protecting group for ribonucleosides has been questioned be-cause the 2′- and 3′-*O*-TBDMS groups (**115** and **116**) migrate in methanolic solutions (Jones & Reese, 1979). At 36°C, the interconversion between **115** and **116** had a half-life of about 1 hr. In anhydrous pyridine, the conversion of **115** to **116** is considerably slower ($t_{1/2}$ = 1140 min at 36°C). This slow conversion rate is only apparent, because **116** interconverted to **115** with a half-life of 380 min under identical conditions (Jones & Reese, 1979). No detectable isomerization of TBDMS between O2′ and O3′ occurs under acidic conditions (80% acetic acid and 0.1 M methanolic hydrochloric acid) or in pure solvents (chloroform and anhydrous *N,N*-dimethylformamide) within 24 hr at 20°C (Ogilvie et al., 1978; Ogilvie & Entwistle, 1981). In addition, less than 1% of isomerization of the TBDMS group between the 2′- and 3′-OHs occurs in either tetrahy-drofuran or acetonitrile within 24 hr (Ogilvie & Entwistle, 1981). The potential posi-tional isomerization of TBDMS during the preparation of ribonucleoside phospho-ramidites **118a** or **118b** from **117** has also been investigated. Specifically, the reaction of **117** with either *N,N*-diisopropyl-(2-cyanoethyl)phosphonamidic chloride or (*N,N*-di-isopropyl)methylphosphonamidic chloride in the presence of 4-dimethylaminopyridine and *N,N*-diisopropylethylamine in dry tetrahydrofuran leads to the exclusive formation of **118a** or **118b** according to ^{1}H and ^{31}P NMR spectroscopy (Damha & Ogilvie, 1993; Wu & Ogilvie, 1990).

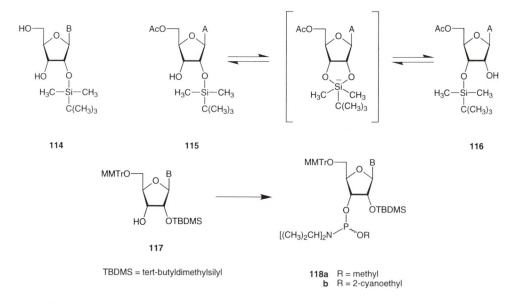

Ribonucleoside phosphoramidites structurally related to **118a** were activated with 1*H*-tetrazole and admixed with 3′-*O*-TBDMS ribonucleosides covalently attached to LCAA-CPG through a 2′-*O*-succinate linkage. Using a 15-min condensation step, this approach led to the preparation of oligoribonucleotides (43- and 77-mers) with an average stepwise yield of 98% (Ogilvie et al., 1988). The sequence corresponding to an *E. coli* tRNA[fMet] analogue (77-mer) demonstrated only modest methionine acceptance activity, perhaps because the 2'-*O*-TBDMS group was not entirely stable under the basic conditions required for deprotection of the nucleobases. For example, treatment of ribonucleotide **119** with concentrated ammonium hydroxide at 55°C resulted in the partial removal of the 2'-*O*-TBDMS group (27% after 8 hr) and cleavage of the internucleotide linkage (3% after 8 hr) (Stawinski et al., 1988b, c). However, the addition of ethanol to concentrated ammonium hydroxide has been shown to decrease the 2'-*O*-TBDMS cleavage rate significantly. Typically, treatment of **119** with concentrated ammonium hydroxide-ethanol (3:1) at 55°C produced only about 4% cleavage of TBDMS after 8 hr and complete stability of the phosphodiester linkage (Stawinski et al., 1988b,c). For *N*-deacylation of larger oligoribonucleotides (>12-mers) the use of concentrated ammonium hydroxide-ethanol (3:1) at room temperature (2–3 days) is recommended. These conditions minimize chain cleavage which results from premature removal of the 2'-*O*-TBDMS group (Damha & Ogilvie, 1993). The removal of the 2'-*O*-TBDMS group from **119** with tetra-*n*-butylammonium fluoride (TBAF) occurs smoothly within 4 hr without cleavage or isomerization of the phosphodiester function (Stawinski et al., 1988b,c), but elimination of TBDMS groups from larger oligoribonucleotides is carried out at ambient temperature for 16 hr using 50 equivalents of TBAF per TBDMS group (Damha & Ogilvie, 1993).

One important issue is the effect of steric hindrance, as generated by the 2'-*O*-protecting group, on the coupling rates of phosphoramidite monomers. In a competitive condensation reaction involving ribonucleoside phosphoramidites **120a–d** with deoxythymidine bound to a solid support, analysis and quantitation of the dimers **121a–d** after treatment with ammonium hydroxide revealed that the coupling efficiency of **120a–d** decreased in the order **120a** > **120b** > **120c** > **120d** (Kierzek et al., 1987). Thus, the steric bulk of the 2'-*O*-protecting group represents an additional factor in the design of such groups. In this context, the 2'-*O*-(2-nitrobenzyloxymethyl)ribonucleoside phosphoramidites **122** have been tested in the solid-phase synthesis of oligoribonucleotides. Given a condensation time of 2 min, the average coupling yield was greater than 98% (Schwartz et al., 1992). This efficiency may be attributed to a reduction of steric crowding in the vicinity of the phosphoramidite function when compared to 2'-*O*-TBDMS ribonucleoside phosphoramidite monomers.

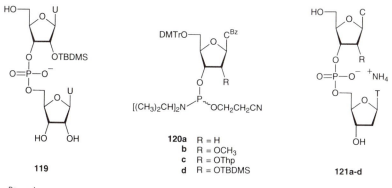

120a R = H
  b   R = OCH$_3$
  c   R = OThp
  d   R = OTBDMS

119

121a-d

C^{Bz} = N^4-benzoylcytosin-1-yl

The ribonucleoside phosphoramidites **123** have been used in the total chemical synthesis of *E. coli* tRNA^{Ala}. Coupling yields greater than 98% during a 2-min condensation time were reported. Triethylamine hydrofluoride was found more effective than tetra-*n*-butylammonium fluoride for the complete deprotection of 2'-*O*-silyl protecting groups (Gasparutto et al., 1992). Although the use of 2'-O-TBDMS groups in conjunction with the phosphoramidite approach has been very popular in the synthesis of oligoribonucleotides, ribonucleoside *H*-phosphonates having 5'-*O*-DMTr and 2'-*O*-TBDMS have also been prepared without significant isomerization of the TBDMS group (less than 0.5%) (Stawinski et al., 1988b,c). Such monomers were used successfully to synthesize a 21-mer (Garegg et al., 1986).

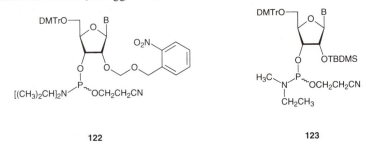

122   **123**

Applications of Oligodeoxyribonucleotides and Oligoribonucleotides

Oligodeoxyribonucleotide Phosphorothioates and Phosphorodithioates. Over the past several years, the development of high-yielding, rapid methods for synthesizing DNA have led to a large number of applications for these molecules (Agrawal, 1992; Englisch & Gauss, 1991; Goodchild, 1990; Uhlmann & Peyman, 1990). These applications include total gene synthesis for cloning experiments (Gröger et al., 1988), primers for sequencing DNA and initiating PCR studies (Arnhelm & Erlich, 1992), site-specific gene mutagenesis (Leatherbarrow & Fersht, 1986), the study of protein–DNA interactions (Harrison & Aggarwal, 1990), and probing nucleic acid structure (Kennard & Hunter, 1989). The exciting results from these investigations have further stimulated the development of oligonucleotide analogues for additional applications in basic research, DNA probe diagnostics, and the use of DNA, potentially, as a therapeutic drug.

In this context, the incorporation of internucleotide phosphorothioate linkages into oligodeoxyribonucleotides markedly enhances their resistance to hydrolysis by nucleases (Eckstein, 1985). Because of such properties, oligodeoxyribonucleotide phosphorothioates have been tested extensively as antisense molecules in the inhibition of gene expression. The inhibitory mechanism presumably results from hybridization of a specific messenger RNA (the sense sequence) with complementary antisense DNA. Translation of the message is thus impaired by either formation of a DNA–RNA duplex or degradation of the heteroduplex by RNase H. As an example of this approach, antisense phosphorothioates complementary to the messenger RNA of the HIV-1 *rev* gene have been shown to inhibit the viral cytopathic effect of chronically infected H9 cells (Matsukura et al., 1989). By contrast, neither the corresponding sense phosphorothioate analogue nor the unmodified antisense oligodeoxyribonucleotide was effective in inhibiting viral expression under identical conditions.

The automated synthesis of phosphorothioates can be achieved using the *H*-phosphonate approach. Once synthesized, oligodeoxyribonucleotide *H*-phosphonates are converted to phosphorothioates by a single sulfurization reaction using elemental sulfur (Froehler, 1986). Unfortunately, this methodology cannot be easily used to produce oligodeoxyribonucleotides carrying predetermined combinations of natural and phosphorothioate linkages. The automated synthesis of oligodeoxyribonucleotide phosphorothioates can also be performed via the phosphoramidite approach. In this strategy, the stepwise aqueous iodine oxidation step, which is responsible for converting phosphite triesters to phosphotriesters, is replaced with a relatively slow sulfurization reaction requiring elemental sulfur (S_8). Because elemental sulfur is sparingly soluble in most organic solvents, the sulfurization reaction is difficult and requires carefully controlled conditions. Recently, five sulfur-transfer reagents which eliminate this problem have been described. For example, low concentrations (0.05–0.2 M) of the thiosulfonate **124** in acetonitrile converts the dinucleoside phosphite triester **125** to the corresponding phosphorothioate dimer **126** in yields exceeding 99% within 30 sec at 20°C (Iyer et al., 1990; Regan et al., 1992), from which deblocked dimer **127** may be obtained. This reagent has, therefore, led to a rapid, efficient, and reliable automated synthesis of phosphorothioate oligomers carrying exclusively, or carrying a predetermined number of, phosphorothioate functions without detectable modification of the nucleobases (Iyer et al., 1990). Other reagents such as dibenzoyl tetrasulfide (**128**) and bis-(*O*,*O*-diisopropoxyphosphinothioyl) disulfide (**129**) have also been used to rapidly sulfurize oligodeoxyribonucleotide phosphite triesters to afford the phosphorothioate derivatives. These reagents, like the thiosulfonate **124**, have rapid sulfurization kinet-

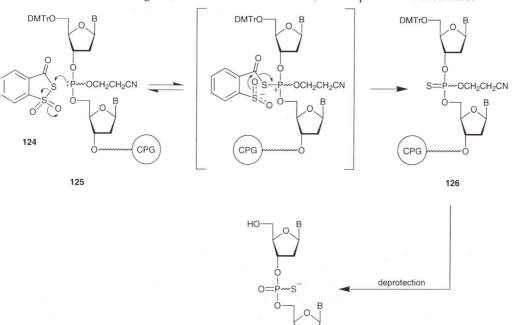

ics and satisfactory solubility properties. A 0.4 M solution of **128** in tetrahydrofuran converts internucleotide phosphite triesters to phosphorothioates within 1 min in almost quantitative yields (Rao et al., 1992). Alternatively, a 0.2 M solution of **129** in pyridine can be used to complete the sulfur transfer reaction within about 4 min (Stec et al., 1993). Phenylacetyl disulfide (**130**) and *N,N,N',N'*-tetraethylthiuram disulfide (**131**) have also been used successfully in the solid-phase synthesis of oligodeoxyribonucleotide phosphorothioates. Typically, a 5% solution of **130** in 1,2-dichloroethane-*sym*-collidine (4:1) enabled sulfurization of a dinucleoside phosphite triester within 5 min at 20°C (Kamer et al., 1989). Relative to **124**, the sulfurization kinetics with **131** are sluggish, because 15 min is required to completely sulfurize phosphite triesters (Vu & Hirschbein, 1991). It must be noted that the sulfur transfer reagent **124** has also been used successfully in the solid-phase synthesis of oligoribonucleotide phosphorothioates (Morvan et al., 1990).

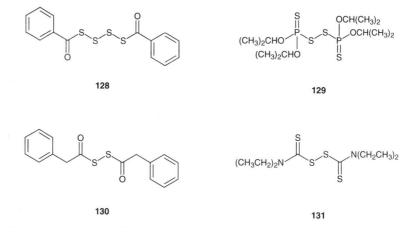

Although oligodeoxyribonucleotide phosphorothioates have potential as antisense DNA analogues, other properties may limit their usefulness. For example, this analog inhibits human DNA polymerases and RNase H in vitro. Specifically, the phosphorothioate homopolymer of deoxycytidine (S-dC$_{28}$) was shown to be a competitive inhibitor of DNA polymerases α and β with respect to DNA template and to also competitively inhibit RNase H$_1$ and H$_2$ with respect to the RNA–DNA duplex. The effect was not sequence-specific but was length-dependent (Gao et al., 1992). However, the relevance of these observations to the potential pharmacological utility of phosphorothioate DNA must still be determined. In vivo, the effective antiviral activity of this oligomer has been obtained in the concentration range 10^{-8}–10^{-6} M without noticeable cytotoxicity up to 10^{-4} M (Agrawal, 1992). Perhaps due to poorly understood cellular biochemistry, the observed inhibition of certain enzymes is irrelevant to their potential as DNA therapeutic drugs. For example, although oligodeoxyribonucleotide phosphorothioates inhibit DNA polymerases, the nuclear location of these enzymes in very specific complexes with cellular DNA and many other proteins may render them inaccessible to inhibition by these analogs. Thus, even if some of the biochemical data suggests caution, research with this analogue should continue because it has been shown to represent a potential new class of chemotherapeutic agents against AIDS (Agrawal & Sarin, 1991), hepatitis B (Goodarzi et al., 1990), and other infectious diseases (Cohen, 1991).

It has been demonstrated recently that the excess of activated deoxyribonucleoside phosphoramidites used during solid-phase oligodeoxyribonucleotide synthesis can be converted to the parent phosphoramidites **61** upon treatment with anhydrous N,N-diisopropylamine (Scremin et al., 1994). Chromatographically purified **61** can be reutilized efficiently in the solid-phase synthesis of oligodeoxyribonucleotides. This simple regeneration and recovery of deoxyribonucleoside phosphoramidites may have a profound impact on the economic considerations stemming from large-scale syntheses of, for example, oligodeoxyribonucleotide phosphorothioates that are required for clinical evaluation of these potential therapeutic agents.

The presence of the larger sulfur atom in oligodeoxyribonucleotide phosphorothioates produces a chiral phosphate backbone, alters the charge distribution around phosphate, and leads to subtle conformation changes (Latimer et al., 1989). The chirality at phosphorus can be abolished by replacing the remaining nonbridging oxygen atom of the internucleotide phosphodiester function with a sulfur atom. The resulting oligodeoxyribonucleotide phosphorodithioate exhibits complete resistance to nuclease hydrolysis (Wiesler et al., 1993). Like oligodeoxyribonucleotide phosphorothioates, the phosphorodithioate oligodeoxyribonucleotide analog forms less stable complexes with natural DNA and activates endogenous RNase H in HeLa cells. Oligodeoxyribonucleotide phosphorodithioates also bind to reverse transcriptases with very high affinity (Marshall & Caruthers, 1993; Marshall et al., 1992; Wiesler et al., 1993).

Synthesis of oligodeoxyribonucleotides having phosphorodithioate linkages can be achieved using several approaches. The currently most useful method is based upon the utilization of deoxyribonucleoside phosphorothioamidite synthons. These were prepared by phosphitylation of properly protected 2'-deoxyribonucleosides (**23**) with tris-(pyrrolidino) phosphine **132** and 1H-tetrazole. The resulting phosphorodiamidite intermediates **133** were converted to the phosphorothioamidites **134** upon treatment with monobenzoylethanedithiol in the presence of 1H-tetrazole. Automated solid-phase synthesis of oligodeoxyribonucleotide phosphorodithioates was accomplished by activation of nucleoside phosphorothioamidites **134** with 1H-tetrazole followed by oxidation of the phosphorothioites **135** with elemental sulfur. Upon completion of the synthesis, phosphorodithioate oligomers are deprotected by treatment with ethanolic ammonia containing 15% benzene. In addition to deblocking nucleobases, this formulation leads to debenzoylation of the phosphorodithioate blocking group, which induces elimination of ethylene sulfide via intramolecular nucleophilic displacement, and produces the phosphorodithioates **137**. Phosphorothioate contamination (2% to 5%) is observed during ammonolysis, presumably because of nonspecific hydrolysis of the phosphorothioate triester (Wiesler et al., 1993). Several other methods have been explored in attempts to synthesize these oligodeoxyribonucleotide analogues. These include (1) the phosphotriester approach (Beaton et al., 1991) and (2) the use of deoxyribonucleoside H-phosphonothioates (Stawinski et al., 1989) and deoxyribonucleoside H-phosphonodithioates (Brill et al., 1989; Porritt & Reese, 1989) as synthons. Because studies have only begun on the biophysical and biochemical properties of phosphorodithioate oligomers, it is difficult to fully assess the potential applications of these analogues in basic research or biotechnology.

Oligodeoxyribonucleotide Methyl Phosphonates and Lipophilic Oligodeoxyribonucleotide Analogs. If oligodeoxyribonucleotides are to become useful as biological probes, their transport into cells must be understood and methods for enhancing this

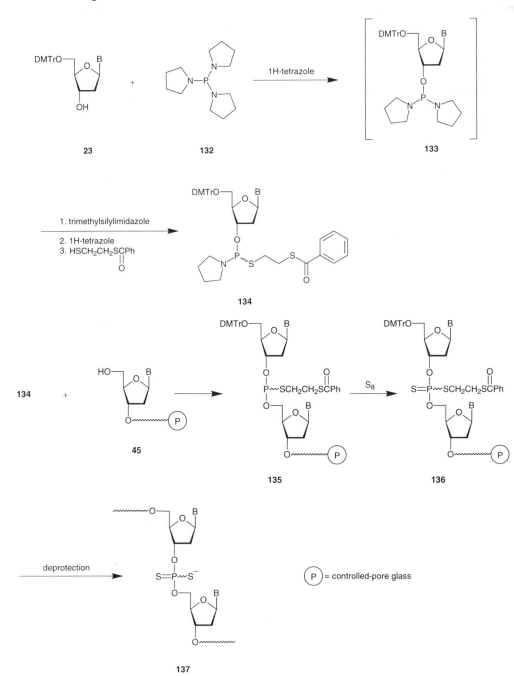

process must be developed. Nonionic oligomers, such as oligodeoxyribonucleotide methylphosphonates, are taken up by cells through a fluid phase/adsorptive route (Akhtar et al., 1991; Shoji et al., 1991). This pathway is distinct from the endocytic mechanisms used by oligodeoxyribonucleotide phosphodiesters or phosphorothioates for entering cells (Akhtar et al., 1991; Loke et al., 1989; Shoji et al., 1991).

Oligodeoxyribonucleotide methylphosphonates are totally resistant to nuclease hydrolysis (Miller & Ts'O, 1987) both in culture media and in cells (Shoji et al., 1991) and have been shown to exhibit biological activity (Laurence et al., 1991). Of practical importance, the incorporation of two contiguous methylphosphonate linkages at each terminus of an oligonucleotide increased its resistance to hydrolysis by exonucleases (Tidd & Warenius, 1989). Unlike native oligodeoxyribonucleotides and their phosphorothioate analogues, oligodeoxyribonucleotides having exclusively or alternating methylphosphonate linkages were unable to form RNA–DNA hybrids that were substrates for *E. coli* RNase H activity (Quartin et al., 1989). However, methylphosphonate oligomers having three or more contiguous natural phosphodiester linkages did stimulate RNase H activity. Based upon these results, it may be possible to design methylphosphonate oligomers that are transported into cells via an adsorptive process and retain the ability to stimulate RNase H to degrade specific cellular or viral RNAs (Giles & Tidd, 1992; Quartin et al., 1989). One potential complication, however, is that this linkage is chiral at phosphorus and hybridizes poorly to natural DNA. Perhaps this problem is due to conformational distortions imparted by the methyl group in its interactions with the 3′- and 5′-*O-P*-linkages (Maher & Dolnick, 1988). In an attempt to address some of these complications with a new analog, the 3′-phosphoramidite derivative of the dinucleoside 5′-deoxy-5′-methylphosphonate **138** was prepared and incorporated into an oligothymidylate analog (Böhringer et al., 1993). However, because the duplex stability, nuclease sensitivity, and RNase H activity of this analogue have not been reported, its biochemical and biological utility is unknown at this time.

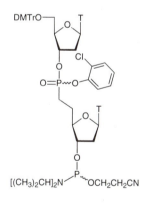

138

Synthesis of methylphosphonate-linked oligodeoxyribonucleotides is most often completed on a solid support using deoxyribonucleoside phosphonamidites (**139**). Activation with 1*H*-tetrazole and incorporation into DNA usually occurs within 30–120 sec with a yield of 96% to 97% (Löschner & Engels, 1988; Viari et al., 1987). Unlike phosphite triesters, methylphosphonite intermediates (**140**) are more susceptible to hydrolysis via catalysis with 1*H*-tetrazole (Hogrefe et al., 1993). Because of this potential problem, imidazole has been suggested as an activating agent for nucleoside methylphosphonamidites (Dorman et al., 1984). However, because 1*H*-tetrazole is widely used for activation, the problem can perhaps be circumvented by rapidly oxidizing methylphosphonite to methylphosphonate diesters (**141**) with a reagent con-

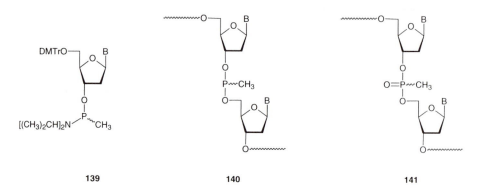

139   140   141

taining a minimal amount of water. Another complication is the sensitivity of methylphosphonate linkages to the 1-methylimidazole that is present in the capping reagent during solid-phase synthesis. For the preparation of this analogue, it is also important to substitute *N,N*-dimethylaminopyridine for 1-methylimidazole which reduces the cleavage rate of methylphosphonate linkages.

Over the past few years, the design and synthesis of dinucleoside 3'-*O*-phosphoramidites related to **142**, where X, Y, and Z represent various atoms or groups of atoms, has attracted considerable attention in an effort to provide oligodeoxyribonucleotide analogues with nuclease resistance and lipophilicity without compromising hybridizing abilities. Part of this work has been reviewed (Beaucage & Iyer, 1993) and recently published (Caulfield et al., 1993; Chur et al., 1993; Idziak et al., 1993; Kawai et al., 1993; Lebreton et al., 1993; Saha et al., 1993).

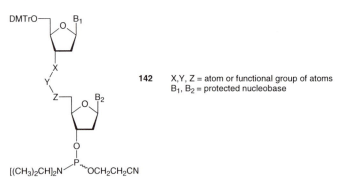

142   X,Y, Z = atom or functional group of atoms
B$_1$, B$_2$ = protected nucleobase

α-Oligodeoxyribonucleotides. The search for novel synthetic oligodeoxyribonucleotides as potential inhibitors of gene expression has focused not only on the chemical modification of the atoms linked to phosphate but also on chemical modifications of the nucleoside per se. One particular analogue that has attracted considerable attention is called α-DNA. This derivative has the sugar and nucleobase linked by an α-glycosidic bond as opposed to the natural β-glycosidic linkage. From work published some time ago (Séquin, 1974), snake venom and bovine spleen phosphodiesterases are known to hydrolyze this derivative at rates considerably slower than those for natural oligodeoxyribonucleotides. This observation prompted a major effort to synthesize and study the utility of α-DNA as a potential antisense drug. Initially α-oligodeoxyribonucleotides were prepared by the phosphotriester methodology (Morvan et al., 1986), but more recently the α-deoxyribonucleoside phosphoramidites **143** and **144** have been used for this purpose (Morvan et al., 1988).

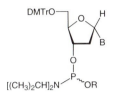

143 R = methyl
144 R = 2-cyanoethyl

In order to evaluate the suitability of α-oligodeoxyribonucleotides as antisense molecules, an α-hexadecadeoxyribonucleotide has been prepared, radiolabeled, and incubated in undiluted fetal bovine serum. The half-life of the α-oligodeoxyribonucleotide under these conditions is about 24 hr. In contrast, natural β-oligomers are completely hydrolyzed within 15 min. NMR and ultraviolet studies indicate that, unlike native DNA, α-oligodeoxyribonucleotides hybridize with complementary β-strands in a parallel orientation (Imbach et al., 1989). Furthermore, duplexes composed of α-DNA and complementary β-RNA strands are not substrates for *E. coli* RNase H or *Drosophila* embryo RNase H. Consequently, hydrolysis of the β-RNA strand of these duplexes does not occur (Imbach et al., 1989). The parallel annealing of α-DNA to β-RNA is presumably responsible for the protection of RNA against RNase H-mediated hydrolysis. Of interest, α-oligodeoxyribonucleotide phosphorothioates complementary to the splice acceptor site of the pre-mRNA encoding the Tat protein or targeted against the translation initiation site of *rev* mRNA exhibit anti-HIV activity (Morvan et al., 1993; Rayner et al., 1990).

The preparation of alternating α,β-oligodeoxyribonucleotides with (3′ → 3′)- and (5′ → 5′)-internucleotidic linkages (**145**) has recently been reported (Koga et al., 1991, 1993, 1995). The design of these analogues is based on the assumption that oligodeoxyribonucleotides having alternating (3′ → 3′) and (5′ → 5′)-internucleotidic phosphodiester linkages may not be hydrolyzed by nucleases as rapidly as natural oligomers. When tested, these α,β-oligomers were found to be less resistant to snake venom and calf spleen phosphodiesterases than β-oligodeoxyribonucleotide phosphorothioate congeners, but considerably more stable than unmodified oligomers under similar conditions (Koga et al., 1991, 1995). As expected, α,β-oligodeoxyribonucleotide phosphorothioates were more resistant than β-oligodeoxyribonucleotide phosphorothioates to the nucleolytic activity of both snake venom and calf spleen phosphodiesterases (Koga et al., 1995). Hybridization studies performed with

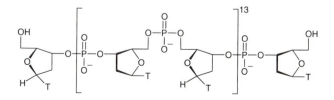

145

α,β-oligodeoxyribonucleotides and complementary unmodified DNA sequences indicated that the stability of these complexes was comparable to that obtained with similar fully phosphorothioated β-oligodeoxyribonucleotides under the same conditions (Koga et al., 1991, 1993, 1995). Thus, α,β-oligodeoxyribonucleotides with alternating $(3' \rightarrow 3')$ and $(5' \rightarrow 5')$-internucleotidic phosphodiester linkages may represent a new class of antisense molecules.

Oligoribonucleotides. Perhaps the most extensively studied RNA analogue is the 2'-*O*-methyl derivative. Its attractive features include resistance to ribonucleases and stability toward alkaline hydrolysis. This analogue can be synthesized on a urethane-linked aminopropylated CPG support using the phosphoramidites **146** and 5-(*p*-nitrophenyl)-1*H*-tetrazole as an activator (Sproat et al., 1989). A 6-min condensation time, presumably due to the steric bulk of the 2'-*O*-methyl group, is required to ensure high coupling efficiency (99%). 2'-*O*-Methyl RNA oligomers are resistant to many RNA and DNA nucleases, but sensitive to P1 nuclease, snake venom phosphodiesterase, and *Bal* 31 nuclease (Sproat et al., 1989). These oligomers are useful for probing the structure and function of small ribonucleoprotein particles (snRNPs) which are subunits of functional spliceosomes (Sproat & Lamond, 1991).

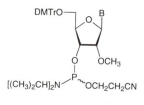

146

Because 2'-*O*-methyl RNA oligomers are resistant to nucleases and form stable heteroduplexes with complementary, unmodified RNA (Inoue et al., 1987), the suitability of these oligomers as antisense molecules has been investigated. For example, oligo-(2'-*O*-methyl)ribonucleotide phosphorothioates inhibit the replication of the HIV-1 virus in cultured MT-4 cells (Shibahara et al., 1989). Particularly interesting is the observation that oligoribonucleotides having five phosphorothioate linkages at both 3'- and 5'-termini exhibit almost as much activity as fully phosphorothioated oligomers. In contrast, both 2'-*O*-methylribo- and oligodeoxyribonucleotides fail to inhibit replication of this virus. These results indicate that fully phosphorothioated oligonucleotides are not necessary to obtain anti-HIV activity and that oligo-(2'-*O*-methyl)ribonucleotides are perhaps sensitive to the nucleases present in the cell line used for this study (Shibahara et al., 1989). More recently, 2'-*O*-allylribonucleotides have been prepared from 2'-*O*-allyl-3'-*O*-phosphoramidites (**147**). The advantages of 2'-*O*-allyloligoribonucleotides over their 2'-*O*-methyl congeners have been discussed (Iribarren et al., 1990). Perhaps the most significant advantage is the reduction of nonspecific interactions, which can be a very important consideration for antisense applications.

The incorporation of modified ribonucleosides into oligoribonucleotides has also been used to provide a better understanding of the mechanisms whereby RNA catalyzes sequence-specific chain cleavages. As an example, the catalytic activity of a RNA enzyme (ribozyme) derived from the group I *Tetrahymena* self-splicing intron

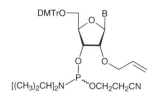

147

depended on a wobble base pair rather than a Watson–Crick base pair at the 5′-splice site. This conclusion was derived from experiments involving the substitution of U·I for the native U·G wobble base pair (Green et al., 1991). Other work with chimeric deoxyribo- and ribooligonucleotides has shown that catalytic activity requires a 2′-hydroxyl group adjacent to the cleavage site in the substrate (Perreault et al., 1990). Several additional 2′-hydroxyl groups located elsewhere in the catalytic core were also shown to be important for catalysis (Perreault et al., 1991). Research directed toward understanding catalysis in the hammerhead ribozymes has also benefited from using these chimeric oligomers and 2′-*O*-methyloligoribonucleotides. It was discovered that ribozymes containing predominantly deoxynucleotides, but also four to seven critically placed ribonucleotides, were still weakly active (Yang et al., 1992). Other ribozymes having 2′-*O*-methylribonucleotides or deoxyribonucleotides flanking sequences had increased catalytic activity and superior stability toward nucleases (Goodchild, 1992; Taylor et al., 1992).

The significance of the 2′-hydroxyl group for catalysis was also studied by incorporating 2′-fluoro- or 2′-amino-2′-deoxyribonucleotides into key sites on hammerhead ribozymes. These ribozymes were prepared chemically from the appropriate phosphoramidites (**148** and **149**). Ribozymes having every adenosine replaced with 2′-deoxyadenosine or 2′-fluoro-2′-deoxyadenosine showed significantly lower catalytic activity compared to unmodified ribozymes. However, no single substitution was responsible for the decrease in activity. It was concluded that the 2′-hydroxyl group of adenosine was not essential for catalysis or for proper formation of the tertiary structure of hammerhead ribozymes (Olsen et al., 1991). Conversely, replacement of two guanosine 2′-hydroxyl groups, located in the conserved central core region of the ribozyme, with 2′-fluoro and 2′-amino groups reduced the catalytic activity of the corresponding ribozymes by factors of at least 150 and 15, respectively. These results suggest that the 2′-amino group can partially fulfill the role of the 2′-hydroxyl group in the catalytic core (Williams et al., 1992). It is noteworthy that ribozymes containing 2′-fluoropyrimidines at all uridine and cytidine positions were stabilized against nu-

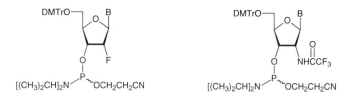

148   **149**

cleolytic degradation in rabbit serum by factors of at least 1000 relative to those of unmodified ribozymes (Pieken et al., 1991). Furthermore, experiments aimed at cleaving the long terminal repeat RNA of HIV-1 with hammerhead ribozymes indicated that replacing the pyrimidines of a ribozyme with the corresponding 2'-fluorocytidines and 2'-flurorouridines, together with the incorporation of phosphorothioate linkages at both termini, caused only a seven-fold decrease in catalytic efficiency (Heidenreich & Eckstein, 1992). The modified ribozyme also exhibited a 50-fold increase in stability toward hydrolysis by nucleases. These results demonstrated the possibility of increasing ribozyme resistance to nucleases without severely affecting catalytic activity. The design and application of ribozymes as antisense and therapeutic agents has been reviewed (Parker et al., 1992).

Summary

The recent advances in the chemical synthesis of deoxyribonucleoside and ribonucleoside building blocks along with their analogs have resulted in the creative application of modified oligodeoxyribonucleotides and oligoribonucleotides to drug discovery. However, many hurdles must be overcome before the use of modified oligomers as therapeutic agents becomes a reality. These include cell permeation properties and in vivo stability, target specificity, and the economics associated with large-scale synthesis and purification. These stringent requirements should, therefore, provide the impetus to develop simple and efficient technologies for the preparation of analogues having therapeutic value.

Enzymatic Methods for the Preparation and Manipulation of Nucleic Acids

Mark Van Cleve

Enzymatic methods for the preparation of nucleic acids benefit from the characteristics of the tools: enzymes combine high specificity with catalytic efficiency. Over the past 25 years, the research community has assembled a formidable array of nucleic acid modifying enzymes, resulting in the rapid growth of a family of technologies known collectively as "biotechnology" or "genetic engineering" (Kornberg & Baker, 1992). Several factors have contributed to the production of this enzymatic armamentarium.

First, the maintenance, mobilization, and replication of genetic material within the living cell require many enzymatic activities, which, acting in concert, give each organism access to the informational content of its genes (Lewin, 1990; Watson et al., 1987). Thus, DNA polymerases and ligases are active in genome replication and repair, RNA polymerases synthesize cellular RNAs, restriction endonucleases protect against invasion by foreign DNA, and so forth. Each enzyme has evolved for a particular purpose, and with particular activities and specificity. We capture these enzymes through sophisticated purification techniques (Boyer, 1982), and their abilities can be harnessed to our experimental purposes (Ausubel et al., 1989; Sambrook et al., 1989); for example, DNA polymerases are used for sequencing, and ligases and restriction endonucleases are used to make recombinant DNA molecules.

The considerable power of the overall technology, which reflects the efforts of thousands of researchers, can probably best be understood by noting that our technical ability has become such as to require the development of new ethical guidelines for its management (Dyson & Harris, 1994).

The examples in this chapter are intended to illustrate how the characteristics of nucleic acid modifying enzymes are used in the pursuit of experimental goals. The fascination of this story lies in understanding how the detailed specifics of the functioning of these remarkable biological catalysts, once the specialized pursuit of the dedicated enzymologist, have now become the working parts of a technological revolution (Kornberg & Baker, 1992). This chapter begins with a discussion of the different types of vectors, then proceeds to the salient features of those enzymes most useful for producing nucleic acids of defined properties.

Vectors: Plasmid and Virus

Genomic DNA molecules from most organisms other than viruses are enormous, double-stranded nucleic acids millions of base pairs long (Kornberg & Baker, 1992). If laid end-to-end, the total DNA in all the chromosomes of each human cell would reach a length that can be calculated at several *feet* (Kornberg & Baker, 1992). The technology for handling such molecules intact is in its infancy. The genetic elements used in the vast majority of molecular biology experiments are orders of magnitude smaller, typically only thousands of base pairs in length, yet contain all the sequence information necessary for their replication and maintenance by host cells. These genetic elements, called *vectors* (Wu & Grossman, 1987), can be purified, characterized, and manipulated according to well-established experimental protocols (Ausubel et al., 1989; Sambrook et al., 1989).

Phage, short for bacteriophage, are viruses that infect bacteria (Watson et al., 1987). They are often used as vectors (Ausbel et al., 1989; Sambrook et al., 1989; Wu & Grossman, 1987). Their study has been central to the development of molecular biology (Cairns et al., 1966; Portugal & Cohen, 1980). The DNA of phage, and also that of eukaryotic viral vectors, is relatively small and easy to manipulate (Watson et al., 1987), and viruses readily enter living cells. The passage of vector DNA into and out of living cells is necessary in many experimental protocols for two reasons. First, cells are capable of repairing and replicating the vector, and this is often the last step in producing a recombinant molecule. Second, it is within a living system that the functioning of a gene is ordinarily observed and studied. So a vector can be thought of as an experimental platform that can hold a gene of interest and can easily be moved back and forth from cell-free to cellular environments. In the test tube, a gene present within a vector can be sequenced, altered, or placed under a variety of control elements. When the vector is placed in a cell, the biological effects of any changes can be observed, or an enzyme or RNA encoded by the gene can be expressed and studied. Obtaining a particular gene from the genomic DNA of an organism and inserting it into a vector is called *cloning*. Moving genes or other DNA fragments from vector to vector is called *subcloning*.

Plasmids are small, circular genetic elements, usually only a few thousand base pairs in length, that are harbored by a wide variety of host cells (Kornberg & Baker, 1992). They occur naturally and confer various phenotypes on the host cell: resistance to antibiotics, for example (Mitsuhashi et al., 1977). Plasmid DNA can be introduced into host cells by a number of techniques, and it can easily be purified from bacterial culture (Ausubel et al., 1989; Sambrook et al., 1989; Walker, 1984). Plasmids are very popular vectors because they are easy to work with and are readily designed to accommodate specific experimental goals; there are presently dozens of plasmids intended for all sorts of purposes and designed with the convenience of the experimenter in mind (Ausubel et al., 1989; Balbas et al., 1986; Sambrook et al., 1989; Wu & Grossman, 1987).

Figure 3-1 shows a diagram of a commercially available vector that can be used for a variety of purposes. Plasmids are often represented as circular maps this way in order to provide a summary of their features. The ampicillin resistance gene (*amp*) is a *selectable marker gene* (Ausubel et al., 1989; Sambrook et al., 1989). Such genes are useful in *transformation* experiments (Ausubel et al., 1989; Sambrook et al., 1989) in which the experimenter attempts to introduce a plasmid into a bacterial cell culture such

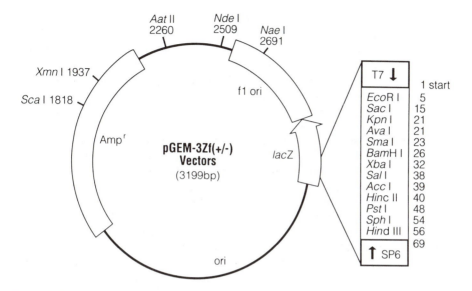

Figure 3-1. Circular map of plasmid pGEM-3Zf. To the right of the circle is a list of the restriction enzyme sites in the polylinker, or multiple cloning site, of the plasmid. Other restriction sites are shown at various places around the plasmid, together with the position of cleavage (base number). Reprinted with permission of Promega Corporation.

that some of the individual bacteria incorporate the plasmid into their cytoplasm. Transformation is generally an inefficient process, with only a small fraction of the individuals in the culture accepting the plasmid. However, if a bacterial population transformed with pGEM-3Zf is subsequently grown in media containing ampicillin (i.e., *selection media*) (Ausubel et al., 1989; Sambrook et al., 1989) the result will be a pure culture of transformants, since all the untransformed bacteria will be killed by the antibiotic.

The *polylinker*, also called the *multiple cloning site* (MCS) (Ausubel et al., 1989; Sambrook et al., 1989), is a short region containing the recognition sites of several restriction enzymes (13, in the case of pGEM-3Zf) commonly used in molecular biology experiments. The polylinker makes it possible for this plasmid to receive inserts prepared by any one or any pair of these restriction enzymes (see "subcloning").

The polylinker is positioned in the middle of the *lacZ* gene, which encodes the enzyme β-galactosidase. This allows easy identification of bacteria harboring recombinant plasmids (Norrander et al., 1983; Viera & Messing, 1982; Yanisch-Perron et al., 1985). The enzyme causes bacterial colonies to appear blue in color when they are grown in the presence of the dye 5-bromo-4-chloro-3-indolyl-β-galactoside (Xgal). Xgal (Horwitz et al., 1964) is normally colorless, but it turns deep blue when hydrolyzed by β-galactosidase. Thus, bacterial colonies that bear the plasmid pGEM-3Zf will be blue unless a DNA fragment or gene has been cloned into the polylinker, in which case the *lac* gene is disrupted. If the gene is disrupted, functional β-galactosidase is not made, and the colony will have its normal white color. This "blue-white" test makes it easy for the experimenter to distinguish colonies bearing recombinant plasmid, which has an insert of new DNA in the polylinker (white) from colonies bearing the parent plasmid, pGEM-3Zf that did not receive an insert (blue).

The T7 and SP6 *promoters* are short DNA sequences recognized as polymerization start sites by RNA polymerases encoded by bacteriophages T7 and SP6 (Butler & Chamberlin, 1982; Melton et al., 1984; Moffatt et al., 1984). Placement of these promoters near the polylinker enables the experimenter to obtain large quantities of transcript mRNA from a gene cloned into the polylinker (Krieg & Melton, 1987; Milligan & Uhlenbeck, 1989). Because the two promoters are "aimed" in opposite directions (one is in the Watson strand, the other in the Crick strand), a transcript can be obtained from genes cloned in either orientation.

The *fl origin of replication* allows the experimenter to produce single-stranded plasmid, which is useful for DNA sequencing and for site-directed mutagenesis experiments. The replication origin is recognized by bacteriophage M13 (Dotto & Zinder, 1983; Dotto et al., 1984). If bacteria bearing this plasmid are infected with M13, the phage will produce single-stranded pGEM-3Zf and package it in phage particles that will bud from the surface of the infected cells (Dente et al., 1983; Dotto et al., 1981; Geider et al., 1985; Levinson et al., 1984; Mead et al., 1985; Zagursky & Berman, 1984). Single-stranded plasmid can then be isolated from these particles (Ausubel et al., 1989).

Thus, the features that have been incorporated into pGEM-3Zf allow its convenient use as a general-purpose cloning vehicle. Other plasmids have been designed for a multitude of specialized purposes (Wu & Grossman, 1987): There are plasmids designed for gene regulation studies and for heterologous gene expression (for example, expression of mammalian genes in *E. coli*) and plasmids that can shuttle back and forth between eukaryotic and prokaryotic hosts. They all have in common the ability to accept cloned or subcloned DNA, providing the researcher with a specialized experimental platform that facilitates study of the properties of the DNA both in the test tube and in the living cell. The tools commonly used in these experiments (i.e., enzymes) are discussed below.

Basics of Enzymatic Synthesis and Manipulation of Nucleic Acids

Polymerization: DNA-Dependent DNA Polymerases. DNA-dependent DNA polymerases catalyze a variety of reactions in in vivo DNA repair and replication (Friedberg, 1985; Kornberg & Baker, 1992). The reaction for which they are named is shown in Fig. 3-2. The polymerase activity adds nucleotides to the 3'-end of a primer strand, which can be RNA or DNA (Karkas, 1973). The newly synthesized strand is complementary to the template; the *fidelity* of a polymerase is its ratio of errors to correctly incorporated bases. The incorporated nucleotides are derived from deoxynucleoside 5'-triphosphates in an enzyme-catalyzed reaction that proceeds by nucleophilic attack by the primer 3'-hydroxyl group on the α-position of the triphosphate (Kornberg, 1969).

In addition to the above reaction, DNA polymerase I of *E. coli* has two exonuclease activities. A $3' \to 5'$ exonuclease can remove bases from the strand being synthesized, with the effect of reversing the polymerase activity. This $3' \to 5'$ exonuclease mediates the enzyme's "proofreading" function, because it is most active on bases that are not correctly paired with the template strand—that is, bases that the polymerase activity has incorporated incorrectly (Brutlag & Kornberg, 1972; Kunkel, 1988). A $5' \to 3'$ exonuclease activity degrades a DNA strand being displaced ahead of the polymerizing enzyme (see Fig. 3-3) (Cozzarelli et al., 1969; Deutscher & Kornberg, 1969; Kelly et al., 1969; Klett et al., 1968).

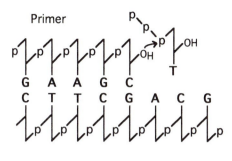

Template

Figure 3-2. Base incorporation by DNA polymerase. DNA-dependent DNA polymerases incorporate nucleotides into DNA that are complementary to a template strand. In the reaction shown, thymidine is added to the 3′-end of a growing strand, a reaction that occurs by enzyme-catalyzed nucleophilic attack by the 3′-terminal hydroxyl oxygen on the α-phosphorus of a deoxynucleotide triphosphate (arrow). In this example, thymidine is added to the 3′-end of the primer strand, base-pairing with adenosine in the template strand. Next, the 3′-hydroxyl group of the newly added T residue will attack the α-phosphate of a deoxyguanosine triphosphate molecule, resulting in the incorporation of G.

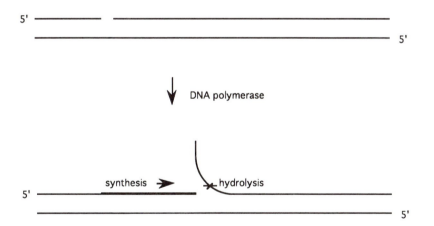

Figure 3-3. Strand displacement by DNA polymerase. When the substrate is a DNA duplex with a nick in one strand, some DNA polymerases can incorporate nucleotides starting at the 3′-end of the nick and can synthesize a new strand complementary to the template, displacing the previous strand and simultaneously hydrolyzing it. The end result is a repaired, nick-free duplex. Not all polymerases have the strand-displacement activity, and some that do lack the 5′ → 3′ exonuclease. The ability to initiate polymerization at a single-stranded nick forms the basis of a method for radioactively labeling DNA called "nick translation"; in this method the polymerase is supplied with deoxynucleoside triphosphates carrying a radioactive phosphate at the α-position.

DNA polymerases are found wherever DNA is found (Kornberg, 1981a). A number of different enzymes, from different sources, find application in molecular biology, including the *E. coli* enzyme Pol I (polymerase I) (Kornberg & Baker, 1992; Lehman, 1981a), a derivative of Pol I called the *Klenow fragment* (Klenow & Henningsen, 1970) which lacks the 5′ → 3′ exonuclease activity, and polymerases encoded by bacteriophages T4 and T7 (Lehman, 1981b). Each of these enzymes has subtly different characteristics, and there is a "best" polymerase for each of the many uses to which polymerases are put, including radioactive labeling of DNA, site-directed mutagenesis, construction of complementary DNA (cDNA), and so on. Special mention should also be made of the "thermostable" polymerases (Herbert, 1992). These enzymes are isolated from bacteria that live under extreme environmental conditions and are very stable to high temperatures. This property has been exploited in the development of a number of techniques, particularly the polymerase chain reaction (PCR) (Saiki et al., 1988), an innovation which has revolutionized molecular biology (Erlich, 1989; Erlich et al., 1989). PCR is described in detail in Chapter 14.

Total Gene Synthesis. Recently it has become possible to synthesize large (>100 bp) double-stranded DNAs de novo, and several complete genes have been constructed in the laboratory without recourse to cloning (Frank et al., 1987, Khorana, 1979). This method offers the researcher complete control over gene sequence, codon usage, and control sequences, thus obviating many of the challenging techniques associated with making defined changes in cloned DNA. Figure 3-4 illustrates the process of total gene synthesis. First, single-stranded oligodeoxyribonucleotides are synthesized. Automated DNA synthesizers are used to perform this task; the oligomers can be up to 100 bases long, or even longer, depending on the limitations of the individual synthesizer. The oligomers are made to base-pair with each other, when annealed, in an arrangement like that shown in the figure, so that an extended structure of overlapping oligos is formed. At this stage, before polymerization, the structure can be viewed as the "skeleton" of the desired gene: All the sequence information necessary to specify the complete gene is present, in one "strand" or the other, even though neither strand is complete. The constituent oligomers of the annealed structure can then be extended, using the ability of DNA polymerase to fill gaps in duplex DNA. For this purpose, the DNA polymerase encoded by phage T4 is a good choice, since it lacks $5' \rightarrow 3'$ exonuclease activity and will not displace DNA strands ahead of polymerization (Ausubel et al., 1989; Sambrook et al., 1989). After polymerization the structure is a double-stranded DNA molecule with four nicks, each indicated by a "p" in the figure; only ligation, using T4 DNA ligase (see below), is necessary to complete the synthesis of the gene. The ligase forms phosphodiester bonds between the $3'$-end of the oligomers extended by polymerase and the $5'$-end of the neighboring oligomers (Ausubel et al., 1989; Sambrook et al., 1989).

DNA Sequencing. The manipulations of genetic engineering permit the nucleotide sequence of DNA to be changed. The ability to determine the sequence of DNA experimentally was instrumental in the development of genetic engineering, because it is by sequencing that the results of nucleotide sequence manipulation are ascertained. DNA sequencing can confirm the structure of every new clone, subclone, or construct.

One method of DNA sequence analysis is the Sanger ("dideoxy") method, which employs a DNA-dependent DNA polymerase to synthesize radioactive DNA strands of varying lengths (Fig. 3-5) (Messing et al., 1977; Sanger et al., 1977). After these strands have been separated by denaturing polyacrylamide gel electrophoresis (Sambrook et al., 1989), they are visualized by autoradiography (Sambrook et al., 1989), and DNA sequence information can be "read" from the resulting pattern of bands (Fig. 3-6).

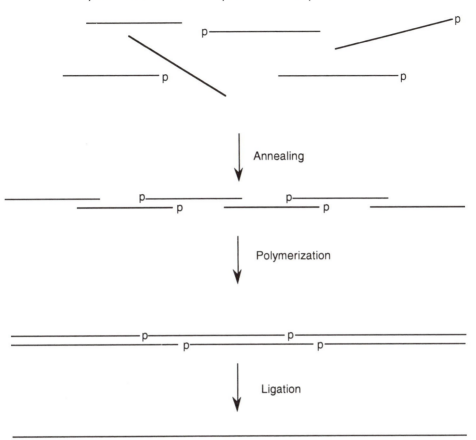

Figure 3-4. Total gene synthesis is a method for making long, synthetic DNA duplexes. In the annealing step, a mixture of 5′-phosphorylated oligodeoxyribonucleotides is heated to 95°C, then allowed to cool slowly. The oligomers are designed so that their sequences will base-pair to form a unique, overlapping structure like the one shown. This structure contains all the sequence information of the desired product duplex in one strand or the other. In the polymerization step, DNA polymerase uses the 3′-ends of annealed oligomers as primers and "fills in" the gaps in the structure, using the opposite strand as template. The result is a nicked duplex, which can be treated with DNA ligase to complete the synthesis. In this way, biologically functional DNA duplexes hundreds of base pairs in length can be made.

To synthesize DNA, the polymerase must be supplied with deoxynucleoside 5′-triphosphates of deoxynucleosides A, T, G, and C. These are incorporated into the growing primer strand in complementarity with the template strand; for example, at every position where a G occurs in the template strand, a C will be incorporated into the new strand.

Dideoxynucleotides (ddNTPs), which have neither a 2′- nor a 3′-hydroxyl group, can also be incorporated into the growing strand by DNA polymerase, but once this is done the strand cannot be further elongated because there is no 3′-OH group to form the next phosphodiester bond. Incorporation of a ddNTP thus results in termination of

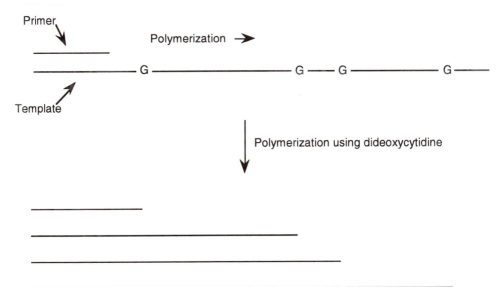

Primer

Polymerization →

Template

G ———————————————————— G —— G ———————— G ——

Polymerization using dideoxycytidine

Figure 3-5. DNA sequencing by the "dideoxy" method. The primer strand is extended by DNA polymerase in a reaction that includes, in addition to the four natural *deoxy*nucleoside triphosphates dATP, dTTP, dGTP, and dCPT, a *dideoxy*nucleotide, in this case ddCTP. Dideoxynucleotides have neither 2'- nor 3'-hydroxyl groups. In the polymerization reaction, the ddCTP is added at much lower concentration than dCTP, so that, occasionally, the polymerase will incorporate ddCTP rather than dCTP. Because the incorporated ddC has no 3'-OH group, polymerization cannot proceed past this point and the strand terminates. Thus, at each G in the template strand, a certain number of primer strands will terminate as a result of ddC incorporation, and the result of the reaction shown in this figure will be a set of elongated primer strand populations, each terminating with a ddC residue. In this figure, each of the lines of different lengths represents one such population. The newly synthesized strands are made radioactive by incorporation of isotope-containing nucleotide, so that they can be visualized by autoradiography following size separation by polyacrylamide gel electrophoresis.

the growing strand. Figure 3-5 shows what happens if a reaction is performed in which there is a small amount of ddCTP in addition to dCTP. In such a reaction, at each position where a G occurs in the template, there is a small probability that the polymerase will incorporate dideoxy C, resulting in chain termination. The result is a collection of DNA strands of various lengths, each terminated with ddC.

In order to determine a DNA sequence, four such reactions are performed. In each, one of the four dNTPs is supplemented with a small amount of the corresponding ddNTP. In addition, a radioactive dNTP is included in each reaction so that the polymerized strands can be visualized by autoradiography after separation by gel electrophoresis. Such an autoradiogram is shown in Fig. 3-6; the DNA sequence can easily be read directly from the figure.

For use in dideoxy sequencing, the ideal DNA polymerase should have several attributes. First, it must accept nucleotide analogues, including the dideoxynucleoside triphosphates and the $[\alpha\text{-}^{35}S]$ nucleotide used for radioactive labeling. In addition, the polymerase should have a high rate of DNA synthesis and high *processivity* (Kornberg & Baker, 1992). Processivity is defined as the average number of nucleotides incorporated in a single binding event. Premature dissociation leads to termination of the growing strand, and it will produce "background" on the sequencing gel. Lastly, the polymerase should lack exonuclease activities, which are not needed.

Several polymerases have at least some of these features. The Klenow fragment of Pol I as well as several thermostable polymerases can be used, depending on the characteristics of the sequence to be obtained. In addition, researchers have produced (Tabor & Richardson, 1987a,b) a modified bacteriophage T7 DNA polymerase that is tailored for use in dideoxy sequencing. As stated above, the unique specificities and characteristics of nucleic acid-modifying enzymes have given rise to many powerful techniques. In the case of dideoxy sequencing, the techniques are giving rise to more powerful enzymes!

RNA Polymerase. The "central dogma" of molecular biology is that the flow of genetic information in living systems is from DNA to RNA to protein: the information is *transcribed* from DNA to RNA and then *translated* to protein (Watson et al., 1987).

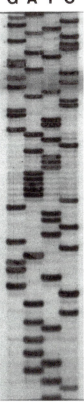

Figure 3-6. Autoradiogram of a denaturing polyacrylamide gel used to separate the products of dideoxy sequencing reactions. The DNA sequence can easily be "read" in the 5′ → 3′ direction from the figure, starting at the bottom: TCTATAATACACG. Autoradiogram courtesy of Patricia A. Spears, Becton Dickinson Research Center, Research Triangle Park, North Carolina.

Thus, the enzymes used by the cell (and the researcher) to make RNA are, for the most part, dependent on DNA templates. RNA polymerases (Chamberlin, 1982; Chamberlin & Ryan, 1982; Lewis & Burgess, 1982) recognize DNA sequences called *promoters* (Chamberlin, 1974) which are necessary for polymerase binding and appear to direct the polymerase to the *transcription start site*, where the polymerase begins making an RNA strand complementary to the DNA template (von Hippel et al., 1984).

The polymerase unwinds (von Hippel et al., 1984) a short region of the DNA duplex, and the new RNA strand, growing in the 5′ to 3′ direction, is temporarily base-paired with the template strand, so that the nucleotide sequence of the DNA is "copied" into RNA (see Fig. 3-7) (Lewis & Burgess, 1982). The result is an RNA strand with a sequence complementary to the template strand of the DNA (the bottom DNA strand in the figure) and identical to the top, or "coding," strand. To further complicate the issue, the two strands of the DNA duplex are sometimes referred to as the "sense" (top) and "antisense" (bottom) strands.

Uses of RNA Polymerases. In the cell, transcription initiation and termination are complex events, requiring the interplay of many macromolecules (Kornberg & Baker, 1992; Watson et al., 1987), but several bacteriophage RNA polymerases are capable of synthesizing RNA in vitro in comparatively simple systems (Chamberlin & Ryan, 1982). In particular, bacteriophages T7 and SP6 encode RNA polymerases that can be used by the researcher to make milligram quantities of RNA when supplied with little more than template DNA and nucleoside triphosphates (Krieg & Melton, 1987; Milligan & Uhlenbeck, 1989). These enzymes readily synthesize RNA molecules hundreds of nucleotides in length, as well as RNAs only a few nucleotides long. The template requirements of the T7 and SP6 polymerases are somewhat relaxed as well; only the promoter region of the DNA need be a duplex; the rest of the coding strand is not required (Milligan & Uhlenbeck, 1989). Also, termination signals are not required; the polymerases will "run off" the end of a substrate and then associate with another DNA promoter, with the result that many "run-off" transcripts are made from each DNA sub-

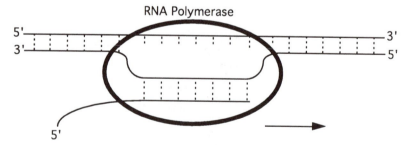

Figure 3-7. RNA synthesis by RNA polymerase. The RNA polymerase molecule (heavy oval) unwinds a short stretch of duplex DNA in a process sometimes called "melting in." For simplicity, the melted region shown is only a few bases long; the actual number varies from polymerase to polymerase. Beginning at a particular base on the DNA called the *transcription start site*, an RNA molecule is synthesized in complementarity to the "template strand" of the DNA, resulting in copying or "transcription" of the DNA nucleotide sequence into the new RNA molecule. As the RNA is synthesized, the polymerase moves along the DNA duplex in the direction shown, transiently melting the duplex as it goes.

strate. This capability has provided researchers with biologically functional messenger RNAs (Krieg & Melton, 1984b) and tRNAs (Noren et al., 1989; Sampson & Uhlenbeck, 1988) as well as molecules needed for structural studies (Krieg & Melton, 1984a) and studies of the catalytic properties of RNA (Butcher & Burke, 1994). An alternative to RNA polymerase is automated chemical synthesis of RNA (Lyttle et al., 1981) which can presently provide multimilligram amounts of RNA of length up to about 30–40 nucleotides.

As is the case with DNA polymerases, RNA polymerase can incorporate radiolabeled nucleotides when, for example, nucleoside triphosphates labeled at the α-position of the triphosphate are used as substrates. The resulting uniformly labeled transcripts find application in experiments requiring high specific activity RNAs (Sambrook et al., 1989).

Polymerization: Reverse Transcription. Enzymes capable of "transcribing" DNA from RNA templates (reverse transcription) were first discovered in virus-infected cells (Varmus & Swanstrom, 1985b). Such viruses contain an RNA genome which is copied into a duplex DNA molecule upon infection of a host cell, after which the resulting DNA viral genome may incorporate into the host genome (Varmus & Swanstrom, 1985b). Viruses whose life cycle includes a reverse transcription step are called *retroviruses*. One member of this family is the human immunodeficiency virus, the causative agent of acquired immune deficiency syndrome (AIDS) (Wong-Staal, 1988). Reverse transcriptases are the object of intense study: A drug or other agent that inhibits their activity might be an effective treatment for the viruses that use them (Haseltine, 1988).

The reverse transcriptases commonly used for the synthesis and manipulation of nucleic acids derive from either the avian myeloblastosis virus (AMV), the murine Moloney leukemia virus (MMLV), or the thermophilic bacterium *Thermus thermophilus* (*Tth*) (Varmus & Swanstrom, 1985a). These enzymes synthesize (Kornberg & Baker, 1992) a DNA strand, starting from a primer, that is complementary to either RNA or DNA templates; also, they possess what is known as *ribonuclease H (RNase H) activity*, which degrades the RNA strand of an RNA/DNA duplex starting from either the 5'- or 3'-direction. Polymerization by these enzymes can be primed by either DNA or RNA oligomers (in vivo, the primer is a tRNA molecule) and results in a DNA/RNA hybrid duplex, the RNA strand of which is subject to degradation by the RNase H activity. Reverse transcriptases do not possess the $3' \rightarrow 5'$ exonuclease activity that appears to function as a "proofreading" activity in other polymerases; perhaps as a result, their rate of misincorporation is high. This low accuracy is not necessarily a drawback; the high mutation rate may provide the virus with a means of evading host immune systems (Preston et al., 1988; Roberts et al., 1988).

Uses of Reverse Transcriptases. Reverse transcriptases are of great importance in molecular biology because they have facilitated the cloning of complex, "interrupted" genes (Ausubel et al., 1989; Sambrook et al., 1989). Unlike the genes of bacteria, the genes of higher organisms are often interrupted by *introns*. These are lengths of DNA that do not code for protein sequence, and they must be removed from the RNA message prior to translation. In the cell, messenger RNAs are *spliced*, after transcription, to remove the introns; the resulting mature mRNA contains the complete coding sequence for the protein to be translated, without any extraneous information.

The problem for researchers wishing to clone a gene is that the bacterial expression systems commonly used in cloning lack the machinery for recognizing and removing introns from eukaryotic genes. Thus, when cloning a eukaryotic gene, researchers often choose to start from the mature mRNA, and they reverse transcribe it into DNA for cloning. The cloned, intronless version of the gene can then be expressed by bacteria to yield the desired protein. The DNA made by reverse transcriptase is complementary to an RNA template, and thus it is called *complementary DNA* (*cDNA*) (Berger & Kimmel, 1987).

The DNA polymerase activity of reverse transcriptase can also be used for 3'-end labeling of DNA or RNA molecules, by the "filling-in" of recessed 3'-ends with nucleoside triphosphates radioactively labeled at the α-position. Also, since reverse transcriptase can synthesize DNA from a DNA template as well, it can be used for dideoxy sequencing.

Template-Independent Polymerization. Terminal deoxynucleotidyltransferase (Ratliff, 1981) catalyzes the incorporation of deoxynucleotides onto the 3'-end of single-stranded DNA or duplex DNA with a 3'-overhang; blunt-ended DNA is a less efficient substrate. This template-independent nucleotide incorporation adds bases in proportion to their relative concentration. Terminal deoxytransferase has been used to add lengths of homopolymeric DNA "tails" to the ends of DNA fragments to facilitate cloning. In this methodology, the cloning vector may be "tailed" with G residues, and the DNA insert with C residues, so that the two will anneal with each other prior to ligation with DNA ligase. The transferase can also accept non-naturally occurring bases as substrates for incorporation, including bases derivatized with biotin (Kumar et al., 1988). The resulting biotin-labeled DNA binds tightly to streptavidin and can thus be immobilized or purified from solution by binding to a streptavidin-coated material. Streptavidin-coated magnetic beads, for example, have proven very useful for purification of biotinylated DNA (Olsvik et al., 1994). A biotinylated probe oligomer can quickly "fish out" a desired sequence from a complex mixture of DNA: The probe base-pairs to a single-stranded DNA sequence, then the probe–DNA duplex is removed from solution by application of a magnetic field.

Template-independent RNA polymerases include poly(A)polymerase and polynucleotide phosphorylase. The template-independent enzymes are polymerases in the sense that they polymerize nucleic acids, but they are outside the generally accepted definition of "polymerase" as an enzyme that copies an existing nucleic acid molecule.

Ligation: RNA Ligase. RNA ligase (Brennan et al., 1983; Romaniuk & Uhlenbeck, 1983) was discovered (Silber et al., 1972) during a purification of DNA ligase from bacteriophage T4-infected *E. coli*, as an activity catalyzing the formation of intramolecular phosphodiester bonds in ribonucleic acid. Subsequent investigations have revealed it to be what Kornberg (1981b) called the "most versatile ligase"; its usefulness results from thorough investigations of its mechanism and careful optimization of its various reactions. RNA ligase catalyzes the formation of phosphodiester bonds between single-stranded nucleic acid molecules (see Fig. 3-8).

RNA ligase can use ATP or dATP as its energy source. The enzyme exhibits some preferences as to the bases near the 3'-end of the acceptor. A wide variety of naturally occurring and synthetic nucleotide analogues are accepted as donors; this has allowed synthesis of oligonucleotides containing synthetic analogues of adenine, thymine, cytidine, and guanine (Brennan & Gumport, 1985).

...NpNpN   +   pMp...   →   ...NpNpNpMp...

    acceptor   donor   product

Figure 3-8. Single-base addition by T4 RNA ligase. When describing RNA ligase-catalyzed reactions, "donor" refers to the 5'-phosphorylated substrate and "acceptor" refers to the 3'-hydroxylated substrate. Both donor and acceptor can be either DNA or RNA. The substrates shown are the minimum lengths accepted as substrates; the series of dots indicates that there is no upper length limit. A reaction in which the donor is a nucleotide monomer results in a product that is one base longer than the acceptor, this is termed "single base addition." If the donor is longer, the reaction is termed "oligomer–oligomer joining."

..G G   ..G G

..C C A C   +   CA-amino acid   →   ..C C A C C A - amino acid

   acceptor   donor   product

Figure 3-9. Misacylation of tRNA by RNA ligase. "Acceptor" here represents the amino acid acceptor stem of a partially truncated tRNA having a cytidine at the 3'-end. "Donor" is an aminoacylated dinucleotide of the form pCpA-amino acid, which forms a mature, full-length aminoacylated tRNA when joined to the acceptor by RNA ligase.

...CpCpA   +   [^{32}p]Cp   →   ...CpCpA[^{32}p]Cp

Figure 3-10. Radioactive labeling at the 3'-end of a nucleotide acid molecule by the use of T4 RNA ligase.

Experimental Uses. Many applications have been found for the oligomer–oligomer joining reaction, including the synthesis of base analogue-containing deoxyribonucleotides (Brennan & Gumport, 1985; Bruce & Uhlenbeck, 1982a,b), synthesis of biologically active tRNAs, and a recent innovation termed "ligation anchored PCR" (Troutt et al., 1992) which enables the production of full-length cDNA for a gene of interest, starting from only partial sequence information.

As described later in this chapter, Hecht and coworkers (Roesser et al., 1986) have developed a technique for chemically misacylating tRNA, which involves ligating partially truncated tRNAs to aminoacylated dinucleotides (Fig. 3-9). Here, the "acceptor" is a tRNA from which the terminal C and A bases have been removed, leaving a 3'-hydroxyl terminus. The donor is the dinucleotide pCpA, chemically acylated with an amino acid. When acceptor and donor have been joined by RNA ligase, the product is an aminoacylated tRNA that can participate in protein biosynthesis.

RNA ligase can also add a single base to the 3'-end of a nucleic acid (Brennan et al., 1983). A common use for this reaction is termed "3'-end labeling." In this procedure, a nucleic acid is radiolabeled by RNA ligase-catalyzed addition of a radioactive nucleoside 3',5'-bisphosphate (Fig. 3-10). In the reaction illustrated, cytidine 3',5'-[^{32}P]bisphosphate is ligated to the 3'-end of an oligomeric nucleic acid molecule ending with the sequence CCA. The result is a radioactively labeled oligomer that is one base longer and that carries a 3′-phosphate. This reaction is particularly useful for labeling nucleic acids that are not substrates for T4 polynucleotide kinase.

As noted above, the single-base addition reaction can also be used to incorporate synthetic analogues of the naturally occurring bases into nucleic acids, since RNA ligase will accept a wide variety of nucleoside 3′,5′-bisphosphates as donor substrates.

Ligation: T4 DNA Ligase. The enzyme T4 DNA ligase (Engler & Richardson, 1982) catalyzes the formation of phosphodiester bonds between 5′-phosphate and 3′-hydroxyl groups in double-stranded nucleic acids, with concomitant hydrolysis of ATP to AMP and pyrophosphate (Kornberg & Baker, 1992). Despite its name, it accepts both DNA and RNA substrates (Kleppe et al., 1970; Moore & Sharp, 1992).

In Fig. 3-11, the phosphodiester bonds between adjacent bases are represented as dashes. The nucleic acid duplex on the left side contains a nick between the T and the A in the upper strand, which is represented in the 5′ → 3′ direction. At the nick, the T has a 3′-OH terminus and the A has a 5′-phosphate. Although the nicked substrate shown is perfectly base-paired, DNA ligase has tolerance for mismatches at or near the ligation point (Harada & Orgel, 1993); its specificity is for a *duplex* rather than correct base-pairing. Though incorrectly base-paired duplexes are not very stable, they can transiently form and be bound by the ligase.

The enzyme also catalyzes a reaction with DNA referred to as "blunt-end ligation." Here, two separate duplexes are ligated together, with formation of phosphodiester bonds between the T and the A in both strands of the new duplex. As usual, the upper strand is represented in the 5′ → 3′ direction, while the lower strand is shown in the 3′ → 5′ direction. In both strands, the new phosphodiester bond is formed between a 5′-phosphate and a 3′-hydroxyl group.

Experimental Uses. DNA ligase has been used to construct long single-stranded molecules; in these experiments, short oligonucleotides are used as "splints" (Fig. 3-12) (Moore & Sharp, 1992). In these experiments, two RNA or DNA strands are joined to make a new, longer strand. In the first step, the two strands are annealed with a "splint" strand to form the "annealed substrate" for DNA ligase. Theoretically, either DNA or

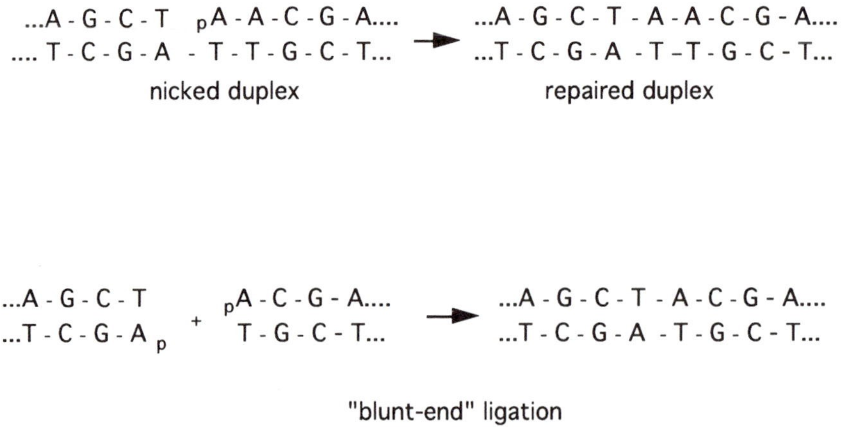

```
...A - G - C - T  pA - A - C - G - A....        ...A - G - C - T - A - A - C - G - A....
.... T - C - G - A - T - T - G - C - T...   →   ...T - C - G - A - T –T - G - C - T...
            nicked duplex                              repaired duplex



...A - G - C - T       pA - C - G - A....        ...A - G - C - T - A - C - G - A....
...T - C - G - A p  +    T - G - C - T...    →   ...T - C - G - A - T - G - C - T...
```

"blunt-end" ligation

Figure 3-11. Reactions performed by bacteriophage T4 DNA ligase.

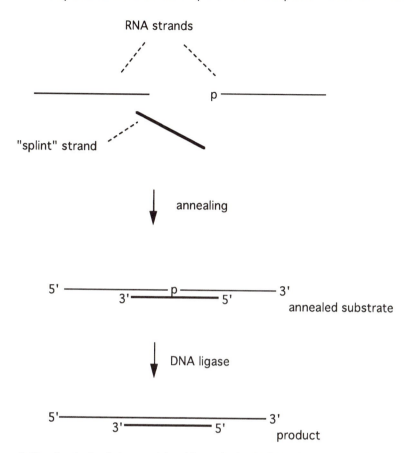

Figure 3-12. Synthesis of a long nucleic acid strand using T4 DNA ligase and a "splint." The three strands shown at the top have been synthesized so that, when annealed by heating and slow cooling, they will form the second structure shown, a nicked duplex. After nick-sealing by ligase, two of the strands have been joined to form a longer strand. In this way, long single strands can be made.

RNA could be used for the splint; in practice, DNA is generally used. This annealed substrate is effectively a nicked duplex of the type described earlier, and DNA ligase will catalyze the formation of a phosphodiester bond at the "nick"; the junction between the ends of the two RNA molecules. The new RNA strand can then be separated from the splint by denaturation. By this method, long RNAs can be made from shorter constituent strands. As described, RNA ligase can also be used for oligomer–oligomer joining; the advantage of using DNA ligase is that the splint strand organizes the substrates and prevents unwanted ligations.

Perhaps the most common experimental use (Ausubel et al., 1989; Sambrook et al., 1989) for DNA ligase is ligation of "sticky ends," formed by the action of certain restriction endonucleases. As described in detail in a later section, restriction endonucleases cleave both strands of duplex DNA at specific sites. Each of the many restriction endonucleases has a characteristic "cleavage site"; for example, the *Eco*RI restriction endonuclease, the first such enzyme obtained from *E. coli* strain R, recognizes the sequence GAATTC and then cleaves it as shown in Fig. 3-13. The two strands of the duplex are cleaved in a "staggered" fashion. The new termini are called "sticky ends" because, under the right conditions, they can anneal to form a duplex with two nicks. This annealed duplex with two nicks is, of course, a substrate for DNA ligase, which will regenerate the original duplex DNA.

In the example shown, a site is cut, annealed, then religated. This may not seem a very interesting enterprise until one realizes that *any* two sticky ends generated by *Eco*RI can be joined with ligase, not just two from the same molecule, as below. Thus, by using restriction enzymes and DNA ligase, researchers can "cut and paste" DNA

```
......NNNNNNGAATTCNNNNNNNN........
......NNNNNNCTTAAGNNNNNNNN........
```

↓ *Eco*RI cleavage

```
......NNNNNNG    pAATTCNNNNNNNN........
......NNNNNNCTTAAp    GNNNNNNNN........
```

↓ transient annealing

```
......NNNNNNG.AATTCNNNNNNNN........
......NNNNNNCTTAA.GNNNNNNNN........
```

↓ ligation

```
......NNNNNNGAATTCNNNNNNNN........
......NNNNNNCTTAAGNNNNNNNN........
```

Figure 3-13. Ligation of "sticky ends" by T4 DNA ligase. DNA can be cut and resealed at restriction enzyme sites because many restriction enzymes leave "sticky ends"; that is, they cut DNA in a "staggered" fashion, leaving overhanging single-stranded ends as shown. Though only a few bases in length, these ends can transiently reanneal to each other under the right conditions. Here, the *Eco*RI endonuclease produces a double-stranded break in a DNA duplex at the sequence GAATTC. The ends thus generated can transiently reanneal, forming the third structure shown. The dots between G and A in the structure indicate that it is not a covalently closed duplex, but a "temporary" duplex held together only by base-pairing between the A and T residues, the "sticky ends." This temporary duplex can be bound by T4 DNA ligase, which will nick-seal the structure to regenerate the original covalent duplex.

$$5' \text{ OH-NpNpNpN } 3' \quad + \quad \text{ATP} \quad \rightarrow \quad 5' \text{ pNpNpNpN } 3' \quad + \quad \text{ADP}$$

Figure 3-14. Phosphorylation of a nucleic acid molecule by T4 polynucleotide kinase. The enzyme transfers inorganic phosphate from the γ-position of adenosine triphosphate to the 5'-hydroxyl end of a nucleic acid, leaving as products the 5'-phosphorylated oligonucleotide and adenosine diphosphate.

at specific sites, an ability that is central to genetic engineering, as discussed later in the chapter.

Terminal Phosphates: T4 Polynucleotide Kinase. Initially purified from bacterio-phage T4-infected *E. coli*, polynucleotide kinase (Lillehaug, 1977; Richardson, 1981) has become one of the indispensable tools of molecular biology (Ausubel et al., 1989; Sambrook et al., 1989). It catalyzes the transfer of phosphate from the γ-position of nucleoside 5'-triphosphates to the 5'-hydroxyl groups of a variety of nucleic acids. A typical reaction is shown in Fig. 3-14. The reaction shown is reversible; thus nucleic acids already bearing a 5'-phosphate can be dephosphorylated, or radioactively end-labeled by an exchange reaction (Ausubel et al., 1982; Sambrook et al., 1989).

Polynucleotide kinase will readily phosphorylate single-stranded DNAs, double-stranded DNAs with overhanging 5'-hydroxyl termini, and many RNAs. The enzyme shows no base-specificity. In the example shown, ATP is used as the phosphate donor; other nucleoside triphosphates can also serve. At pH 6, polynucleotide kinase exhibits a 3'-phosphatase activity which releases inorganic phosphate from deoxynucleoside 3'-monophosphates and 3'-phosphorylated ribo- and deoxyribooligonucleotides (Cameron & Uhlenbeck, 1977). The physiological function of polynucleotide kinase has not been determined, but there is evidence both from studies of T4-infected *E. coli* (David et al., 1982) and by analogy to similar enzymes from other organisms (Pick & Hurwitz, 1986) that the enzyme is involved in alteration of host tRNA, perhaps acting in con-cert with T4 RNA ligase.

In addition to preparing nucleic acid substrates for other enzymes that require a 5'-phosphate group, such as T4 RNA ligase and DNA ligase, polynucleotide kinase is commonly used for labeling the 5'-ends of nucleic acids with radioactive phosphate, either ^{32}P or [33]P (Ausubel et al., 1989; Sambrook et al., 1989). Current highly sensi-tive radioactivity quantification methods allow detection of femtomolar quantities of ^{32}P-labeled nucleic acids (Berger, 1987), and 5'-end-labeled DNAs and RNAs have found utility in a wide variety of molecular biological applications (Ausubel et al., 1989; Greene & Guarente, 1987; Sambrook et al., 1989). In fact, radioactivity is prob-ably the primary means of detection of nucleic acid molecules at present, although the situation is rapidly changing with the advent of fluorescence and other nonradioactive detection methods (Kessler, 1993).

Terminal Phosphates: Phosphatases. When it is necessary to remove terminal phos-phates from a nucleic acid molecule, several enzymes are available; the most com-monly used are bacterial alkaline phosphatase and calf intestinal alkaline phosphatase (Ausubel et al., 1989; Sambrook et al., 1989). As their names imply, both enzymes work at alkaline pH. Neither is specific for the position of the phosphates they remove, and both work on either DNA or RNA. Both enzymes are also heat-stable, and thus can be difficult to inactivate, though the calf intestine enzyme is somewhat better in this regard.

The phosphatases are the basis of some nonradioactive detection methods due to their ability to hydrolyze phosphate substrates other than nucleic acids (Kessler, 1993). A number of dye molecules, in particular, yield bright colors when treated with phosphatase.

Nucleases. Enzymes that catalyze the hydrolysis of nucleic acid phosphodiester bonds are called *nucleases* (Linn & Roberts, 1982). A great variety of nucleases, with a wide range of specificities, are available to the researcher. Some digest only RNA, others only DNA, and some are active against both. *Exonucleases* hydrolyze only the outermost phosphodiester bond of their substrate, thus removing one base at a time from either the 5'- or the 3' terminus. *Endonucleases* hydrolyze interior phosphodiester bonds, thereby causing rapid change in the length of their substrate. *Restriction endonucleases* produce double-stranded breaks in duplex DNA at specific recognition sites, as described below. Finally, several multifunctional nucleic acid metabolic enzymes have nuclease activities as part of their repertoire—for example, DNA polymerase I and reverse transcriptase, described above.

Nucleases hydrolyze phosphodiester bonds by catalyzing a nucleophilic attack at phosphorous (Boyer, 1982) (Fig. 3-15). This may involve direct attack by a water molecule; however, in the case of some ribonucleases, cleavage is achieved via a "cyclization" reaction in which the 2'-hydroxyl group is the nucleophile, leading initially to a 2',3'-cyclic phosphate and then typically to the 3'-phosphate (Blackburn & Moore, 1982). The location of the phosphate group after hydrolysis of the phosphodiester bond is a defining characteristic of nuclease action. Ribonucleases responsible for processing or maturing cellular RNAs, and most deoxyribonucleases, yield 5'-phosphate and 3'-hydroxyl termini (Linn & Roberts, 1982). Ribonucleases responsible for degrading RNA typically yield 3'-phosphate and 5'-hydroxyl termini (Linn & Roberts, 1982).

Hundreds of nucleases, of almost every conceivable specificity, are available to the researcher; a small sampling is given below. The creative use of nuclease activities has been an invaluable part of structural studies of nucleic acids. For example, nuclease S1 has been used in many studies designed to locate single- and double-stranded regions of nucleic acid molecules; it is active only against single stranded regions, so the structure can be deduced from the S1 digestion pattern (Reyes & Wallace, 1987). A comprehensive listing of known nuclease activities is available (Adams et al., 1986).

Exonucleases. Spleen phosphodiesterase (Bernardi & Bernardi, 1971) removes nucleoside 3'-monophosphates from single-stranded RNA or DNA. It requires a 5'-hydroxyl terminus. Its counterpart enzyme is *snake venom phosphodiesterase* (Laskowki, 1971), isolated from snake venom, which removes nucleoside 5'-monophosphates from single-stranded DNA or RNA and requires a 3'-hydroxyl terminus. These enzymes have been useful in many studies of nucleic acid structure. In one example, snake venom phosphodiesterase has been used in conjunction with T4 polynucleotide kinase in the design of an ingenious assay for several types of DNA damage associated with cell transformation and cancer. In this work, Liuzzi and Paterson (1992) took advantage of the specificities of both enzymes in such a way as to achieve detection of the damaged DNA bases with femtomolar sensitivity. Irradiation of DNA with ultraviolet (UV) light or other ionizing radiations causes several types of damage, one of which is the formation of structures called *pyrimidine dimers*, in which two adjacent pyrim-

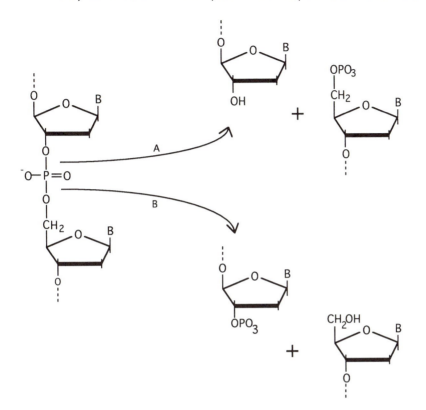

Figure 3-15. Nuclease digestion of nucleic acids. Depending on which bond is cleaved during the course of phosphodiester bond hydrolysis, the products of the reaction can have 5'-phosphate and 3'-hydroxyl groups (A) or 3'-phosphate and 5'-hydroxyl groups (B). Different nucleases use different strategies. In the cleavage of RNA, the 2'-hydroxyl group is often the attacking nucleophile; such reactions proceed through a 2',3'-cyclic phosphate intermediate, which typically hydrolyzes to the products shown (B). In other enzyme-catalyzed hydrolyses, a water molecule is the attacking species.

idine bases on the same strand become directly linked to one another by covalent bonds. These pyrimidine dimers are sometimes denoted by the symbol T <> T.

Snake venom phosphodiesterase removes bases processively from the 3'-end of DNA until it encounters a damaged base, at which point it cleaves a trinucleotide oligomer from the damaged strand and continues. The basis of this "reaching over" capability is unknown, but it nicely meshes with the specificity of T4 polynucleotide kinase to provide the basis for this assay. The product of codigestion of a pyrimidine dimer-containing DNA sample with snake venom phosphodiesterase and bacterial alkaline phosphatase includes nucleosides, which *are not* substrates for T4 polynucleotide kinase, and trinucleotides of the form TpT <> T, which *are*. After codigestion, radioactive labeling with T4 polynucleotide kinase will reveal the presence of the characteristic trinucleotide with very high sensitivity.

RNase H (Crouch & Dirksen, 1982) is the name given to enzymatic activities catalyzing the degradation of the RNA strand of RNA–DNA hybrid duplexes. RNase H activity is present in many cell types and appears to function in DNA replication, perhaps in the removal of RNA primers. Reverse transcriptase includes an RNase H activity that degrades the RNA template strand after using it to synthesize cDNA, used in cloning. RNase H has also been implicated in the mechanism of action of some *antisense therapeutic agents* (Agrawal, 1992). These are typically oligodeoxynucleotides complementary to a sequence segment of a targeted message—for example, the human immunodeficiency virus. Upon entering an infected cell, the antisense oligodeoxynucleotide binds the RNA target, forming a substrate for RNase H, which destroys that segment of the RNA target involved in the complex.

Bal 31 (Shishido & Ando, 1982), isolated from *Alteromonas espejiano*, can digest both DNA and RNA. It acts endonucleolytically against single-stranded substrates and exonucleolytically against double-stranded substrates. Bal 31 has proved very useful for creating deletions of controlled size in cloned DNA, allowing researchers to identify the boundaries of genes and gene control regions.

DNAse I (Moore, 1981), isolated from pancreas, hydrolyzes double- and single-stranded DNA, leaving 5'-phosphate termini. DNAse I has proved very useful in studies of protein–nucleic acid interaction through the technique known as *footprinting* (Galas & Schmitz, 1978), by which a researcher can locate where on a long DNA molecule a protein has bound by the "footprint" of the protein—that is, that area of DNA protected from DNase I digestion by the bound protein.

Nuclease T1 (Takahashi & Moore, 1982), obtained from *Aspergillus oryzae*, hydrolyzes phosphodiester bonds 3' to G residues, leaving 3'-phosphate termini. Nuclease T1 is used along with other sequence-specific RNA endonucleases for RNA sequence analysis (Donis-Keller, 1980).

Micrococcal nuclease (Cotton & Hazen, 1971) nonspecifically digests single- and double-stranded DNAs and RNAs.

RNase A (Blackburn & Moore, 1982) cleaves RNA 3' to pyrimidine residues, leaving 3'-phosphate termini.

RNase VI (Lockard & Kumar, 1981) cleaves only double-stranded RNA.

Restriction/Modification Systems: Characteristics. In the same way that a road map gives the traveler access to an unknown country, the availability of restriction endonucleases has given researchers enormously enhanced access to the informational content of DNA (Endlich & Linn, 1981; Kannan et al., 1989; Wells et al., 1981). The importance of this large and growing collection of enzymes to the biotechnology revolution is difficult to overstate.

Restriction/modification systems are found in prokaryotes. Each includes two enzymes, a restriction endonuclease and a modification methylase, both of which recognize the same short base sequence of double-stranded DNA. For example, the *Eco*RI system, so named because it was the first restriction/modification system discovered in the R strain of *E. coli*, recognizes the site shown in Fig. 3-16.

The *Eco*RI restriction endonuclease catalyzes the hydrolysis of the phosphodiester bonds indicated by arrows, the result being a double-stranded break in the DNA. The *Eco*RI modification methylase transfers methyl groups from *S*-adenosylmethionine to the marked adenine base in each strand. Both enzymes are quite site-specific: They perform their respective actions *only* at the sequence GAATTC.

The actions of these enzymes are mutually exclusive. A GAATTC site that has been methylated in one or both strands is refractory to cleavage by the endonuclease; thus the activity of the methylase protects the cellular genome from damage by the en-

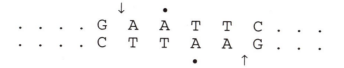

Figure 3-16. The sequence recognized by the *Eco*RI restriction/modification system. *Eco*RI endonuclease cuts the phosphodiester strands of duplex DNA at any occurrence of the sequence shown. The arrows show the location of cutting in each strand. The *Eco*RI methylase modifies the adenine base marked with a darkened dot (·) in each strand, adding a methyl group to the exocyclic N6 atom. An *Eco*RI site modified in either or both strands is resistant to cleavage by the endonuclease.

donuclease. In this way, restriction/modification systems provide the bacterial cell a measure of protection against viral or other foreign DNA. Any foreign DNA that may penetrate the cell wall of an individual of *E. coli* strain R will be subject to double-stranded cleavage at each occurrence of the sequence GAATTC. The cell's own DNA is protected from the endonuclease by virtue of being methylated at all GAATTC sites. Working together in this way, restriction/modification systems can be thought of as a primitive kind of immune system, distinguishing self from nonself and destroying the latter.

Over the last 20 years, hundreds of restriction enzymes have been identified, from many sources. A small sample is shown in Table 3-1. In addition to their usefulness in restriction mapping and cloning (see below), restriction enzymes have served as model systems for the study of protein–DNA interactions in general and, specifically, for the phenomenon of DNA sequence recognition by enzymes (Brennan et al., 1986; George et al., 1985; Kim et al., 1990; King et al., 1989; Lesser et al., 1990; Modrich & Zabel, 1976; Van Cleve & Gumport, 1992). As a result of intensive study, many intriguing characteristics of restriction/modification systems have come to light, in some cases providing researchers with new experimental tools. The investigation of this large and diverse family of enzymes will be a challenging topic of research for some time to come.

Restriction/Modification Systems: Uses. The human genome, the genome of yeast or bacterial cells, and even some bacteriophage genomes are too large to manipulate as complete molecules (Kornberg & Baker, 1992). The DNA of bacteriophage lambda, for example, is nearly 40,000 base pairs in length. Such molecules are easily broken during isolation, resulting in a heterogeneous preparation. In addition, they contain hundreds or thousands of genes, a huge library of information. Restriction endonucleases allow the experimenter to study such a library one "book" at a time. Not only are the large genomes cut reproducibly at exact sites by restriction endonucleases, they are cut into pieces that can easily be inserted into vectors for further study (Berger, 1987).

In the early days of molecular biology, most DNA sequencing was done by the Maxam–Gilbert method (Maxam & Gilbert, 1977), which provides sequence data beginning at the 5'-end of a DNA molecule and continuing for perhaps 100 bases. The Maxam–Gilbert method requires microgram amounts of the DNA to be sequenced, meaning that to produce a readable sequencing gel, the experimenter needs at least 10^{10} DNA molecules, all terminated at the same base (Ambrose & Pless, 1987). The products of restriction enzyme digesion meet this criterion, because restriction enzyme

Table 3-1 Restriction Enzyme Names, Sites, and Sources

Enzyme[a]	Recognition Sequence[b]	Source Organism[c]
BamHI[d]	..G|G A T C C..	*Bacillus amyloliquefaciens*
	..C C T A G|G..	
BsrI[d]	..A C T G G N|N..	*Bacillus stearothermophilus*
	..T G A C|C N N..	
BsaB[d]	..G A T N N|N N A T C..	*Bacillus stearothermophilus* B
	..C T A N N|N N T A G..	
DdeI[d]	..C|T N A G..	*Desulfovibrio desulfuricans*
	..G A N T|C..	
HpaII[d]	..G T T|A A C..	*Haemophilus parainfluenzae*
	..C A A|T T G..	
NotI[d]	..G C|G G C C G C..	*Nocardia otitidiscaviarum*
	..C G C C G G|C G..	
SphI[d]	..G C A T G|C..	*Streptomyces phaeochromogenes*
	..C|G T A C G..	
XhoI[d]	..C|T C G A G..	*Xanthomonas holcicola*
	..G A G C T|C..	

[a]Restriction enzyme names are acronyms derived from the name of the source organism.

[b]The vertical lines represent the sites at which each enzyme produces strand breaks. The letter "N" represents a site at which the identity of the base is not specified; for example, the enzyme *Dde*I will accept any base pair at the center of its recognition site.

[c]The original source for each enzyme is given; many restriction enzymes are now purified from bacterial strains of *E. coli* engineered to produce large quantities of the enzyme.

[d]*Bam*HI (Wilson & Young, 1975), *Bsr*I (Polisson & Morgan, 1988); *Bsa*BI (Chen, Z., & Kong, H., unpublished data); *Dde*I (Makula & Meagher, 1980); *Hpa*II (Sharp et al., 1973); *Not*I (McClelland & Nelson, 1985); *Sph*I (Fuchs et al., 1980); *Xho*I (Gingeras et al., 1978).

cleavage is so specific. Thus the combination of restriction enzyme cleavage and Maxam–Gilbert sequencing gave researchers their first real access to the huge informational content of DNA.

A graphic representation of the restriction sites appearing on a DNA molecule is termed a *restriction map*. Figure 3-17 shows the map of restriction enzyme *Hin*dIII cuts on bacteriophage lambda. Equipped with a map of restriction sites, the researcher interested in studying some segment of a DNA molecule (e.g., a particular gene) needs only identify restriction sites on either side of the gene. Following the digestion with the enzyme that utilizes these sites, the DNA fragment of interest can be isolated and subcloned into an appropriate vector. Thus, restriction sites are the signposts used by researchers to demarcate large genomes. Digestion of such a molecule by a restriction endonuclease produces a characteristic pattern of bands. Figure 3-18 shows a photograph of an agarose gel that has been used to separate the restriction fragments derived from bacteriophage lambda.

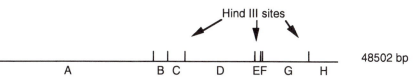

Figure 3-17. Digestion of bacteriophage lambda DNA by restriction endonuclease *Hin*dIII. The *Hin*dIII recognition site, AAGCTT, occurs seven times in the genome of bacteriophage lambda (vertical lines). The enzyme cleaves the 48502 base-pair lambda genome at these sites, forming eight shorter DNAs. These digestion products are labeled A through H. The products of a lambda *Hin*dIII digest are shown in Fig. 3-18.

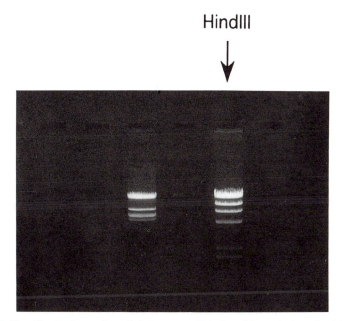

Figure 3-18. The products of restriction enzyme digestion of bacteriophage lambda. Digestion of lambda with different restriction enzymes produces different distributions of products. These products can be separated from each other by electrophoresis through an agarose "gel." When stained with ethidium bromide, the DNA fragments will fluoresce and can be photographed. The leftmost lane of the gel shows undigested lambda DNA.

The "staggered" ends produced by many restriction enzymes facilitate subcloning (Fig. 3-19) (Linn & Roberts, 1987). As explained above in the section on DNA ligase, the staggered ends can be religated to produce a covalently closed duplex. In fact, any two DNA molecules that are the products of digestion by a restriction endonuclease that produces staggered or "sticky" ends can easily be joined end-to-end by T4 DNA ligase; under suitable conditions, the cleaved ends can anneal to each other, forming a ligase substrate. Much of the work of molecular biology involves construction of plasmids and phage with novel sequences and properties, and this work is done through the creative use of DNA ligase and the wide variety of restriction enzymes now available.

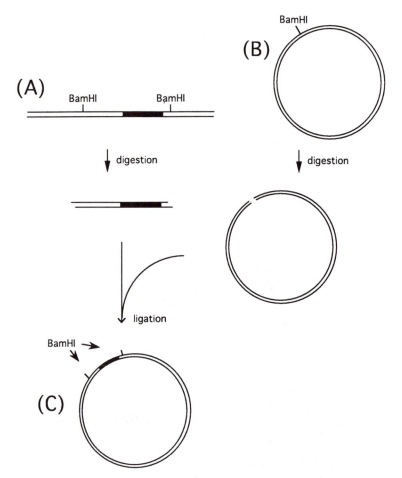

Figure 3-19. Subcloning. A sequence of interest (**A**) is found to be located between two *Bam*HI recognition sites. This sequence can be subcloned into a plasmid (**B**) having a single *Bam*HI site as diagrammed here. First, both the sequence of interest and the plasmid are digested with restriction endonuclease *Bam*HI. The digested plasmid and the gene fragment can be joined by ligation, since any ends produced by *Bam*HI digestion are compatible. The resulting recombinant plasmid (**C**) contains the sequence of interest and two *Bam*HI sites. Note that the gene fragment can anneal to the digested plasmid in either orientation, since its two sticky ends are equivalent; only one orientation is shown. In addition, it is possible for the two ends of the digested plasmid to rejoin to each other without accepting the gene fragment. In practice, this is avoided by treating the digested plasmid with alkaline phosphatase.

Protein Engineering

As mentioned in the introduction to this chapter, our present degree of sophistication in the synthesis and manipulation of nucleic acids owes much to the careful, painstaking work of enzymologists. As we have seen, a great variety of experimental designs have arisen from our detailed knowledge of the functioning of these biological catalysts. In the same way that an endless variety of houses can be built with the same set of carpenter's tools, a large and growing collection of techniques for the study of nucleic acids has been developed over the years. Some of these techniques are described briefly above. Collections of standardized protocols are available (Ausubel et al., 1989; Sambrook et al., 1989). Used singly and in combination, these techniques have revolutionized most areas of the life sciences and have created entirely new fields as well. One of the latter is protein engineering (Robson & Garnier, 1988).

The ability to manipulate genetic information affords, by extension, the ability to manipulate protein structure. A multitude of studies have appeared in the scientific literature during the last decade in which researchers have systematically varied the amino acid sequence of a protein of interest by techniques of molecular biology that employ many of the enzymes described in this chapter. The results of these *site-directed mutagenesis* experiments are obtained by kinetic or structural studies of the altered proteins, resulting in new insights and, in some cases, new enzymes with properties different from those of the original, or "wild-type," enzymes (Oxender & Fox, 1987). Very recently, our knowledge of the relationship between protein structure and function has advanced to the point that researchers have successfully designed enzymes "de novo"—that is, "from scratch" (Handel et al., 1993; O'Neil et al., 1990) All of these studies together constitute the new field of protein engineering.

Site-directed mutagenesis (Zoller & Smith, 1987) is a core technology of protein engineering. It is a technique used most profitably for the further study of enzymes that are already well characterized. One must have access to the gene before mutagenesis can be carried out, and an *expression system* (Bitter, 1987; Cullen, 1987; Shatzman & Rosenberg, 1987) is required to translate the mutated gene. The enzyme must be understood sufficiently that meaningful questions can be asked about the functioning of individual amino acid residues within its structure; thus a structural study, typically X-ray crystallographic analysis (Kim et al., 1990), is often performed. The amino acids critical for the functioning of the enzyme are usually clustered in one location, often somewhere on the surface of the enzyme, termed the *active site*. Location of the active site often allows testable predictions to be made concerning the functioning of the various amino acid residues located there, so these residues are prime targets for site-directed mutagenesis (King et al., 1989).

Thus one can consider a typical sequence of events for enzymological research. First, an activity is discovered, then the enzyme responsible for this activity is isolated. Next, kinetic and structural studies are performed, and hypotheses are advanced to explain various aspects of the structure of the enzyme, as well as its stability and catalytic mechanism. At this point, site-directed mutagenesis becomes useful. In other words, the systematic replacement of amino acid residues within a given enzyme can only be planned, and will only be useful, if the experimenter is starting from rather detailed knowledge of the structure and function of the enzyme.

It should be noted that the above trajectory is only one example of the routes taken in identifying proteins of interest. Our knowledge of the relationship between protein structure and function is becoming sufficiently advanced that isolation of the enzyme under study has lost some importance. In the case of a heritable disease, genetic analysis of families afflicted with the disease may lead to location of the responsible gene and identification of the protein it encodes prior to measurement of the enzymatic activity. For example, the cloning and sequencing of the gene associated with cystic fibrosis led immediately to testable hypotheses about the structure and function of the protein it encodes (Roberts, 1990).

"Semisynthetic" Enzymes. At present, several laboratories are developing methodologies that refine site-directed mutagenesis and may be seen as the "next step" in the technology (Hecht, 1992). In one such methodology, in vitro simulation of each step in the central dogma has been achieved: Genes are created by total synthesis, transcribed in vitro by phage polymerases, and translated in vitro using artificially charged transfer RNAs; every step of the process is achieved through concerted use of the enzymes of nucleic acid synthesis and manipulation. The result is an ability to synthesize functional enzymes bearing synthetic, non-naturally occurring amino acids at predetermined sites. For example, an aspartic acid residue might be replaced with β-chloroaspartate, which has an altered side chain pK_a (Fig. 3-20).

The ability to make this kind of replacement allows students of enzyme structure and mechanism a level of control over functional group chemistry that is much finer than that offered by traditional site-directed mutagenesis, the latter of which is limited to the 20 naturally occurring amino acids. For our last example of the applications of nucleic acid enzymology, we will consider in detail the technology that leads to these "semisynthetic" enzymes.

In in vivo protein biosynthesis, genetic information that has been transcribed from DNA to RNA is translated into protein (Fig. 3-21) (Lewin, 1990). The translation takes place at the ribosome, a large complex of RNA and protein components that provides the framework and catalytic site for assembly of new proteins. At the ribosome, the base triplets (codons) of messenger RNA are base-paired with the complementary anticodons of tRNAs activated with their cognate amino acids. The tRNA is the translator molecule, transferring meaning from one language (its anticodon base triplet) to another (the amino acid corresponding to the anticodon). That each tRNA is aminoacylated only with the amino acid corresponding to its anticodon is ensured by the high specificity of the aminoacyl-tRNA synthetases, a fascinating group of enzymes (Schimmel, 1987) that activate tRNAs by forming ester bonds between the terminal 2'- or 3'-hydroxyl group of mature tRNAs and the carboxyl group of the amino acid. Misacylation of a tRNA during in vivo protein biosynthesis will

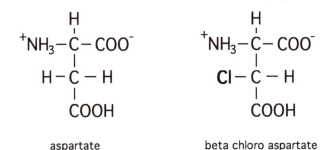

aspartate   beta chloro aspartate

Figure 3-20. Structures of the amino acids aspartate and β-chloroaspartate. The chlorine atom is electron-withdrawing, thus lowering the pK_a of the side-chain carboxyl group. Site-specific substitution of amino acid analogues in enzymes allows structural and mechanistic investigations beyond the reach of classical site-directed mutagenesis, which is limited to the 20 naturally occurring amino acids.

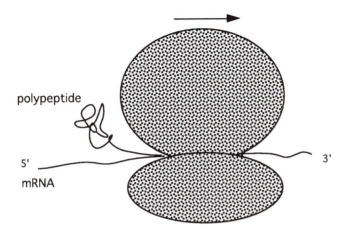

Ribosome

Figure 3-21. Protein biosynthesis. The ribosome, composed of two subunits each of which contains ribosomal proteins bound together within a ribosomal RNA framework, moves along messenger RNA in the $5' \rightarrow 3'$ direction. Aminoacylated tRNAs (not shown) bring amino acids to the site of peptide bond formation, where base-pairing between the tRNA anticodons and the messenger RNA codons specifies the sequence of the newly synthesized polypeptide.

result in insertion of an incorrect amino acid in the protein being synthesized (Chapeville et al., 1962).

If a researcher were to devise a way to misacylate, say, all of the alanine tRNAs in a protein biosynthetic reaction with valine, then the resulting protein would have valine residues at each position normally occupied by alanine. Likewise, if all the alanine tRNAs were misacylated with a non-naturally occurring, synthetic amino acid such as β-chloroaspartate, then this amino acid would appear in place of alanine. Much more desirable is the ability to replace a *single* amino acid within the structure of an enzyme, and this is achieved through creative use of the genetic code.

Expanding the Genetic Code. In addition to specifying each of the 20 naturally occuring amino acids, the genetic code contains "stop" signals that mark the position at which protein translation is to terminate (Lewin, 1990). These "stop codons" cause a translating ribosome to disengage from messenger RNA and release the completed protein. In the technology developed by Hecht and coworkers (Hecht, 1992), the gene for an enzyme is engineered so as to encode a messenger RNA containing a stop codon in place of the codon of the amino acid of interest. Under normal circumstances, such a messenger RNA will not encode functional enzyme; rather, the polypeptide will be truncated at the position corresponding to the stop codon. However, a special kind of tRNA, called a *suppressor* because it suppresses stop codons, can "read through" such a messenger RNA to make full-length enzyme, because it carries an anticodon that is complementary to the stop codon (Fig. 3-22).

In this way, the codon used by living cells as a stop signal for translation can be adapted for use as an amino acid codon. A protein biosynthetic reaction supplied with aminoacylated suppressor tRNA will insert its amino acid at every occurrence of the

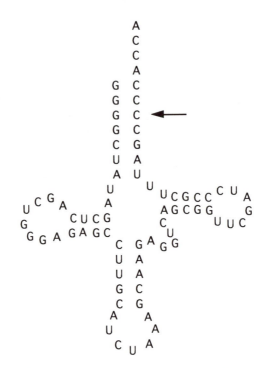

Figure 3-22. Transfer RNA cloverleaf structure. This transcribed tRNA is a "suppppressor" because its anticodon, 5'-CUA-3', base-pairs with the mRNA sequence UAG, a stop codon. This has the effect of "suppressing" the stop codon and allowing the ribosome to continue along the mRNA. The anticodon is at the bottom of the diagram; the amino acid acceptor site (CCA) at the top. Transfer RNA isolated from natural sources contains modified bases. The arrow points to a base (number 70) that is implicated in recognition of this tRNA by its aminoacyl-tRNA synthetase. When the base is a C, as shown, the tRNA is not charged; when U, it is charged with alanine. For the incorporation of unnatural amino acids, it is important that the tRNA not be chargeable in vivo; otherwise it could be re-used by the biosynthetic reaction, with resultant incorporation of alanine at stop codons.

stop codon to which it is complementary. If the suppressor tRNA is activated with an unnatural amino acid, the result of the protein biosynthetic reaction will be a "semi-synthetic" enzyme, with a modified amino acid at one or more predetermined sites in the polypeptide chain (Fig. 3-23).

Transfer RNA isolated from biological sources contains many modified bases, the functions of which have been the object of much study (Schimmel et al., 1979). Transcribed tRNA contains only A, G, C, and U, yet has been found to function effectively in in vitro translation (Noren et al., 1989). Thus it is possible to produce the suppressor tRNA necessary for the Hecht technology by an in vitro transcription reaction of the kind described above under "RNA polymerase." Because tRNA structural genes are only 70 to 80 bases in length, they are well within the range of total gene synthesis, and large quantities of the tRNA can be made by transcription from a strong promoter, such as the one recognized by bacteriophage T7 polymerase.

The chemical activation of the suppressor tRNA is a multistep process involving enzymatic ligation of a partially truncated tRNA molecule to an aminoacylated dinucleotide (Hecht, 1992). The result is an activated tRNA, biologically indistinguishable from charged tRNAs produced by aminoacyl-tRNA synthetases. When introduced into a protein biosynthesis reaction programmed with messenger RNA bearing a stop codon, the misacylated suppressor tRNA will deliver its synthetic amino acid to the growing peptide chain at the desired position, and no other.

Unnatural Enzymes. The technology for introducing non-naturally occurring amino acids at defined positions in enzymes has allowed studies of enzyme stability, structure, and mechanism that would not have been possible before. In one study (Mendel et al., 1991), a photoactivatable protein was made by incorporating an aspartyl β-nitrobenzyl ester at the active site of phage T4 lysozyme. This enzyme was inactive until irradiated with a high-pressure mercury lamp, after which its activity was similar to that of wild-type enzyme. Such "caged" enzymes may be useful in time-resolved studies of enzymatic reactions and protein folding. In another study (Chung et al., 1993) the deletion of a single methyl group from isoleucine by substitution of the unnatural amino acid norvaline abolished the activation by GTPase-activating protein of the onco-gene product Ha-Ras p21 protein, which is implicated in cell transformation, shedding

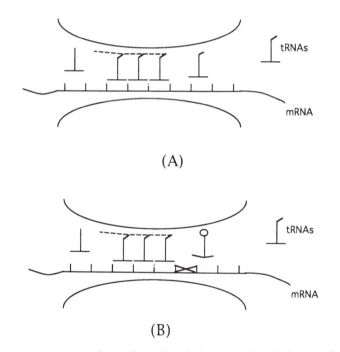

(A)

(B)

Figure 3-23. Protein biosynthesis. (**A**) Shown here is the normal situation in a protein biosynthetic reaction. Aminoacylated tRNAs conduct amino acids to the site of peptide bond formation. The tRNAs bind in an order determined by the messenger RNA sequence to which their anticodons are complementary. Peptide bonds are formed between the amino acids, and the new polypeptide chain dissociates from the ribosomal assembly. (**B**) In this figure a suppressor tRNA activated with an unnatural amino acid (small circle) enters the site to base-pair with a stop codon, shown by crossed lines. The unnatural amino acid will be incorporated into the growing polypeptide uniquely at the sites of such stop codons.

light on the nature of this protein–protein interaction. In the same study, substitution of threonine-35 with allothreonine inactivated the enzyme, demonstrating the importance of the stereochemical orientation of this amino acid in the functioning of Ras. Unnatural enzymes have also been employed in protein stability studies. Position 133 in T4 lysozyme was replaced with norvaline and ethylglycine as part of a study (Mendel et al., 1992) designed to measure the cost of stepwise removal of methyl groups from the hydrophobic core.

These studies demonstrate the potential power of the semisynthetic enzyme technology, one of the latest "offsprings" of nucleic acid enzymology. Refinement of the technology must address (1) the considerable technical difficulty of producing the misacylated tRNAs and (2) the relatively low yield of in vitro translation reactions.

Summary

The material in this chapter will be outdated fairly soon: Nucleic acid enzymology continues to produce new methods for the study of macromolecules at a dizzying rate. For example, one of the most exciting innovations of this decade involves the artificial "evolution," in the test tube, of nucleic acid molecules designed for specific tasks of binding or catalysis (Ellington & Szostak, 1990). These combinatorial techniques are discussed in Chapter 14.

If one can accept the proposition that nucleic acids are the most "powerful" substances known, it follows that our developing ability to employ them for our purposes is of tremendous potential. Clear benefits in the diagnosis and treatment of human diseases, both infectious and genetic, can be seen, as can benefits to agriculture. What was once the specialized pursuit of the dedicated enzymologist has become an academic–industrial complex that seems to produce, with each new innovation, not only a flurry of scientific papers but also new commercial opportunities. Things have come a long way; the responsible use of this technology seems likely to take us much farther.

4

Structure of DNA and RNA

M. Sriram and Andrew H.-J. Wang

The key roles played by nucleic acids in the molecular processes associated with life, especially the storage and transfer of genetic information, have focused our attention on these molecules. They have been studied over the years with every available technique, and a wealth of information has been obtained. Not only is the amount of information accumulated enormous, but the rate of acquisition of information is also increasing, with the major thrust being placed on genome projects and superior research techniques.

The biological, chemical, and physical properties of nucleic acids and their roles are enumerated in the other chapters of this volume. However, some repetition is unavoidable, albeit with a change in emphasis in this chapter to the structure of nucleic acids.

The notion that the three-dimensional structure of a molecule determines its function is universally accepted. The premise that structure determines function has been the foundation in the explanation of the function and regulation of biologically important molecules in terms of their atomic structures. As our understanding of the structural basis for the function of biologically important molecules increases, so does our ability to design molecules that contribute to improve longevity and the quality of life.

Nucleic acids are molecules found in relative abundance in the nucleus of cells, where they constitute 5% to 15% of the dry weight of the cell. Their acidic character was reflected in their propensity to undergo staining by basic dyes; hence they were called *nucleic acids*. Nucleic acids are of two types, namely DNA and RNA, as described in Chapter 1. These are shown schematically in Fig. 4-1.

As shown in the figure, DNA and RNA are comprised of nucleotides linked by $3' \rightarrow 5'$ phosphodiester bonds to form a polymer. The polynucleotide chain has directionality due to this linkage, and a nucleotide chain is described as going from 5' to 3' by convention. The repetitive linking of sugar-phosphate-sugar-phosphate . . . forms the "backbone" of the nucleic acid and different bases are attached by the β-glycosidic linkage to the C1' atoms of the sugar. A dodecanucleotide sequence 5'-cytidine-guanosine-cytidine-guanosine-adenosine-adenosine-thymidine-thymidine-cytidine-guansine-cytidine-guanosine-3' is written in short notation as 5'-d(CGCGAATTCGCG)-3', where the prefix "d" indicates that the sugar is 2'-deoxyribose for this sequence. Similarly, the sequence 5'-r(GGCCGGCC)-3' denotes an RNA octanucleotide. Frequently the 5' and 3' markers are omitted because the chain direction is fixed by convention, reducing the short notation for these two sequences to d(CGCGAATTCGCG) and r(GGCCGGCC).

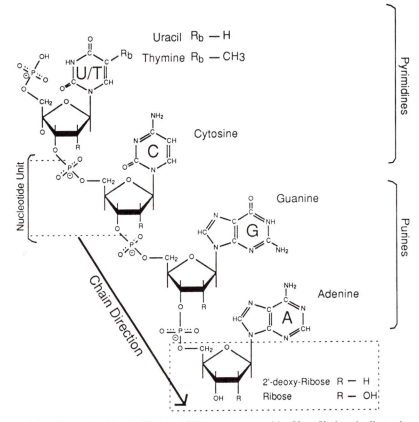

Figure 4-1. The nucleotides in DNA and RNA are connected by $3' \rightarrow 5'$ phosphodiester bonds. A representative tetranucleotide fragment of DNA/RNA is shown. The sugar and phosphate form the backbone to which different bases are attached. Because of its asymmetry, the nucleotide chain has a direction which by convention is defined from $5'$ to $3'$. Uracil is found in RNA only, replacing thymine which occurs in DNA.

Key Structural Features of Nucleic Acids

Base Pairs in Nucleic Acids. As illustrated in Fig. 4-1, the nucleic acid backbone is very polar and negatively charged while the bases provide large aromatic surfaces and are hydrophobic. In the cell when DNA or RNA occur as single strands, they are most likely stabilized by bound proteins. In cells, DNA occurs predominantly in double-stranded form with antiparallel strands. The bases of the two strands pair with each other to form "rungs" of a ladder, with the backbone forming the sides. The two strands are helically coiled; this coiling maximizes the exposure of the charged and polar back-bone to water, while the aromatic bases lying within the middle are shielded. The base-pairing observed is quite specific, with adenine pairing to thymine and guanine to cytosine. This specific base pairing was proposed by Watson and Crick in 1953, and it played a crucial role in determining the structure of DNA.

Figure 4-2 shows a detailed view of the Watson–Crick base pairs. The key element producing the specificity of base-pairing is the formation of hydrogen bonds. Hydrogen bonds have an element of directionality such that the donor (the hydrogen attached to an electronegative atom such as oxygen or nitrogen) and the acceptor (oxygen or nitrogen) have to be aligned almost linearly for a good hydrogen bond to form.

The complementary arrangement of donors and acceptors allow the Watson–Crick base-pairing of cytosine with guanine and adenine with thymine. This makes the two C1′ atoms of the sugars of a C·G base pair and an A·T base pair essentially equidistant, and it minimizes the strain in the backbone of the DNA double helix. The Watson–Crick arrangement also maximizes the number of hydrogen bonds possible in the two base pairs.

The sugars and phosphate are located toward one side of the C·G and A·T base pairs. The edge of a base pair from the sugar C1′ along the N3 edge of a purine (pu), and C2 edge of a pyrimidine (pyr) to the sugar C1′, is located in the minor groove. The other edge of the base pair from the sugar C1′ along the C6 of a pu and C4 edge

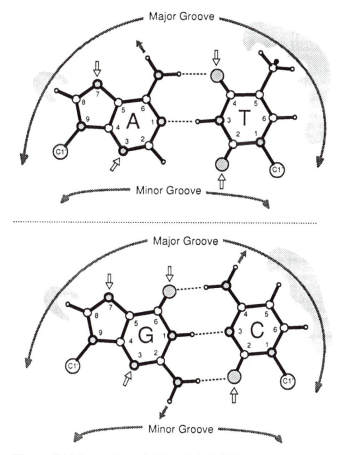

Figure 4-2. Watson–Crick base-pairing of A·T and G·C. Hydrogen bonds are shown as dashed lines. Hydrogen bond donors (gray arrows) and acceptors (white arrows) of the bases are shown, and it is obvious that the bases can pair in more than one way. Hydrophobic edges of the base pairs are emphasized by gray-shaded clouds. The major and minor groove edges of the base pairs are shown.

of a pyr to its sugar C1′ is located in the major groove. This is clearly indicated in Fig. 4-2 by the broad shaded curved arrows. The grooves are formed when the two polynucleotide chains are coiled into a double helix, as the phosphate and sugar form ridges on helix surface and the two edges of the bases form grooves. More than one type of helix can be formed by DNA and the "major groove" and "minor groove" vary in depth and width, depending on the type of helix formed. In the B-DNA helix correctly predicted by Watson and Crick in 1953, the major groove is indeed the bigger groove as illustrated in Fig. 4-2 (cf. Fig. 1-8). Significantly, the arrangement of hydrogen bond donors and acceptors in the major groove and minor groove provides a means for the recognition of DNA sequences by other molecules, especially DNA-binding proteins.

DNA Double Helices. The polymorphism of DNA and the different helical forms that it adopts under the influence of different environmental conditions are described in Chapter 1. At this point it would be useful to summarize the general features and emphasize some of the structural details that are used in the next part of this chapter to analyze the actual atomic structures of oligonucleotides as solved by single crystal X-ray diffraction.

Table 1-1 lists some of the structural parameters of ideal helices of A, B, and Z types of DNA helices, while Fig. 1-9 shows a ball-and-stick representation of these types of helices. Figure 4-3 shows the skeletal representation of B, A, and Z-type of DNA helices in a view down the helix axis. The disposition of the bases and nucleic acid backbone with respect to the helix axis is obvious, and because the double helices are drawn to scale it is easy to recognize the differences in diameter among different helices.

As shown in the figures, A-DNA helices are short and fat with the two DNA strands wrapped around the helix axis, and with the base pairs and backbone far away from the helix axis (Figs. 4-3 and 1-9). The bases are tilted with respect to the helix axis, and the helix formed by the two strands has a right-hand twist. Double-stranded RNA forms A-type helices invariably. In the B-DNA helix, the base pairs are perpendicular to the helix axis which passes through the center of the base pairs. The B-DNA helix is also right-handed and is the predominant type of DNA structure under physiologi-

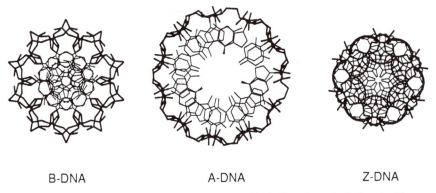

B-DNA   A-DNA   Z-DNA

Figure 4-3. End view of the B-DNA, A-DNA, and Z-DNA helices down the helix axis. The sugar-phosphate backbone is shown as darker lines. Note the change in diameter as well as the relative positions of the base pairs and backbone with respect to the helix axis.

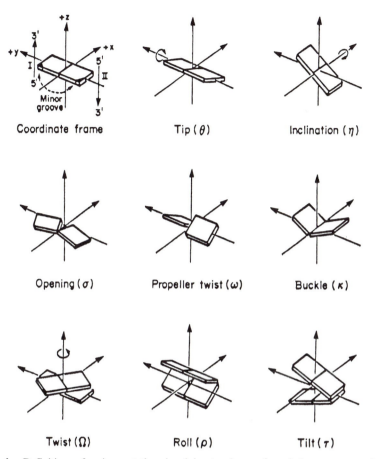

Figure 4-4. Definitions of various rotations involving two bases of a pair (upper two rows) or two successive base pairs (bottom row). In the top row the motions of the bases are coordinated, and in the middle row their motions are opposed. The left, center, and right columns describe rotations about the z, y, and x axes, respectively. The standard coordinate frame is defined at the upper left.

cal conditions. The Z-DNA helix is left-handed, and is stretched tall and thin for a given number of base pairs, as shown. Z-DNA helices have been demonstrated to exist in polytene chromosomes of drosophila by fluorescent antibodies (Nordheim et al., 1981); structurally, Z-DNA would serve to relieve the strain on DNA when it becomes negatively supercoiled and packed into chromosomes (Klysik et al., 1981; Singleton et al., 1982). More recently, Z-DNA-specific binding proteins have been discovered in chromatin (Zhang et al., 1992), and it has been established that Z-DNA occurs in metabolically active nuclei and the level is dependent on DNA torsional strain and increases with transcription (Wittig et al., 1992).

In Fig. 4-4 a series of schematic examples define the parameters used to quantify the distortions in the base pairs of the DNA double helix. For a more complete definition, the reader is referred to the Cambridge Convention in Dickerson (1989). The standard coordinate frame is defined in the top left illustration, and the other figures define the terms and the symbols commonly used to describe the structural parameters. Figure 4-5 shows the two preferred orientations (*anti* or *syn*) of the base with respect to the sugar as determined by steric restrictions during the torsion (χ) about the C1′—N glycosyl bond. The definitions of torsion angles of the nucleoside backbone are shown in Fig. 4-6. The sugar is nonplanar, and two of its common "puckering" modes, C2′-endo and C3′-endo, are shown in Fig. 4-7 along with the atom numbering scheme. Two parameters, the sugar pseudorotation phase angle and its maximum torsion angle, as shown in Fig. 4-8, are used to describe quantitatively all of the accessible puckering modes available in terms of the sugar torsion angles.

Analogous to coins stacked on top of each other to form a cylinder, the bases in a DNA double helix can be visualized as stacking on top of each other. This interaction of bases is stabilized by the electronic π–π interaction of the aromatic rings; it also minimizes the exposure of the bases to the solvent. The base stacking force plays a key role in stabilizing the helix due to the heteroatomic nature of the bases, because they possess characteristic dipole distributions which determine the preferred patterns of vertical base–base interactions. The stacking interactions vary with the helix type and whether it is a pu–pu step, a pu–pyr step or a pyr–pu step. Generally, the greater the degree of overlap of the heteroatom aromatic surfaces, the greater the stability introduced by base-pair stacking. Figures 4-9, 4-10, and 4-11 show the stacking interaction of the bases observed in ideal A-, B-, and Z-DNA helices, respectively. The location of the helix axis is shown by a dark circle to give an indication about the position of a base pair with respect to the helix axis. This view perpendicular to the base pairs for A-, B-, and Z-DNA shows that the helix axis shifts from *above* the base pairs in A-DNA (i.e., in the major groove), to being *central* to the base pairs in B-DNA, to *below* the base pairs for Z-DNA. It can be seen in Fig. 4-9 that the pu–pu step has the best stacking interaction for A-DNA; it is not surprising that nucleotide sequences with successive guanines have a greater tendency to form A-DNA-type helices. The stacking interaction seen in B-DNA helices (Fig. 4-10) is reasonable for all three possible base pair steps, but is least in the pyr–pu base-pair step. Would it be a fair guess to

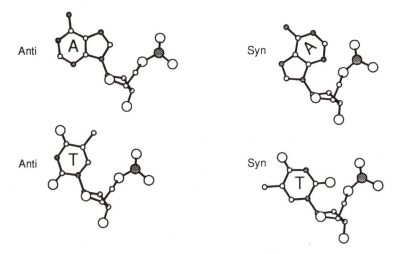

Figure 4-5. The orientation of the base with respect to the sugar is determined by the torsion angle about the C1′—N glycosyl link. The *syn* and *anti* orientations for a purine and a pyrimidine base are shown.

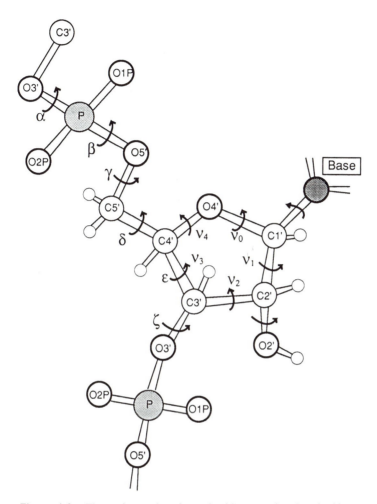

Figure 4-6. The torsion angles of a nucleotide sugar-phosphate backbone.

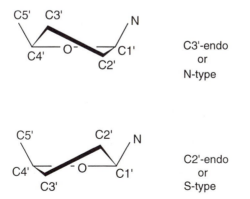

Figure 4-7. The sugar is nonplanar or "puckered." The puckering of the sugar is described as the relative displacement of C2′ and C3′ atoms with respect to the C1′—O4′—C4′ plane. The two common forms of puckering are C2′-endo and C3′-endo.

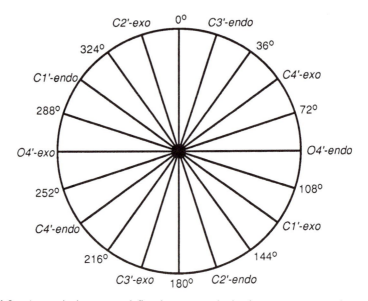

Figure 4-8. A quantitative way to define the sugar puckering is to use sugar pseudorotation phase angle P and maximum torsion angle ν_{max}, which are defined below:

$$\nu_{max} = \frac{\nu_2}{\cos P}$$

$$\tan P = \frac{(\nu_4 + \nu_1) - (\nu_3 + \nu_0)}{2\nu_2(\sin 36° + \sin 72°)}$$

The entire sugar pseudorotation cycle and the names for sugar puckers are shown. This nomenclature was used for cyclopentane, where all conformations are equally populated. In a sugar, however, there are preferred sugar puckering modes.

say that the oligonucleotide sequences with successive pyr–pu steps would be deformed more readily when the sequence adopts an B-DNA conformation? The stacking interaction seen in Z-DNA is unique for the pyr–pu step, where the sugar O4' atom of cytosine (Fig. 4-11) stacks onto the following guanine. It should be noted that DNA sequences with successive repeats of CG steps have been observed to have a greater tendency to form Z-DNA. Therefore, differences in stacking interactions play a major role in sequence-dependent conformational preference.

The A-type helix (Fig. 4-3) is the broadest of the three, with a helical diameter of ~28 Å; it is also the shortest, with a helical pitch of 25 Å. A-form helix is right-handed with ~11 base pairs per helical turn. The predominant sugar pucker is C3'-endo and the glycosyl conformation is *anti*. The base pairs have a 19° tilt from the normal to the helix axis, which lies displaced ~4.5 Å towards the major groove. The rise per base pair is 2.3 Å and the helical twist between successive base pairs is 32°. The major groove has a width of ~3 Å and a depth of ~13 Å, while the minor groove has a width of 11 Å and a depth of 3 Å. The base pairs also display a minor propeller twist.

The B-type helix (Fig. 4-3) is a right-handed helix with a diameter of ~20 Å and a helical pitch of 33–34 Å. There are 10 base pairs per helical turn, giving a 3.3- to 3.4-Å rise per base pair and a helical twist of 36° between successive base pairs. The sugar puckers range from O4′-endo to C2′-endo (Fig. 4-8) and the glycosyl conformation is *anti*. The base pairs are normal to the helix axis which is minimally displaced toward the minor groove (0.2–1 Å). The major groove and minor groove are equivalent in depth, 8.5 Å and 7.5 Å, respectively; the width of the major groove is ~12 Å, while that of the minor groove is ~6 Å.

The Z-type helix (Fig. 4-3) is a left-handed helix with a diameter of ~18 Å and a helical pitch of ~45 Å, making it a tall and thin helix. This helix type is preferred by alternating pyr–pu sequences, especially $(CG)_n$. The glycosyl conformation of the cytosine is *anti* while that of guanine is *syn*. The sugar pucker for cytosine is C2′-endo, while that for guanine is predominantly C3′-endo but varies in range from C2′-exo to C1′-exo. A dinucleotide repeat forms the unit for this helix type. The base pairs are inclined by −9° from normal to the helix axis and there are 12 base pairs per helical turn. The rise per dinucleotide repeat is ~7.4 Å and the rotation per dinucleotide repeat is −60°.

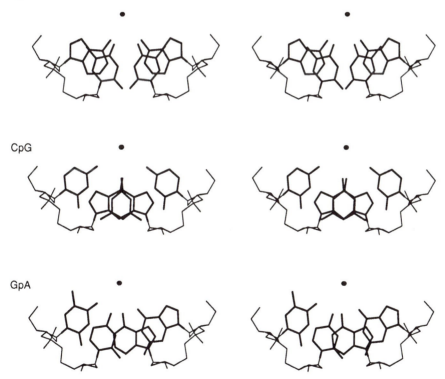

Figure 4-9. Stacking interactions in A-DNA showing (**top**) purine–pyrimidine step-stacking, (**middle**) purine–purine step-stacking, and (**bottom**) pyrimidine–purine step-stacking. The darkened dot (·) shows the position of the helix axis.

ApT

CpG

GpA

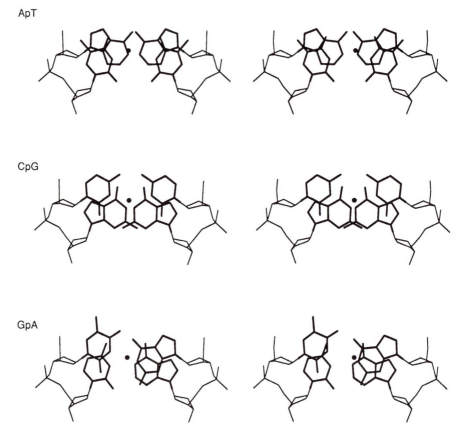

Figure 4-10. Stacking interactions in B-DNA showing (**top**) purine–pyrimidine step-stacking, (**middle**) purine–purine step-stacking, and (**bottom**) pyrimidine–purine step-stacking. The darkened dot shows the position of the helix axis.

Hydration Around DNA. The hydration environment around a DNA double helix plays an important role in determining the type of conformation adopted and in determining other properties. Recent structural work has suggested that some proteins (e.g., Trp repressor) recognize DNA sequences through direct hydrogen bonds, nonpolar contacts, indirect structural effects and, surprisingly, water-mediated interactions (Schevitz et al., 1985). Interestingly, many of the water molecules involved in the water-mediated interactions are preserved in the crystal structure of the free DNA molecule (Shakked et al., 1994). It is now generally recognized that the hydration shell of the DNA molecule is likely to play an important role in its recognition by proteins and other ligands.

Structure Determination of Oligonucleotides by X-Ray Crystallography

Single-crystal X-ray diffraction is a very powerful technique for determining the three-dimensional structure of a molecule. DNA and RNA structures determined by this method are discussed in this chapter, and it seems appropriate to give a short overview of the technique. A brief flow diagram indicating the steps involved in structure determination is shown in Fig. 4-12.

The electrons of atoms in a molecule scatter X-rays. In a crystal, the molecules are arranged in a highly regular manner forming a three-dimensional lattice. This highly periodic and oriented arrangement of molecules in space enables them to scatter X-rays collectively. The constructive interference of scattered X-rays results in a set of diffracted spots (*reflections*) of X-ray intensities. The diffraction of X-rays from a crystal can be thought of as reflection of incident X-ray beams at an angle θ from a set of Miller planes. When the interplanar spacing of the Miller planes, d_{hkl}, is such that the path difference for the X-rays from each plane is an integral multiple of the incident wavelength λ, constructive interference (a reflection) results. This is described more precisely by the Bragg equation

$$n\lambda = 2d_{hkl} \sin \theta \tag{4-1}$$

where n is an integer, d_{hkl} is interplanar distance, hkl are the Miller indices, and θ is the incident and reflected angle. The smallest d_{hkl} limit to which data can be obtained from each crystal is referred to as the *resolution*. The aim is to obtain the correct electron density distribution, from which a model of a molecule can be traced, refined, and interpreted. The electron density distribution $\rho(x, y, z)$ in a crystal lattice at a position (x, y, z) is expressed by the equation

$$\rho(x, y, z) = \frac{1}{V}\sum_h\sum_k\sum_l |F(hkl)|\, e^{-i[2\pi(hx+ky+lz)-\Phi(hkl)]} \tag{4-2}$$

where $|F(hkl)|$ and $\Phi(hkl)$ are the magnitude and phase angle, respectively, of the structure factor for each reflection and V is the volume of the unit cell. Note that both $|F(hkl)|$

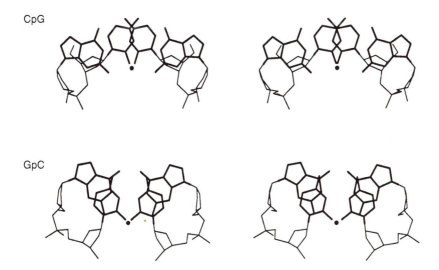

Figure 4-11. Stacking interactions in Z-DNA showing (**top**) pyrimidine–purine step-stacking, (**bottom**) purine–pyrimidine step-stacking. The darkened dot shows the position of the helix axis.

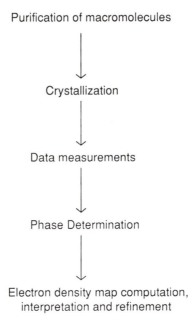

Purification of macromolecules

↓

Crystallization

↓

Data measurements

↓

Phase Determination

↓

Electron density map computation,
interpretation and refinement

Figure 4-12. A flow chart of the steps involved in structure determination by X-ray crystallography.

and $\Phi(hkl)$ need to be determined in order to calculate the electron density, but that only $|F(hkl)|$ can be directly obtained from the diffraction data. This is termed the "phase problem." The steps involved in the process of crystal structure determination are (1) purification of macromolecules, (2) obtaining suitable crystals, (3) X-ray diffraction data measurement, (4) phase determination and calculation of the electron density, and (5) refinement of the molecular model built to fit the electron density. Purification of macromolecules will not be discussed here, but a brief description of the last four steps is presented below. The aim is to give a noncrystallographer sufficient background to be better equipped for interpreting crystallographic results. Comprehensive treatises (Blundell & Johnson, 1976; Drenth, 1994; Ladd & Palmer, 1993; Stout & Jensen, 1989; Wycoff et al., 1985) and recent reviews (Eisenberg & Hill, 1989; Wang & Robinson, 1992) provide more detailed information.

Crystallization. The basic principle of crystallization is to precipitate the molecule(s) of interest from the solution ("mother liquor") slowly, thereby maximizing the chance for the molecules to pack in a regular three-dimensional array (McPherson, 1982). A large single crystal with dimensions approaching ~0.2–1.0 mm in all dimensions is the aim. The procedure for crystallization of oligonucleotides and their complexes with drugs is quite empirical and involves considerable trial and error. The pH, cations (both metal ion and organic) and precipitants are the major factors screened by trial and error, to obtain useful crystals. Though dust and temperature play an important role, the amount of dust is only minimized and the temperature is kept constant. The crystallization experiments are usually done in duplicate, both at room temperature and at 4°C, to see if lower temperature can provide better crystals. Occasionally, higher temperature is tried when prompted by the stability and native condition of the sample. A practical overview of the procedures used has been reviewed (Wang & Gao, 1990).

The number of parameters influencing the nucleation and growth of biological macromolecules is large. It is important to try to identify the potentially important factors by setting up efficient screening trials. Overall, the best experiments for this purpose are the factorial designs used by generations of industrial chemists (Carter, 1992). Factorial experiments have been applied very successfully to the problem of protein crystallization (Carter, 1992) and to date with less success for the crystallization of nucleic acids. Using the crystallization conditions for 200 proteins, peptides and oligonucleotides, a factorial crystal screen kit incorporating 50 conditions is now commercially available from Hampton Research, Inc.; it has been used successfully for several new macromolecules.

Collection of X-Ray Diffraction Data. When suitable crystals have been obtained, a single crystal is selected and mounted in a glass or quartz capillary along with a separate drop of mother liquor. The ends of the capillary are sealed with dental wax, followed by a coat of fingernail polish. In the sealed capillary environment the macromolecule crystals are generally relatively stable for the duration of data collection. The oligonucleotide crystals contain large solvent channels in the crystal lattice and would disintegrate upon exposure to the atmosphere. The capillary with the precious crystal is glued onto a brass pin and placed on a goniometer, which has translation and angular (arc) adjustment components. The crystal is visually centered on an instrument designed to measure X-ray diffraction intensities, such that it can be irradiated with a uniform and parallel beam of X-rays. Measurement of X-ray diffraction intensities can be done by photographic techniques, by diffractometry, or by using an electronic or image plate area-detector instrument. The preliminary step is the characterization of the space group of the crystal in question and the determination of the resolution limit [d_{hkl} of Eq. (4-1)] to which it diffracts. Then the scattered intensity data are collected.

The collected intensity data I_{hkl} is related to the structure factor F_{hkl} in the following way:

$$I_{hkl} = k \cdot (\text{Lp})^{-1} \cdot |F_{hkl}|^2 \qquad (4\text{-}3)$$

where k is the factor used to make corrections for absorption, crystal decay, and so on, and $(\text{Lp})^{-1}$ is a geometric/physical correction factor called the Lorentz polarization factor. It can be seen from Eq. (4-3) that after data processing and reduction, we obtain the amplitude of the structure factor $|F_{hkl}|$ is obtained. The phase information $\Phi(hkl)$ associated with each reflection is lost during the data collection. The phase information is recovered via indirect means.

Solving the Phase Problem. There are four major ways to solve the phase problem. For molecules with less than 100 nonhydrogen atoms, the direct method (Hauptman, 1986; Karle, 1986) is used. For larger molecules, the classical technique is the multiple isomorphous replacement (MIR) method where several heavy atom derivative crystals isomorphous to the native crystal are grown and used to extract the phase information (Blundell & Johnson, 1976; Drenth, 1994; Ladd & Palmer, 1993; Stout & Jensen, 1989; Wang & Robinson, 1992; Wyckoff et al., 1985). A recent technique for large molecules is the multiple anomalous dispersion (MAD) method, which extracts the phase information using the anomalous scattering of certain heavy metals near their absorption edge (Hendrickson et al., 1985a; Smith, 1991). A significant achievement of MAD method has been obtained on a study of the regulatory enzyme of de novo synthesis of purine nucleotides, glutamine 5-phospho-ribosyl-1-pyrophosphate (PRPP) amidotransferase from *Bacillus subtilis*, having a molecular weight of ~200,000 (Smith et al., 1994). It should be noted that this method requires several wavelengths of X-rays which can be obtained effectively only on a synchrotron.

The molecular replacement (MR) method is another popular technique used for solving the phase problem of large molecules and is a particularly useful technique for solving structures of oligonucleotides. The MR method is useful when an appreciable portion of the structure of the molecule is known in advance. The first step is to construct a model of a molecule which is assumed to be close to the real structure. Then the model molecule is rotated in the unit cell to determine the three correct rotation parameters. The properly oriented molecule is then systematically moved along the three unit cell directions to determine the correct translation parameters. Hence, this method is often called the *rotation–translation search method*. Several programs packages like ULTIMA (Rabinovich & Shakked, 1984), MERLOT (Fitzgerald, 1988), and XPLOR (Brunger, 1992a) can be used for this method. In practice, one must consider the top 50 or more rotation solutions, using each as the object of a translation search, and then test the several possible translation solutions by intermolecular contact criteria and subsequent refinement. It is a laborious process but is particularly suitable for nucleic acids alone, because they often adopt a structure in one of the three families, A-DNA, B-DNA, or Z-DNA. For drug–DNA complexes, this method becomes more complicated, because it is more difficult to build an initial model that is close to the real structure.

A powerful application of the molecular replacement method is in the investigation of DNA lesions or drug–DNA interaction. When a good crystal lattice is available, it is possible to generate isomorphous crystals of a derived but similar molecule. The onus is now on growing isomorphous crystals. Here, care is taken to design new sequences that would retain the packing property of the original molecule in its crystal lattice. This type of sequence engineering has been frequently applied to the dodecamer d(CGC-GAATTCGCG) in the B-DNA conformation (Drew & Dickerson, 1981a). Several related sequences and their complexes with minor groove binding drugs have been crystallized in the isomorphous $P2_12_12_1$ lattice and their structure solved by molecular replacement (Sriram et al., 1992a,b). In these studies, once an isomorphous crystal can be grown, the initial phases for the new crystal are derived from the original crystal.

Refinement. The initial model structure can be used to calculate the theoretical structure factor $|F_c|$. Because the initial model is a crude representation of the actual model, the $|F_c|$ and the experimentally observed structure factor $|F_o|$ are not in good agreement. The initial model is adjusted and a new set of $|F_c|$ is calculated so that the discrepancies between $|F_o|$ and $|F_c|$ are reduced. This step is repeated until the values of $|F_o|$ and $|F_c|$ agree as closely as possible. This highly iterative process is called *refinement*. The treatment of this type of problem was discussed by Legendre in 1806, in a paper describing the calculation of the orbits of comets from a number of observations. Legendre's principle states that the best and most plausible parameters are those for which the sums of the squares of differences between the observed and calculated values is a minimum. From this arises the name *least-squares refinement procedure*.

The overdeterminacy—that is, $n < m$ (n unknown parameters, m observations)—is what enables X-ray diffraction to estimate the position of atoms to an accuracy well beyond that of the resolution [d_{hkl}-spacing limit, see Eq. (4-1)] of the structure. For a small molecule (with less than 100 nonhydrogen atoms), the crystals usually diffract well and the ratio of observations per variable is high enough to allow direct refinement of the three positional parameters (x, y, z) and the six thermal parameters (temperature factors) describing the anisotropic motion of individual atoms in a molecule. Crystals of biological macromolecules usually diffract X-rays to lower Bragg angle θ [Eq. (4-1)], and this results in substantially fewer observed reflections. Consequently, the observations/parameter ratio is too low to allow refinement of individual atom parameters. Because biological macromolecules are made up of small building units, it is assumed that the geometric parameters (bond lengths and bond angles) are very similar to those of the individual building units. The geometric parameters of the individual building units are used as stereochemical constraints, and the torsion angles are varied to minimize the difference between $|F_o|$ and $|F_c|$. This ingenious method was developed to overcome the problem of underdeterminancy ($n > m$). The program PROLSQ developed originally for proteins by Hendrickson et al. (1985b), and later modified for nucleic acids (NUCLSQ) by Gary Quigley and Eric Westhof (Westhof et al., 1985), incorporates the following function Ω_{total} to be minimized by a least-squares procedure:

$$\Omega_{total} = \Omega_F + \Omega_{distance} + \Omega_{vdW} + \Omega_{planar} + \Omega_{chiral} + \Omega_B \qquad (4\text{-}4)$$

The structure factor term (Ω_F) is minimized together with constraint terms, distance constraints ($\Omega_{distance}$), van der Waals contact (Ω_{vdW}), planarity (Ω_{planar}), stereoconfiguration of chiral centers (Ω_{chiral}), and temperature factor (Ω_B). Additional optional terms for constraint of occupancy (Ω_Q), torsion angle (Ω_{tor}), and sugar pseudorotation (Ω_{pseudo}) can also used when needed. This kind of stereochemically constrained refinement reduces the number of parameters to be refined by a factor of about 5. It becomes possible to refine a biological macromolecule structure even at a relatively low resolution (~ 3 Å).

The principal shortcoming of the least-squares refinement process is that it is easily trapped in local minima and cannot correct residues that are misplaced by more than 1 Å, making human intervention necessary. Progress has been reported in the nonlinear optimization of a function of many parameters by the introduction of simulated annealing (Kirkpatrick et al., 1983). The method involves simulating the many-parameter system by a Monte Carlo algorithm (Metropolis et al., 1953) with the temperature being kept high; then the system is "annealed" by reducing the temperature slowly. The process simulates the transition from a glass (high-energy state) to a crystal (low-energy state) state of the same substance. Brunger et al. (1987) introduced simulated annealing into crystallographic refinement, where molecular dynamics (Karplus and McCammon, 1983) is used to search the conformational space (MD-refinement) of a molecule in regions allowed by the diffraction data through the introduction of an effective potential energy which includes the crystallographic residuals. It was shown that MD-refinement has a much larger radius of convergence than does conventional least-squares restrained refinement, and it reduces the need for manual corrections. In general, MD-refinement yields an improved R-factor, a small decrease in the average difference between MIR phases and calculated phases, and a slight improvement in the stereochemistry of the model. A powerful package called X-PLOR for such "simulated annealing" procedure has been developed and is widely used (Brunger, 1992a).

The initial model used for the refinement of macromolecules does not account for everything in the unit cell. The residual error (R-factor, see below for definition) at this stage is about 25% to 30%. The additional material in the unit cell (e.g., the drug molecule, chemical adducts on DNA) is located using a difference Fourier electron density map ($|F_o| - |F_c|$ map). Such a map is calculated based on the equation

$$\Delta \rho(x, y, z) = \frac{1}{V} \sum_h \sum_k \sum_l (\left||F_o(hkl)| - |F_c(hkl)|\right|) e^{-i[2\pi(hx + ky + lz) - \Phi_c(hkl)]} \qquad (4\text{-}5)$$

where $|F_o(hkl)|$ and $|F_c(hkl)|$ are the observed and calculated (based on the current model) structure factors, respectively, and Φ_c is the calculated phase angle. The ($|F_o| - |F_c|$) map is particularly useful in locating missing portions of the model. Another closely related type of Fourier map is a $2|F_o| - |F_c|$ map, called a *sum map* because it contains the sum of a regular $|F_o|$ map and the $|F_o| - |F_c|$ difference map. The solvent molecules and ions present in the unit cell are often located from a sum map.

Assessing the Refinement and the Structure. The refinement of a macromolecular structure along with the location of all missing portions of the model is continued iteratively until it converges. The progress of a refinement is monitored most commonly by the crystallographic residual factor (R-factor). The R-factor is defined as

$$R = \frac{\sum \left||F_o| - |F_c|\right|}{\sum |F_o|} \qquad (4\text{-}6)$$

The R-factor for small molecules can often be as low as 5% due to the very reliable measurements of intensity data. Macromolecular crystals contain large solvent channels and the disordered solvent molecules cannot be accounted for. In addition, the flexible regions of the molecule add to the uncertainty of the model. A macromolecule which has been refined to an R-factor between 15% and 20% with good constraints is considered satisfactory. The constraints are usually presented as root mean square deviation (RMSD) of bond lengths from ideal values in angstroms, and sometimes together with the RMSD of bond angles from ideal values in degrees. Because of the weighting scheme usually used to impose constraints during refinement, the RMSD of bond lengths from ideal values is a critical indicator of the overall constraint. When the RMSD of bond lengths is less than 0.025 Å, the constraints are considered satisfactory.

The accuracy of a structure cannot be judged by the R-factor and constraints alone. The visual inspection of an electron density map must be done to see that there is no obvious discrepancy between the model and the electron density in the map. The electron density map provides a good way to judge both the quality of the data as well as the degree to which the model fits the electron density. A poorly refined structure produces a noisy electron density map. The accuracy of a structure is directly dependent upon the resolution to which the crystal diffracts. Figure 4-13 shows the electron density maps of four structures drawn at different resolutions. It is obvious by comparison that the uncertainty associated with a atom location increases as the resolution of the structure decreases.

The free R-factor provides a more quantitative and objective way of assessing the quality of a structure by using the statistical concept of cross validation in reciprocal

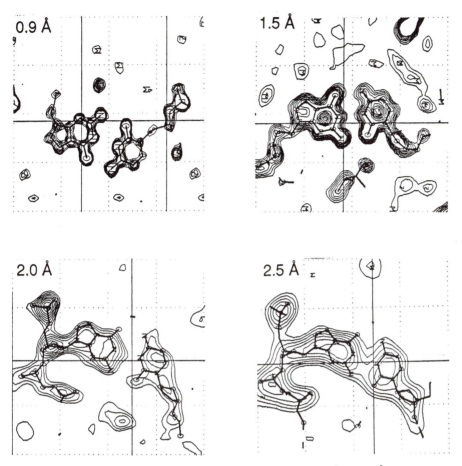

Figure 4-13. The electron density maps of (**top left**) d(CGCGCG)·Mg²⁺ at 0.9-Å resolution, (**top right**) d(CGT(2-NH₂A)CG at 1.5-Å resolution, (**bottom left**) d(CGCGAATTCGCG)·netropsin at 2.0-Å resolution, and (**bottom right**) d(CGCGAATTCGCG)·Hoechst 33342 at 2.5-Å resolution.

space (Brunger, 1992b, 1993). The crystallographic residual factor (R-factor) can be made arbitrarily small by increasing the number of model parameters; that is, the diffraction data can be overfitted without changing the information content of the atomic model. Fortunately, crystallographic data are redundant and a small portion of the data can be omitted without seriously affecting the result. The observed reflections are partitioned into two disjoint sets, a test set T and a working set A (Brunger, 1992a,b) such that their conjunction is the full set of observed reflections. The *free R value* (R_T^{free}) corresponds to the crystallographic residual factor computed for the T-set reflections which is omitted from the model refinement process. R_T^{free} would be less prone to overfitting. Both R_T^{free} and the room mean square difference between the model refined against the complete data set and the model refined against A increase more or less monotonically as a function of the percentage of omitted data (Brunger, 1992a,b). In practice, as a compromise, the T set is obtained from a random selection of 10% of the observed reflections. R_T^{free} provides a way of assessing when the parameters be-

Table 4-1 A Representative List of A-DNA Structures

Oligonucleotide	Space Group	Unit Cell (Å)	Resolution (Å)	R-Factor (%)	Number of Waters	Number of Reflections	Reference
ACCGGCCGGT	$P6_122$	$a = b = 39.23$ $c = 78.0$	2.0	18	36	1434	Frederick et al., 1990
CCCCGGGG	$P4_32_12$	$a = b = 43.36$ $c = 24.83$	2.25	15	82	1018	Haran et al., 1987
CCCGGCCGGG 1993	$P2_12_12_1$	$a = 24.88$ $b = 44.60$ $c = 46.97$	1.65	18.5	103	6033	Ramakrishnan & Sundaralingam,
CCGTACGTACGG	$P6_122$	$a = b = 46.2$ $c = 71.5$	2.5	15	40	1664	Bingman et al., 1992
CpsCGTACGTACGG	$P6_122$	$a = b = 46.2$ $c = 71.5$	2.5	13.6	71	2415	Bingman et al., 1989
CTCTAGAG	$P4_32_12$	$a = b = 42.53$ $c = 24.33$	2.15	14.7	35	931	Hunter et al., 1989
GCCCGGGC	$P4_32_12$	$a = b = 43.25$ $c = 24.61$	1.8	17.1	34	1359	Heineman et al., 1987
GGATGGGAG	$P4_3$	$a = b45.29$ $c = 24.73$	3.0	33	0	1032	McCall et al., 1986
GGbr5UAbr5UACC	$P6_1$	$a = 45.05$ $c = 41.72$	1.7	14	84	4669	Kennard et al., 1986
GGCCGGCC	$P4_32_12$	$a = b = 40.51$ $c = 24.67$	2.25	15.8	84	874	Fujii et al., 1982
GGCCGGCC	$P4_32_12$	$a = b = 42.06$ $c = 25.17$	2.25	17.4	75	920	Wang et al., 1982a
GGGATCCC	$P6_1$	$a = b = 46.83$ $c = 44.49$	2.5	16.6	9	1428	Lauble et al., 1988

Sequence	Space group	Cell dimensions (Å)					Reference
GGGCGCCC	$P4_32_12$	$a = b = 42.74$, $c = 24.57$	1.7	21	62	1694	Eisenstein et al., 1988
GGGCGCCC	$P4_32_12$	$a = b = 43.28$, $c = 24.66$	1.8	15.9	88	2082	Rabinovich et al., 1988
GGGCGCCC	$P6_1$	$a = b = 43.62$, $c = 41.49$	1.9	16	100	2827	Shakked et al., 1989
GGGCGCCC	$P6_1$	$a = b = 46.96$, $c = 44.31$	2.9	12	45	1068	Shakked et al., 1989
GGGGCCCC	$P6_1$	$a = b = 45.32$, $c = 42.25$	2.5	14	106	1283	McCall et al., 1985
GGGGCTCC	$P6_1$	$a = b = 45.20$, $c = 42.97$	2.25	13.6	52	1924	Hunter et al., 1986c
GGGGTCCC	$P6_1$	$a = b = 44.71$, $c = 42.40$	2.1	14.5	104	2486	Kneale et al., 1985
GGGTGCCC	$P6_1$	$a = b = 45.62$, $c = 40.99$	2.5	15.2	63	1362	Rabinovich et al., 1988
GGTATACC	$P6_1$	$a = b = 45.01$, $c = 41.55$	1.8	19.8	66	3251	Shakked et al., 1983
GTGTACAC	$P4_32_12$	$a = b = 42.43$, $c = 24.75$	2.0	11.5	43	1214	Jain et al., 1989
GTGTACAC	$P6_122$	$a = b = 32.4$, $c = 79.5$	2.0	12.5	50	1561	Sundaralingam & Jain, 1989
iCCGG	$P4_32_12$	$a = b = 41.10$, $c = 26.70$	2.0	16.5	86	1486	Conner et al., 1984
r(GCG)d(TATACGC)	$P2_12_12_1$	$a = 24.20$, $b = 43.46$, $c = 49.40$	2.0	16	179	2521	Wang et al., 1982b
UUAUAUAUAUAUAA	$P2_12_12_1$	$a = 34.11$, $b = 44.61$, $c = 49.11$	2.25	13.1	91	2492	Dock-Bregeon et al., 1988

ing added to the refinement are reducing the *R*-factor without increasing the information of the atomic model.

Crystal Structures of Oligonucleotides

The crystallographic details of a representative list of oligonucleotides which form the A-type helix is shown in Table 4-1, the B-type helix in Table 4-2, and the Z-type helix in Table 4-3. The list shows several dozen structures in the A-DNA conformation, the B-DNA conformation, and the Z-DNA conformation. While the list appears long, the variety of structures is narrower because several structures are in the same space group and their structures can be solved readily by molecular replacement.

The extent to which crystal packing forces affect the outcome of these structures is an intriguing problem and there are two ways to distinguish the extent of the problem. One of the successful approaches, as is evident from the list, is to grow isomorphous crystals of different sequences. Then crystal packing becomes a common factor and all the differences seen are due to the sequence-dependent variations. The second approach relies on the ability to grow crystals of the same sequence in more than one space group. Then all conserved structural features are independent of the crystal packing forces and probably strongly influenced by the sequence, but all the differences can be due to the crystal packing or due to the hydration changes. From this list, it is possible to glean that several high-resolution structures of oligonucleotides are available. These oligonucleotide structures can be analyzed in great detail to accumulate structural details at a molecular level for different types of DNA helices.

B-DNA. The first crystal structure of B-DNA was that of a dodecamer sequence d(CGCGAATTCGCG) (Wing et al., 1980) which contains the *Eco*RI restriction site GAATTC. The crystals of this dodecamer sequence can be grown readily; this is due to the type of intermolecular packing adopted by the molecules in the crystal. The terminal two base pairs of the helix are hydrogen bonded via the minor groove edges of the base to the first two base pairs of the next helix. The minor groove of the terminal base pairs are hence occluded and the helix is also bent at the end to allow such a crystal packing arrangement. A ball-and-stick representation of this structure is shown in Fig. 4-14 along with the water molecules in the minor groove.

The dodecamer helix d(CGCGAATTCGCG) exhibits a distinct bend at the end of the helix. This is achieved by rolling of one base pair, atop the next, along their long axis toward the major groove and compressing it. The bend is localized at the junction between the CGCG end and the AATT center. The other end of the helix is not bent and though the sequence is self-complementary, the two ends of the helix occupy different environments. In other related sequences such as d(CGCAAATTTGCG), the bend in the helix is present but moves to the junction between the GC and the AT regions.

The sugar pucker of the deoxyribose ring favors the C2'-endo conformation though a range of conformations from C1'-exo to O4'-endo is present. The sugar pucker observed in B-DNA crystal structures in general tends to show a greater spread than in A-DNA crystal structures. This may reflect the higher flexibility associated with the B-DNA structures. The base pairs in the central AATT region of the helix have a pattern of distinctly larger propeller twists when compared to those of the CGCG ends.

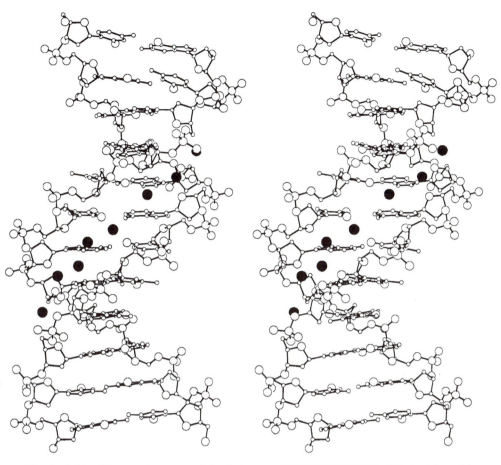

Figure 4-14. The crystal structure of the dodecamer d(CGCGAATTCGCG). This is stereoscopic ball-and-stick representation with water molecules in the minor groove shown as black circles.

The propeller twisting enhances the stacking of the bases along each strand of the double helix and the A·T base pairs are in general less restrictive to propeller twisting because of fewer hydrogen bonds than G·C base pairs. The propeller twist is high in regions of all the dodecamers where there are successive runs of two or more adenines.

The minor groove is distinctly narrower at the AATT region of the double helix than at the CGCG ends. The narrow minor groove is filled by a spine of water molecules at the AATT region which form hydrogen bonds to both the O2 of the thymines and N3 of the adenines (Fig. 4-14). This spine of water disappears in the wider CGCG region where the ends of the helix pack into each other.

The narrow minor groove region of the dodecamer structure forms an excellent binding site for several minor groove binding drugs (e.g., Hoechst 33258, Hoechst 33342, netropsin, distamycin, DAPI) and crystals of the complex of the related dodecamer sequences with minor groove binding drugs are readily obtained. The drug replaces the spine of hydration in the minor groove and the overall conformation of both the DNA helix and the drug is close to that of the native structures. The detailed skeletal view of the d(CGCAAATTTGCG)–distamycin complex is shown in Fig. 4-15. Though the overall conformation is conserved, detailed analysis reveals that both the DNA and the drug adapt their conformation to form a stable complex. The narrow minor groove associated with the central AATT base pairs is essential for the binding of the minor groove drugs. A portion of the binding energy derives from the gain in entropy associated with the displacement of the water molecules. The sequence preference for AT regions by minor groove binding drugs is due to the greater negative electrostatic potential at the bottom of the minor groove at AT regions and the absence of N2 amino groups of guanine which provide both a charge and steric hindrance to drug binding.

The crystal structures of B-DNA decamers listed in Table 4-2 in general have two advantages over that of the dodecamers. The decamer helices tend to stack end-over-end in the crystal mimicking infinite B helices. Secondly, the data from decamer crystals are of a much higher resolution, ranging from 1.8 Å to 1.3 Å and thus affording

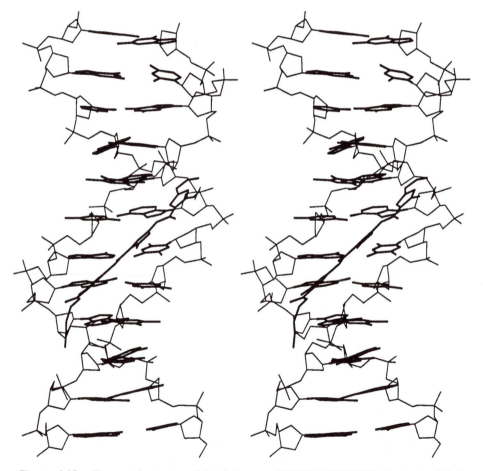

Figure 4-15. The crystal structure of the dodecamer d(CGCAAATTTGCG)-distamycin. This is stereoscopic skeletal representation with drug displacing the waters in the minor groove.

Table 4-2 A Representative List of B-DNA Structures

Oligonucleotide	Space Group	Unit Cell (Å)	Resolution (Å)	R-Factor (%)	Number of Waters	Number of Reflections	Reference
ACCGGCGCCACA	R3	$a = 65.90$ $b = 65.90$ $c = 47.10$	2.8	15.3	20	1101	Timsit et al., 1989
CATGGCCATG	P2$_1$2$_1$2$_1$	$a = 36.60$ $b = 42.29$ $c = 34.69$	2.0	19.6	49	3217	Goodsell et al., 1993
CCAACGTTGG	C2	$a = 32.25$ $b = 25.49$ $c = 34.65$ $\beta = 114.0°$	1.4	17.9	71	4707	Prive et al., 1991
CCAACGTTGG	C2	$a = 32.52$ $b = 26.17$ $c = 34.30$ $\beta = 118.9°$	1.4	17.8	52 + 3Mg	5072	Prive et al., 1991
CCAACITTGG	C2	$a = 31.87$ $b = 25.69$ $c = 34.21$ $\beta = 114.1°$	1.3	15.2	72	5026	Lipanov et al., 1993
CCAACITTGG	P3$_2$21	$a,b = 33.23$ $c = 94.77$ $\gamma = 120.1°$	2.2	16.4	36	1725	Lipanov et al., 1993
CCAAGATTGG	C2	$a = 32.25$ $b = 25.53$ $c = 34.38$ $\beta = 113.4°$	1.3	16.0	56 + 4Mg	4402	Prive et al., 1987

continued

Table 4-2 (continued)

Oligonucleotide	Space Group	Unit Cell (Å)	Resolution (Å)	R-Factor (%)	Number of Waters	Number of Reflections	Reference
CCAAGCTTGG	C2	$a = 32.52$ $b = 26.17$ $c = 34.30$ $\beta = 118.9°$	1.3	16.4	68	4016	Prive et al., 1988
CCAGGCCTGG	C2	$a = 32.15$ $b = 25.49$ $c = 34.82$ $\beta = 116.71°$	1.6	16.9	—	2420	Heineman & Alings, 1985
CGATATATCG	$P2_12_12_1$	$a = 38.76$ $b = 40.06$ $c = 33.73$	1.7	17.8	95	3683	Yuan et al., 1992
CGATCGATCG	$P2_12_12_1$	$a = 38.93$ $b = 39.63$ $c = 33.30$	1.8	16.1	142 + 2Mg	5107	Grzeskowiak et al., 1991
CGATTAATCG	$P2_12_12_1$	$a = 38.60$ $b = 39.10$ $c = 33.07$	1.5	15.7	108	3793	Quintana et al., 1992
CGCAAAAAGCG	$P2_12_12_1$	$a = 25.40$ $b = 40.70$ $c = 65.80$	2.5	20	27	2003	Nelson et al., 1987
CGCAAAAATGCG	$P2_12_12_1$	$a = 25.54$ $b = 40.32$ $c = 65.86$	2.6	20.1	23	1975	DiGabriele et al., 1989
CGCAAATTCGCG	$P2_12_12_1$	$a = 25.37$ $b = 41.44$ $c = 65.20$	2.5	19.0	82	2029	Hunter et al., 1986b

Sequence	Space group	Cell dimensions					Reference
CGCAAATTGGCG	P2₁2₁2₁	$a = 25.23$ $b = 41.16$ $c = 65.01$	2.25	16	94	2262	Brown et al., 1989
CGCAAATTTGCG + Distamycin	P2₁2₁2₁	$a = 25.20$ $b = 41.07$ $c = 64.65$	2.2	19.6	70	2369	Coll et al., 1987
CGCAGAATTCGCG	C2	$a = 78.48$ $b = 42.85$ $c = 25.16$ $\beta = 99.36°$	2.8	15	50	—	Joshua-Tor et al., 1988
CGCATATATGCG	P2₁2₁2₁	$a = 23.54$ $b = 38.86$ $c = 66.57$	2.2	18.7	43	1915	Yoon et al., 1988
CGCGAAATTTACGCG	I222	$a = 37.00$ $b = 53.70$ $c = 101.6$	3.0	26.3	—	1943	Miller et al., 1988
CGCGAATTAGCG	P2₁2₁2₁	$a = 25.69$ $b = 41.96$ $c = 65.19$	2.5	17.0	83	2028	Hunter et al., 1986b
CGCGAATT^{br5}CGCG	P2₁2₁2₁	$a = 24.20$ $b = 40.09$ $c = 63.95$	2.3	17.3	115	2919	Kopka et al., 1983
CGCGAATT^{br5}CGCG	P2₁2₁2₁	$a = 24.71$ $b = 40.56$ $c = 65.62$	3.0	17.3	44	1515	Kopka et al., 1983
CGCGAATT^{br5}CGCG + Netropsin	P2₁2₁2₁	$a = 24.27$ $b = 39.62$ $c = 63.57$	2.2	21.1	75	2528	Kopka et al., 1983

continued

Table 4-2 (continued)

Oligonucleotide	Space Group	Unit Cell (Å)	Resolution (Å)	R-Factor (%)	Number of Waters	Number of Reflections	Reference
CGCGAATTCGCG	P2₁2₁2₁	a = 23.44 b = 39.31 c = 65.26	2.7	15.1	83	1051	Drew et al., 1982
CGCGAATTCGCG	P2₁2₁2₁	a = 24.87 b = 40.39 c = 66.20	1.9	17.8	80	2725	Drew and Dickerson, 1981a
CGCGAATTCGCG	P2₁2₁2₁	a = 24.87 b = 40.39 c = 66.20	1.9	18.8	64	3979	Westhof, 1987
CGCGAATTCGCG	P2₁2₁2₁	a = 24.87 b = 40.39 c = 66.20	1.9	14.9	80	2728	Holbrook et al., 1985
CGCGAATTCGCG + Berenil	P2₁2₁2₁	a = 25.29 b = 41.78 c = 64.76	2.5	—	—	—	Brown et al., 1992
CGCGAATTCGCG + Cisplatin	P2₁2₁2₁	a = 24.16 b = 39.93 c = 66.12	2.6	11.2	128	2088	Wing et al., 1984
CGCGAATTCGCG + DAPI	P2₁2₁2₁	a = 25.35 b = 40.71 c = 66.53	2.4	21.5	24	2428	Larsen et al., 1989
CGCGAATTCGCG + Hoechst 33258	P2₁2₁2₁	a = 25.04 b = 40.33 c = 65.85	2.2	14.0	175	1569	Pjura et al., 1987

Sequence	Space group	Cell parameters					Reference
CGCGAATTCGCG + Hoechst 33258	$P2_12_12_1$	$a = 25.23$ $b = 40.58$ $c = 66.08$	2.25	15.7	126	2000	Teng et al., 1988
CGCGAATTTGCG	$P2_12_12_1$	$a = 25.53$ $b = 41.22$ $c = 65.63$	2.5	13	71	2000	Hunter et al., 1987
CGCGA^{me}ATTCGCG	$P2_12_12_1$	$a = 25.62$ $b = 40.70$ $c = 67.32$	2.0	16.9	86	2379	Frederick et al., 1988
CGCGATATCGCG + Netropsin	$P2_12_12_1$	$a = 25.48$ $b = 41.26$ $c = 66.88$	2.4	20.1	60	1848	Coll et al., 1989a
CGCTTAAGCG	C2	$a = 77.27$ $b = 25.97$ $c = 79.17$ $\beta = 104.5°$	—	—	—	—	Larsen, unpublished
GpsCGpsCGpsC	$P2_12_12_1$	$a = 34.90$ $b = 39.15$ $c = 20.64$	2.17	14.5	72	1327	Cruse et al., 1986

a greater precision of analysis. The crystal structure of the decamer d(CCAGGCCTGG) (Heinemann & Alings, 1985) is presented below.

The stereoscopic skeletal figure of d(CCAGGCCTGG) is shown in Fig. 4-16. The minor groove width of the decamer in general is wider but some variations do exist. The central GC step in the decamer shows a slight narrowing of the minor groove. Though this narrowing of the minor groove is similar to dodecamer structures, other decamers have a narrow or wide minor groove depending on the sequence. In general the AT regions have a narrower minor groove. The minor groove width is strongly dependent on the phosphate conformation. The hydration structure in the groove is dependent on the groove width. A single spine of waters along the floor of the minor groove is associated with a narrow minor groove, whereas a ribbon of *two waters* along the walls of the minor groove bridges the base edge N or O atom to the O4′ atoms of the sugar ring (Fig. 4-17). This seems to be generally true of all the B-DNA structures.

A-DNA and RNA. There is increasing evidence that the A-form of DNA may play an important role in processes such as protein recognition (Drew & Travers, 1984; Suck & Oefner, 1986) and transcription regulation (Kim et al., 1993a,b; McCall et al., 1986;

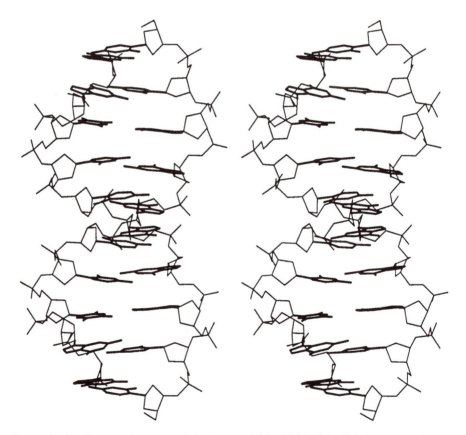

Figure 4-16. The crystal structure of the decamer d(CCAGGCCTGG). This is stereoscopic skeletal representation.

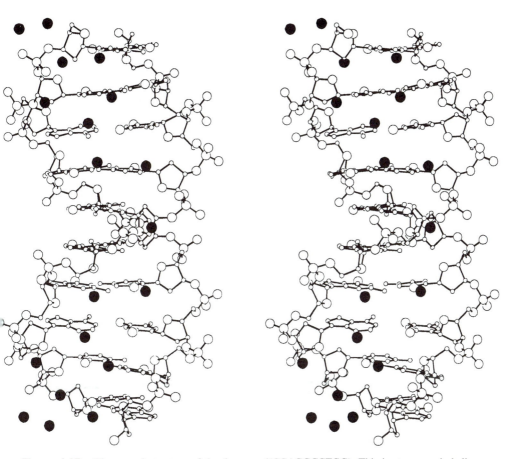

Figure 4-17. The crystal structure of the decamer d(CCAGGCCTGG). This is stereoscopic ball-and-stick representation with the waters bound to the minor groove shown as black circles.

Rhodes & Klug, 1986). Fiber X-ray diffraction analysis has shown that B-DNA converts to A-DNA under conditions of low hydration and that the process is reversible (Saenger et al., 1986). It has also been shown that there is a sequence-specific propensity to form A-DNA with guanine-rich regions readily forming A-DNA while stretches of adenine resist the formation of A-DNA (Benevides et al., 1986). The guanine-rich regions show a characteristic intrastrand guanine–guanine stacking interaction in A-DNA helix which may explain the propensity of these sequences to adopt the A-DNA conformation (McCall et al., 1985; Wang et al., 1982a).

Table 4-1 shows a list of oligonucleotide crystal structures adopting the A-DNA conformation, many of which are octamers. The study of these short DNA octanucleotide structures revealed that although the overall conformation resembles the idealized A-DNA helix derived from fiber diffraction, there are significant differences in details. The tilt of base pairs with respect to the helix axis has a significantly smaller value (8° to 10°) in the crystal structures than the 19° from fiber diffraction studies. Furthermore, the major groove is wider and the helical pitch is longer than those in the fiber studies.

The crystal structure of the decamer d(ACCGGCCGGT) (Frederick et al., 1990) was found closer to the fiber A-DNA structure. A sequence-specific stacking pattern with the very low helical twist of 24° at the CpG step is seen in the decamer structure. This may be a property of the CpG dinucleotide step in A-DNA conformation. The overall conformation of the decamer A-DNA helix is shown in Fig. 4-18, and the characteristic features such as the wide flat minor groove and reduced helical pitch is obvious. The stereoscopic skeletal views as shown in Fig. 4-18 allow a thorough examination of several structural details.

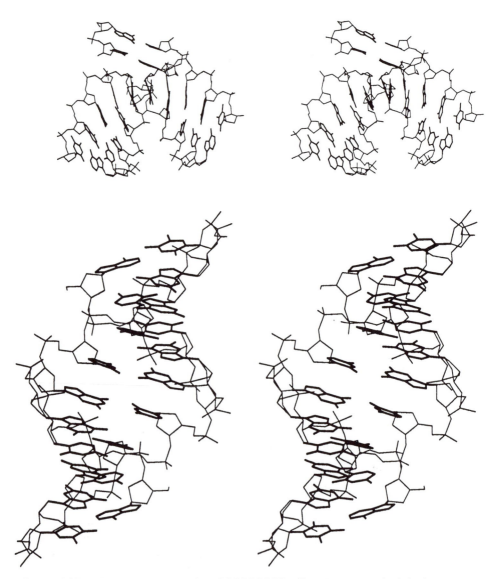

Figure 4-18. The crystal structure of d(ACCGGCCGGT). (**Top**) A stereoscopic skeletal representation of crystal packing of the terminal two base pairs of each helix into the minor groove of the neighboring helix. (**Bottom**) A stereoscopic skeletal representation of decamer DNA duplex.

The end view of the helix (not shown) is not perfectly circular due to structural irregularities and also the fact that it represents less than one helical turn. In the end view of a decamer helix a clear open cylindrical channel of approximately 10 Å extends down the center of the molecule as the base pairs are displaced out toward the perimeter of the molecule. The nucleotide-backbone torsion angles of this decamer A-DNA helix, as well as those of the other octamer helix, are in general quite similar to those of the fiber A-DNA double helix and are fairly uniform with small standard deviations when compared to those of crystal structures of B-DNA dodecamers. The glycosyl torsion angle χ is also very uniform. The α torsion angle is the most variable in the decamer A-DNA structure, ranging from $-31.5°$ to $-104.1°$ with an average of $-65.5°$ and a standard deviation of $23°$.

When the central eight base pairs of the d(ACCGGCCGGT) decamer A-DNA helix are superposed on the octamer d(GGCCGGCC) (Wang et al., 1982a), the RMSD of the sugar-phosphate backbone is 1.78 Å. The octamer has a lower tilt angle of 12° and a higher rise per residue of 3 Å, while the decamer has an average tilt angle of 18° and a correspondingly smaller rise per residue of 2.64 Å. The eight central base pairs of the decamer represent a shorter helical segment than the octamer. The major groove width in the octamer structure is more like a B-DNA helix and is significantly wider; for example, the phosphate P3 to phosphate P13 distance across the major groove is 7.9 Å instead of 3.0 Å in the decamer crystal structure, compared to 2.7 Å in an ideal fiber A-DNA model. The crystal structures of the several octamers which have formed the A-type helix represent something between an A-DNA and B-DNA. The minor groove of these two A-DNA crystal structures are very consistent and vary by 1 Å only, because this value is dependent on individual base pair geometry.

The A-DNA model structure for the identical sequence d(ACCGGCCGGT) derived from fiber diffraction data is close to the crystal structure with a RMSD of 1.15 Å for all the atoms of the two decamers. The local helical parameters of these two decamer helices are similar. Note that there is a relatively high propeller twist angle (avg. 15.2°) associated with the GC base pairs in the decamer crystal structure. This has been observed in some individual base pairs in B-DNA dodecamers where it reduces sequence-specific cross-strand clashes (Calladine, 1982). The functional advantage of a high propeller twist in A-DNA is as yet unclear. The base-stacking pattern in the crystal structure, though not uniform, clearly adopts an overall structure similar to the fiber A-DNA structure. The packing of the helices in the crystal lattice minimizes the accessibility of the solvent to the wide minor groove by having the terminal base pairs on one helix abutting into the minor groove of the neighboring helix. Because a low humidity environment favors the formation of A-type helix, the displacement of surface solvent molecules by hydrophobic base pairs provides a driving force in stabilizing short oligonucleotides in A-DNA conformation.

Finally, the crystal structure of r(GCG)d(TATACGC) as a self-complementary DNA–RNA hybrid helix has been chosen as an example of a typical A-DNA helix (Wang et al., 1982b). Figure 4-19 shows the stereoscopic skeletal representation of the DNA–RNA hybrid helix. The molecule adopts a conformation close to that seen in the 11-fold RNA double helix. An almost perfect A-type geometry is adopted by the molecule with all the ribose and 2'-deoxyribose sugars in the C3'-endo conformation. The 2'-hydroxyl groups of ribose are involved in different types of hydrogen bonding to the adjacent nucleotides in the chain. The cytidine 2'-hydroxyl groups are hydrogen-bonded through a bridging water to the O2 of the same cytidine residue. The 2'-hydroxyl group of the riboguanosine residues are hydrogen-bonded to waters that form

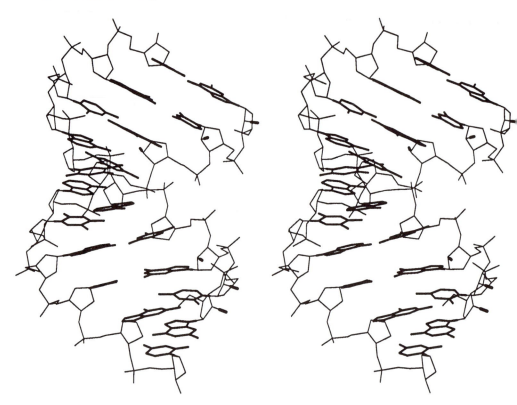

Figure 4-19. The crystal structure of r(GCG)d(TATACGC), a self-complementary DNA–RNA hybrid. There are two DNA–RNA hybrid segments in this stereoscopic skeletal plot. The O2′–ribose oxygens are shown as thicker lines.

hydrogen bonds with adjacent molecules in the lattice. The hybrid DNA duplex contains no structural discontinuities though the central TATA step has some structural irregularities.

Z-DNA. The first crystal structure of an extended DNA oligonucleotide helix was that of d(CGCGCG); to the surprise of all concerned, it was a left-handed extended helix (Wang et al., 1979). The DNA hexamer duplexes stacked on each other, forming infinite helices in the crystal lattice. The unprecedented high resolution of 0.9 Å allowed an unambiguous determination of the atomic structure of the molecule. It is ironic that the infant field of single-crystal diffraction of oligonucleotides was started with an extreme view of DNA polymorphism. Subsequent studies with macromolecular DNA showed that DNA can undergo reversible conformational change between the right-handed B-DNA and left-handed Z-DNA (see Table 4-3 for a list of Z-DNA structures). The conformational equilibrium is influenced by negative supercoiling, changes in the ionic environment, and the presence of Z-DNA interacting proteins (Rich et al., 1984). Several other oligonucleotides have been crystallized in the Z-DNA conformation, which include the tetramer d(CGCG) (Crawford et al., 1980; Drew et al., 1980), fragments containing A·T base pairs (Brennan et al., 1986; Fujii et al., 1985; Wang et al.,

Table 4-3 A Representative List of Z-DNA Structures

Oligonucleotide	Space Group	Unit Cell (Å)	Resolution (Å)	R-Factor (%)	Number of Waters	Number of Reflections	Reference
br5CGATbr5CG	P2₁2₁2₁	a = 18.30 b = 31.10 c = 44.10	1.54	19.3	88	2,386	Wang et al., 1985
br5CGbr5CGbr5CG	P2₁2₁2₁	a = 17.93 b = 30.83 c = 44.73	1.4	12.5	83	3,765	Chevrier et al., 1986
br5CGbr5CGbr5CG	P2₁2₁2₁	a = 18.01 b = 30.88 c = 44.76	1.4	13.3	61	2,919	Chevrier et al., 1986
br5UGCGCG	P2₁2₁2₁	a = 17.94 b = 30.85 c = 49.94	2.25	15.6	64	1,105	Brown et al., 1986
CGCATGCG	P6₅	a,b = 30.90 c = 43.14	2.5	16	—	—	Fujii et al., 1985
CGCG	C222₁	a = 19.50 b = 31.27 c = 64.67	1.5	21	84	1,900	Drew et al., 1981b
CGCG	P6₅	a,b = 31.25 b = 31.25 c = 44.06	1.5	19.3	85	—	Wang, unpublished
CGCGCG	P2₁2₁2₁	a = 18.01 b = 31.03 c = 44.80	1.0	17.5	88	10,893	Gessner et al., 1989

continued

Table 4-3 (continued)

Oligonucleotide	Space Group	Unit Cell (Å)	Resolution (Å)	R-Factor (%)	Number of Waters	Number of Reflections	Reference
CGCGCG/spermine	$P2_12_12_1$	a = 17.88 b = 31.55 c = 44.58	0.9	13	68	15,000	Wang et al., 1979
CGCGCGCG	$P6_6$	a,b = 31.27 c = 44.06	1.6	19	85	—	Fujii et al., 1985
CGCGCGTTTTCGCGCG (hairpin)	C2	a = 57.18 b = 21.63 c = 36.40 β = 95.22°	2.1	20.0	70	1,509	Chattopadhyaya et al., 1988
CGCGf5UG	$P2_12_12_1$	a = 17.38 b = 31.06 c = 45.39	1.5	20.0	64	3,008	Coll et al., 1989b
CGCGTG	$P2_12_12_1$	a = 17.45 b = 31.63 c = 45.56	1.0	19.5	91	10,964	Ho et al., 1985
CGTACGTACG	$P6_5$	a,b = 17.93 c = 43.41	1.5	25.5	7	506	Brennan et al., 1986
m5CGm5CGm5CG	$P2_12_12_1$	a = 17.76 b = 30.57 c = 45.42	1.3	15.6	98	4,208	Fujii et al., 1982
m5CGTAm5CG	$P2_12_12_1$	a = 17.91 b = 30.43 c = 44.96	1.2	16	98	5,412	Wang et al., 1984a

1984a), nonalternating pu–pyr sequences (Ho et al., 1987; Wang et al., 1985), and wobble G·T base pairs (Brown et al., 1986; Ho et al., 1985).

The stereoscopic skeletal plot of the Z-DNA structure of d(CGCGCG)/Mg^{2+} form is shown in Fig. 4-20. The helix axis of hexamer duplex coincides with the crystallographic two-fold screw axis parallel to c-axis at $a = 0.75$ and $b = 0.5$. The helical twist of each dinucleotide repeat in Z-DNA is $-60°$, leading to the required $-180°$/hexanucleotide. The packing imitates a DNA molecule of almost infinite length extending along the c-axis throughout the crystal. Hydrated magnesium ion complexes interact with the DNA duplex and also form hydrogen bonds to the symmetry-related DNA molecules, stabilizing the lattice packing. A row of water molecules are in the deep groove connecting O2 keto groups of successive cytosine residues. The residues along each chain alternate with all cytidines in the *anti* conformation and all guanosines in the *syn* conformation, making a dinucleotide step the repeat unit in the helix and a zigzag appearance of the backbone. The N7 of guanine lies protruding on the surface of the Z-type helix, making it readily available (for example) for coordination to metal ions or reaction with mutagens. The torsion angles of the nucleic acid backbone and sugars of the d(CGCGCG) duplex show a small RMSD (1°–6°) indicative of structural regularity. The dG residue sugars, with the exception of the terminal residue, are in the C3'-endo pucker while all the cytosine sugars adopt the C2'-endo pucker.

The Z-DNA helix has a lot more internal consistency, and the base pairs show little tilt or roll. The buckle and propeller twist of base pairs are much less relative to that seen for an average base pair in the A- or B-DNA helix. The small differences between base-pair orientation from one structure to another suggest that these small differences are influenced by ions packing around them. The helical twist for the CpG

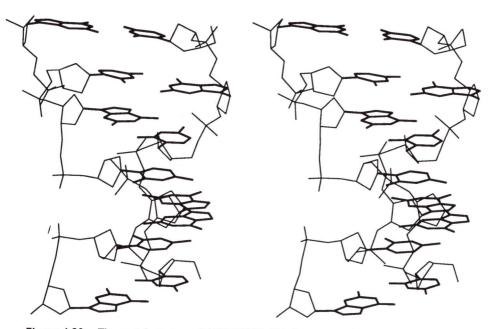

Figure 4-20. The crystal structure of d(CGCGCG). This is stereoscopic skeletal representation.

step is small ($-7°$ to $-11°$), and the base pairs are displaced by 7 Å with respect to each other. The helical twist for the GpC step is a much larger value ($-49°$ to $-53°$). The average helical twist angles are $-8.6°$ for the CpG and $-51.4°$ for the GpC with a sum of $-60°$ for the dinucleotide repeat unit. This would give rise to a helix with six dinucleotide steps per helical turn.

The single groove of Z-DNA as shown in Fig. 4-20 is very narrow and deep. When the van der Waals radius of the phosphate group is also included, the space available for access to the groove is made up of openings 6–7 Å in width. The major groove becomes very shallow, in fact almost nonexistent, and lies exposed to the solvent. The hydrated metal ions interact with the phosphate, and occasionally one of the hydration shell waters is replaced by a direct coordination of the metal ion to N7 of guanine. The metal ion coordination is discussed in detail below. The spermine and hydrated metal ions are seen interacting with the base edge atoms and the phosphate group atoms (Bancroft et al., 1994). The interaction of spermine and metal ions with the phosphate groups of the DNA plays a role in charge neutralization. Several spermine molecules were located pointing perpendicularly away from the DNA helix, and these play a major role in crystal packing rather than interacting with any one specific DNA molecule. Some spermine molecules bridge the minor groove, interacting with phosphate groups on both sides, and play a major role in stabilizing the Z-DNA helix.

The stacking interaction between the base pairs is very ordered and plays a major role in stabilizing Z-DNA. At the CpG step, the cytosines stack on top of each other, and guanosines are stacked upon the O1′ oxygen atoms of the preceding deoxyribose residues. This is allowed by the low helical twist angle associated with this base pair as well as the 7-Å translation seen. The stacking interaction at this step is unique to Z-DNA. At the GpC step there is considerable stacking overlap of bases on each other in a manner similar to that of B-DNA.

Crystallographic Studies of Ligand–DNA Interactions

Metal Ion–DNA Interaction. The negative charges associated with the phosphate groups of DNA make it a polyelectrolyte (Saenger, 1984). The charge neutralization of DNA is mostly attained from the positive charges of metal ions in solution, and it is not surprising to find nucleic acids bound with metal ions in vivo. The coordinated metal ions are important for the biological action of nucleic acids. It is believed that metal ions serve to screen the negative charges by direct interactions with the phosphate oxygen atoms. The most common intracellular metal ion is the potassium ion. Magnesium is another important metal ion for the function of nucleic acids. For example, magnesium ion plays an integral role in the folding of transfer RNA (Holbrook et al., 1977; Jack et al., 1977; Quigley et al., 1978). It has also been shown to be essential for ribozyme action (Piccirilli et al., 1993). High-resolution X-ray diffraction studies of oligonucleotides offer reliable answers about how these metal ions bind to nucleic acids. For example, the structures of several Z-DNA structures (Gessner et al., 1989; Wang et al., 1979, 1981), B-DNA structures (Gao et al., 1991b; Heinemann & Hahn, 1992; Prive et al., 1987; Yanagi et al., 1991) and drug–DNA complexes (Frederick et al., 1988; Gao et al., 1991a; Wang et al., 1984b, 1987, 1991, 1992) revealed the locations of magnesium and sodium ions. In crystals, both sodium and magnesium ions have been found to interact with various sites on nucleic acids, including phosphate oxygen and N7 of guanine. Interestingly, the interactions may be either through direct metal ion coordination or mediated through water molecules of the hydration shell of the metal ion.

Transition metal ions such as cobalt(II) or copper(II) have different binding characteristics toward DNA (Marzilli, 1977). They bind almost exclusively by coordinating to the N7 position of purines, especially of guanine. While there have been extensive structural studies of heavy metal ion interactions with nucleic acid components (e.g., bases, nucleosides) (Swaminathan & Sundaralingam 1979; reviewed in Saenger, 1984), little information is available on how they interact with macromolecular nucleic acids at the molecular level. A limited number of studies on the interaction of Pb(II) with tRNA indicated that it attacked the phosphate group, causing backbone breakage at neutral pH (Behlen et al., 1990).

More recently, the availability of oligonucleotide crystals has made it possible to diffuse different metal ions into the crystal lattice. The reasonably open solvent channels facilitated the diffusion of metal ions. The first CGCGCG Z-DNA structure (spermine form) was solved by the MIR method using Ba(II) and soaked Co(II) as heavy atoms (Wang et al., 1979). Co(II) was found to coordinate to N7 of the G6 residue. Subsequently, studies on the interactions of $Ru(NH_3)_6^{3+}$ (Ho et al., 1987), $Co(NH_3)_6^{3+}$ (Gessner et al., 1985) and Cu(II) (Geierstanger et al., 1991; Kagawa et al., 1991) with Z-DNA have been carried out. In a study of the cisplatin–CGCGAATTCGCG complex, cisplatin [i.e., *cis*-diamminedichloroplatinum(II)] was diffused into the B-DNA crystals. The N7 of guanine was observed as one of the square-planar Pt ligands (Wing et al., 1984).

A recent X-ray diffraction analysis of the interactions of the two transition metal ions Co(II) and Cu(II), and the alkaline earth metal ion Ba(II), with DNA of different conformations revealed several novel features (Gao et al., 1993). The binding modes of Co(II) as well as Ba(II) ions were compared with that of the Cu(II) ion (Kagawa et al., 1991; Geierstanger et al., 1991). The goal was to understand the different preferred coordination geometries of Co(II), Ba(II), and Cu(II) to DNA. The approach took advantage of the excellent crystal quality of Z-DNA and that of daunorubicin–DNA complex (Wang et al., 1987) and afforded a reliable determination of the metal ion locations and the coordinated water molecules.

In crystals, Co(II) ion binds *exclusively* to the N7 position of guanine bases by direct coordination. The coordination geometry around Co(II) is octahedral, although some sites have an incomplete hydration shell. The averaged Co—N7 bond distance is 2.3 Å. The averaged Co—N7—C8 angle is 121°, significantly smaller than the value of 128° if the Co—N7 vector were to bisect the C5—N7—C8 bond angle. Model building of Co(II) binding to guanine N7 in B-DNA indicates that the coordinated waters in the axial positions would have a van der Waals clash with the neighboring base on the 5′-side. In contrast, the narrow but deep major groove of A-DNA does not have enough room to accommodate the entire hydration shell. This suggests that Co(II) binding to either B-DNA or A-DNA may induce significant conformational changes. The Z-DNA structure of Cu(II)-soaked CGCGTG crystal revealed that the Cu(II) ion is bis-coordinated to N7 position of G10 and [#]G12 ([#] denotes a symmetry-related position) bases with a trigonal bipyramid geometry, suggesting a possible N7—Cu—N7 crosslinking mechanism. A similar bis-coordination to two guanines has also been seen in the interaction of Cu(II) in m^5CGUAm^5CG Z-DNA crystal and of Ba(II) with two other Z-DNA crystals.

A more detailed analysis of metal ion-DNA interaction is provided in Chapter 9.

Drug–DNA Interaction. Nucleic acids are key targets for pharmaceutical agents presently being used as well as those being designed for use in the treatment of cancers. These drugs interact with DNA or RNA molecules in diseased cells, inhibiting further cell growth. The need to design more specific and efficient drugs is increasing. It has become imperative to understand how these molecules bind and interact with DNA at an atomic level, because this should enable newer and more effective therapies to be designed. The types of interactions of nucleic acid binding drugs with DNA can be broadly classified as (1) intercalation of the drug between base pair, (2) covalent bonding of the drug to specific groups on DNA, and (3) binding of the drugs in the grooves of DNA. There are a multitude of drugs available for the interaction of drugs with DNA, and a recent comprehensive review on this topic has been provided by Propst and Perun (1992). Additional information is also available in Chapters 6 and 10 of this volume.

Protein–DNA Interaction. The interaction of DNA with protein is one of the most active areas of research in structural biology because of its direct importance to our understanding of cellular control mechanisms. Several principles govern the interaction of proteins with nucleic acids. Direct hydrogen bond formation and van der Waals interaction between protein and DNA atoms provide the major forces for sequence-specific recognition of DNA by protein. The proteins bind primarily to the major groove of DNA and to a lesser extent to the minor groove of DNA. The arrangement of hydrogen bond donors and hydrophobic groups on the DNA as a function of sequence provide a means for recognition of the sequence of DNA. The bendability or deformability of DNA as a function of sequence also provides a means for sequence-specific recognition of DNA, because it allows some sequences to take up a particular structure required for binding to protein at a lower free energy cost than other sequences. The phosphate groups have been found to play vital roles in the stabilization of protein–DNA complexes through charge interactions and hydrogen bond formation, though their role in sequence-specific recognition is unclear. Several buried water molecules have been found at the protein–nucleic acid interface. These waters apparently play a major role in the stability and specificity of certain DNA–protein complexes (Schevitz et al., 1985), and many of the water molecules involved in water-mediated interactions are preserved in the crystal structure of the free DNA molecule (Shakked et al., 1994). Steitz (1990) has provided an excellent review of protein–DNA interaction. The current opinion is that proteins recognize DNA sequences through direct hydrogen bonds, nonpolar contacts, indirect structural effects, and water-mediated interactions (Shakked et al., 1994).

Unusual DNA Structures

Certain sequences of DNA form 3D structures that are not the common A-, B-, or Z-form of double helices. These sequences form other, higher-order structures such as hairpin loops, triple-stranded structures, cruciforms, and pseudoknots. Specific functions have already been identified for some of these higher-order structures. One such example is the structure of telomeric DNA which plays an important role in stabilizing chromosomes. Telomeres refer to the terminal protein–DNA complexes of the chromosomes, and the DNA sequences found in telomeres are rich in G and contain a short, highly repetitive sequence such as $(TTGGGG)_n$. The structure of a fragment of a telomeric sequence from a protozoa "Oxytricha", d(GGGGTTTTGGGG), has been crystallized and solved (Kang et al., 1992). The oligonucleotide was found to form a hairpin, and two of these hairpins joined to make a four-stranded helical structure with the thymines forming loops at either end. The four-stranded structure was formed by the guanine residues being held together by a cyclic hydrogen bonding network and a metal ion (e.g., K^+) located at the center. The four guanine residues were found to have an glycosyl conformation that alternates between *anti* and *syn*. The four-stranded telomeric DNA is very stable. A detailed review of the quadruplex structure may be found in Williamson (1993).

Other unusual DNA structures have been studied in solution by two-dimensional nuclear magnetic resonance (2D-NMR) and other spectroscopic techniques. These include DNA hairpin and triple helices. Recently, a hairpin sequence was used to grow crystals of the complex of TATA-box binding protein (TBP) with TATA box DNA (Kim et al., 1993b). The TBP was also crystallized and solved with a nonhairpin sequence (Kim et al., 1993a). In both cases the TATA box DNA helix was found to be bent through an extreme angle of 90° over a distance of eight base pairs. It is an elegant example of the extreme distortion some sequences of DNA can undergo while binding to proteins with high affinity.

A recent addition to unusual DNA structures has been the discovery that some sequences form parallel-stranded DNA (Robinson et al., 1992). The parallel-stranded motif was discovered in a low pH study of the sequence d(CGATCG). The structure, solved by NMR, revealed non-Watson–Crick homo base-paired parallel-stranded DNA, called Π DNA. In a subsequent study (Robinson & Wang, 1993) it was discovered that the d(CGA) sequence is a strong motif for forming Π-DNA and that a sequence d(CGACGAC) can form Π-DNA at a pH close to neutral (pH 6.8). The CGA motif can also induce other sequences such as $(GA)_n$ in Π-conformation (Robinson et al., 1994). Interestingly, low pH induces DNA sequences that are rich in cytosines (e.g., TCCCCC) into highly unusual mutually intercalated tetra-stranded structures (Gehring et al., 1993).

Summary

One important direction for study of nucleic acid structures clearly lies in investigating the structures associated with new and significant biological functions. Higher-order structures associated with DNA sequences found in telomeres and centromeres of chromosomes, as well as other unusual structures such as cruciforms, hairpins, or pseudoknots, will undoubtedly be studied with much interest. The structure of RNA molecules has not been studied in detail; considerable emphasis will be placed on this species in the immediate future. One area where a lot of work is being done is the development of a reliable method for obtaining suitable crystals for X-ray crystallography. This would remove the principal bottleneck between obtaining a nucleic acid sequence and determining its structure.

Chemical Mapping of Nucleic Acid Conformation

Thomas D. Tullius

One lesson that can be taken from this text is that nucleic acids exist in a bewildering variety of conformations. Despite the limited range of building blocks (the four ribonucleotides and four deoxyribonucleotides) available to RNA and DNA, structures can form that rival the folds of proteins in complexity.

This diversity of structure was not appreciated until recently, though, especially for DNA. Most scientists envisioned the highly symmetrical Watson–Crick double helix when thinking about DNA (Watson and Crick, 1953), and the only three-dimensional structure known for RNA was the compactly folded L-shaped tRNA (Robertus et al., 1974; Suddath et al., 1974).

The X-ray crystal structures of DNA that began to emerge in the late 1970s widened the perception of the structural possibilities available to nucleic acids (Dickerson, 1992). More recently, nuclear magnetic resonance (NMR) has been used to determine three-dimensional structures of DNA and RNA in solution (Feigon et al., 1992; Wemmer, 1991). These methods, and the fruits of their application to nucleic acid molecules, are covered in other chapters in this text. Nucleic acids with one, two, three, or four strands, and with right- or left-handed double helices, have been investigated at atomic resolution by X-ray crystallography (Dickerson, 1992; DiGabriele & Steitz, 1993; Kang et al., 1992; Robertus et al., 1974; Suddath et al., 1974; Wang et al., 1979; Wing et al., 1980) or NMR (Cheong et al., 1990; Schultze et al., 1994; Sklenar & Feigon, 1990; Smith & Feigon, 1993; Wemmer, 1991).

We are fortunate to have these high-resolution structures, so that we may understand better the formation of, for example, DNA triplexes or left-handed Z-DNA. But high-resolution structural methods have their limitations, particularly if we are interested in how unusual nucleic acid structures participate in the functioning of the cell. Biologically active DNA and RNA molecules often are too large to be studied by NMR or X-ray crystallography at atomic resolution. Consider that while X-ray structures have been solved for DNA molecules having one or two turns of the double helix (10–20 base pairs) in length, hundreds or thousands of base pairs of DNA constitute a region of the chromosome that participates in gene regulation. We have very little idea of how long segments of duplex DNA fold into functional structures, or how unusual DNA structures might be involved in gene transcription. Similarly, it is becoming evident that ribosomal RNA is an active participant in protein synthesis, and not just a scaffold for the proteins of the ribosome. We do know something of the secondary structure that these large RNA molecules adopt, mainly due to phylogenetic comparisons of sequences from a large number of organisms. We also have an idea of the structural possibilities for small segments of RNA based on high-resolution structures. But still, the three-dimensional fold (the tertiary structure) of ribosomal RNA is unknown.

Our ultimate goal in the study of nucleic acids is to understand how nature employs these molecules in living organisms. A key question is whether the variety of nucleic acid structures that we now know can form (duplexes, triplexes, quadruplexes, and so on) do in fact participate in cellular function. To do this we will need methods for observing the structures of nucleic acids inside cells, to put into perspective the structural information we have collected in vitro.

Chemical Probes of Nucleic Acid Structure

How can we work out the structure of a large and complicated nucleic acid molecule, whether pure in the test tube or functioning in a cell? One way to approach this problem is to forego solving the structure by measuring the three-dimensional position of every atom in the molecule, as is done by X-ray crystallography or NMR. Instead, we could collect information on many different aspects of the structure, and put these pieces of data together to make an "image" of the structure, although possibly not at atomic resolution. For example, we could try to find out which individual nucleotides are on the "outside" and which are on the "inside" of the molecule, or which are in duplexes and which are single-stranded. Such an experimental approach recalls the fable of the blind men describing the elephant, and it might lead to the same problems of determining an accurate and unique structure based on fragmentary and sometimes conflicting data. But if we are trying to solve a structural problem that is too difficult for conventional high-resolution methods, the information we obtain from this "piecemeal" approach might still be the best available.

In recent years an experimental approach based on chemical transformations has been developed that attempts to put together such a piecemeal "image" of a nucleic acid. These methods, sometimes called *solution probe* or *chemical probe* experiments (Tullius, 1991), depend on having a battery of small molecules that are sensitive in some way to the structural details of nucleic acids. For example, a consequence of the presence of a single-stranded region in a DNA molecule is the exposure of functional groups on the DNA bases that would normally be protected due to participation in hydrogen bonding. A chemical reagent that reacts with these exposed functional groups can then be used to make a map of which segments of the DNA strand are single-stranded. Unreactive bases, on the other hand, are assigned as occurring in a duplex.

Many chemical probes have been developed in recent years to monitor different features of nucleic acid structure, including the shape of the surface of the molecule, distortions from ideal structures, and the occurrence of single-, double-, or triple-stranded regions. Examples of small molecules that have proven useful as structural probes include dimethyl sulfate (Maxam & Gilbert, 1980), iron(II)·EDTA (Price & Tullius, 1992), osmium tetroxide (Palecek, 1992a), bis(o-phenanthroline)copper(I) (Yoon et al., 1988), Rh(phen)$_2$(phi)$^{2+}$ (phen = o-phenanthroline; phi = phenanthrenequinonediimine) (Chow & Barton, 1992), and the uranyl ion (Nielsen et al., 1988). More details on these and other reagents is available. An excellent starting point is a recent volume of *Methods in Enzymology* (Lilley & Dahlberg, 1992). Some of these molecules, and others, are discussed later in this chapter. Each chemical probe molecule has particular strengths and weaknesses as a tool for structure determination. The most powerful experimental approach combines information from more than one reagent to piece together the structure.

Advantages of Solution Probe Methods. A key advantage of these new methods is that they can be used to study large nucleic acid molecules of complicated structure. The individual experiments are rapid, especially in comparison to the time needed to determine the three-dimensional structure of an oligonucleotide by X-ray crystallography or NMR, and require very small amounts of nucleic acid (perhaps picomole to femtomole quantities).

The results of solution probe experiments are complementary to the information obtained by high-resolution methods. For example, a common approach is to calibrate the chemistry of a new reagent by applying it to a nucleic acid molecule of known structure. Once the chemistry of the reagent is understood by means of such control experiments, nucleic acids of unusual structure can be studied and the results interpreted with more confidence.

In addition, it often is useful to compare the structure of a nucleic acid molecule in solution to the structure determined by X-ray crystallography in the solid state. A problem that has bedeviled nucleic acid crystallographers is determining which structural features are inherent to the molecule under study, and which are the result of crystal packing forces (Dickerson, 1992). It is not commonly appreciated by newcomers to macromolecular crystallography that extensive intermolecular interactions are present in a crystal. What is presented most often is a graphical representation of the structure of the individual macromolecule, but a crystal packing diagram shows that a substantial proportion of the surface of the macromolecule is in contact with other molecules in the crystal. Proteins are less subject to this experimental caveat, since the interesting part of a protein molecule is often the interior, where packing interactions have little influence on the structure. Nucleic acids are different, since intermolecular crystal contacts can have a sizable effect on the parts of the structure of most interest. Solution probe experiments might thus be used to determine whether unusual structural features of a nucleic acid molecule that are revealed by X-ray crystallography persist in solution.

A further advantage of the solution probe approach lies in the ability of the experimenter to explore a wide range of conditions systematically. The type of counterion and its concentration, temperature, pH, and even pressure may be varied in the experiment. This is not so easy to do with X-ray crystallography, where very specific conditions are usually necessary for crystallization, or with NMR spectroscopy, where the solubility properties of the nucleic acid often limit the solution conditions that can be studied.

The nucleic acid itself can be obtained from any of a variety of sources. Natural DNA sequences can be isolated in the form of a bacterial plasmid. This is an inexpensive and convenient source of large DNA molecules of a single sequence for preliminary assessment of the DNA modifying ability of a new reagent. Because a wide variety of sequences in many contexts is present in a plasmid, which might consist of about 4000 base pairs, any sequence preference or specificity of the reagent is easily surveyed. Plasmid DNA is also the ideal vehicle for studying the effect of supercoiling on structural transitions in DNA, since any DNA sequence can be cloned into a plasmid, and straightforward methods are available for manipulating the supercoil density of a plasmid.

RNA may be prepared by in vitro transcription of a DNA sequence that has been cloned in a plasmid. DNA or RNA of essentially any sequence (but of limited length) may be prepared by automated chemical synthesis. A more recent method of preparation, the polymerase chain reaction (PCR), in principle allows the isolation of any DNA sequence directly from a genome (Saiki et al., 1988). Direct chemical synthesis (and in some cases PCR) may be used to incorporate modified nucleotides into a DNA molecule (Diekmann et al., 1992; Mazzarelli et al., 1992).

Straightforward and relatively inexpensive methods are used to analyze the reactions of the reagent with the nucleic acid. The most common strategy involves mapping the positions of cleavage of the nucleic acid backbone that are the result of the interaction of the reagent with the nucleic acid molecule. Where the initial chemical modification does not produce strand scission, other methods can often be used to produce cleavage at the site of modification. The sites of cleavage are detected by denaturing polyacrylamide gel electrophoresis of the nucleic acid reaction products. Because this is the same kind of gel that is used for DNA sequencing, the methods are well known to many laboratories and the equipment is widely available (Maxam & Gilbert, 1980). A further advantage of DNA sequencing technology for analysis is the use of radiolabeled nucleic acid molecules. Because very high specific activity can be achieved by incorporation of ^{32}P-labeled phosphate at the end of one strand of a nucleic acid molecule, vanishingly small quantities of the nucleic acid (typically femtomoles) are sufficient for an experiment. This means that rare or difficult-to-prepare nucleic acid molecules may be studied easily in solution probe experiments.

Disadvantages of Solution Probe Methods. As discussed above, chemical probe experiments cannot be expected to produce atomic-resolution structural information. No reagent, or collection of reagents, provides a complete structural picture of a nucleic acid molecule. A further difficulty, which is the subject of continuing investigation, is that the specific structural feature of a nucleic acid molecule that is recognized by a given chemical probe often is not known for certain. In part this is due to a paucity of model nucleic acid molecules having known structures in solution, especially for complicated DNA structures and for RNA in general. As more RNA and DNA structures are determined, and the reactivities of chemical probes toward these different structures are characterized, more detailed and reliable interpretation of chemical probe experiments will result.

Basic Principles of Conformational Mapping by Chemical Probing

Design of the Experiment. The aim of a chemical probe experiment is to determine where in a nucleic acid molecule the probe reagent binds (or reacts). Because the probe reagent was chosen to recognize some aspect of nucleic acid structure, mapping its loci of interaction with the nucleic acid indicates where this structural feature occurs. The most direct way of mapping the interaction of the chemical probe is to somehow cause cleavage of the nucleic acid backbone where the probe interacts. A diagram showing the steps of a chemical probe experiment is shown in Fig. 5-1. Each of the steps is discussed in more detail below.

Cleavage may be a direct result of the reaction of the probe reagent with the nucleic acid, for example due to oxidative scission of the sugar-phosphate backbone. The hydroxyl radical (·OH) reacts in this way; a hydrogen atom of a deoxyribose residue is abstracted by the hydroxyl radical. The resulting sugar-based radical undergoes further chemistry, leading ultimately to degradation of the deoxyribose and a single-nucleoside gap in the DNA strand. Many metal-based chemical probe reagents follow this general outline of oxidative, radical-based attack on a backbone sugar (Stubbe & Kozarich, 1987).

Alternatively, cleavage can be induced at a later time at sites that are chemically modified by the reagent. For example, the reagent ethylnitrosourea ethylates a back-

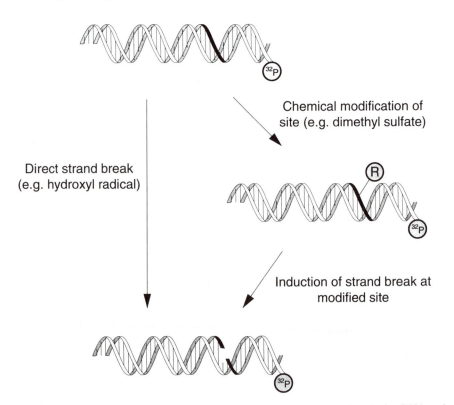

Figure 5-1. The steps of a chemical probe experiment. Depicted at the top is a duplex DNA molecule, with a radioactive phosphorus atom attached to one end of one strand. Colored black is a segment of one DNA strand that will be targeted by a chemical probe reagent. If the chemical probe causes a strand break upon its initial interaction with DNA, the experiment proceeds as on the left. If a subsequent step is required to convert a site, of chemical modification (denoted by R) to a strand break, the experiment follows the path on the right. The result in both cases is a DNA molecule with one strand that was broken following interaction with the chemical probe reagent. The experiment is analyzed by separating the strands of the DNA molecule by heat treatment and then running the single-stranded DNA fragments on a denaturing electrophoresis gel. Only radioactive DNA fragments are observed. The site of cleavage can be mapped by this method to single-nucleotide resolution.

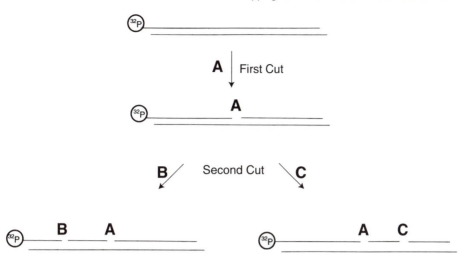

Figure 5-2. Single versus double-hit cleavage. The first cleavage event is marked A. If the experiment is not properly controlled, a DNA molecule can receive more than one cut. If the second cut occurs between the first cut and the radiolabel (cut B), the situation on the left obtains. If the second cut is to the other side of the initial cut (cut C), the result is as shown on the right. In either case, information on one of the cuts is lost, since it is not connected to the radiolabel and thus cannot be detected by autoradiography after gel electrophoresis of the products.

bone phosphodiester (Siebenlist & Gilbert, 1980). The resulting phosphotriester is labile to attack by base, resulting in strand cleavage in a step subsequent to the initial alkylation reaction. Dimethyl sulfate, which methylates the N7 and N3 positions of the bases guanine and adenine, respectively, also works in this general fashion (Maxam & Gilbert, 1980). The alkylated guanine or adenine nucleotide is easily depurinated. Alkali is then employed to specifically degrade the abasic sugar that remains. The result is a break in the nucleic acid strand. The two-step approach of modification and cleavage has the advantage that the topological properties of supercoiled DNA are not changed by the initial modification reaction, so processes that depend on supercoil density (protein binding, for example) are not affected by the initial reaction of the chemical probe. The sites of modification can then be determined later, for example, once the DNA is isolated from the cell and purified. This feature can be advantageous for in vivo chemical probe experiments.

Another dimension is added to the experiment when the modification or cleavage reaction is accomplished photochemically. Several DNA cleaving reagents are activated by light, including the uranyl ion and $[Rh(phi)_2bipy]^{3+}$ (bipy = 2,2′-bipyridine) (Chow & Barton, 1992). Such photoactivatable chemical probes could be used for time-resolved experiments or for experiments in vivo, but few such investigations have been reported as yet.

Once the chemical probe reagent has reacted with the nucleic acid molecule, the positions at which it has induced cleavage must be determined. The experiment is analyzed by performing denaturing gel electrophoresis on the radiolabeled nucleic acid that has been cleaved. The extent of reaction is generally controlled so that on average any individual nucleic acid molecule undergoes cleavage once at most. This is an important point; if cleavage occurs more than once, then information is lost. One may think about the experiment in the following way (Fig. 5-2). Suppose a duplex DNA molecule, with a radioactive phosphorous atom attached to the 5′-end of one of its strands, is cleaved at one nucleotide. This cleavage site is marked A in Fig. 5-2. The initial cleavage splits the strand into two pieces: One piece carries a radioactive label, the other does not. We find out where in the strand the cleavage reaction occurred by measuring the length of the broken, radioactive fragment using denaturing gel electrophoresis. DNA molecules migrate strictly by size on a denaturing gel, so it is possible to measure the length of a piece of DNA exactly by running standard length DNA fragments in parallel. These standards are usually generated by DNA sequencing reactions (Maxam & Gilbert, 1980). Now, consider what happens if the original DNA strand is cleaved a second time before gel analysis. There are two possibilities: Either the second cut occurs between the first cut and the radioactive label, or the second cut occurs away from the label. If the former is true (Fig. 5-2, cleavage B), then all information concerning the original cut is lost, since it is now detached from the radiolabel and thus is invisible to detection. If the latter is the case (Fig. 5-2, cleavage C), then the second cut is not observed, again because the fragment containing that cut has no radiolabel.

There are two ways to overcome this problem. Both depend on assuming that DNA cleavage is essentially randomly distributed, even if not every nucleotide is cleaved by the particular reagent. One approach is to calculate a factor that is applied to band intensities to correct for the underrepresentation of longer fragments in DNA samples which have been cleaved more than once per strand. This calculation was first described in connection with DNase I cleavage experiments on the nucleosome (Lutter, 1978). A second approach is to ensure by the design of the experiment that (almost) every DNA molecule suffers either 1 or 0 cleavages. This can be achieved by performing the cleavage reaction so that no more than about 30% of the DNA molecules are cleaved; most of the DNA in the sample therefore has not reacted with the reagent, and it remains completely intact. A calculation using the Poisson equation shows that the probability is very small that any DNA molecule in this sample was cleaved twice or more (Brenowitz et al., 1986). This second approach, sometimes called "single-hit kinetics," is the preferred way to perform a chemical probe experiment.

One limitation inherent in the analysis described above is that the signal observed in the experiment, a band on a DNA sequencing gel, must by necessity be kept low in intensity so that we do not encounter the problem of multiple cutting. The signal is low because not all DNA molecules in the experiment contribute to the cleavage pattern, since a majority of the DNA is never cleaved. For example, consider a 100-nucleotide-long DNA strand. If our chemical probe reagent cleaves equally at every nucleotide, and we have made sure that only 30% of the DNA molecules have been cleaved at all, then each individual band in the gel contains only 0.3% of the radioactivity that was in the original intact DNA strand.

New methods of DNA amplification (the PCR and its variants, like primer extension) offer a way to increase the experimental signal linearly (Sasse-Dwight & Gralla, 1988) or even exponentially (Mueller & Wold, 1989). Using DNA amplification, each cleaved DNA molecule in the sample is observed multiple times, leading to a substantial increase in the intensity of the experimental signal. A further advantage of amplification is that the DNA sample does not need to be pure. The oligonucleotide primers that are used in the reaction direct amplification only to the sequence of interest, even in the presence of an entire genome's worth of DNA. Thus these methods can be used on single-copy sequences in a large genome. The constant stream of innovative variants of PCR (rivaling the pace of development of new NMR pulse sequences!) makes it likely that, even if a particular experiment seems to be impossible today, someone soon will introduce a new PCR protocol that will do the job.

Choice of the Chemical Probe. As mentioned briefly above, the best way to use chemical probe experiments to study the structure of a nucleic acid molecule is to use several probes, each of which monitors a different structural feature. In this section an overview of some of the available chemical probes is presented, as well as a discussion of the information each provides.

For an idea of the overall shape of nucleic acid, a probe that reports on the accessibility of the surface of the molecule would be useful. The hydroxyl radical is such a reagent. This highly reactive radical reacts with organic molecules by abstracting a hydrogen atom from a C—H bond, to give a molecule of water and a carbon-centered radical (Walling, 1975):

$$\cdot OH + R_3CH \rightarrow H_2O + R_3C\cdot \tag{5-1}$$

This new radical can then undergo any of a number of subsequent reactions depending on its structure. If the organic molecule is a nucleic acid, the (deoxy)ribose backbone offers a number of C—H bonds that in principle can react with the hydroxyl radical. For at least some of the resulting sugar radicals the final product is a broken strand. Because the hydroxyl radical is so reactive, it hardly discriminates with which backbone sugar it reacts in a nucleic acid molecule, at least for nucleic acids of uniform structure such as a DNA or RNA duplex. The cleavage pattern of such a uniformly structured nucleic acid molecule is thus exceedingly simple, consisting of a ladder of bands of equal intensity with one band per nucleotide.

If the DNA or RNA molecule is not structured so simply, the cleavage pattern changes to reflect the more complicated fold of the nucleic acid. An example that is easy to understand is a DNA–protein complex. When the protein is bound to its DNA site, it "covers" several nucleotides. These nucleotides then become inaccessible to the hydroxyl radical and thus do not react; consequently, the DNA is not cleaved at these sites. The result is a "footprint" of the protein bound to DNA (Dixon et al., 1991), which appears as a series of blanks in the electrophoresis gel pattern (cf. Fig. 1-26).

How can we understand the footprint in terms of the structure of the DNA–protein complex, and how much quantitative detail of the structure is present in the footprint pattern? The Tullius laboratory has compared the hydroxyl radical footprint of a structurally characterized DNA-binding protein, the bacteriophage lambda repressor, with a calculation of the solvent-accessible surface area of the deoxyribose hydrogen atoms of the DNA-binding site (Dixon et al., 1991). There is a satisfying correlation between the exposed surface area of the complex and the nucleotides most reactive to the hydroxyl radical (Fig. 5-3). We like to think of the hydroxyl radical as a "reactive water molecule." The cleavage pattern generated by the hydroxyl radical then provides a map of the relative accessibility to solvent of each nucleotide of an RNA or DNA molecule.

Other structural features of a nucleic acid that might be interesting to detect include discontinuities in an otherwise uniform structure. Discontinuities in structure have been proposed to represent landmarks in a DNA or RNA molecule that are recognized by

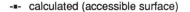

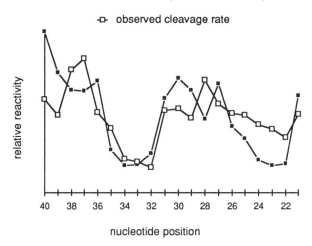

nucleotide position

Figure 5-3. Relationship of the hydroxyl radical cleavage pattern to the accessible surface of a DNA–protein complex. Shown is a plot of the extent of cleavage by the hydroxyl radical at each nucleotide of the binding site of the bacteriophage lambda repressor (open symbols), for the complex of repressor with DNA. This plot is a representation of the footprint of the lambda repressor. Sites where repressor contacts the DNA (nucleotides 32–35, for example) are cleaved to a lesser extent than sites not contacted by protein. The solvent-accessible surface area of each C3′ deoxyribose carbon atom, calculated by the Lee and Richards algorithm from the co-crystal structure, is denoted by the solid symbols. The correlation that is apparent between the two plots is evidence that the hydroxyl radical cleavage pattern reports on the solvent-accessible surface area of a nucleic acid molecule. Reprinted from Dixon et al., 1991, with permission.

proteins. For example, a kink, bulge, or unstacked base might direct the binding of a protein to a particular site in a nucleic acid molecule. This mode of recognition is thought to be of particular importance in RNA–protein complexes.

How might such discontinuities in structure be detected using a small molecule chemical probe? Some intercalators are known to bind preferentially to bulges in DNA or RNA, which are the result of the presence of a stretch of noncomplementary bases within a region that is otherwise perfectly paired in a duplex. Binding of the intercalator can be detected by attaching a reactive group to it that cleaves the nucleic acid. The classic example of such a bifunctional molecule is the iron(II) complex of methidiumpropyl EDTA (MPE) (Dervan, 1986). The methidium heterocycle intercalates into DNA or RNA, and the iron EDTA-containing side chain produces a flux of hydroxyl radicals near the binding site. MPE has proven useful in mapping bulges in RNA and DNA and has also proven useful in mapping kinks induced in DNA by bound proteins (Hatfull et al., 1987).

Unstacked thymine bases can be detected by their enhanced reactivity toward potassium permanganate (Borowiec et al., 1987) or osmium tetroxide (Palecek, 1992a). If a thymine is well-stacked with adjacent bases in a regular duplex, access of the reagent is limited and little reactivity is seen. But if the thymine is not stacked on its neighboring bases, due to a kink in the duplex or in the limiting case because the DNA is single-stranded, the reagent can easily approach the thymine base from the preferred direction and react efficiently. Osmium tetroxide has the added benefit of being amenable to introduction into living cells, so that unusual DNA structures can be seen in vivo (Palecek, 1992b).

Other reagents provide information on the exposure of the various functional groups of the nucleic acid bases. As mentioned above, dimethyl sulfate methylates the N7 position of guanine, which is exposed in the major groove and is the most nucleophilic site in DNA. In normal duplex DNA all guanine N7's will be methylated; in fact, this reaction is the Maxam–Gilbert guanine-specific chemical sequencing reaction (Maxam & Gilbert, 1980). If the electron pair of a guanine N7 is otherwise occupied, though, perhaps because it is participating in a hydrogen bond, dimethyl sulfate will not react. The dimethyl sulfate reaction proved crucial in refining initial models for the structure of the quadruplex that is formed by the guanine-rich ends of telomeric DNA (Sundquist & Klug, 1989; Williamson et al., 1989).

Details of Recent Studies

The first part of this chapter describes the design of the chemical probe experiment and introduces a few of the chemical probe reagents that are now in common use. The remainder of the chapter focuses on a few specific examples of how a chemical probe has been used to determine structural details of RNA and DNA molecules in solution. The emphasis is on what sort of information can be obtained from these experiments and how these results compare to data accessible from other methods. The common feature of the experiments described is the use of the hydroxyl radical as the chemical probe, both because of its versatility and to provide a sense of the way in which chemical probe data are interpreted for several structurally diverse systems.

To produce the hydroxyl radical for use as a chemical probe, a convenient method makes use of the Fenton reaction of iron(II)·EDTA with hydrogen peroxide (Price & Tullius, 1992; Tullius & Dombroski, 1985):

$$[Fe(EDTA)]^{2-} + H_2O_2 \rightarrow [Fe(EDTA)]^- + OH^- + \cdot OH \qquad (5\text{-}2)$$

In this reaction the iron(II)·EDTA complex is oxidized to iron(III)·EDTA. The electron from the metal complex reduces hydrogen peroxide, causing the O—O bond to break. The oxygen-containing products are the hydroxide ion and the hydroxyl radical. We chose iron(II)·EDTA as the metal complex for this reaction because of its negative charge. Because DNA and RNA are polyanions, the anionic iron complex should have little, if any, binding affinity for the nucleic acid molecule. The hydroxyl radical is neutral in charge, though, and can diffuse through solution to the nucleic acid, where it reacts with the C—H bonds of the sugar-phosphate backbone [see Eq. (5-1)]. Thus, the ultimate probe of nucleic acid structure in this system is not the metal complex, but the very small and highly reactive hydroxyl radical (Tullius, 1987).

While iron(II)·EDTA currently is the most widely used chemical probe reagent for producing the hydroxyl radical, other methods have been developed recently. Peroxynitrite can be used as a source of the hydroxyl radical (King et al., 1993), as can the irradiation of aqueous solutions by γ- or X-rays (Hayes et al., 1990a).

Ribozyme Structure. It is fair to say that the discovery that RNA can act as a catalyst fundamentally changed our perception of the role of nucleic acids in biology. This fascinating story is described in Chapter 13; here, the use of a chemical probe experiment to study the structure of the *Tetrahymena* intervening sequence ribozyme (Latham & Cech, 1989) is described. Cech and his colleagues were interested in the shape of the *Tetrahymena* ribozyme: how it is folded, and whether the molecule has an "inside" and "outside" like a protein enzyme. To answer these questions, they used iron(II)·EDTA to make a map of the accessible surface of the ribozyme. Before these experiments were done, a fairly detailed model had been proposed for the structure of the ribozyme, based on sequence conservation in the family of group I introns (of which the *Tetrahymena* ribozyme is a member). To calibrate the chemical probe method for tertiary structure determination of RNA, Latham and Cech cleaved phenylalanyl-tRNA using iron(II)·EDTA. While they found that nearly every nucleotide in unfolded tRNA[Phe] could be cut, when the tRNA molecule was correctly folded in the presence of magnesium ions the cleavage pattern changed. Cleavage was greatly inhibited in three discrete regions of the molecule. These three sites of decreased cleavage corresponded to places in the tRNA strand that become inaccessible to solvent when the molecule folds upon itself to form the native L-shaped structure. Because this cleavage pattern agreed well with the known crystal structure of tRNA[Phe], and in particular with calculations of the solvent-accessible surface area, Latham and Cech concluded that the chemical probe method could be used to study the tertiary structure of the ribozyme.

Treatment of the ribozyme with iron(II)·EDTA in the absence of Mg^{2+} again showed nearly uniform cleavage of all nucleotides, consistent with the RNA molecule being unfolded. Both the experiments with tRNA and with the ribozyme established that the iron(II)·EDTA probe was insensitive to the presence of secondary structure (i.e., single-stranded versus duplex), since RNA molecules lacking tertiary structure were cleaved uniformly. However, when magnesium ion was added and the ribozyme became folded into the catalytically active conformation, the cleavage pattern produced by iron(II)·EDTA was quite different. The cleavage patterns in the absence and presence of magnesium, along with a plot of the difference between the two patterns, are shown in Fig. 5-4. Cleavage was inhibited throughout much of the RNA molecule, to a larger extent than for tRNA[Phe]. The catalytic core region of the ribozyme was found to be especially well protected from cleavage. This result provides evidence that the "active site" of the ribozyme is in the interior of the folded RNA molecule, much like the active site of a protein enzyme.

Another segment of the ribozyme 100 nucleotides in length showed a highly periodic cleavage pattern, indicative of alternating exposure and occlusion of the RNA strand. Such a pattern could result from helices that are contacted on one side by other parts of the ribozyme. Periodic cleavage patterns have also been observed for DNA in situations where the structure is better understood. These include DNA in the nucleosome core particle (Hayes et al., 1990b) and DNA bound to a crystalline precipitate of calcium phosphate (Tullius & Dombroski, 1985). In both cases, one side of the DNA duplex is bound to a surface: the histone octamer of the nucleosome or the flat inorganic crystal surface. The other side of the DNA molecule is exposed to the solvent. Because DNA is helical, each strand is periodically exposed to, and then hidden from, the solvent as it winds around the other strand along the surface. The access of a cleavage reagent (like the hydroxyl radical) is also periodically modulated. A diagram showing a DNA molecule bound to a surface that is being attacked by the hydroxyl radical is shown in Fig. 5-5. Experimental cleavage patterns of DNA bound to a calcium phos-

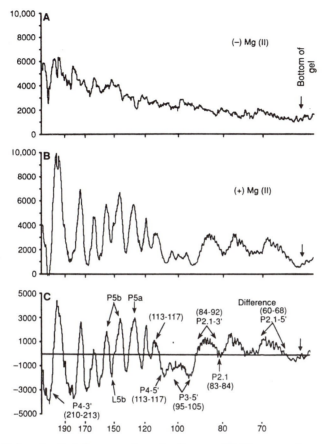

Figure 5-4. Hydroxyl radical cleavage pattern of a ribozyme. (**A**) Cleavage pattern of the *Tetrahymena* ribozyme in the absence of Mg²⁺. (**B**) Cleavage with Mg²⁺ present. (**C**) Difference plot (**B−A**). Reprinted from Latham & Cech, 1989, with permission.

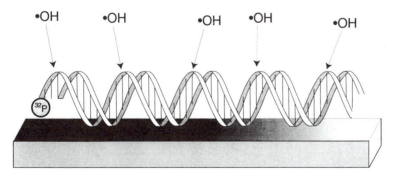

Figure 5-5. Cleavage of DNA bound to a surface. One side of the DNA is accessible to a cleavage reagent (here, the hydroxyl radical), and the other is blocked. The most intense sites of cleavage occur once each helical turn.

phate precipitate and to a nucleosome are shown in Fig. 5-6. The cleavage pattern has the appearance of a sine wave, with a repeat equal to the helical repeat of the DNA. In fact, the helical twist of DNA has been measured by such experiments (Hayes et al., 1990b; Tullius & Dombroski, 1985).

From Latham and Cech's experiments a detailed view of the accessible surface of the ribozyme could be constructed. The data agreed in large part with the model proposed previously from sequence comparisons, although the relative orientation of some segments in the model needed to be changed based on the results of the chemical probe experiment. Celander & Cech (1991) later used iron(II)·EDTA to follow the pathway of folding of the ribozyme and to determine which parts of the molecule interact with each other in the tertiary fold.

Bent DNA. DNA that is bent (or curved) is perhaps the most commonly occurring of the unusual DNA structures. In fact, it might be argued that perfectly straight DNA actually is the most unusual form of DNA.

About 10 years ago, experiments on the kinetoplast minicircle DNA of trypanosomatid parasites provided the first definitive evidence for stably curved DNA (Marini et al., 1982). The chief method of detection of bent DNA is the retardation in electrophoretic mobility of bent, as compared to straight, DNA. While the theory of gel electrophoresis is still not sufficiently developed to provide a detailed explanation of this phenomenon, the experiment is simple and has allowed a large number of DNA sequences to be assayed for bending. The general rule that has emerged from such work is similar to that derived from the original studies on kinetoplast DNA: Short runs of adenine (four

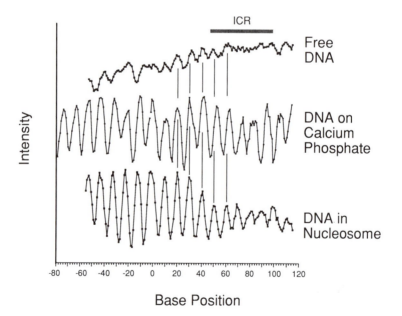

Figure 5-6. Cleavage patterns of nucleosomal DNA. Shown are hydroxyl radical cleavage patterns of DNA from a 5S ribosomal RNA gene of *Xenopus*, either free in solution (**top**), bound to a calcium phosphate precipitate (**center**), or incorporated into a nucleosome (**bottom**). Periodic cleavage patterns are seen for this DNA when it is bound to a surface, either protein or inorganic. Reprinted from Hayes et al., 1990b, with permission.

to six adenines is best), interspersed with mixed sequence, repeated once each helical turn, produce substantial curvature of DNA (Crothers et al., 1990).

The gel mobility experiment gives an idea of the overall topological shape of DNA. Indeed, different sequences (varying in the number of adenines, their spacing, and so forth) are retarded to different extents in gel electrophoresis and presumably are more or less curved. But the structural origin of bending, what is happening at the individual nucleotides, cannot be determined directly by such experiments.

Either X-ray crystallography or NMR would be the logical choice for a method to answer the question of the structural basis of DNA curvature. It might come as a surprise, then, that no clear explanation of DNA bending at the atomic level has yet emerged. The limitations of NMR in this regard are not hard to understand (Metzler et al., 1990; Wemmer, 1991). The power of NMR for structure determination of proteins lies in large part in the ability to observe close spatial contacts between amino acids distant in the sequence, via the nuclear Overhauser effect. DNA, on the other hand, being a linear polymer (even if it might be curved), does not have such long-range interactions. We can use NMR to obtain structural information on individual nucleotides and their neighbors, but the overall shape of the DNA molecule (straight or curved) is inaccessible by this technique.

X-ray crystallography has its own difficulties in addressing the structural basis of bending, although there have been a number of structures recently that may lead to a model. However, there has been a fundamental disagreement in the literature between the direction of bending of DNA observed in X-ray structures (DiGabriele & Steitz, 1993) and that derived from gel mobility experiments (Crothers et al., 1990), which still has not been resolved.

Solution probe experiments can contribute to the effort to understand DNA curvature, by providing information on the structural environments of individual nucleotides in solution. The hydroxyl radical cleavage pattern of curved DNA is particularly striking and provides clues to the structural details of bent DNA. We performed our initial experiments (Burkhoff & Tullius, 1987) on a kinetoplast DNA sequence from *Crithidia fasciculata* that has 18 adenine tracts and that was established both by gel mobility and electron microscopy to be highly curved. The cleavage pattern of this DNA molecule is remarkably wavelike in character, quite unlike the even pattern we have observed many times for mixed-sequence DNA that is not curved (Fig. 5-7). The pattern is perfectly phased with the adenine tracts: Each minimum in the cleavage pattern coincides with the adenine at the 3'-end of a run of adenines, and each maximum is found at or near the adenine at the 5'-end of each adenine tract. The adenines in between are cleaved to intermediate extents that vary smoothly depending on their place in the run of adenines. The cleavage pattern on the complementary, thymine-rich strand gave us the first structural clue. This strand also had a periodic cleavage pattern, but it was shifted relative to the pattern on the other strand. That is, the minima in the thymine-strand pattern did not occur at the thymines that were base-paired to the adenines at the 3'-ends of the adenine tracts. Instead, the minima were found two or three nucleotides in the direction of the 3'-end of the thymine-rich strand.

What does this shift tell us about the structure? The correlation in cleavage across strands gives information on which of the DNA grooves, major or minor, is involved in whatever phenomenon is being studied. This is because the closest distance between strands is not directly across a base pair. Because of the way B-form DNA winds, nucleotides that are closest to each other across the minor groove are two or three nucleotides apart in the sequence, shifted relative to one another toward the 3'-end of

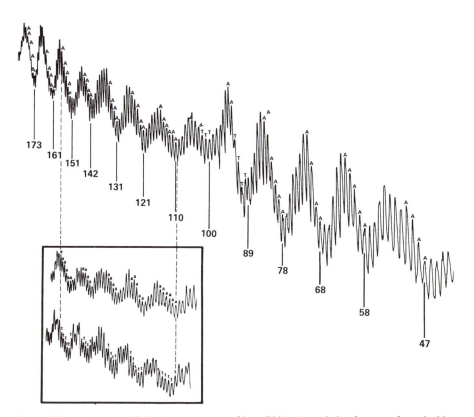

Figure 5-7. Hydroxyl radical cleavage pattern of bent DNA. A restriction fragment from the kinetoplast minicircle DNA of the trypanosomatid *Crithidia fasciculata* was cleaved by the hydroxyl radical. The densitometer scan of the cleavage pattern of the adenine-rich strand is shown, with the peaks representing the adenine nucleotides marked. (**Inset**) A comparison of the cleavage patterns of a segment of the adenine- and thymine-rich strands, showing the shift in the pattern from one strand to the other. Reprinted from Burkhoff & Tullius, 1987, with permission.

each strand (Fig. 5-8). Correspondingly, nucleotides directly across the major groove from one another are shifted in the 5′-direction relative to each other in the sequence. So, noting the offset in cleavage pattern between the strands is a powerful tool to determine which DNA groove is being monitored in a chemical probe experiment.

Recall that the offset observed for curved DNA was in the 3′-direction. This indicated to us that nucleotides in equivalent structural environments on each strand were directly across the minor groove from one another. We concluded that the hydroxyl radical cleavage experiment was revealing that the structure of the minor groove in bent DNA was smoothly changing along the adenine tracts. Furthermore, this change in structure has a polarity: At the 5′-end of a tract, an adenine is cleaved to the same extent as it would be in normal, mixed-sequence, unbent DNA, while an adenine at the 3′-end is cleaved to a much lesser extent, perhaps 60% of the usual extent of cleavage.

Our idea has been that the hydroxyl radical monitors the solvent-accessible surface area of a nucleic acid molecule. This is clearly seen to be true for tRNA and for the lambda repressor–DNA complex, as discussed above. How well does this notion work in the case of curved DNA (Price & Tullius, 1992)? Both NMR and X-ray crystallographic studies on adenine tract-containing oligonucleotides also show the $5' \rightarrow 3'$ polarity of adenine tract structure. Several crystal structures of adenine tracts have been published. A common feature of most of these structures is a decided narrowing of the minor groove in the adenine tract. The groove is normal in width at the 5'-end of the tract, and it smoothly decreases in width to a minimum at the 3'-adenine. NMR experiments detect an unusually strong cross-strand NOE signal at the 3'-end of an adenine tract, consistent with the minor groove being especially narrow at this point. Other experiments, including drug binding studies and chemical probe experiments, also point to an unusual structural feature at the 3' (and not 5')-ends of adenine tracts.

Holliday Junction. A key intermediate in a DNA recombination reaction is the four-stranded Holliday junction. This unusual DNA structure arises in the process of strand exchange between two duplex DNAs. While a four-stranded structure was postulated several years ago to occur during recombination, until recently little was known of its structure. In Chapter 7, David Lilley discusses a variety of experimental evidence that recently has led to a detailed understanding of the structure of the four-way DNA junction. The use of a chemical probe experiment to determine the symmetry of a model Holliday junction, and to limit the possible models for its structure, is shown here.

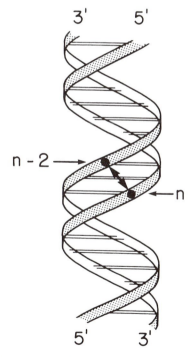

Figure 5-8. The closest distance across the minor groove of right-handed B-form DNA is not directly across a base pair, but offset in the 3'-direction by two (or three) nucleotides. Reprinted from Burkhoff & Tullius, 1987, with permission.

A serious difficulty in studying "real" Holliday junctions is that by definition they are unstable. The four-way junctions that occur in recombination reactions form because two segments of DNA duplex are identical, or nearly so, in sequence. This sequence identity allows a strand to base-pair either with its original partner strand or with a strand of the second duplex. The point of crossover, where a strand switches from one duplex to the other, clearly is not constrained. This fluidity in structure makes the Holliday junction difficult to study by biophysical or structural methods.

A breakthrough in how this problem could be approached came from the work of Nadrian Seeman, who showed how to make a stable four-way junction by lowering the symmetry of the sequences of the four strands (Kallenbach et al., 1983). The resulting immobile junction had only one possible crossover position. A depiction of the strands making up an immobile junction is shown in Fig. 5-9.

Even though the design of the immobile junction requires that the strands be base-paired as depicted, the higher-order structure of the molecule still needs to be determined. For example, are the arms of the junction extended 90° to each other as in the figure, or are they directed toward the corners of a tetrahedron? Or do two arms be-

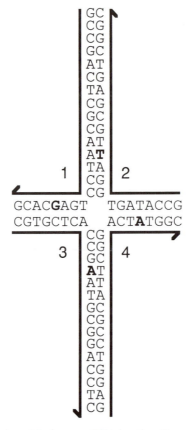

Figure 5-9. Sequence of an immobile four-way DNA junction. The numbering convention (see Fig. 5-10) is that nucleotides 8 and 9 on each strand flank the junction. The nucleotide at position 12 in each strand is indicated by boldface. Reprinted from Price & Tullius, 1992, with permission.

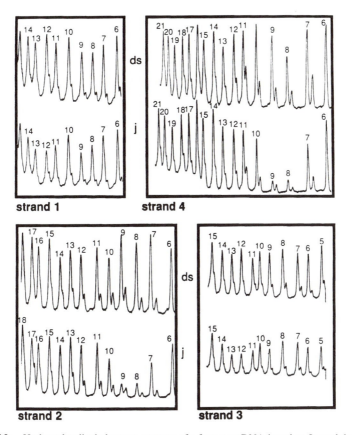

Figure 5-10. Hydroxyl radical cleavage patterns of a four-way DNA junction. In each box is shown the cleavage pattern of one strand, incorporated either in a duplex (top) or in the four-way junction (bottom). Nucleotides 8 and 9 of strands 2 and 4 (which are at the crossover of the junction) are cleaved to a very low extent; most other nucleotides are cleaved nearly the same as in the duplex control. Reprinted from Price & Tullius, 1992, with permission.

come coaxial due to base-stacking across the junction? If this latter case were true, which strands would cross over and which would remain helical?

We used the hydroxyl radical cleavage method to study the structure of an immobile four-way junction (Churchill et al., 1988). Each of the four strands was radiolabeled in turn, allowed to hybridize to the other three strands, and then subjected to reaction with the hydroxyl radical. As a control experiment, each strand also was allowed to form a duplex with its complementary strand and then cleaved. We compared the cleavage patterns of the four strands when incorporated either in a four-way junction or in a duplex. The cleavage patterns are shown in Fig. 5-10. The results were rather surprising. The cleavage patterns showed a remarkable symmetry. Cleavage was strongly inhibited at nucleotides abutting the crossover only for strands 2 and 4. In contrast, strands 1 and 3 were cleaved readily around the crossover position. The cleavage pattern thus exhibited twofold symmetry: Strands 1 and 3 had similar patterns, as did strands 2 and 4, but the two patterns were different.

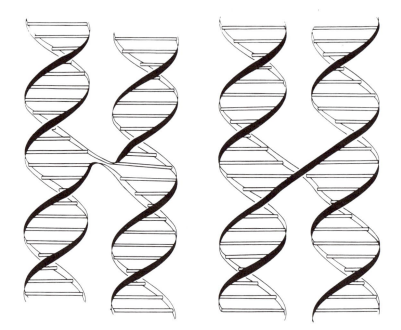

Figure 5-11. Possible twofold symmetric structures for a four-way junction. The structures differ in how the strands cross over at the junction: The left-hand structure is H-shaped, while the right-hand structure is X-shaped. Reprinted from Price & Tullius, 1992, with permission.

This symmetry argument immediately allowed us to exclude any structure with four-fold symmetry, like the square-planar structure shown in Fig. 5-9 or the related tetrahedral structure. Structures of the type depicted in Fig. 5-11 could easily be accommodated by our data, since there are two different types of strands in these structures, crossover strands and helical strands. The crossover strands would be expected to be protected from hydroxyl radical cleavage near the junction, while the helical strands would be cleaved to the same extent as if they were in a normal duplex. Our data did not allow us to decide between these two possible structures, but other experiments eventually showed that the H-shaped structure was correct for the immobile four-way junction (Cooper & Hagerman, 1987; Duckett et al., 1988).

Summary

This chapter shows how chemical probe experiments can be used to determine structural information for nucleic acid molecules in solution. While atomic-resolution structures cannot be obtained by these methods, a surprising level of detail is nonetheless accessible from these rapid and relatively simple experiments. Chemical probe methods have the advantage of "scalability." That is, they can be performed on nucleic acid molecules having a very wide range of sizes, so DNA and RNA molecules capable of biological function can be studied. Finally, a most exciting possibility is to use chemical probe experiments to determine the structure of nucleic acid molecules while they are functioning in living cells. While this might sound fanciful, there are already examples of such experiments in the literature (Palecek, 1992b), and more will surely follow.

Introduction to ¹H NMR Spectroscopy of DNA

Kenneth J. Addess and Juli Feigon

¹H nuclear magnetic resonance (¹H NMR) spectroscopy is currently the most practical experimental method available to study the structure of nucleic acids in solution. Although nucleic acids contain only four different bases, the ring current shifts due to stacking of the aromatic rings in the generally helical structures of DNA and RNA usually result in reasonable chemical shift dispersion for the protons on the bases and sugars. Early studies focused on observation of the hydrogen-bonded imino protons of the base pairs and provided evidence for base-pairing interactions in transfer RNA (tRNA) and synthetic RNA polymers (Kearns et al., 1971). With the development of convenient methods for the chemical synthesis of DNA oligonucleotides, it became possible to obtain well-defined samples of DNA suitable for detailed structural studies. This coincided with the development of two-dimensional NMR methods for the study of protein structures (reviewed in Wüthrich, 1986). Two-dimensional NMR spectra of double-stranded DNA were first published in 1982 (Feigon et al., 1982), and methods for the sequential assignment of B-DNA oligonucleotides were published soon afterward (Feigon et al., 1983; Hare et al., 1983; Scheek et al., 1983). Although the methods used to make ¹H NMR assignments of double-helical DNA are straightforward, determining three-dimensional structures of DNA from the data is more difficult than for proteins. This is due to two major factors: (1) The proton density of nucleic acids is lower than that of proteins, leading to fewer possible nuclear Overhauser effect (NOE) constraints per residue, and (2) because of the generally helical structure of DNA, there are no long-range NOE constraints. Both factors provide critical information in defining secondary and tertiary structural features of proteins studied by NMR. Partially because of this, structure determination of DNA by NMR has lagged significantly behind proteins. It is only in the last year or two that methods have improved to the point where at least in some cases reasonably well-determined structures of DNA can be calculated from constraints derived from NMR data.

In this chapter, the basic approach to assigning the ¹H NMR resonances of DNA oligonucleotides is presented. Methods for using the data from two-dimensional NMR data for qualitative and quantitative structure determination are discussed. As a practical example of the application of these methods, NMR studies on the interaction of [*N*-MeCys³, *N*-MeCys⁷]TANDEM (CysMeTANDEM), a bis-intercalating antibiotic, with DNA oligonucleotides are reviewed.

Assignment Strategies

Assignments of B-DNA Resonances. In proton NMR spectroscopy of proteins, resonances are assigned by the sequential assignment method first proposed by Wüthrich
and coworkers (reviewed in Wüthrich, 1986). This sequential assignment method relies on correlated spectroscopy (COSY, TOCSY, etc.) to identify amino acid spin systems and NOE spectroscopy (NOESY) to correlate neighboring amino acids along the
peptide chain and is model-independent. A similar approach was developed for sequential assignment of nonexchangeable base and sugar protons in DNA, except that
correlation of base to sugar within the same nucleotide as well as in the neighboring
nucleotide relies solely on NOESY spectroscopy (Chazin et al., 1986; Feigon et al.,
1983; Haasnoot et al., 1983; Hare et al., 1983; Reid, 1987; Scheek et al., 1983, 1984;
reviewed in Wüthrich, 1986) and requires an A- or B-DNA-type helix. Figure 6-1 illustrates the base pairs and deoxyribose moiety of DNA including the numbering
scheme for the protons. As a consequence of the structure of DNA (see Chapter 1),
there should be a series of short interproton distances that give rise to NOEs which

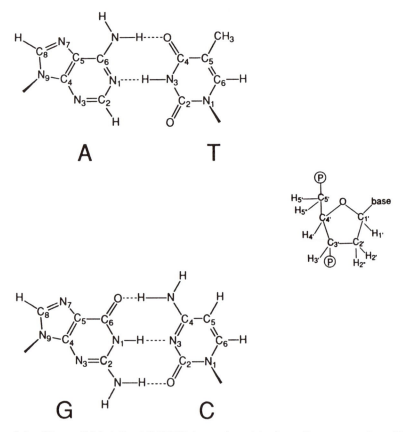

Figure 6-1. Watson–Crick A·T and G·C DNA base pairs and the deoxyribose sugar moiety of DNA.
Reprinted from Feigon et al., 1992, with permission.

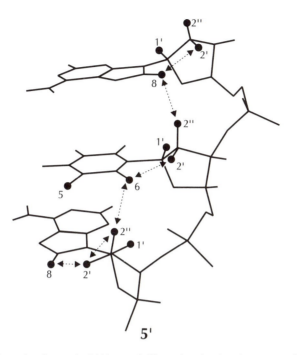

Figure 6-2. Schematic of part of a DNA strand, illustrating the short interproton distances in DNA that afford NOEs in the NOESY spectra. The H1′, H2′, and H2″ protons are all close enough to the base H6 and H8 protons to produce NOEs; the shortest interproton distances are denoted by dashed lines. Reprinted from Feigon, 1993, with permission.

permit the sequential connectivities to be determined. Thus, in the NOESY spectrum of a DNA duplex that has an A- or B-type helical structure, it is possible to do sequential assignments because the H1′, H2′, and H2″ protons of the sugar ring of one residue have a priori NOE cross-peaks to the aromatic H8 or H6 proton of the same (n) nucleotide and of the 3′($n + 1$) nucleotide. These sequential connectivities are illustrated in Fig. 6-2. The spin system assignments of the sugar protons can be obtained from correlated spectra (e.g., phase-sensitive COSY and TOCSY), some details of which are briefly described below.

As a consequence of having all *anti*-glycosidic torsion angles, the sequential assignments of B-DNA can reliably be expected to involve strong H6,H8–H2′,H2″ NOE cross-peaks, as well as weaker H6,H8–H1′ NOE cross-peaks (Fig. 6-2). Thus, at least two different regions of the NOESY spectrum can be used to make sequential assignments. Although the shortest intranucleotide base–sugar connectivity is H6,H8–H2′ and the shortest intermolecular one is H6,H8–H2″, sequential assignments are usually made most readily by means of the H6,H8–H1′ cross-peaks. This is because sequential connectivities are much easier to trace in the base–H1′ region of the spectrum, since it is more well-resolved than the base–H2′,H2″ region and the sequential pathway involves only one (base–H1′) rather than two (base–H2′ and base–H2′) sets of connectivities.

The assignment process is facilitated by first identifying as many of the aromatic resonances as possible to proton types particular to A, C, G, or T. This gives check points for the sequential assignment. For pyrimidines, the cytidine H5 and H6 and the thymidine methyl and H6 resonances have characteristic COSY cross-peaks in unique parts of the two-dimensional spectrum and can thus be distinguished from the H8 resonances of adenosine and guanosine. For the purines, AH8 and GH8 resonate at the low-field end of the aromatic spectrum ($\sim$7–8 ppm), with AH8 generally at lower field than GH8s. The adenosine H2 resonances can be identified by the strong imino-AH2 cross-peak in Watson–Crick base pairs, observable in the NOESY spectra of H_2O solutions of DNA. Assignment of the AH2s is important, because it precludes the misassignment of occasional AH2–H1$'$ cross-peaks as sequential connectivities. In principle, sequential assignment of resonances can start anywhere in the structure, and proceed in either the 3$'$- or 5$'$-direction. Thus, one does not have to be able to identify the nucleosides at the ends of the molecule in order to begin the assignment process.

The assignment of exchangeable amino and imino protons is made from NOESY spectra recorded in H_2O (Boelens et al., 1985; Rajagopal et al., 1988; Sklenár & Feigon, 1990a). Sequential imino–imino connectivities between adjacent base pairs permit the assignment of imino proton resonances. The strong T imino-AH2 cross-peak of Watson–Crick base pairs is then used to assign the AH2 resonance. The amino resonances are also identifiable via cross-peaks with the imino protons. Because they have different exchange rates and chemical shifts, the amino resonances from the individual types of bases can be distinguished. The cytidine amino protons exchange slowly on the NMR time scale; accordingly, they give sharp cross-peaks that are readily identified. The adenosine amino protons are intermediate to slow in their exchange rates; the resonances can generally be assigned. The exchange of guanosine amino protons is usually fast; the resulting resonances are often too broad to identify.

Assignments of Non-B-DNA Structures and Larger Oligonucleotides

Some of the most interesting DNA structures—for example, those involving non-B-DNA structures or complexes of DNA with low molecular weight ligands such as drugs—cannot be assigned using the usual NOESY sequential assignment strategies. This is a consequence of changes in the distances of intra- and internucleotide base-sugar connectivities, such as those caused by the presence of bases that are in the *syn* conformation. A good example of this is Z-DNA, in which the alternate bases are in the *syn* conformation, resulting in disruption of the sequential connectivities between the *anti* base and the sugar protons on the *syn* nucleotide to its 3$'$-side (Feigon et al., 1984b). For larger oligonucleotides, the greater complexity of the spectra often results in overlap of resonances, which may preclude the assignment of sequential connectivities. Accordingly, other strategies must be used on a case-by-case basis, to identify the types of base protons and then to permit them to be used in determination of sequential connectivities. Some of these strategies are outlined below.

Variation of Experimental Conditions. Alterations in experimental conditions represent a very simple, and often quite effective, method for addressing spectral ambiguities due to overlap of resonances. For example, variations in salt conditions can be important, because the stability (and sometimes structure) of DNA is a strong function of cation concentration. Divalent cations such as Mg^{2+} can stabilize DNA structure, making it possible to take spectra at higher temperature with a corresponding increase in resolution. The simple procedure of acquiring two-dimensional spectra at a number of different temperatures can often resolve ambiguities due to overlap of resonances. Likewise, the acquisition of NOESY spectra at different mixing times can be informative. Long mixing times facilitate the identification of small cross-peaks due to base–H1′ interactions, while short mixing times improve resolution in the base–H2′,H2″ region, thus allowing interproton distances to be determined more accurately. A slightly different approach to variation of experimental conditions involves the study of similar molecules whose structures differ in some small way likely to resolve ambiguities—for example, two oligonucleotides differing by a single base at one position or substitution of inosine for guanosine.

Base-Pairing Scheme. Identification and assignment of non-B-DNA structures usually begin by analysis of one-dimensional and NOESY spectra of the exchangeable proton resonances obtained on samples in H_2O. Observation of imino proton resonances between 10 and 16 ppm is generally indicative of base-pair formation, although some unpaired iminos may be observed at lower pH, usually around 10 ppm. The chemical shifts and number of imino proton resonances will be determined by the DNA structure formed. For example, DNA triplex formation results in the formation of additional base pairs, and additional imino resonances due to these base pairs are observed (Rajagopal & Feigon, 1989; Sklenár & Feigon, 1990a). The protonation of the C nucleotides in the third strand also results in low-field shifted imino and amino resonances, relative to unprotonated C iminos and aminos in Watson–Crick G·C base pairs (Rajagopal & Feigon, 1989; Sklenár & Feigon, 1990a). Intercalation of the aromatic chromophore of drugs into DNA results in substantial upfield shifts of imino resonances at positions in the DNA adjacent to the position at which intercalation has occurred (Feigon et al., 1984a).

As discussed above, the sequential assignment of imino resonances in canonical B-DNA can be made using imino–imino NOEs, and the amino cross-peaks to the imino protons can then be used to identify specific amino resonances. The strong T-imino–AH2 cross-peak, which is used to assign AH2 resonances in B-DNA, can be misleading where the actual DNA structure is non-B-DNA, or simply of unknown conformation. For example, there is a strong cross-peak from the T imino to AH8 in Hoogsteen A·T base pairs (Gilbert et al., 1989), so that differentiating between AH2 and AH8 is essential for establishing whether one is dealing with an A·T base pair of the Watson–Crick or Hoogsteen type. One method to accomplish this involves deuteration of the H atom at C8.

Purine Deuteration at C8. The protons at the C8 position of purines are susceptible to exchange at elevated temperature and pH. This can be exploited for assignment purposes, by exchanging the protons for deuterons by heating samples in D_2O for an appropriate period of time. Complete exchange of purine H8s occurs in 2–3 days at pH 8 and 60°C. The guanosine H8s exchange more rapidly than adenosine H8s, which can potentially be used to advantage as well. By comparison of the cross-peaks present in the NOESY spectra (in D_2O and H_2O) before and after deuterium exchange of the purine nucleotides, it is often possible to identify the exchangeable purine resonances unambiguously (Schweizer et al., 1964). This would then allow differentiation between T-imino cross-peaks with AH2 versus AH8, as discussed above. Clearly, because deuteration results in simplification of the proton spectrum, it may also be useful in resolving ambiguities due to overlap of cross-peaks.

Chemical Exchange. The differences in energy between B-DNA and non-B-DNA structures are frequently small enough so that both can exist in significant concentrations under a given set of experimental conditions. Where these conditions can be defined and where the two species are interchanging slowly on the NMR time scale, the exchange cross-peaks (Jeener et al., 1979) may be useful in assigning resonances in the non-B-DNA structure. This presumes, of course, that the resonances in the B-DNA structure have been assigned. This technique has been used to assign the resonances in the Z-form of the self-complementary DNA hexanucleotide m^{5}CGm^{5}CGm^{5}CG (Feigon et al., 1984b). Drug–DNA and drug–protein complexes are also potentially interpretable by application of this strategy (see, e.g., Klevit et al., 1986; Pelton & Wemmer, 1989, 1990). An obvious potential problem with this technique is distinguishing between NOE and exchange cross-peaks. However, this is straightforward in ROESY experiments (Bothner-By et al., 1984) in which exchange cross-peaks are positive and NOE cross-peaks are negative.

Characterization of CH6 Resonances and Cross-Peaks by Selective Excitation. While their strong COSY cross-peaks generally permits the identification of the CH5 and CH6 resonances, spectral overlap with other base–sugar proton cross-peaks can obscure the CH6–sugar cross-peaks. By the use of the HOENOE experiment (Sklenář & Feigon, 1990b), it is possible to observe the CH6 cross-peaks to other resonances selectively. The HOENOE experiment involves transfer of magnetization from CH5 to CH6 resonances via selective excitation of the region containing the CH5 resonances; this is done before the t_1 evolution period of a regular two-dimensional NOE experiment.

Three-Dimensional NMR Spectra. Three-dimensional NMR experiments can be used to facilitate assignments by resolving overlapping cross-peaks. This is especially the case for H5' and H5″ resonances because the region containing these resonances in a two-dimensional NOESY experiment is severely overlapped. In some cases, it is possible to assign the H5' and H5″ resonances on the basis of NOE connectivities to the H3' and H4', which are easier to assign, but it is rare to be able to assign all the H5' and H5″ resonances in this manner. To overcome this challenge, it is possible to spread the NOE information in a third proton dimension in either a NOESY–NOESY (Boelens et al., 1989) or TOCSY–NOESY experiment (Mooren et al., 1991; Piotto & Gorenstein, 1991). In the case of the latter experiment for example, the NOE transfer from the H5' to the H3' during the NOESY mixing period of the experiment contains an H1' frequency label as a result of isotropic mixing between the H1' and H3' during the spin lock period of the TOCSY portion of the experiment. Therefore, in a plane that slices through the H3' resonance (located in the second dimension) of the three-dimensional spectrum, one can observe a resolved NOE between the H3' and the H5' resonance (located in the third dimension) at the chemical shift of the H1' resonance (located in the first dimension).

Future Directions. A major problem in determining three-dimensional structures of unusual and larger DNA oligonucleotides is the inability to assign the DNA without relying on some structural assumptions in making sequential assignments. Although theoretically assignments can be made from proton to phosphorus along the phospho-diester backbone (Pardi et al., 1983), in practice this method often fails due to the small chemical shift dispersion of the phosphate and the 3′,4′,5′,5″ sugar resonances as well as the low sensitivity of these experiments (reviewed in Gorenstein, 1992; see also Feigon, 1993; Kellogg et al., 1992). Clearly, NMR spectroscopy of DNA could bene-fit from the current advances in macromolecular NMR of proteins that rely on three- and four-dimensional double and triple resonance experiments on uniformly ^{15}N,^{13}C-labeled proteins. For example, overlapping cross-peaks in the NOESY spectra of some DNA molecules, particularly in the base–H2′,H2″ region, pose too much difficulty to allow for complete sequential assignments. Many of these difficulties could be sur-mounted by spreading the unresolved cross-peaks of a two-dimensional NOESY over a third ^{13}C dimension in a three-dimensional (^{1}H, ^{13}C, ^{1}H) HMQC-NOESY. Thus far, these methods have been applied to DNA in only a very limited way, due to the un-availability of labeled DNA samples. However, several laboratories are working on ways to obtain fully ^{15}N,^{13}C-labeled DNA oligonucleotides, and it is certain that once samples are obtained, these methods will begin to be fully exploited for studying DNA in solution. It is worth noting that methods for obtaining fully labeled RNA oligonu-cleotides have already been worked out (Batey et al., 1992; Nikonowicz, 1992), and NMR methods development on these labeled samples has advanced rapidly in the past year (reviewed in Dieckmann & Feigon, 1994).

DNA Oligonucleotide Structure Determination

Qualitative Analysis of DNA Structure. The assignment procedures discussed above provide secondary structure information about DNA oligonucleotides, at least at a qual-itative level. One- and two-dimensional spectra obtained in water afford information concerning base-pairing schemes and stability. In general, the helical structure (A-DNA, B-DNA, or Z-DNA) can be deduced from the NOESY and COSY spectra (Wüthrich, 1986). Also accessible from such spectra is information concerning the DNA sugar moieties, such as glycosidic torsion angle and sugar pucker. The latter, which can be obtained from phase-sensitive correlated spectroscopy through an analy-sis of derived coupling constants and also less directly from NOE spectra, has been described in detail (van Wijk et al., 1992).

NMR experiments can also be used to determine the formation of more complex DNA tertiary structures, including such features as triplexes and quadruplexes, hair-pins, and bulges. As noted above in a limited fashion and discussed in the sections be-low in detail, NMR spectroscopy is also useful for determining the nature of drug–DNA interactions. In favorable cases where there is tight, sequence-specific binding of a drug to DNA, a high-resolution three-dimensional structure can be determined. The quali-tative and quantitative information that can be derived by two-dimensional ^{1}H NMR for drug–DNA complexes is discussed in detail below for the specific example of bind-ing of the bisintercalator CysMeTANDEM to DNA oligonucleotides. Thus, much use-ful information can be deduced using relative NOE cross-peak data and coupling con-stants. By combining the NOE data and information concerning dihedral bond angle with distance geometry or molecular dynamics approaches, a more sophisticated struc-ture analysis can be carried out.

Determination of Three-Dimensional Structures. As mentioned in the introduction, the determination of the molecular structures of DNAs with the precision of atomic resolution has some unique problems not encountered in the determination of protein structure. Experimental NOEs alone, if sufficient in number, are enough to define all the atomic positions in a given protein macromolecule while the number of experimental ^{1}H NOEs alone, no matter how abundant, can define the atomic positions of the base and deoxyribose portions of any given oligonucleotide with only marginal precision and cannot define the atomic positions of the phosphodiester backbone. Early attempts to calculate three-dimensional structures of B-DNA oligonucleotides were not very successful; the structures were often underwound and had unusual groove widths and bends that have not passed the test of time. In general, the NOE constraints used were insufficient to give a structure that was any better than a model structure based on a B-DNA helix, and deviations from this could easily be overinterpreted to be meaningful.

However, the situation has improved considerably in the last few years, especially for folded structures such as intramolecular triplexes and drug–DNA complexes, which provide additional constraints for the structure calculations. Improvements in the analysis of NOE data, addition of dihedral bond angle constraints of the deoxyribose rings derived from analysis of ^{3}J(H,H) coupling constants, and improvements in computational analysis of the data have all contributed to this progress. Furthermore, experimental information provided by ^{3}J(H,P) couplings, which can be measured from (^{1}H, ^{31}P) heteroCOSY, is beginning to be used in determining the torsional angles of the phosphodiester backbone. However, one must still be cautious when evaluating reported structures; as with crystallography, their accuracy and precision depends on the quality of the input data, the quality of the data analysis, and the method of structure calculation.

In general, structures are refined by energy minimization from initial starting structures. These starting structures can be models, such as an A- or B-DNA helix (when a right-handed helical structure is the expected outcome), or they can be generated by metric matrix distance geometry. In distance geometry calculations, a matrix consisting of upper and lower bounds of distances, both inherent to the primary structure of the molecule such as covalent bonds lengths and derived from experimental NOE crosspeaks, is used to generate a set of atomic coordinates representing the starting structure of the complex. The latter is the preferred method, because no model bias goes into the calculation. The metric matrix distance geometry starting structures can be generated with an incomplete set of distance constraints if desired.

Refinement of the model or distance-geometry-generated starting structures should include as many constraints as possible, including distance constraints from both exchangeable and nonexchangeable protons and dihedral bond angle constraints from analysis of coupling constants. The distance constraints are obtained by integration of the NOE cross-peaks, since to a first approximation the intensity of the NOE is proportional to the inverse sixth power of the interproton distance. This holds for a single correlation time and assumes interactions between the two protons being considered only (two-spin approximation). It is usually necessary, although not the preferable approach, to explicitly include hydrogen bond constraints from the base pairs. Once the energy refined structures are obtained, they are subject to a final NOE-based refinement step. This involves calculation of the full relaxation matrix or direct NOE refinement, using data from at least two different NOESY mixing times. The reason for doing this is that distances derived from cross-peak intensities in a single NOESY spectrum are subject to error due to spin diffusion. This is because the intensity of an NOE between a proton pair in a DNA oligomer results both from the interproton distance between the proton pair and from indirect dipolar cross-relaxation through other protons that are in the vicinity of the proton pair. This indirect transfer (spin diffusion) and its effects can be significant in NOESY spectrum acquired with longer mixing times. This last step usually does not result in large changes in the atom positions but does significantly improve the R-factor.

Analysis of the Binding of CysMeTANDEM to DNA Oligonucleotides Using ¹H NMR Spectroscopy

[N-MeCys$_3$, N-MeCys$_7$]TANDEM (CysMeTANDEM) is a cyclic octadepsipeptide that exhibits antibiotic activity. Structurally, the molecule has a twofold axis of symmetry and contains a disulfide bridge between the two cysteine residues, which contain N-methyl substituents; there are two quinoxaline rings attached to D-serine residues (Fig. 6-3). CysMeTANDEM is structurally related to the naturally occurring antibiotic triostin A, which contains N-methylated cysteines and valines, and the synthetic TANDEM, which contains no N-methylated residues. In addition to these antibiotics, which are called *triostins*, there is a second family of quinoxaline antibiotics called *quinomycins*, whose members include echinomycin (Katagiri et al., 1975; Waring, 1979). These quinoxaline antibiotics also exhibit potent antitumor activity (Katagiri et al., 1975).

The binding of triostin A and echinomycin to DNA have been studied by footprinting with MPE-Fe(II) (van Dyke & Dervan, 1984) and with DNase I (Low et al., 1984a,b). Both techniques indicated that the preferred binding sites for the compounds were localized around CpG. X-ray crystallographic analyses and NMR studies have now defined the interactions of triostin A and echinomycin with DNA at high resolution (Gao & Patel, 1988; Gilbert et al., 1989; Gilbert & Feigon, 1991; Ughetto et al., 1985; Wang et al., 1984, 1986). Binding involves bis-intercalation of the quinoxaline

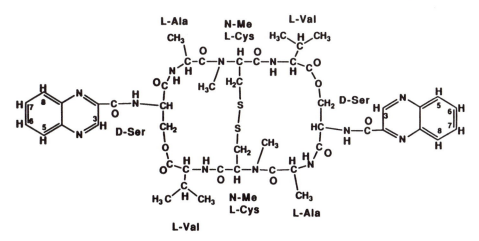

Figure 6-3. Chemical structure of the quinoxaline antibiotic CysMeTANDEM.

moieties on either side of a CpG structural element; the peptide moiety is oriented in the DNA minor groove.

Though a detailed picture had emerged that provided insight into the nature of the binding of triostin A and echinomycin to DNA, much less was known about the antibiotics TANDEM and CysMeTANDEM, which preferentially bind to A·T-rich DNA (Low et al., 1984a,b, 1986). Based on the crystal structure of TANDEM, Viswamitra et al. (1981) proposed that TANDEM should be able to bis-intercalate at ApT sequences. However, when the interaction between CysMeTANDEM and the DNA decamer (CCCGATCGGG)$_2$ was studied by NMR, no evidence for intercalation was found (Powers et al., 1989). Futhermore, restriction enzyme inhibition studies of Low et al. (1986) provided evidence that CysMeTANDEM binds preferentially to TpA rather than ApT sequences.

As part of a program to understand the sequence-specific binding of the quinoxaline antibiotics to DNA, we investigated the binding of CysMeTANDEM to the DNA octanucleotides [d(GGATATCC)]$_2$ and [d(GGTTAACC)]$_2$, both of which contain single TpA sites, and [d(GGAATTCC)]$_2$, which contains a single ApT site. As discussed in detail below, we found that CysMeTANDEM binds specifically to TpA as its recognition element in DNA (Fig. 6-4). This conclusion was based on the observation that CysMeTANDEM bound only weakly to [d(GGA<u>AT</u>TCC)$_2$] and gave no evidence of binding by intercalation, consistent with the findings of Powers et al. (1989) for (CC-CG<u>AT</u>CGGG)$_2$. In contrast, CysMeTANDEM bound strongly to the octamers [d(GGA<u>TA</u>TCC)$_2$] and [d(GGT<u>TA</u>ACC)$_2$]. The NMR spectra gave clear indications of bis-intercalative binding, as discussed below. These findings alone established convincingly that CysMeTANDEM recognizes TpA, rather than ApT, sequences in B-DNA (Addess et al., 1992).

Assaying Complex Formation Between CysMeTANDEM and DNA Using ^{1}*H NMR Spectroscopy.* Figure 6-5 shows the one-dimensional ^{1}H-NMR spectra of [d(GGATATCC)]$_2$, recorded in D$_2$O, both in the absence of drug and in the presence of one equivalent of CysMeTANDEM per DNA duplex. In the spectrum of the free [d(GGATATCC)]$_2$, each of the different proton types resonate at frequency ranges characteristic of B-DNA. For example, the aromatic H6, H8, and H2 protons resonate in the frequency range from 8.5 to 6.5 ppm; the cytidine H5 and sugar H1' protons resonate in the 6.5- to 5-ppm range; the sugar H3' protons resonate in the region from 5 to 4.5 ppm; the H4, H5', and H5" protons resonate in the range from 4.5 to 3 ppm; the H2', H2", and thymidine methyl protons resonate at the highest field, between 3.0 and 0 ppm. As shown, the addition of CysMeTANDEM results in the appearance of new resonances corresponding to the bound drug itself, as well as the shift of DNA resonances due to complex formation. The changes in the spectra include a downfield

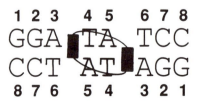

Figure 6-4. Orientation of CysMeTANDEM on the DNA duplex [d(GGATATCC)]$_2$.

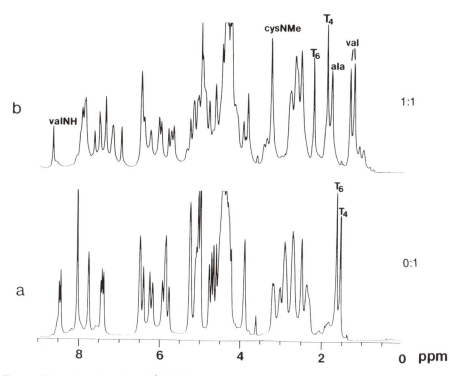

Figure 6-5. One-dimensional ^{1}H-NMR spectra for a 1:1 complex of CysMeTANDEM and [d(GGATATCC)]$_2$ (**b**) and [d(GGATATCC)]$_2$ alone (**a**). The spectra were acquired in D$_2$O at 25°C, using 2 mM DNA duplex and 150 mM NaCl at pH 6.5.

shift in the thymidine methyl resonances, which are labeled in the figure. Also labeled in Fig. 6-5 are the positions of the alanine and valine methyl groups, the cysteine *N*-methyl group, and the valine amide residues of CysMeTANDEM. In the absence of added CysMeTANDEM, the two strands of the DNA duplex are symmetrical, and only a single resonance is observed for each type of proton. Likewise, no doubling of the DNA or CysMeTANDEM resonances is observed when the DNA was completely saturated with drug; this indicates that both retain their twofold symmetry in the formed complex.

Titration of the octanucleotide DNA duplexes [d(GGATATCC)]$_2$ and [d(GGT-TAACC)]$_2$ with less than saturating amounts of CysMeTANDEM results in the appearance of two sets of resonances, one of which corresponds to that of free DNA and the other to the 1:1 CysMeTANDEM–DNA complex. This observation indicates that both CysMeTANDEM–DNA complexes are in slow exchange with the octamer duplexes on the NMR time scale.

Shown in Fig. 6-6 are the imino proton spectra in H$_2$O of [d(GGATATCC)]$_2$ alone and as part of a 1:1 complex with CysMeTANDEM. As expected, the spectrum of the free DNA oligonucleotide (panel a) contains four imino resonances, two each for G·C and A·T, corresponding to the four sets of two equivalent base pairs. The T-imino protons resonate in a range of 13.5–14.5 ppm, which is at lower field than the G-imino protons, which resonate in the range of 12.5–13.5 ppm. This is again characteristic of

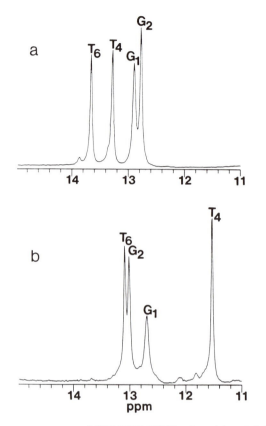

Figure 6-6. Imino proton spectra of [d(GGATATCC)]$_2$ alone (**a**) and 1:1 CysMeTANDEM–[d(GGATATCC)]$_2$ (**b**) at 10°C in 9:1 H$_2$O–D$_2$O. The samples were the same as those in Fig. 6-5.

the free B-DNA octamer. For the CysMeTANDEM–[d(GGATATCC)]$_2$ complex, the spectra indicate large upfield shifts of the imino resonances, consistent with intercalative binding of the drug to the DNA substrate (Feigon et al., 1984a).

Assignment of Nonexchangeable DNA Resonances in CysMeTANDEM–[d(GGATATCC)]$_2$. The nonexchangeable DNA resonances in the complex formed between CysMeTANDEM and d[d(GGATATCC)]$_2$ were made based on the NOESY (Fig. 6-7) and COSY (Fig. 6-8), as well as the HOHAHA spectra recorded on the sample in D$_2$O. Following standard methods (Chazin et al., 1986; Feigon et al., 1983; Hare et al., 1983; Scheek et al., 1983), the COSY and HOHAHA spectra were used to identify the deoxyribose, cytidine H5–H6, and thymidine H6–Me spin systems. Furthermore, previous work on the ^{1}H assignments of several echinomycin–DNA complexes provided confirmation of assignments of analogous resonances in the CysMeTANDEM–[d(GGATATCC)]$_2$ complex (Gao & Patel, 1988; Gilbert et al., 1989; Gilbert & Feigon, 1991). A D$_2$O solution of the complex gave the well-resolved NOESY spectrum shown in Fig. 6-7, from which virtually all of the resonances were assigned.

The inset in the NOESY spectrum, shown in expanded form in panel b, contains

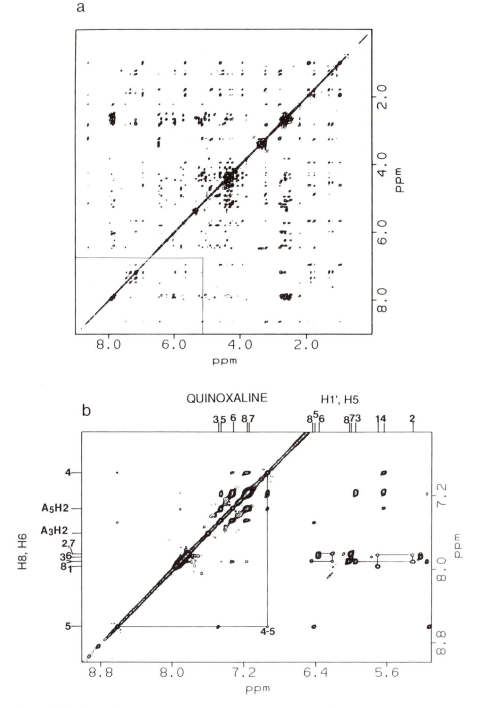

a

QUINOXALINE   H1', H5

b

Figure 6-7. NOESY spectrum of the CysMeTANDEM–[d(GGATATCC)]₂ complex in D₂O at 25°C (**a**). The boxed region is expanded in the panel **b**. The base–H1′ sequential connectivities are indicated by solid lines, as is the cross-peak between T_4H6 and A_5H8. The sample conditions were the same as in Fig. 6-5. Reprinted from Addess et al., 1992, with permission.

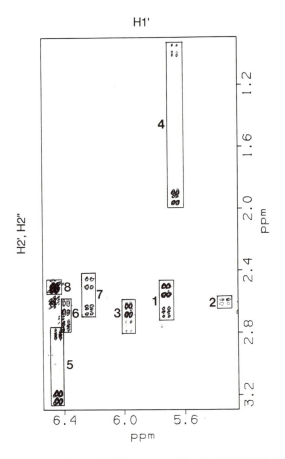

Figure 6-8. Expanded region of the P.COSY spectrum of the CysMeTANDEM–[d(GGATATCC)]₂ complex, illustrating the H1′–H2′ and H1′–H2″ cross-peaks. The cross-peaks for individual sugars are shown in boxes with labels. Adapted from Addess et al., 1992.

the aromatic Hs and the base–H1′ region of the 1:1 drug–DNA complex. In the NOESY spectrum of the unbound octamer, the sequential assignments of the base–H1′ and base–H2′,H2″ cross-peaks from the 5′- to the 3′-end of the DNA were made quite easily using the conventional sequential walk. Because of intercalation of the quinoxaline moieties, there is an interruption in the sequential assignments of the base and H1′, H2′, H2″ protons of the nucleotides flanking the intercalated quinoxaline rings (Gilbert & Feigon, 1991). Sequential connectivities between base and H1′ resonances are observable for nucleotides 1–3 and 6–8 (Fig. 6-7) but not for the central A·T bases, which are sandwiched by intercalation of the quinoxaline rings. However, a new base–base cross-peak (T₄H6-A₅H8), which is not observed the spectrum of the octamer alone, is observed in the spectrum of the complex. This cross-peak, together with the sequential connectivity data, permitted the assignment of all H8, H6, and H1′ DNA resonances. The strong T-imino–AH2 NOE cross-peaks associated with A·T base pairs, discussed above, also permitted the identification of the resonances due to A₃H2 and A₅H2 (see below). Connectivities observed between the base and H2′,H2″ resonances served to confirm the sequential base-H1′ assignments. The intensities of the NOE cross-peaks between H1′ and the H2′ and H2″ resonances permitted the latter to be distinguished from each other, since H1′ is always closer to H2″ than to H2′ (cf. Fig. 6-2).

Interestingly, a few protons resonances in the drug–DNA complexes display unusual chemical shifts. This includes the T_4H2′ and H2″ resonances, which both resonate at higher field ($\sim$1 ppm) relative to the positions of the same protons in the uncomplexed DNA. In previous work on several echinomycin–DNA complexes, the CH2′ and CH2″ resonances of the CpG binding site were also shown to resonate at higher field relative to the resonances of the same protons in the free DNA. This helped us to identify the T_4H2′ and H2″ resonances in the CysMeTANDEM–DNA complex. Also, the A_5H2′ proton resonates at lower field than the A_5H2″, which is reversed from the usual order in DNA (and from the order in free [d(GGATATCC)]$_2$). The relative intensities of the base–H1′ (weak) and base–H2′,H2″ (strong) cross-peaks indicates that all of the nucleotides are in the usual *anti* conformation.

Assignment of Exchangeable DNA Resonances in the Drug–DNA Complex. As noted previously, the two A·T-imino resonances gave strong cross-peaks to adenosine H2, which permitted their identification; the sequence-specific assignment was made from cross-peaks (via spin diffusion) to the corresponding thymidine Me resonance. Specifically, the imino resonance at highest field had a cross-peak to T_4Me, permitting its assignment as due to the T_4-imino proton. Relative to the T_4-imino proton in the free oligonucleotide, this H was shifted upfield by approximately 2 ppm (Fig. 6-6). Furthermore, this T4-imino resonance resonated at higher field than the two G-imino resonances in the complex, in contrast to what was observed in the imino proton spectrum of the free DNA. This finding was entirely consistent with intercalative binding of the quinoxaline moiety in proximity to the A_5·T_4 base pair and thus reinforces the conclusion reached from interruption of sequential connectivities. In the present case, it was also possible to assign the G-imino resonances. One of these had a cross-peak to A_3H2, which permitted its assignment as the G_2-imino H; the (terminal) G_1-imino exhibited a large exchange cross-peak with water. The assignment of the G-imino resonances permitted identification of the C-amino Hs via their cross-peaks to the G-imino resonances.

Assignment of Resonances Due to CysMeTANDEM. As noted, only one set of resonances was observed for bound CysMeTANDEM, indicating that the symmetry of the molecule was not affected by DNA binding. The COSY and HOHAHA spectra, recorded in D$_2$O, were used to identify the four-peptide spin systems. Because of slow exchange, the valine amide H could be observed in spectra taken in D$_2$O. As anticipated due to the presence of the modified amino acids *N*-methylcysteine and D-serine, the "standard" sequential connectivities associated with the peptide ring were absent. Those observed in the NOESY spectra, taken in H$_2$O, included alanine NH to D-serine NH and valine NH to *N*-methylcysteine C$_\alpha$H. The HOHAHA spectrum, taken in H$_2$O, was used to confirm the assignments of the valine NH and alanine NH.

The COSY spectra, taken in D$_2$O, permitted the assignment of quinoxaline resonances H5–H8 to be made. Interestingly, it was found that quinoxaline H3 had a cross-peak to A_5H8, the latter of which had been noted above to have a cross-peak to TH6 in the NOESY spectrum.

Identification of Sequence-Specific Binding and Structural Features of the CysMeTANDEM-[d(GGATATCC)]$_2$ Complex by Analysis of NOE Cross-Peaks. As described earlier, the NOE data provide considerable information regarding the mode of binding of CysMeTANDEM to the DNA oligonucleotide [d(GGATATCC)]$_2$. It is clear that binding occurs in proximity to the TpA sequence within the oligomer, involves bis-intercalation of the quinoxaline moieties, and results in the formation of a single, symmetrical complex. The binding site in the substrate investigated in this study is the sequence ATAT, since intercalative binding occurs between the TpA sequence and the nucleotides flanking this two-base site. As noted previously, evidence consistent with intercalation includes both the interruption in the base–H1$'$ sequential connectivities at the putative intercalation sites and the upfield shift of the T-imino protons, relative to their position in the uncomplexed DNA.

As illustrated Fig. 6-7, there were NOE cross-peaks between quinoxaline protons H5–H7, and A$_3$H8 and T$_4$H6. There were also cross-peaks between the same quinoxaline Hs and T$_4$Me and also between quinoxaline H7, H8, and A$_3$H2. It is clear from these cross-peaks that the quinoxaline rings must be stacked between the base moieties of A$_3$ and T$_4$ (Fig. 6-4). A further NOE cross-peak, between quinoxaline H3 and A$_5$H8, establishes the orientation of the quinoxaline ring, whose long axis must necessarily be parallel to the long axis of the A·T base pair.

Many of the resonance assignments and pattern of NOE cross-peaks in the spectrum of the CysMeTANDEM-[d(GGATATCC)]]$_2$ complex were similar to those in the spectra of several echinomycin–DNA complexes. For example, it was shown for echinomycin–DNA complexes that there are NOEs between methyl groups in the side chains of alanine and valine, as well as protons in the minor groove of the DNA (Gao & Patel, 1988; Gilbert & Feigon, 1991). Analogously, for the CysMeTANDEM–[d(GGATATCC)]$_2$ complex, there were strong NOEs between the alanine and valine methyl groups and H1$'$, H4$'$ of the T$_4$ and A$_5$ moieties. This observation alone provides strong support for the binding of the peptide moiety of CysMeTANDEM in the minor groove of [d(GGATATCC)]$_2$ (Fig. 6-4).

Additional evidence was obtained from the NOESY spectrum taken in H$_2$O. The drug–DNA complex exhibited an ~1-ppm downfield shift for the alanine NH resonance; this is a usual consequence of H-bonding. This resonance also had strong cross-peaks to the A$_5$H1$'$ and A$_5$H2 resonances, consistent with H-bonding to N3 of A$_5$. Again, this finding is closely analogous to observations made for triostin A and echinomycin complexes with DNA, in which the same alanine NH is H-bonded to GN3 at the CpG binding sites of the drug–DNA complexes (Gao & Patel, 1988; Gilbert & Feigon, 1991; Wang et al., 1984, 1986). The role of this H bond in DNA binding was studied by Olsen et al. (1986). They prepared a TANDEM analog in which the alanine was replaced by lactic acid; the resulting analogue was incapable of protecting the substrate DNA from digestion by DNase I, suggesting that this hydrogen bond is important for the site-selective binding of DNA by quinoxaline antibiotics such as TANDEM and CysMeTANDEM.

Another analogy with echinomycin–DNA complexes was found when the sugar conformations of the nucleotide at the site of drug binding were studied. The sugar conformation of T$_4$, in particular, was found to adopt predominantly an N-type (C3$'$-endo-like) conformation. Aside from the evidence from the P.COSY spectrum of the complex, additional evidence for the conformation of the T4 sugar is the large upfield shifts observed for the T$_4$H2$'$ and H2$''$ resonances. These were analogous to findings made for echinomycin, which caused upfield shifts of similar magnitude for the cytidine H2$'$ and H2$''$ resonances at the putative CpG binding sites in DNA. This would be expected to lead to a minor groove which is wider at the drug binding site in the complex than in free DNA itself. In fact this widening would likely facilitate minor groove binding of the peptide moiety of CysMeTANDEM, both in terms of the steric requirements for binding and with regard to stabilization of van der Waals interactions between the DNA sugar and drug peptide moiety. Essentially the same effect on sugar conformation has been noted for the binding of echinomycin to DNA oligonucleotides, specifically involving C_2 in the tetranucleotides [d(ACGT)]$_2$ and [d(TCGA)]$_2$ (Gao & Patel, 1988) as well as in the self-complementary octanucleotides [d(ACGTACGT)]$_2$ and [d(TCGATCGA)]$_2$ (Gilbert & Feigon, 1991).

Also observed in the NOESY spectrum of the complexed DNA was a cross-peak between the T_4H6 resonance and the A_5H8 resonance. Based on a comparison of the integrated intensity of this cross-peak with that of the cross-peak between CH5 and CH6, the T_4H6–A_5H8 distance was calculated to be ~4.5 Å.

Identification of the Sugar Conformations in the Drug–DNA Complex Using $^3J(H,H)$ Coupling Constants. Figure 6-8 illustrates that portion of the P.COSY spectrum of the CysMeTANDEM–d[d(GGATATCC)]$_2$ complex which contains the H1'–H2' and H1'–H2" cross-peaks. With the exception of T_4, all of the residues exhibit similar cross-peak patterns, indicative of a sugar pucker that is predominantly S-type. The P.COSY spectrum was analyzed by computer simulations using the SPHINX and LINSHA programs; the derived $J_{1'2'}$ and $J_{1'2''}$ coupling constants are shown in Table 6-1 (Widmer & Wüthrich, 1987). This analysis indicated that T_4 adopts a conformation that is predominantly N-type (i.e., similar to C3'-endo), while G_1, A_3, T_5, A_6, and C_7 adopted sugar puckers that were predominantly S-type (i.e., similar to C2'-endo). The P.COSY spectrum did not permit the characterization of the conformations of the sugars in G_2 and C_8, due to the lack of resolution between the 2' and 2" resonances. In order to be able to assign the sugar puckers in G_2 and C_8, a NOESY spectrum with a short mixing time was recorded. We found that the base–H3' cross-peaks for these two residues were much less intense than the analogous H6–H3' cross-peak for T_4, which is consistent with a longer H6/8–H3' distance for G_2 and C_8 than for T_4. As noted above, T_4 has been shown to have a C3'-endo-type sugar, suggesting that the greater base–H3'

Table 6-1 ^{1}H–^{1}H Coupling Constants of the Deoxyribose Sugar Rings in the CysMeTANDEM–[d(GGATATCC)]$_2$ Complex

	$J_{1'2'}$ (Hz)	$J_{1'2''}$ (Hz)	$\Sigma 1'$	%S ($pS \times 100\%$)
G_1	8.6	6.6	15.2	92
G_2				
A_3	9.2	4.6	13.8	68
T_4	3.5	7.7	11.2	24
A_5	8.9	5.6	14.5	80
T_6	8.9	5.3	14.2	75
C_7	7.7	6.5	14.2	75
C_8				

Source: Addess et al., 1992.

distances in G_2 and C_8 probably reflects a C2′-endo-type (i.e., S-type) conformation for the sugars. This analysis is consistent with the finding that in A-DNA, which is known to have C3′-endo conformation, the thymidine H6–H3′ distance is shorter than the corresponding distance in B-DNA (Wüthrich, 1986).

Solution Structure of a Complex Between CysMeTANDEM and a DNA Oligonucleotide

The foregoing studies defined the sequence-specific binding of CysMeTANDEM to DNA and resulted in a qualitative structure (model) for the CysMeTANDEM–[d(GGATATCC)]$_2$ complex. To obtain more detailed information concerning the mode of binding of CysMeTANDEM to DNA oligonucleotides containing TpA binding sites, a three-dimensional solution structure of CysMeTANDEM with a DNA oligonucleotide was subsequently determined using the hexanucleotide [d(GATATC)]$_2$ (Addess et al., 1993). We chose to use this hexanucleotide for determination of a three-dimensional structure, rather than d[d(GGATATCC)]$_2$, because of the superior resolution of the cross-peaks in the two-dimensional NOESY spectra. Accurate integration of a majority of the cross-peaks was thus possible, which facilitated the structure determination. The three-dimensional structure described below represents the first one carried out for any quinoxaline antibiotic specific for TpA sequences in DNA.

Assignment of Resonances. The complex between CysMeTANDEM and [d(GATATC)]$_2$, as well as the free hexanucleotide, was analyzed in the same fashion as the complex involving [d(GGATATCC)]$_2$ (Addess et al., 1992) using NOESY, HOHAHA, and P. COSY spectra to assign the proton resonances in the 1:1 drug–DNA complex (Addess et al., 1993). As expected, the overall patterns of cross-peaks for CysMeTANDEM and the DNA hexanucleotide were similar to those observed for the 1:1 CysMeTANDEM–[d(GGATATCC)]$_2$ complex (Addess et al., 1992). The NOESY spectrum of the complex was sufficiently well resolved to allow assignment of almost all DNA proton resonances. Importantly from the perspective of structure calculation, virtually all of the cross-peaks were amenable to integration.

Starting Structures and Initial Refinement. Interproton distances were determined from cross-peaks intensities. The standard used for calibration was the C_6H5–H6 cross-peak, since the distance between the CH5 and CH6 protons is known to be 2.44 Å. The accuracy of the calibration was checked by measuring the H2′–H2″ interproton distances for the two best resolved H2′–H2″ cross-peaks. The calculated distances were 1.8 ± 0.2 Å for A_4 and 1.9 ± 0.2 Å for T_3. These values compared favorably with the known H2′–H2″ distance of 1.77 Å.

The strategy employed for the three-dimensional structure determination involved initial structure generation of the complex via metric matrix distance geometry. This strategy thus avoids the use of initial DNA models or docking of the drug onto the DNA prior to final energy refinement. A combination of energy minimization, restrained molecular dynamics, which alleviates any covalent bond and angle violations that are associated with distance geometry structures, and direct NOE refinement, which takes into account spin diffusion, was then used to produce the final refined structures. The well-resolved spectra made it possible to obtain a large number of intraresidue and

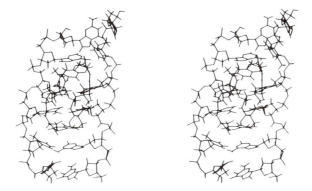

Figure 6-9. Stereoview of one of the 10 starting structures of CysMeTANDEM–[d(GATATC)]$_2$. Reprinted from Addess et al., 1993, with permission.

drug–DNA distances constraints. This contributed to the high quality of the structures obtained.

One of the 10 starting structures is shown in Fig. 6-9. Here it is examined to illustrate qualitatively how well the initial embedding actually worked. In common with the other nine starting structures, the structure in the figure was in reasonably good agreement with the features noted from our earlier NMR study (Addess et al., 1992). For example, all 10 starting structures existed in a right-handed double helical conformation in which CysMeTANDEM was bound via bis-intercalation. Likewise, all of the complexes had the quinoxaline moieties flanking the TpA sequence, with the cyclic peptide in the minor groove of the duplex. Not unexpectedly at this point in the structure calculation, each of the 10 starting structures has several covalent bonds and dihedral bond angles whose values were clearly unreasonable. The quinoxaline rings and some of the bases were also distorted and nonplanar. Excluding the terminal base pairs of the complex, the average pairwise root mean square difference was calculated to be 2.28 ± 0.32 Å for the heavy atoms of all starting structures.

The structure refinement was carried out using the X-PLOR program, following standard protocols. In the first step, 10,000 cycles of energy minimization were run. Twenty picoseconds of restrained molecular dynamics was then followed by further cycles of minimization sufficient to obtain convergence of the gradient of all structures. This step in the refinement resulted in quite significant improvement in the total energy of the structures.

Relaxation Matrix Refinement. In the last step of the refinement, the distance constraints were replaced with NOE volume constraints for direct NOE refinement employing the full relaxation matrix. In the case of the calculation of the structure of the complex, this procedure resulted in a significantly better R-factor, although the atoms actually displayed little movement. The observation that the atomic positions were altered only minimally indicates that the distance restraints used in the initial stages of structure determination (i.e., distance geometry and molecular dynamics) defined the positions of the atoms of the refined structures with reasonable accuracy. The average pairwise root mean square difference for all atoms between each of the eight final refined structures and these eight structures after molecular dynamics and energy mini-

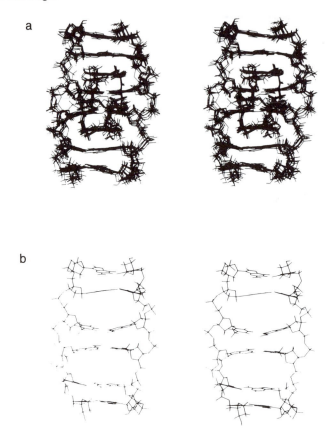

Figure 6-10. (**a**) Stereoview of the eight superimposed final structures of CysMeTANDEM-[d(GATATC)]$_2$. Minor groove of the DNA is in front. (**b**) Stereoview of one of the eight best structures, showing the DNA constituent of the 1:1 complex. Major groove of the DNA is in front. Reprinted from Addess et al., 1993, with permission.

mization was calculated to be only 0.72 ± 0.07 Å. Obviously, the significant change in R-factor without large changes in atomic positions reinforces the finding that R-factor can be quite sensitive to small changes in atomic position (Nilges et al., 1991). This is typical of results we have obtained for other DNA structure calculations. However, this may not be the case for larger molecules in which the effects of spin diffusion on NOE intensities becomes much greater. For such molecules, NOE-based refinement may produce much more dramatic changes in atomic positions.

A stereoview of the superimposed eight final structures of the CysMeTANDEM–[d(GATATC)]$_2$ complex is shown in Fig. 6-10a. These eight refined structures have an average pairwise root mean square difference of 1.1 Å for all heavy atoms except the terminal base pairs. The backbone conformation of CysMeTANDEM varied relatively little from one structure to another; the same was also true of the four internal base pairs. In contrast, the greatest interstructure variation was for the phosphate backbone, due principally to the lack of experimental constraints for the phosphates and possibly also to the motion of the backbone. For CysMeTANDEM alone, the root mean square difference was 1.01 ± 0.12 Å. The least well defined portion of the drug was the side chains of valine; omission of these atoms from the analysis reduced the root mean square difference to 0.63 ± 0.24 Å. Comparison of values for the best and least well defined structural elements of the DNA is also instructive. The average pairwise root mean square difference for the bases and sugars of the four internal nucleotides was 0.75 ± 0.21 Å, while the value for the phosphate backbone was 1.44 ± 0.24 Å. During the course of the refinement, the average R-factors for the eight structures diminished from 0.48 ± 0.09 for the starting structures to 0.21 ± 0.01 for the final structures.

In comparing the precision of the various structural element of the complex, we can understand one of the reasons why structure determination in proteins has advanced much further than in nucleic acids. Though the number of experimental NOEs used to calculate the structure was sufficient to produce a well-defined backbone for the peptide ring of the drug, NOEs alone were not able to provide us precise ranges for all the phosphodiester backbone torsional angles of the DNA. Significant improvement in the precision of the phosphodiester backbone could be made with constraints derived from heteronuclear ^{3}J(H, P) coupling measured from a $(^1H,^{31}P)$ heteroCOSY.

CysMeTANDEM–[d(GATATC)]$_2$ Three-Dimensional Structure. A minor groove view of the eight final refined structures, showing the quinoxaline rings intercalated on either side of the central T·A base pairs and the cyclic peptide moiety of CysMeTANDEM bound in the minor groove, is shown in Fig. 6-10. The complex has an approximate twofold axis of symmetry between the central T·A base pairs. Because no symmetry constraint was imposed during the refinement, there are variations in the two symmetrical halves of the complexes, effectively doubling the number of structures. As discussed below, the intercalation of the quinoxaline moiety of CysMeTANDEM causes DNA helix unwinding; the distance between the affected base pairs increases by about 1.5 Å (from ~ 3.5 Å to ~ 5.0 Å).

Shown in Fig. 6-11 is a space-filling model of the complex of CysMeTANDEM with d[d(GATATC)]$_2$. From the view into the major groove of the complex, shown on the left, the edges of the quinoxaline rings are clearly visible, intercalated on either side of the T_3·A_4 base pairs. The view into the minor groove, shown on the right, indicates the orientation of the cyclic peptide moiety of CysMeTANDEM. In addition, as described below, the figure illustrates a number of van der Waals interactions between the central four base pairs and the peptide side chains. An H bond between the alanine NH group and adenosine$_4$ N3 is also in evidence.

The Structure of the DNA in the Complex. Figure 6-10b is a stereoview of the DNA oligonucleotide d[d(GATATC)]$_2$ as it exists within the drug–DNA complex. The DNA within the complex contains the normal Watson–Crick base pairs, with the individual nucleotides all having the *anti* configuration. The major structural changes from the free DNA (B-form) that accompany CysMeTANDEM binding are clear from this figure; they include unwinding of the helix, a separation of A_2·T_5 and T_3·A_4 base pairs to permit intercalation of the quinoxaline moieties, and buckling of the base-pairing adjacent to the site of intercalation.

The pattern of DNA unwinding illustrated here is fairly typical of the pattern that is observed for drugs that intercalate between base pairs and also contain some structural element, such as a peptide, that is essentially perpendicular to the long axes of

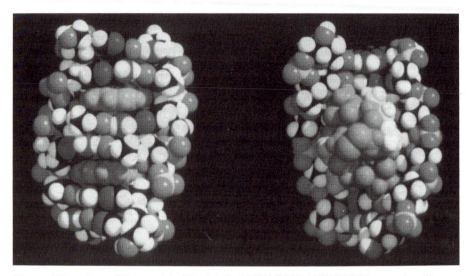

Figure 6-11. Space-filling model of the 1:1 CysMeTANDEM–[d(GATATC)]₂ complex, illustrating the major groove (**left**) and minor groove (**right**). Reprinted from Addess et al., 1993, with permission.

the DNA bases. Other such "perpendicular intercalators" include daunomycin and triostin A, for which there are crystal structures, and nogalamycin, which has been studied both crystallographically and by solution methods. Like CysMeTANDEM, each of these drugs effects DNA unwinding not at the site of intercalation per se, but rather at the adjacent internucleotide step (Egli et al., 1991; Gao et al., 1990; Nunn et al., 1991; Ughetto et al., 1985; Wang et al., 1984, 1986; Zhang & Patel, 1990).

The inward buckling of the central T·A base pairs is quite evident in Fig. 6-10b. This structural feature was present in each of the eight final structures for the Cys-MeTANDEM–[d(GATATC)]₂ complex and had an average value of $18.2 \pm 3.6°$ for all structures. Correspondingly, there was an outward buckle of the $A_2 \cdot T_5$ base pair with an average value of $14.2 \pm 6.0°$. Again, this type of pattern is typical of the effects of perpendicular intercalators and has been noted for DNA complexes of triostin A, nogalamycin, and daunomycin (Wang et al., 1984; Ughetto et al., 1985). This type of buckling is not noted for "parallel intercalators" such as ethidium, suggesting that its formation must occur to create a cavity in the DNA optimal for binding of perpendicular intercalators. It seems likely that this would include enhanced opportunities for van der Waals interactions (Williams et al., 1992).

Drug–DNA Interactions Within the Complex. As noted above, there is a hydrogen bond connecting the alanine NH and N3 of A_4. For all of the structures, the average distance between the two N atoms was 2.26 ± 0.07 Å. This H bond must be essential for binding of the drug to its DNA substrates because, as noted above, the replacement of alanine by lactic acid diminished drug binding (Olsen et al., 1986). The analogous H bond between alanine NH and guanosine N3 of CpG-specific quinoxaline antibiotics has been noted in DNA complexes of triostin A, as well as those involving echinomycin (Gao & Patel, 1988; Gilbert & Feigon, 1991; Gilbert et al., 1989; Ughetto et al., 1985; Wang et al., 1984, 1986).

As illustrated in Fig. 6-11, the CysMeTANDEM–[d(GATATC)]$_2$ complex contains a number of significant van der Waals interactions. Notable among these are those between the alanine Me in the peptide moiety with thymidine$_3$ C1′, C2′, and A$_4$O4′. Earlier studies of triostin A have indicated that the analogous group in the cyclic peptide has similar van der Waals interactions with the sugars of the cytidine and guanosine moieties at the site(s) of drug binding (Ughetto et al., 1985; Wang et al., 1984). One of the clear differences between the CysMeTANDEM and triostin A complexes is a strong van der Waals interaction between one valine methyl group and T$_3$O2. This is absent in the 2:1 triostin A–[d(CGTACG)]$_2$ complex, apparently because the valine methyl groups in the latter complex are rotated too far from the cytidine O2.

T·A Versus C·G Specificity of CysMeTANDEM. The large numbers of analogous features in the complexes of TpA- versus CpG-specific quinoxaline antibiotics raises the issue of the actual source of sequence selectivity for CysMeTANDEM and TANDEM versus triostin A and echinomycin. In order to investigate this further, we have also calculated the structure of the analogous triostin A–[d(GA_CG_TC)]$_2$ complex and have studied the complexes formed by both triostin A and CysMeTANDEM to [d(GGACITCC)]$_2$. The results of these studies are discussed in detail in Addess and Feigon (1994a,b).

Summary

The continuing increases in the sensitivity of NMR spectrometers and the number of pulse sequence techniques available for structure determination have combined to make NMR spectroscopy an extremely powerful tool for the analysis of nucleic acid structures. This chapter has focused on the techniques employed for the analysis of B-form DNA and its interaction with a class of peptide-derived antibiotics that bind to DNA by an intercalative mechanism.

Future directions include the development of strategies for determining the three-dimensional structures of DNA using uniformly ^{15}N,^{13}C-labeled DNA oligonucleotides. Although the discussion in this chapter has been limited to DNA, parallel efforts in the area of RNA structure determination are ongoing and have already produced exciting results. The great variety of secondary and tertiary structures assumed by RNA molecules, which should be amenable to definition using appropriate NMR techniques, argues that NMR spectroscopy will have an important role to play in the area of RNA structure determination.

Acknowledgments

This work was supported by grants from NIH (R01 GM 37254) and NSF Presidential Young Investigator Award with matching funds from Amgen, Inc., Monsanto Co., Dupont/Merck, Inc., and Sterling Winthrop, Inc. to J.F.

The Formation of Alternative Structures in DNA

David M. J. Lilley

Under the normal conditions that prevail in the cell, most DNA exists as a right-handed double-stranded species in the B-conformation, in which two antiparallel strands are held together by complementary base-specific hydrogen bonding. However, all these facets are capable of alteration under certain conditions, and it is the purpose of this chapter to review how the geometric properties can vary and what controls the structural repertoire. At the end, the elucidation of one folded DNA structure is discussed in greater detail.

Interactions That Control the Structure of DNA

The structure of DNA comprises a series of flat aromatic heterocyclic bases attached to a hydrophilic, negatively charged sugar-phosphate backbone. There are seven degrees of torsional freedom in the nucleotide unit: Six are on the backbone itself, and the seventh sets the glycosidic bond between the base and the sugar. This means that the backbone is relatively flexible and can relax to accommodate structural alteration caused by variation in base–base interactions. Three main kinds of interactions govern most of the structural changes observed in DNA. First, the charged phosphate groups of the backbone are mutually repulsive, and this sets limits on the extent of structural variation that is possible. These electrostatic interactions normally require screening by metal cations, and ion interactions are critical to the formation of many structures. Second, hydrogen bonding between bases is the driving force for the formation of multistranded structures; these interactions have directional character and their vectorial nature generates base-specificity. Third, the base pairs exert short-range attractive interactions on each other, leading to stacking interactions between the flat surfaces. These properties lead to the adoption of the normal double-stranded structure of DNA with a right-handed twist, comprising the charged backbone on the outside and the stacked base pairs on the inside. But they also determine the extent to which the structure may be altered, and some themes are recurrent. For example, the requirement to reduce electrostatic repulsion, as well as the tendency for base pairs to undergo stacking, is found in many rearranged DNA structures.

Sequence-Dependent Modifications to B DNA; the Variable Properties

The Helical Repeat of Double-Stranded DNA. A simple geometric calculation based on the requirement to stack base pairs (Calladine & Drew, 1992) suggests that DNA should be helical with a repeat of around 11 base pairs/helical turn. An average helical repeat for B-DNA is about 10.6 bp/turn (Peck & Wang, 1981; Rhodes & Klug, 1980), but local twist is quite sequence-dependent and certain sequences adopt structures in which the repeat is significantly different from this value. For example, stretches of oligo dA·oligo dT adopt a structure called B′ in which there is a pronounced propeller twist of the A·T base pairs such that an adenine amino proton is shared between two thymine bases to form three-centered hydrogen bonds (Coll et al., 1987; Nelson et al., 1987). This structure, which is the basis of sequence-directed curvature of DNA (Diekmann & Wang, 1985; Hagerman, 1985; Koo et al., 1986; Marini et al., 1982; Wu & Crothers, 1984), is relatively overwound, with a helical repeat of around 10.0 bp/turn. By contrast, alternating adenine–thymine tracts $\{d(ApT)_n\}$ appear to be of particularly low torsional stiffness and are capable of being either underwound (McClellan et al., 1986) or overwound (McClellan & Lilley, 1991). Certain (G + C)-rich sequences exhibit a propensity for adopting a different helical structure, called A-DNA. As described in Chapter 1, this is characterized by a movement of the base pairs away from the axis of the helix, creating a pronounced hole down the center and a very deep and narrow major groove (Connor et al., 1982; Shakked et al., 1983). There is a change in deoxyribose pucker from a predominantly south to north (i.e., around C3′-endo) conformation, and the bases take up a significant inclination to the axis. The A-structure is underwound relative to B-DNA, with a helical repeat of 11–12 bp/turn.

The Handedness of DNA. The variants on normal B-DNA structure discussed so far are all right-handed, but the handedness of DNA is not invariant. The most famous example of left-handed DNA is the Z-conformation adopted by alternating pyrimidine-purine sequences, especially by $d(CpG)_n$ repeats (Wang et al., 1979). The alteration to the left-handed geometry requires a major reorganization of the backbone conformation, as described below.

Alternatives to Watson–Crick Base-Pairing. Normal DNA structure is based upon the complementary base pairs of adenine with thymine, and guanine with cytosine, which are the most stable base pairs thermodynamically (Breslauer et al., 1986; Gotoh & Tagashira, 1981a,b). However, single mispaired bases can be incorporated into the overall geometry of B-DNA with very little disruption in many cases (Bhattacharyya & Lilley, 1989b; Brown et al., 1986; Gao & Patel, 1987, 1988; Hunter et al., 1986a; Kneale et al., 1985). Mispairs are invariably destabilizing, the degree of which depends on the nature of the mismatch and its sequence context (Aboul-ela et al., 1985; Werntges et al., 1986). Rearranged base-pairing may involve hydrogen bond donors and acceptors other than the normal ones. Hoogsteen first recognized that altered base-pairing could be mediated by use of guanine N7 and O6 and of adenine N7 and N6 (Hoogsteen, 1963). These positions are the basis of altered base-pairing adjacent to certain sites of intercalation (Quigley et al., 1986), accommodation of a third strand in the H-triplex structure (Htun & Dahlberg, 1988; Johnston, 1988; Kohwi &

Kohwi-Shigematsu, 1988; Lyamichev et al., 1986; Mirkin et al., 1987), and the formation of guanine tetrads in tetraplexes formed from telomere-related sequences (Aboul-ela et al., 1992b, 1994; Gellert et al., 1962; Jin et al., 1992; Laughlan et al., 1994; Sen & Gilbert, 1988; Sen & Gilbert, 1990; Smith & Feigon, 1992; Sundquist & Klug, 1989; Wang et al., 1991; Wang & Patel, 1992; Williamson et al., 1989). Base pairs may also be broken, leaving single-stranded bases. In relaxed DNA base pairs are very stable; nuclear magnetic resonance (NMR) measurements of imino proton exchange rates show that opening of base pairs is infrequent (Guéron et al., 1987). However, in supercoiled DNA, $(A + T)$-rich regions can undergo cooperative opening at low ionic strength and elevated temperature, to generate stable bubbles of significant size (Bowater et al., 1991; Furlong et al., 1989; Kowalski et al., 1988; Lee & Bauer, 1985).

Deformation of the Helix Axis. To a first approximation, the axis of normal B-DNA can be considered as straight, but the axial trajectory can be altered quite readily in a number of ways. The deformation can be discontinuous as a kink, or the curvature may be less localized. An example of the former behavior is the kinking that is introduced by base bulges in DNA or RNA (Bhattacharyya & Lilley, 1989a; Bhattacharyya et al., 1990; Hsieh & Griffith, 1989; Rice & Crothers, 1989; Riordan et al., 1992; Wang & Griffith, 1991). Sharp kinking of the DNA may also be introduced as a result of adduct formation at a particular site (Husain et al., 1988; Rice et al., 1988). Local sequence may also result in deformation of the helix axis. The most extreme case of this is the pronounced curvature of DNA associated with oligoadenine tracts that are phased with the helical repeat (Diekmann & Wang, 1985; Hagerman, 1985; Koo et al., 1986; Marini et al., 1982; Wu & Crothers, 1984), but lesser distortions are clearly found in many sequences and truly straight segments of DNA are probably rare. In addition to intrinsic curvature, certain sequences may exhibit enhanced bendability in response to protein binding.

The Number of Strands. Standard DNA is double-stranded. However, the width of the major groove in B-DNA is sufficient to accommodate a third strand, and thus triplex DNA can be formed, chiefly by oligopurine·oligopyrimidine sequences (Htun & Dahlberg, 1988; Johnston, 1988; Kohwi & Kohwi-Shigematsu, 1988; Lyamichev et al., 1986; Mirkin et al., 1987). Four-stranded DNA has been described for guanine-rich sequences, based on the guanine tetrad (Sen & Gilbert, 1988, 1990; Sundquist & Klug, 1989; Williamson et al., 1989).

Relative Orientation of Strands. Normal DNA is a double helix in which the two strands have an antiparallel orientation. This is clearly the thermodynamically more stable arrangement, although parallel-stranded duplexes can be forced together chemically (van de Sande et al., 1988). In the H-triplex structure, the incoming oligopyrimidine strand is oriented parallel to the oligopurine strand of the duplex and antiparallel to the complementary oligopyrimidine strand (Htun & Dahlberg, 1988; Johnston, 1988; Kohwi & Kohwi-Shigematsu, 1988; Lyamichev et al., 1986; Mirkin et al., 1987). Both parallel (Aboul-ela et al., 1992b, 1994; Jin et al., 1992; Laughlan et al., 1994; Sen & Gilbert, 1988, 1990; Wang et al., 1991) and antiparallel (Smith & Feigon, 1992; Sundquist & Klug, 1989; Wang & Patel, 1992; Williamson et al., 1989) forms of the guanine tetraplex have been observed. When the structure is formed from four unlinked strands, they assemble in a parallel orientation, but the formation of hairpin or foldback forms (see below) requires antiparallel association. This is consistent with the greater thermodynamic stability of the parallel tetraplex structure, in the absence of the entropic advantage of intramolecular association (Lu et al., 1993).

Formation of Novel Structures by Rearrangement or by Strand Association. Some structures are formed by rearrangement of normal double-stranded DNA, such as the formation of left-handed Z-DNA or cruciforms. Others may require intramolecular folding of a single strand, such as the foldback tetraplex, or intermolecular association of strands, such as the parallel tetraplex or certain junction structures.

A Brief Review of Structural Variants

It is clear that at the local level, all DNA exhibits sequence-dependent structural variability (e.g., see Drew & Travers, 1984). In this section, however, the discussion is restricted to that of the more grossly rearranged DNA structures.

Base Mismatches. Locally disrupted base-pairing can take a number of different forms (Fig. 7-1):

   Single-base mismatches—a single non-Watson–Crick base opposition flanked by
      normal base pairing.
   Multiple-base mismatches—a group of consecutive non-Watson–Crick base op-
      positions. This is rather similar to a locally melted "bubble" or internal loop.
   Bulged bases—one or more extra bases that are unopposed on the complemen-
      tary strand of the duplex.

The overall degree of disruption caused is different for these different forms of mismatches. There have been a number of crystallographic (Brown et al., 1986, 1990; Hunter et al., 1986a,b; Kneale et al., 1985; Privé et al., 1987; Webster et al., 1990) and NMR (Gao & Patel, 1987, 1988) studies of oligonucleotides containing single-base mismatches. In each case the disruption to the overall geometry of the double helix is quite small, with the mismatched bases remaining stacked into the helix and hydrogen-bonded. This is in agreement with gel electrophoretic experiments indicating that

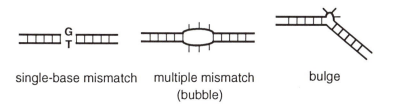

single-base mismatch   multiple mismatch   bulge
                          (bubble)

Figure 7-1. Schematic illustration of the three classes of base mismatches in DNA. Single-base mismatches are single non-Watson–Crick base oppositions flanked by normal base-pairing. A group of successive non-Watson–Crick base oppositions constitute a multiple mismatch, or bubble. One or more extra unopposed bases on one strand introduces a bulge into the duplex.

all possible single-base mismatches cause very little disruption to helical axis trajectory (Bhattacharyya & Lilley, 1989b). However, in many cases the local structure of the mismatch leads to an enhanced chemical reactivity (Bhattacharyya & Lilley, 1989b; Cotton et al., 1988). For example, the thymine base of a G·T mismatch tends to be reactive to addition by osmium tetroxide; this can be understood in terms of the crystal structure (Kneale et al., 1985), where the wobble base pair leads to exposure of the 5,6-bond of the thymine to out-of-plane electrophilic attack in the major groove. All single-base mismatches result in a destabilization of the helix (Aboul-ela et al., 1985; Werntges et al., 1986).

Multiple mismatches and base bulges are of quite different character. In general, multiple mismatches may introduce a point of flexibility, but no anisotropic bending (Bhattacharyya & Lilley, 1989a). The bases exhibit strong reactivity toward enzyme and chemical probes (Bhattacharyya & Lilley, 1989a). Thus multiple mismatches of a significant size behave like amorphous bubbles in the DNA, although there is no doubt that many base–base interactions exist. Similar features can be generated in (A + T)-rich sequences in supercoiled DNA at low ionic strength and elevated temperature, when a bubble can be generated from complementary sequences by a local melting process (Bowater et al., 1991; Furlong et al., 1989; Kowalski et al., 1988; Lee & Bauer, 1985).

By contrast, base bulges introduce a pronounced kinking of the helix axis, which can readily be observed as an electrophoretic retardation of the bulged molecule (Bhattacharyya & Lilley, 1989a; Hsieh & Griffith, 1989; Rice & Crothers, 1989). The magnitude of this retardation is dependent on the position of the bulge, being greatest when the bulge is located at the center of the fragment. In principle, the retardation could be due to a kinking of the helix axis or to a flexible hinge. This was resolved by introducing two such bulges into a series of molecules of fixed length and varying the spacing between the bulges. The end-to-end distance, and hence the mobility, should vary in a sinusoidal way with the spacing if the bulges cause a kinking of the helix axis, and this was observed experimentally (Bhattacharyya et al., 1990). Kinking by bulges in DNA duplexes has been confirmed by fluorescence resonance energy transfer experiments (Gohlke et al., 1994), and kinking has been observed by conventional electron microscopy (Wang et al., 1992) and by electron microscopy in vitreous ice (J. Bednar, J. Dubochet, A. I. H. Murchie and D. M. J. Lilley, in preparation).

The location of bulged nucleotides—stacked within the double helix or extended outside the helix—has been examined by NMR. An extra single base was found to be intrahelical in most cases (Hare et al., 1986; Nikonowicz et al., 1989, 1990; Patel et al., 1982; van den Hoogen et al., 1988; Woodson & Crothers, 1987, 1988). However, pyrimidine bulges within an (A + T)-rich context were concluded to be extrahelical (Morden et al., 1983, 1990), and in one case a bulged cytosine was found to form both structures dependent upon temperature (Kalnik et al., 1990). A crystal structure has been reported for an oligonucleotide containing a single bulged adenosine (Joshua-Tor et al., 1988, 1992; Miller et al., 1988). More recently, reports have appeared describing structures of molecules containing multiple-base bulges in DNA. NMR studies of AA, AAA, and ATA bulges have indicated that the extra bases are all stacked into the helix, such that the sequential path of NOE connectivities through ribose H1′ and base protons is unbroken throughout the bulges (Aboul-ela et al., 1993; Rosen et al., 1992a,b). Significant distortion was found on the nonbulged strand opposite the location of the bulge however, indicating a discontinuity at this point. This might represent the principal location of the axial kink. In one study (Aboul-ela et al., 1993) the loss of a T-imino resonance suggested that an A·T base pair immediately adjacent to an AAA bulge might be disrupted. Rosen et al (1992b) carried out a full structural determination of an ATA bulge by NMR, concluding that the axis was kinked by 50° to 60°. The insertion of a bulge into one strand of a duplex DNA molecule destabilizes the molecule (Gohlke et al., 1994; LeBlanc & Morden, 1991; Morden et al., 1983; Patel et al., 1982; Woodson & Crothers, 1989), by approximately 3.5–4.5 kcal/mol at 37°C for a single-base bulge. The destabilization depends upon interactions with the base pairs directly neighboring the bulges and also with more distal base pairs (Longfellow et al., 1990).

Z-DNA. The most extreme example of a change to the twist of double-stranded DNA must be the Z-structure adopted by alternating purine–pyrimidine $(RpY)_n$ sequences. This was first observed by Pohl and Jovin (Pohl & Jovin, 1972) as an inversion of the circular dichroism of alternating poly dC-G·poly dC-G at high concentrations of sodium ion. The reason for this later became apparent when the crystal structure of dCGCGCG was solved by Rich, Wang, and coworkers (Wang et al., 1979), whereupon it was observed to form a left-handed conventionally base-paired double helix (Fig. 7-2). The

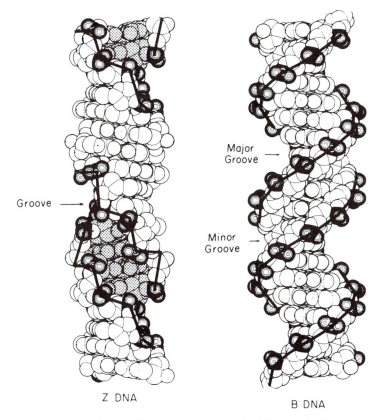

Figure 7-2. Comparison of space-filling models left-handed Z-DNA with B-DNA. Note the zigzag path of the backbone of Z-DNA (heavy lines), showing the dinucleotide repeat. Reprinted from Rich et al., 1984, with permission.

repeating unit of Z-DNA is a dinucleotide, with alternation of sugar pucker and glycosyl torsion angle, giving a pronounced zigzag appearance. Thus while the cytosine bases have normal *anti* glycosyl torsion angles and a C2′-endo deoxyribose conformation, the guanine bases adopt a *syn* conformation with respect to a C3′-endo puckered sugar. In contrast to the A-structure, the helix axis of Z-DNA is displaced into the minor groove. Thus in Z-DNA the minor groove is extremely deep, while the major groove is convex. The exposure of bases in the major groove leads to chemical reactivities not seen in B-DNA (Herr, 1985; Johnston & Rich, 1985), and the structure is also highly immunogenic (Lafer et al., 1981; Möller et al., 1982). The formation of the Z-conformation is not limited to $(GpC)_n$ sequences, and the conformation can be formed by $(GpT)_n$ sequences (Haniford & Pulleyblank, 1983; Nordheim & Rich, 1983). However, the greater the departure from the ideal $(GpC)_n$ sequence, the lower the stability of the left-handed form becomes (Ellison et al., 1985; McLean et al., 1988). Segments of Z-DNA can form within a DNA molecule that is predominantly right-handed (Haniford & Pulleyblank, 1983; Nordheim et al., 1982; Nordheim & Rich, 1983; Peck et al., 1982; Singleton et al., 1982), in which case interfacial B–Z junctions are created; these are susceptible to attack by a number of enzyme and chemical probes (Galazka et al., 1986; Singleton et al., 1984), indicative of a perturbed conformation. Studies (Nordheim et al., 1982; O'Connor et al., 1986; Peck & Wang, 1983) of the energetics of the B-to-Z transition in supercoiled DNA molecules (see below) show that the transition is cooperative, with a significant free energy associated with the formation of two B–Z junctions, in addition to the free energy for conversion of each participating base pair in the Z-conformation.

DNA Curvature. It has been known for some time that the axis of double-stranded DNA can be curved by particular sequences, particularly those found in kinetoplast circles (Diekmann & Wang, 1985; Hagerman, 1985; Marini et al., 1982; Trifonov & Sussman, 1980; Wu & Crothers, 1984). This can be observed readily by the resulting anomalously slow electrophoretic mobility (Diekmann, 1986; Diekmann & Wang, 1985; Hagerman, 1985; Koo et al., 1986; Wu & Crothers, 1984) or by increased rates of self-ligation into circles by relatively short curved fragments (Levene & Crothers, 1986; Ulanovsky et al., 1986). These effects are not simply due to increased flexibility, because two such sequences exhibit a precise phase relationship (Drak & Crothers, 1991). The most pronounced sequence-directed curvature results from runs of oligoadenine·oligothymine having lengths of four base pairs or greater, which are phased with the helical repeat (Fig. 7-3). Inclusion of bases other than adenine into the tract markedly reduced the apparent curvature (Wu & Crothers, 1984), with the exception of a limited number of inosine bases (Diekmann et al., 1987). However, adenine methylation was observed to increase the gel retardation (Diekmann, 1987a). The electrophoretic

CCC**AAAAA**TGTC**AAAAAA**TAGGC**AAAAAA**TGCC**AAAAA**TCCC**AAA**C
GGG**TTTTT**ACAG**TTTTTT**ATCCG**TTTTTTT**ACGG**TTTTT**AGGG**TTT**G

curvature
center

Figure 7-3. The curvature sequence of kinetoplast DNA. Note the runs of oligoadenine (bold), which are repeated with the periodicity of the DNA helix.

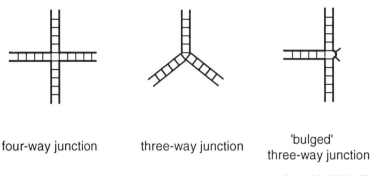

four-way junction   three-way junction   'bulged'
                                                      three-way junction

Figure 7-4. Helical junctions. Three- and four-way junctions may be formed in DNA. These may be perfect (i.e., with full base-pairing), or they may include extra unpaired bases.

anomaly disappears as the temperature is raised (Diekmann, 1987b), suggesting that some critical aspect of the DNA structure is being "melted," and clearly the curvature is associated with the structure of the oligoadenine tracts. This is significantly different from normal B-DNA in a number of respects. A narrow minor groove was indicated by the cleavage of curved DNA fragments by hydroxyl radicals, resulting in a gradient of cleavage frequency decreasing 5′ to 3′ (Milgram-Burkoff & Tullius, 1987). The imino protons of oligoadenine tracts were shown to have unusually slow exchange rates with solvent (Guéron et al., 1988). Crystallography of oligonucleotides with sequences based on oligoadenine runs (Coll et al., 1987; Nelson et al., 1987) has confirmed the existence of a novel structure, generally termed B′. The structure has the expected narrow minor groove and is characterized by a large propeller twist (average 20°) of the A·T base pairs, such that there is bifurcated hydrogen bonding in which the amino proton of a given adenine is shared between O6 acceptors of the opposed thymine and the next one in the 3′ direction. Within the oligoadenine tract there are very small values of roll and tilt, but large roll angles were found at each end of the adenine tract. However, distortion due to crystal packing may be a factor, and the role of bifurcated hydrogen bonding has been questioned (Diekmann et al., 1992).

Helical Junctions. Junctions between helical segments (Fig. 7-4) are formed as intermediates in a number of DNA rearrangements. The most important of these is undoubtedly the four-way junction, the Holliday junction of genetic recombination (Broker & Lehman, 1971; Holliday, 1964; Orr-Weaver et al., 1981; Potter & Dressler, 1976, 1978; Sigal & Alberts, 1972; Sobell, 1972). This structure is discussed in some depth below, but it is subject to some rather simple folding principles. In the presence of sufficient metal cations, the junction folds by pairwise coaxial stacking of helices. Thus the folding process illustrates two very important aspects, which are quite general. First, the folding is driven by the favorable free energy of base-pair stacking. Second, this is not possible at low salt concentrations, because the unfavorable free energy of electrostatic repulsion is greater. When this is reduced by counterion screening, the junction can undergo the folding process.

Three-way DNA junctions (Fig. 7-4) have been studied which test the generality of these principles. The perfect three-way junction comprises three helices linked through the covalent continuity of the strands, with no unpaired bases. Gel electrophoretic experiments suggested that the angles between the three arms of the three-way DNA junction were much closer to being equal than were the equivalent inter-arm angles of the four-way junction (Duckett & Lilley, 1990). This indicated that the arms do not undergo the pairwise coaxial stacking seen in the four-way junction, and this was consistent with the permanent reactivity of thymine bases at the point of strand exchange to osmium tetroxide. This conclusion has been questioned for other sequences (Lu et al., 1991), but while the angles are unlikely to be exactly 120° between each pair of arms, the differences still appear to be smaller than the variation observed between the angles of the four-way junction, and the extended unstacked structure is likely to be broadly correct for most sequences. If we try to construct a three-way junction by breaking a phosphodiester linkage in one strand of a duplex and fusing an additional helix, it is necessary to insert at least the width of the minor groove into the space previously occupied by one phosphate group, and this is not possible. However, if the junction contains extra unpaired bases on one strand (Fig. 7-4), the coaxial stacking of two arms is stereochemically feasible. Such junctions are more thermodynamically stable (Leontis et al., 1991), and it has recently been shown that, like the four-way junction, the bulged three-way junctions undergo ion-dependent folding into a folded conformation based on coaxial stacking of two arms (Leontis et al., 1993; Rosen & Patel, 1993; Welch et al., 1993). Thus when additional conformational flexibility is provided, the same folding principles observed with the four-way junction reassert themselves.

Cruciform Structures. Cruciform structures can be regarded as a special example of a four-way junction. For a more detailed recent review, see Murchie et al. (1992). They are formed from inverted repeat sequences, when the two strands each form intrastranded hairpin structures (Fig. 7-5). They were first discovered in supercoiled DNA about 14 years ago (Gellert et al., 1979; Lilley, 1980; Panayotatos & Wells, 1981). Cruciform structures are unstable, with a positive free energy of formation of about 18 kcal/mol or more for most sequences (Courey & Wang, 1983; Lilley & Hallam, 1984; Mizuuchi et al., 1982; Naylor et al., 1986); they are therefore not observed in relaxed DNA molecules.

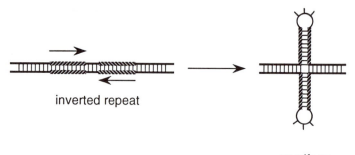

inverted repeat

cruciform

Figure 7-5. Cruciform formation. An inverted repeat sequence (a sequence having a twofold symmetry axis, shown stippled) can undergo a conformational change whereby interstrand base pairing is broken in favor of intrastrand base pairing. The cruciform is equivalent to the formation of hairpin loops on each strand.

C-type extrusion

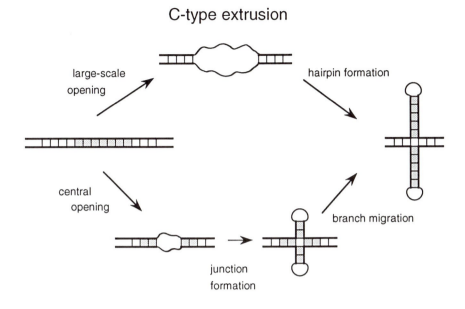

S-type extrusion

Figure 7-6. Two cruciform extrusion mechanisms. The majority of sequences undergo cruciform extrusion by the S-type pathway (**lower**) in which there is a relatively small initial opening of the helix at the center of the inverted repeat, followed by intrastrand base-pairing and formation of a four-way junction. The structure may then undergo branch migration to generate the fully extruded cruciform. This pathway requires salt for optimal extrusion rates. A small number of sequences, in which the inverted repeat is adjacent to (A + T)-rich DNA, may extrude via the C-type mechanism (**upper**). At low salt concentrations the (A + T)-rich DNA may initiate thermal helix opening in a relatively large region, which encompasses the inverted repeat. Intrastrand pairing of the entire inverted repeat may then occur, to generate the fully extruded cruciform in a single event.

One of the more interesting aspects of the cruciform structure concerns the kinetic processes involved in the extrusion from unperturbed DNA. Clearly this must involve a significant disruption to base-pairing as existing interstrand base pairs are broken and new intrastrand base pairs are formed. It has been known for some time that this can be a rather slow process (Courey & Wang, 1983; Gellert et al., 1983; Lilley, 1985; Vologodskii & Frank-Kamenetskii, 1983), and there appear to be two alternative mechanisms according to which the extrusion can proceed (Lilley, 1985) (Fig. 7-6). The choice depends on the sequence context of the inverted repeat that is undergoing the transition. For most DNA sequences, extrusion takes place via the S-type (salt-dependent) mechanism, where there is a local opening at the center of the inverted repeat, formation of a protocruciform, and finally branch migration to the fully extruded cruciform. S-type kinetics are affected by the sequence (Murchie & Lilley, 1987) and methylation state (Murchie & Lilley, 1989) at the very center of the inverted repeat, suggesting that this region is opened at or near the rate-determining step of the extrusion process. Cation concentration and ionic radius are also important (Sullivan & Lilley, 1987), suggesting that a junction is forming in the transition state. By contrast, if the sequence context of the inverted repeat is very (A + T)-rich and the ionic strength is low, there appears to be a much larger cooperative opening of the entire region along with a single-step formation of the cruciform—the C-type mechanism (Bowater et al., 1991; Sullivan & Lilley, 1986). Large-scale opening of DNA is strongly suppressed by moderate salt concentrations (Bowater et al., 1994). The (A + T)-rich sequences are responsible for the cooperative destabilization of the local domain of the super-coiled domain at low ionic strength:

> The effect was dependent on the size, base composition, and state of adenine methylation of the inducing sequence (Murchie et al., 1992).
>
> Short (G + C)-rich blocks of sequence could prevent the effect of the (A + T)-rich sequence, when interposed between it and the inverted repeat (Sullivan et al., 1988).
>
> Helix destabilization by solvents like DMF was found to lead to extrusion by the C-type mechanism, while conversely helix stabilization by distamycin lead to S-type extrusion (Sullivan & Lilley, 1988).
>
> The inducing sequences were chemically reactive to single-strand-selective chemicals when supercoiled (Bowater et al., 1991; Furlong et al., 1989). Many of the properties of the reactivity parallel those of cruciform extrusion.
>
> Statistical mechanical calculations of helix opening potential indicated that the inducing sequences have a propensity for cooperative opening (Schaeffer et al., 1989).

The cruciform extrusion process is probably the most thoroughly studied transition between duplex DNA and a rearranged conformation, and its study has generated many useful data concerning (1) the ability of duplex DNA to undergo large-scale opening and (2) the role of supercoiling, temperature, and ionic conditions.

Triplex Structures. It has been known for a long time that triplex formation can result by uptake of a third strand in the major groove of a duplex (Arnott & Selsing, 1974; Felsenfeld et al., 1957; Morgan & Wells, 1968). This can occur by the uptake of an ex-ogenous third strand or by rearrangement of a duplex structure to generate an intramo-lecular triplex. Triplex formation is based in general on the hydrogen bonding of a pyrim-idine base to the Hoogsteen positions of an existing base pair, to form either an A·T·T or a G·C·C$^+$ base triplet (Fig. 7-7). The requirement for cytosine protonation results in elevated stability of triplex structures at low pH, and the stereochemistry of binding the third strand limits the structure to homopurine·homopyrimidine sequences.

Triplex formation has been observed by the uptake of an exogenous strand into su-percoiled (Lyamichev et al., 1988) and linear DNA (Moser & Dervan, 1987). This forms the basis of a new series of reagents for DNA dissection and genome mapping (Dervan, 1992) and has significant potential for therapeutic application (Hélène, 1991). It has also suggested new ideas for the manner of uptake of a homologous third strand by the RecA protein in genetic recombination (Hsieh et al., 1990; Rao et al., 1991).

Intramolecular triplexes, or H-DNA, have been observed in supercoiled DNA at low pH, mainly by two-dimensional gel electrophoresis and probing experiments. A num-ber of examples of structural perturbations in oligopurine sequences [such as G_n·C_n or $(ApG)_n$·$(CpT)_n$] in supercoiled plasmids had been described (Evans et al., 1984; Hentschel, 1982; Htun et al., 1984; Kohwi-Shigematsu & Kohwi, 1985; Mace et al.,

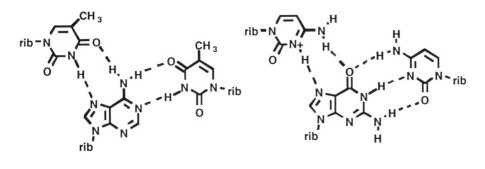

T•A•T triplet   C•G•C⁺ triplet

Figure 7-7. Base triplets. Base triplet formation exploits the purine 6 and 7 positions in the major groove, to introduce an additional pyrimidine base in the manner of a Hoogsteen base pair. In the case of the C·G·C triplet, this requires protonation of the cytosine.

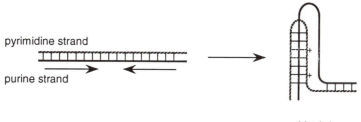

H-triplex

Figure 7-8. The formation of the H-triplex. The H-triplex structure is formed from oligopurine sequences with a central plane of mirror symmetry. In the most common form of the structure, the 3'-end of the pyrimidine strand unpairs with its complementary purine strand and subsequently forms a triple-base interaction in the major groove of the 5'-end of the sequence.

1983; McKeon et al., 1984; Pulleyblank et al., 1985) and had been ascribed to slippage and other structures. The correct explanation was given (albeit at the second attempt) by Frank-Kamenetskii and colleagues (Lyamichev et al., 1985, 1986). They showed that the oligopurine·oligopyrimidine may undergo a large-scale rearrangement at low pH, in which half of the pyrimidine strand forms the third strand of a triplex structure (Fig. 7-8). Thus the pyrimidine strand forms a kind of hairpin structure, while half of the purine strand is left formally single-stranded. A considerable body of work has confirmed the essential details of this structure. Sequence variation showed the requirement for an oligopurine sequence with a mirror repeat (Hanvey et al., 1988; Mirkin et al., 1987), as expected for the H-structure. Chemical probing confirmed the presence of the single-stranded regions (Htun & Dahlberg, 1988; Johnston, 1988; Vojtiskova et al., 1988; Vojtiskova & Palecek, 1987), protection against ultraviolet-induced crosslinking (Lyamichev et al., 1990) and protection against alkylation of purine N7 in the Hoogsteen paired regions (Voloshin et al., 1988).

A number of alternative intramolecular triplex forms are possible and have been observed. In principle, either the 3′ or the 5′ half of the oligopurine strand may be left single-stranded, and Htun and Dahlberg (1989) have studied this possible isomerization. For shorter oligopurine·oligopyrimidine tracts, the form with the 3′-end single-stranded (termed H-y3) is generally observed, while for longer tracts interconversion between different forms is possible (Htun & Dahlberg, 1989; Kang et al., 1992b). Further forms are made possible by the observation of triplex structures based on the uptake of an oligopurine strand—that is, a structure based on R·Y·R triplets. This has been observed for long oligoguanine·oligocytidine tracts at neutral pH (no cytosine protonation is required) in the presence of magnesium (Kohwi & Kohwi-Shigematsu, 1988), as well as for $(GpA)_n \cdot (CpT)_n$ sequences in the presence of zinc ions (Bernues et al., 1989; Kohwi & Kohwi-Shigematsu, 1993).

Initial information on the stereochemistry of triple helices came from fiber X-ray diffraction studies (Arnott & Selsing, 1974), which suggested that the overall structure was of the A-form with C3′-endo deoxyribose pucker. More recent NMR studies (de los Santos et al., 1989; Rajagopal & Feigon, 1989a,b; Sklenar & Feigon, 1990) indicate that the structure is intermediate between A- and B-form, and the presence of critical NOE connectivities has confirmed the base triplet interactions.

Guanine Tetrad Quadruplex Structures. Sequences containing repeated runs of oligoguanine can adopt novel conformations based on association between guanine bases (Henderson et al., 1987). Such sequences are found in chromosome telomeres, which are generally based on a repeated unit of either $T_{2-4}G_4$ or $T_{2-4}AG_3$, with a single-stranded purine-rich 3′-overhang of 12–20 bases (Zakian, 1989). Formation of two different kinds of structure have been observed by guanine-rich sequences in the presence of monovalent cations. These are based on the inter- or intramolecular assembly of four strands, held together by the formation of tetrads of guanine bases that are hydrogen-bonded between N1 and O6 and between N2 and N7, as first proposed by Gellert et al. (1962) (Fig. 7-9). Tetraplex formation may occur either as a parallel intermolecular association of four strands (Sen & Gilbert, 1988, 1990) or as an antipar-

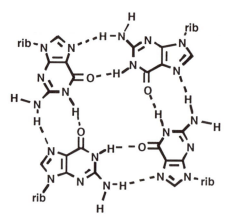

Figure 7-9. Guanine tetrads. Four guanine bases can form a tetrad by hydrogen-bonding from N1 to O6 and from N2 (amino) to N7, in a cyclic manner.

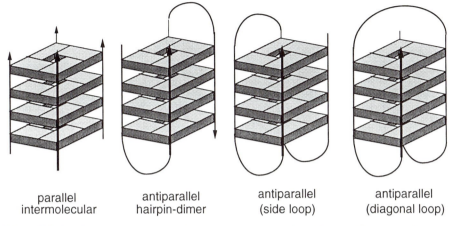

| parallel | antiparallel | antiparallel | antiparallel |
| intermolecular | hairpin-dimer | (side loop) | (diagonal loop) |

Figure 7-10. Tetraplex formation by oligoguanine sequences. A number of different tetraplex structures can be formed by association of oligoguanine sequences, based on the formation of guanine tetrads. Four separate strands containing oligoguanine sequences assemble as a parallel-stranded tetraplex. Oligonucleotides containing two oligoguanine repeats separated by a short region may form a paired hairpin structure; only one example of a number of potential isomeric forms is illustrated. A single strand of a sequence containing four oligoguanine repeats [e.g., $(T_4G_4)_4$] can adopt a foldback structure; two isomers of this structure are possible, as shown.

allel association of either two hairpin-forming strands (Sundquist & Klug, 1989) or a single strand that is folded back to comprise three loops (Williamson et al., 1989). NMR studies show that the parallel structure is based on a backbone conformation in which all glycosyl torsion angles are *anti* (Aboul-ela et al., 1992b, 1994; Jin et al., 1992; Wang & Patel, 1992), while in the antiparallel structure, *syn* and *anti* glycosyl angles alternate along the chain (Smith & Feigon, 1992; Wang et al., 1991). A variety of potential isomeric forms of the antiparallel structures exist, differing in the manner in which the loop regions connect the different oligoguanine blocks of the tetraplex. Structures have been presented in which the loops connect neighboring strands (Kang et al., 1992a) or diametrically separated strands (Smith & Feigon, 1992) (see Fig. 7-10). The structure of a parallel guanine tetraplex has recently been solved by X-ray crystallography to a resolution of 1.2 Å (Laughlan et al., 1994). The structure of dTGGGGT confirms the guanine tetrad structure and reveals the presence of sodium ions on the axis between each pair of tetrads (Fig. 7-11). The parallel tetraplex structure has also been observed in RNA (Kim et al., 1991), where its conformation is closely similar to that of parallel-stranded guanine tetraplex DNA (Cheong & Moore, 1992), and RNA tetraplex formation has been suggested to be important in retroviral genome dimerization (Sundquist & Heaphy, 1993).

The Special Case of $d(ApT)_n$ Sequences—A Structurally Highly Polymorphic Sequence. Alternating adenine–thymine sequences provide an example of a particularly polymorphic structure, which can adopt a variety of conformations depending upon conditions. In principle, $d(ApT)_n$ sequences might adopt a variety of structures, because they are simultaneously 100% A + T (potential melted bubble), alternating purine–pyrimidine sequences (potential Z-DNA), and inverted repeats (potential cru-

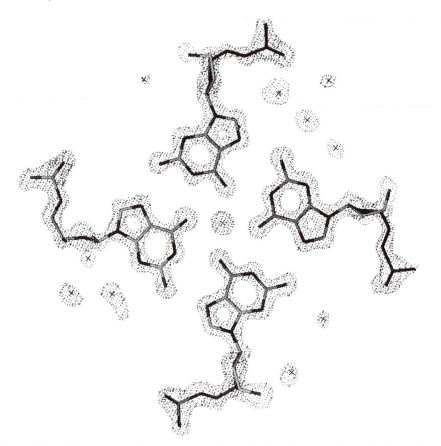

Figure 7-11. The structure of a guanine tetrad determined by X-ray crystallography (Laughlan et al., 1994). Shown here is detail from the 1.2-Å resolution electron density map of a guanine tetrad in the parallel-stranded tetraplex formed by dTGGGGT. Note the presence of the axial sodium ion.

ciform). In addition, extrapolation from the crystal structure of a pApTpApT minihelix (Viswamitra et al., 1978) led to an early suggestion of an alternating B-structure for d(ApT)$_n$ sequences (Klug et al., 1979).

It was found that d(ApT)$_n$ sequences readily adopt the cruciform geometry when present in circular plasmids at sufficient superhelix density (Greaves et al., 1985; Haniford & Pulleyblank, 1985; Panyutin et al., 1985), with the lowest free energy of formation yet observed (13–14 kcal mol^{-1}). Interestingly, no kinetic barrier could be detected, suggesting that extrusion occurs by an extreme example of C-type mechanism, where the d(ApT)$_n$ sequences themselves provide the large-scale helix destabilization. A second abnormality in d(ApT)$_n$ sequences was observed in linear DNA fragments, when it was found that the sequences were uniformly hyperreactive (termed the U-pattern) to a number of chemical probes (McClellan et al., 1986) and enzymes (McClellan et al., 1986; Suggs & Wagner, 1986). Reactivity of thymine bases to osmium tetroxide was strongly increased when the tracts were contained within supercoiled DNA (McClellan & Lilley, 1987). Further analysis of the topology of the extrusion transition by two-dimensional gel electrophoresis suggested that at superhelix densities lower than those that stabilize the cruciform conformation, the $d(ApT)_n$ tracts are preferentially underwound (with a helical repeat of 11.7 bp/turn prior to cruciform formation) (McClellan et al., 1986). The alternating B-structure (Klug et al., 1979) provides a rationale for these observations. This structure is based on an alternating dinucleotide repeat, in which the base stacking at TpA steps (with a high twist of 39°) is sacrificed to improve that of the ApT step (twist of 33°). An alternation of this kind has been observed in the crystal structure of CGCATATATGCG (Yoon et al., 1988). This structure might be expected to have two consequences. First, the very poor stacking at the TpA step cannot be significantly worsened by further unwinding, and thus the $d(ApT)_n$ tracts might act as a torsional "weak point" leading to the observed preferential unwinding. Second, the abnormal twist would lead to exposure of the thymine 5,6 double bond in the major groove, where it would be more accessible to out-of-plane electrophilic addition by osmium tetroxide. The torsional stiffness of the tract might be expected to be low to both under- and overwinding, and it has been shown that the thymines become hyperreactive to osmium tetroxide in *positively* supercoiled DNA (McClellan & Lilley, 1991).

A structural equilibrium exists in $d(ApT)_n$ tracts contained within supercoiled plasmids (McClellan & Lilley, 1987). While the cruciform conformation is stabilized by the presence of salt and low temperature, the uniformly reactive conformation appears at low salt and elevated temperatures. In the presence of nickel ions, a further conformation appears, that of left-handed Z-DNA (Adam et al., 1986; Nejedly et al., 1989). Thus the $d(ApT)_n$ tracts provide a clear example of conformational variability and the effects of prevailing conditions upon the equilibria.

Factors Affecting the Formation of Alternative DNA Structures

As we have seen, DNA structure can potentially exhibit considerable polymorphism, and transitions between conformation variants are possible. A number of factors determine the extent of structural variability that is possible.

Base Sequence. This is probably the most important factor. In general terms the criteria are clear: Inverted and mirror repeats are required for cruciform and H-triplex formation respectively, alternating purine–pyrimidine sequences for Z-DNA, phased adenine tracts for curvature, and so on. Within these requirements there is normally some flexibility, but departure from the ideal sequence will carry an energetic penalty. Thus deviation from perfect twofold symmetry will effectively add the unfavorable free energies of mismatches to the (already significantly unfavorable) free energy of cruciform formation. The ideal Z-forming sequence is a $d(GpC)_n$ tract; departure from this by inclusion of adenine or thymine bases raises the free energy of Z-formation, and breaking the phase of RpY alternation leads to the formation of an unfavorable Z–Z junction. Base modification can have a profound influence on the stability of perturbed DNA structures, exemplified by the stabilization of Z-DNA by cytosine methylation (Behe & Felsenfeld, 1981). We have also seen that the sequence context of regions undergoing structural transitions may exert an important influence (Sullivan & Lilley, 1986).

DNA Supercoiling. Many of the alternative DNA structures have significantly positive-free energies of formation under normal conditions and might not therefore be expected to be of any potential biological importance. However, the formation of most rearranged structures requires a net unwinding of the DNA locally, and in negatively supercoiled DNA molecules this leads to a negative change in supercoiling free energy that may offset the unfavorable free energy of formation of the new structure. Thus negative supercoiling greatly increases the structural repertoire of DNA.

Supercoiling is a property of covalently closed circular DNA molecules (Vinograd et al., 1965). The linking number (Lk), which measures the number of times one strand is linked with the other, is related to two geometrical properties, the twist (Tw, rotation of the strands about the helical axis) and the writhe (Wr, which measures the path of the helix axis in space). These three properties are related by (Fuller, 1971)

$$Lk = Tw + Wr \qquad (7\text{-}1)$$

A relaxed closed circular DNA molecule has a linking number $Lk°$:

$$Lk° = \frac{N}{h} \qquad (7\text{-}2)$$

where N is the number of base pairs and h is the helical repeat under the experimental conditions. A negatively supercoiled molecule has a linking deficit (ΔLk) relative to the relaxed species; that is,

$$\Delta Lk = Lk - Lk° < 0 \qquad (7\text{-}3)$$

Molecules differing in Lk are called *topoisomers* and can be interconverted by topoisomerases. Lk is constant in the absence of strand breakage, and therefore the sum of twist and writhe changes remains constant during any structural change that maintains strand integrity (e.g., B–Z, or B-cruciform transitions). Thus the linking deficit is partitioned between geometric alterations in the molecule of torsional and flexural character:

$$\Delta Lk = \Delta Tw + \Delta Wr \qquad (7\text{-}4)$$

Both deformations are energetically unfavorable, and a supercoiled DNA molecule has a higher free energy compared to its relaxed isomer (Depew & Wang, 1975; Horowitz & Wang, 1984; Pulleyblank et al., 1975),

$$\Delta G°_{sc} = \frac{1050RT}{N} \Delta Lk^2 \qquad (7\text{-}5)$$

where $\Delta G°_{sc}$ is the free energy of supercoiling, R is the gas constant, and T is the absolute temperature. Thus any process that brings about a relaxation of superhelical stress will be favored by negative supercoiling, and this is the basis of the stabilization of a number of structural variants.

Most perturbed DNA structures are underwound relative to B-DNA; that is, their formation contributes a local negative twist change that therefore brings about a partial relaxation of the superhelical stress. The reduction in the free energy of supercoiling therefore helps stabilize the altered conformation. Because the free energy of supercoiling increases quadratically with linking difference [Eq. (7-5)], there will exist a level of negative supercoiling above which the new structure has a stable existence, and this has been experimentally demonstrated for structures such as cruciforms (Gellert et al., 1979; Lilley, 1980; Panayotatos & Wells, 1981), Z-DNA (Peck et al., 1982; Singleton et al., 1982), and H-triplex DNA (Lyamichev et al., 1986). This threshold level of supercoiling, and hence the free energy of formation, may be determined using two-dimensional gel electrophoresis to separate topoisomers (Wang et al., 1983); an example is shown in Fig. 7-12. These arguments apply in reverse for a structure that brings about an overwinding, which should therefore be stabilised by *positive* supercoiling; the only known example is that of the $d(ApT)_n$ sequences discussed above (McClellan & Lilley, 1991).

Complex conformational equilibria are possible within supercoiled DNA circles, discussed theoretically by Benham (1981). It is possible to have sequences within a circular molecule that can adopt more than one conformation, as discussed above for the $d(ApT)_n$ sequences (McClellan & Lilley, 1987). But different sequences within the same molecule will also affect each other, because the transition of one will affect the

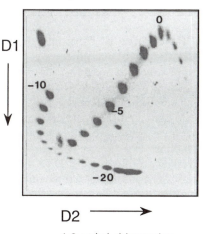

pXG540

D1

D2 ⟶

1.8 µg/ml chloroquine

Figure 7-12. Two-dimensional gel electrophoresis reveals cruciform extrusion by an alternating adenine–thymine sequence in a supercoiled plasmid. A mixture of topoisomers of the plasmid pXG540, which contains an $(ApT)_{34}$ sequence, was electrophoresed in agarose from a single well (dimension D1). Under these conditions the formation of a cruciform by the $(ApT)_{34}$ sequence in topoisomers with sufficient superhelix density relaxes their overall supercoiling, which consequently migrate slower than the equivalent cruciform-free topoisomers with the same linking number. The gel is then electrophoresed in a direction 90° to the first dimension (D2), in the presence of an intercalator. This leads to a partial relaxation, and hence reabsorption of any cruciform structures, leading to a discontinuous "jump" in mobility at the level of supercoiling at which the cruciform became stable in the first dimension (topoisomer −9). From this we can determine the twist change that occurs on extrusion, as well as the free energy of cruciform formation for this sequence. Careful analysis of the data also reveals that the $(ApT)_{34}$ sequence becomes significantly underwound prior to cruciform extrusion.

global topology of the molecule, and hence the probability that the other sequence will undergo a transition. Competitive melting and B–Z transitions that exhibit this kind of coupling have been studied (Aboul-ela et al., 1992a).

Temperature. Any structural transition with a nonzero enthalpy will be affected by temperature. The clearest example of this is the melting of (A + T)-rich regions at low salt concentrations.

pH. If base protonation is required for the formation of a given structure, it will be promoted by low pH. A clear example is provided by the H-triplex, which requires protonation of cytosine to form the required G·C·C$^+$ triplet interaction.

Ionic Conditions. DNA is a polynucleotide in which very high local charge densities may develop. In general, phosphate charge screening by counterions is required, but the folded geometry of DNA structures may also lead to more specific interactions with metal ions. For example, formation of a four-way junction appears to create a site of relatively high affinity for ion interactions, as discussed below. Certain complex metal ions, such as hexammine cobalt (III), are particularly effective in promoting structural transitions in DNA, probably in part due to the ability of the ligands to participate in hydrogen bonding interactions.

The Biological Relevance of Perturbed DNA Structures

When Z-DNA and cruciform structures were first observed, it seemed almost inconceivable that these conformations would not have biological roles. "Anything that DNA can do, it probably does do somewhere" seemed a fairly safe bet. Yet many attempts to ascribe functional roles to perturbed DNA structures have met with failure. After a little more than a decade of investigation we can attempt a more sober appraisal of the role of DNA structure in biology.

The less adventurous variation of the basic B-DNA structure is clearly widely exploited in biology, particularly in terms of protein recognition. DNA curvature is clearly important. But when we move to the more grossly rearranged structures we are on very much weaker biological ground. Z-DNA (Jaworski et al., 1988; Rahmouni & Wells, 1989, 1992; Zacharias et al., 1988), cruciforms (Dayn et al., 1992; McClellan et al., 1990) and H-triplex structures (Karlovsky et al., 1990) have all been detected in plasmid DNA inside bacterial cells, so we know that they can be tolerated to some extent. Yet these sequences were all introduced for the purpose of the experiment, and this is not equivalent to demonstrating a structural transition in the natural DNA of the cell. Of course, one can use these as reporter transitions, to obtain valuable information on local superhelix density in cellular DNA (Dayn et al., 1992; McClellan et al., 1990; Rahmouni & Wells, 1992). But this leaves the assertion that there is as yet no clear biological role for Z-DNA, H-triplex, or cruciforms still unchallenged. It is too early to judge the role of the guanine tetraplex. The selection of telomeres with just those sequences that exhibit a high propensity to form tetraplexes (Murchie & Lilley, 1994) suggests, but clearly does not prove, a biological role for the four-stranded structure. This structure is not a substrate for telomerase (Zahler et al., 1991), leaving possible roles in the protection and/or coherence of chromosome termini.

Perhaps the best argument for biological relevance for a given DNA structure is the existence of proteins that bind and manipulate the structure. Four-way junctions are substrates for structure-selective resolving nucleases and other proteins (see below), implying a biological existence of the structure. In the case of the guanine tetraplex, a series of proteins have been isolated that manipulate the DNA in one way or another. These include (1) a nuclease that is activated by binding to the structure (Liu et al., 1993) and (2) proteins that appear to promote the formation of the tetraplex structure (Fang & Cech, 1993; Giraldo & Rhodes, 1994). The identification of such proteins provides the best hope of establishing the biological role of these structures, because with proteins comes the possibility of applying the powerful methods of genetics.

Non-B-DNA structures may be useful models for biological intermediates. The four-way DNA junction (and thus the cruciform) seems to be an excellent model for the Holliday junction of genetic recombination (Broker & Lehman, 1971; Holliday, 1964; Orr-Weaver et al., 1981; Potter & Dressler, 1976, 1978); such junctions are recognized as substrates for precise cleavage by resolving enzymes for example (Duckett et al., 1992). It has also been suggested that triplex DNA might be a model for heteroduplex formation by RecA protein (Hsieh et al., 1990; Rao et al., 1991), although base substitution experiments appear to weaken this conclusion (Jain et al., 1992).

In general there is much to be learned from studying the full structural repertoire of DNA. Some altered DNA structures are clearly important, and we cannot be sure that any structure is completely excluded from playing a role in the cell. However, the seemingly obligatory claim for an important biological role for every structure that is found should be treated with some caution, and each should be judged on its merits.

Elucidation of the Structure of the Four-Way Junction in DNA

The Stacked X-Structure. The first four-way DNA junctions to be studied were cruciform structures formed by inverted repeat sequences in supercoiled DNA (Gellert et al., 1979; Lilley, 1980; Panayotatos & Wells, 1981). However, these are rather difficult to study, because they require negative supercoiling for their maintenance. Junctions were later constructed from cloned (Bell & Byers, 1979; Gough & Lilley, 1985; Hsu & Landy, 1984) or synthesized (Kallenbach et al., 1983) DNA by hybridizing appropriate sequences, which were incapable of extensive branch migration. Early studies indicated that stable four-way junctions could be formed and that full base-pairing was preserved (Furlong et al., 1989; Gough et al., 1986; Wemmer et al., 1985).

Gel electrophoresis has been very powerful in the study of DNA structures, and it has been particularly useful for the analysis of the structure of the four-way junction. Some years ago it was shown that the creation of a junction at the center of a DNA fragment caused it to exhibit abnormally low mobility in polyacrylamide (Gough & Lilley, 1985), consistent with the introduction of a distinct kink. The electrophoretic mobility was found to be very dependent on the concentration and type of cation present (Diekmann & Lilley, 1987), implying a role for ion interactions in the structure. Cooper and Hagerman (1987) developed a technique based on relative changes in electrophoretic mobility following the pairwise ligation of reporter arms. They concluded that the symmetry of the junction was lower than tetrahedral and that two of the strands were more severely bent than proposed in the model of Sigal and Alberts (1972).

We employed the conceptually equivalent technique of comparing the electrophoretic mobilities of the six possible isomeric junctions with two long and two short arms, created by selective restriction cleavage. In the presence of added cations, we observed a pattern of relative mobilities comprising two fast, two intermediate, and two slow (2:2:2 pattern) species (Fig. 7-13) (Duckett et al., 1988). From this we de-

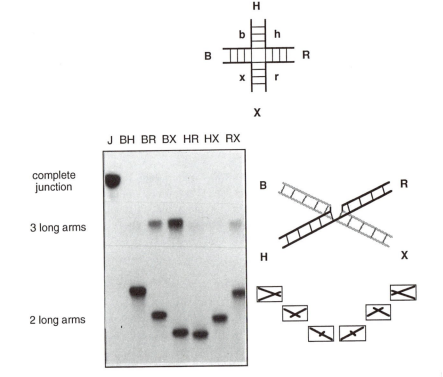

Figure 7-13. Gel electrophoretic analysis of a four-way junction in the presence of magnesium ions. A junction was constructed from four 80-base oligonucleotides (strands b, h, r, and x). Each arm (B, H, R, or X) contained a unique site for a restriction enzyme and could be selectively shortened by enzyme cleavage. The electrophoretic mobility of the six possible double digests of the junction were compared in an 8% polyacrylamide gel in 90 mM Tris-borate (pH 8.3) containing 100 μM $MgCl_2$. The following six tracks contain the junctions after restriction digestion. The digested species are named according to the uncleaved arms (e.g., BH is the junction that has been cleaved in the R and X arms). The digests are incomplete in some cases, and the slower species in some tracks (three long arms) have undergone cleavage in a single arm. Those that have been cleaved in two arms (i.e., containing 2 long arms) exhibit three types of mobility, consistent with a twofold symmetrical X-shape. Cleavage of two arms in an X-shape leads to two species in which the long arms are related by an acute angle, two in which they are related by an obtuse angle, and two linear species; these are indicated by the scheme on the right, assigning the bands to the six possible species with two long arms. Assuming that the mobility will be inversely proportional to the distance between the ends of the long arms, this leads to the antiparallel X-shape as shown. In this junction there is coaxial stacking of the B and X arms and of the R and H arms. By changing the base sequence at the point of strand exchange, it is possible to alter the stacking preference so that there is coaxial stacking of the B and H arms.

duced that the structure of the four-way junction in the presence of metal ions is that of an X-shape.

An X-shape is readily created from a four-way junction by stacking the helical arms in coaxial pairs, followed by a rotation in the manner of opening a pair of scissors. This generates a favorable increase in base-pair stacking interactions, while reducing unfavorable steric and electrostatic interactions between the stacked pairs of arms. The reduction to twofold symmetry generates two classes of strand in the junction; two continuous strands have effectively undeviating axes, while the two exchanging strands pass between the two helical stacks at the point of strand exchange. This is consistent with hydroxyl radical probing of four-way DNA junctions (Churchill et al., 1988); it was found that two strands were protected to a greater extent than the other two.

Two isomers of the stacked X-structure are possible, differing in the selection of helical stacking partners. When the sequence was altered at the point of strand exchange, the electrophoretic pattern of the long-short arm junctions changed, consistent with an exchange of stacking partners (Duckett et al., 1988). This isomerization changes the character of each strand in the structure; exchanging strands become continuous strands, and vice versa. The identity of the most stable isomer will be governed by the thermodynamics of the interactions at the point of strand exchange—that is, probably mainly by the stacking interactions.

Assuming that electrophoretic mobility is determined by end-to-end distance (Lumpkin & Zimm, 1982), the slower species is assigned to those in which the longer arms are related by the smallest angle of the X-structure. This gives an approximately antiparallel alignment of the two continuous strands; that is, the axes of the exchanging strands include an angle significantly smaller than 90°. This is contrary to the normal depiction of Holliday junctions and disagrees with the earlier model of Sigal and Alberts (1972). Therefore, alternative methods that made no assumptions about the relative mobilities of different species in polyacrylamide were sought to analyze the structure.

If the end-to-end distances of a junction with arms of equal length could be compared, it would be equivalent to comparing the angles between arms. Therefore, a method was sought that could provide relative distance information in the range 20–80 Å. The efficiency of fluorescence resonance energy transfer (FRET) is dependent upon distances in this range, and this technique was used for a detailed study of four-way junctions (Clegg et al., 1992; Murchie et al., 1989). A series of junctions of constant base sequence were prepared, with arms of 17 bp each, in which one arm was conjugated with a fluorescein dye at its 5′-end while another was conjugated with rhodamine (Clegg et al., 1992; Murchie et al., 1989). Upon excitation of the donor fluorophore (fluorescein), energy can be transferred to the acceptor (rhodamine) due to dipolar interaction between the two transition moments. This leads to a deexcitation of the donor (i.e., reduced fluorescent quantum yield and shortened lifetime) and an enhanced fluorescent emission from the acceptor. The efficiency of the process depends on the inverse sixth power of the distance between the two dyes (provided that one of the dyes is flexible) and thus can be used to provide distance information. The power and reliability of the method has been demonstrated by studying a series of duplex molecules of varying length, whereupon quantitative agreement was obtained with the theory based on dipolar coupling and the known cylindrical geometry of the double-stranded DNA (Clegg et al., 1993). Comparison of the six possible end-to-end distances of the four-way junction (each of which can be studied in both directions) made it possible to distinguish different models for the structure of the junction. The relative FRET efficiency of a series of labeled junctions was determined in several ways, by measuring (1) the normalized enhancement of acceptor fluorescence, (2) the reduction in donor fluorescent quantum yield, (3) the reduction in the fluorescent nanosecond lifetime of the donor, and (4) the reduction in the measured anisotropy of the acceptor fluorescence (Clegg et al., 1992). All of these techniques confirmed the antiparallel X-structure of the junction. Results for one junction in the presence of magnesium ions are shown in Fig. 7-14. Two of the six end-to-end distances exhibited a greater degree

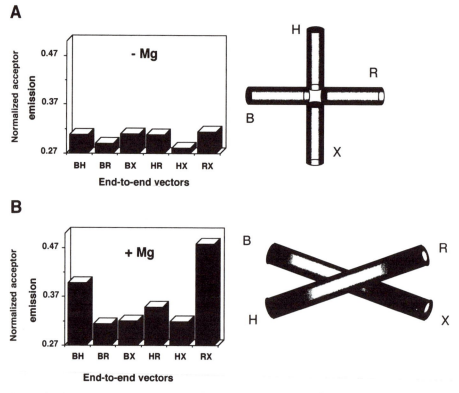

Figure 7-14. Analysis of the structure of a four-way junction by fluorescence resonance energy transfer. The junction chosen for analysis is the same sequence as that analyzed in Fig. 7-13. For these experiments, junctions were assembled from two strands conjugated at their 5′-termini with fluorescein and rhodamine, respectively, and two unlabeled strands, generating a junction carrying the fluorophores at the termini of two arms. The efficiency of FRET between the fluorophores is inversely proportional to the distance between them, raised to the sixth power. Analysis of energy transfer efficiency is performed for each of the six possible end-to-end vectors. The data are presented as normalized acceptor emission values, which contain fluorescent emission from rhodamine due to direct excitation and from FRET. For these experiments the numerical value is 0.27 in the absence of fluorescein (i.e., with direct excitation of rhodamine only); thus under these experimental conditions a value greater than 0.27 indicates acceptor emission due to FRET. (**A**) No added magnesium ions. Under these conditions there is little enhancement of acceptor emission for any of the six end-to-end vectors, indicating an extended structure. (**B**) With added magnesium ions. Two end-to-end vectors exhibit significant increases in FRET efficiency—that is, B to H and R to X. This is fully consistent with the antiparallel stacked X structure in which there is B-on-X coaxial stacking.

of FRET (i.e., shorter distance) than did the remaining four distances, which is clearly consistent with an X-shape. These distances are those expected to be short on the basis of the earlier gel electrophoretic experiments using the same sequence—that is, the antiparallel structure (Duckett et al., 1988). Moreover, upon changing the central sequence of the junction the expected exchange of short distances was observed (Murchie et al., 1989), as expected from the isomerization of the structure deduced from the electrophoretic experiments (Duckett et al., 1988).

Other physical methods agree with this configuration of arms in the four-way junction. Cooper and Hagerman (1989) studied a series of four-way junctions with extended arms by electric birefringence, from which they could calculate angles between selected pairs of arms. They observed that two of the six inter-arm angles were significantly smaller that the other four; these were the two small angles expected for an antiparallel structure. We measured the radius of gyration of a junction comprising two arms of 50 and two of 15 bp by means of neutron scattering. Comparison with values calculated for parallel and antiparallel models showed that the antiparallel structure was in much closer agreement with the experimental value (J. Torbet, A. I. H. Murchie, & D. M. J. Lilley, unpublished data). Indeed, all experimental results known to us indicate that the antiparallel structure is most stable in solution, and there are none consistent with the parallel structure.

Metal Ions and the Folding of the Four-Way Junction. Metal ions play an important role in the structure of the four-way DNA junction. In the absence of added ions the junction cannot fold into the stacked X-structure, but instead remains extended with unstacked arms. Gel electrophoretic experiments (Fig. 7-15) show that the junction adopts a structure with approximately square symmetry in the absence of metal ions (Duckett et al., 1988, 1990), which was confirmed by FRET experiments (Clegg et al., 1994). The folding of the junction into the stacked X-structure creates close phosphate juxtaposition that is destabilizing unless reduced by ion screening, and the fourfold symmetry of this unfolded form indicates that inter-arm stacking cannot occur under these conditions. This is supported by the observation that the thymine bases of A·T base pairs present at the point of strand exchange are strongly reactive to osmium tetroxide in the absence of added ions, but become protected with the folding process (Duckett et al., 1990).

Folding the four-way junction into its stacked form represents a thermodynamic balance between the favorable interactions stabilizing the folded form, among which stacking interactions are probably the most important, and the destabilizing effects of electrostatic repulsion. Unless the latter are reduced by ion screening, the repulsive interactions are dominant and the junction is unable to fold. A variety of ions are able to bring about the folding, with differing efficiencies (Duckett et al., 1990). Group II metals such as magnesium and calcium fold the junction at concentrations higher than about 80 μM. Complex ions and polyamines are more efficient; for example, 2 μM [Co(NH$_3$)$_6$] (III) folds the junction effectively. Group I metal ions, such as sodium or potassium, bring about at least a partial folding of the junction, but only at very high concentrations. FRET experiments suggest that the junction eventually achieves something like the correct global shape in the presence of high sodium concentrations (Clegg et al., 1992), but differences in electrophoretic patterns remain, and junction thymine bases remain reactive to osmium tetroxide (Duckett et al., 1988). The balance between stacking and electrostatic interactions was strikingly revealed by experiments in which selected phosphate groups were replaced by electrically neutral methyl phosphonates:

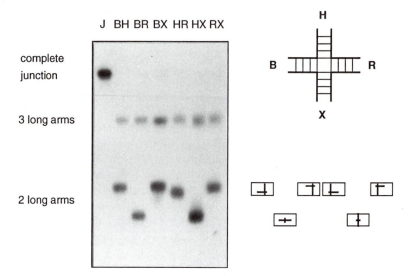

Figure 7-15. Electrophoretic analysis of the structure of a four-way junction in the absence of added metal ions. Long–short arm analysis of a junction in 90 mM Tris-borate (pH 8.3) containing 0.1 mM EDTA. Under these conditions, junctions of any sequence exhibit a pattern of mobilities in which there are four slow and two fast species. This is consistent with a square configuration of helical arms, giving four species where the long arms subtend approximately 90° (slow species) and two species with angles of approximately 180° (faster species), indicated by the scheme on the right.

Different stacking isomers resulted depending on which phosphates were replaced (Duckett et al., 1990). Recently, by means of uranyl photocleavage experiments, a site of high-affinity binding for certain ions has been localized (Møllegaard et al., 1994).

Inclusion of single-base mismatches into the junction can have a destabilizing effect on the folded structure (Duckett & Lilley, 1991). For example, a CC mismatch led to gel electrophoretic patterns indicative of the square configuration of arms in the presence of 100 μM magnesium. In general the effects of these mismatches were dependent both on the nature of the mismatch and on its context. Most junctions, even those containing severely destabilizing mismatches, could be persuaded to fold into the stacked X-structure by increasing ion concentrations, further supporting the putative balance between stacking and electrostatic interactions.

Another perturbation that may lead to alteration of the folded structure of the four-way junction is the presence of discontinuities (nicks) in one of the strands (Pöhler et al., 1994). When these are located at the point of strand exchange, the helices appear to rotate to a position of 90° with respect to each other. Moreover, the nick is placed on the exchanging strands of the junction, even if this means isomerization to the less preferred stacking isomer.

The Stereochemistry of the Four-Way Junction. The above experiments have established the global structure of the four-way DNA junction as an antiparallel stacked X-structure, but this does not address the local stereochemistry at the point of strand exchange. Simple modeling suggests that the most effective way to avoid stereochemical clash is by means of a right-handed, antiparallel structure, as observed experimentally

(Fig. 7-16) (Murchie et al., 1989). This structure allows a relatively favorable juxtaposition of strands and grooves, thereby minimizing steric and electrostatic clash between backbones. This is most effective if the small angle of the X-structure is 60°, which is in good agreement with experimental determinations from FRET, neutron scattering, and electric and magnetic birefringence. Similar packing has been observed in the crystal structures of double-stranded oligonucleotides (Lipanov et al., 1993; Timsit et al., 1989). The anticipated right-handedness of the X-structure was confirmed using FRET (Murchie et al., 1989). The location of the continuous strands in the major grooves 3' to the point of strand exchange leads to a localized protection against cleavage by DNase I (Lu et al., 1989; Murchie et al., 1990).

The collinearity of the helical arms in the coaxial stacks was tested by construction of a junction containing the recognition site for the restriction enzyme *Mbo*II in one arm, oriented such that cleavage required a continuous DNA helix beyond the point of strand exchange (Murchie et al., 1991). Cleavage would therefore imply good collinearity of the arms such that the DNA appears to be continuous from the point of view of the enzyme. Good cleavage was observed, at the expected phosphodiester bonds, suggesting an absence of significant translational, torsional, or flexural distortion at the point of strand exchange.

The antiparallel junction presents two dissimilar sides. When four helices are connected together to form a junction this generates two inequivalent faces, and these are preserved in the formation of the antiparallel stacked X-structure. In this structure, the four base pairs that define the point of strand exchange are oriented in the same di-

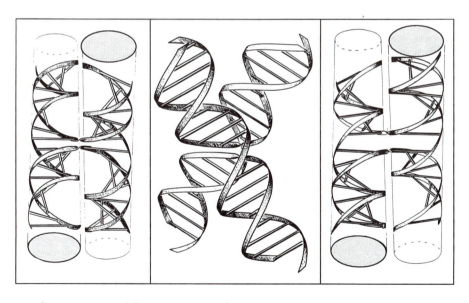

major groove side **face** **minor groove side**

Figure 7-16. A ribbon model of the right-handed, antiparallel stacked X-structure of the four-way DNA junction. **Center:** Face view, showing the X-shape of the folded junction. The two sides of the structure are not equivalent. On one side (**left**) the four base pairs at the point of strand exchange present major groove edges, while on the other side (**right**) the minor groove edges are presented.

rection, leading to sides with major and minor groove characteristics (Fig. 7-16). Thus it may be seen that there is a continuous major groove running down one arm of the major groove side, passing through the point of strand exchange and then continuing on the arm of the opposite helical stack. This asymmetry in the structure has major consequences for interaction with proteins.

A detailed molecular mechanical modeling exercise was performed for the structure of the four-way junction (von Kitzing et al., 1990), and two kinds of structure were identified. One was an extended, unstacked, square structure that closely resembles our view of the junction in the absence of added metal ions. The other was a right-handed, antiparallel stacked X-structure that was in good agreement with all the experimental data. In the folded structure, the stacked helices retained essentially B-DNA conformation, and good stacking was preserved at the point of strand exchange. Two main deviations from normal B-DNA were required at the point of strand exchange to form the four-way junction: (1) a t to g^- change in the ϵ torsion angle and (2) a local widening of the minor groove.

Recognition of the Structure of DNA Junctions by Enzymes. An important and probably ubiquitous class of enzymes that specifically cleave DNA junctions has been isolated from a wide variety of sources including bacteriophage (de Massey et al., 1984; Kemper & Janz, 1976), eubacteria (Connolly et al., 1991; Connolly & West, 1990; Iwasaki et al., 1991), yeasts (Symington & Kolodner, 1985; West et al., 1987), and mammals (Elborough & West, 1990; Jeyaseelan & Shanmugam, 1988; Waldman & Liskay, 1988) (reviewed in Duckett et al., 1992). In addition, there are proteins that selectively bind DNA junctions without inducing cleavage (Lloyd & Sharples, 1993; Parsons et al., 1992), including the interesting class of HMG box proteins (Bianchi et al., 1989; Lilley, 1992). The existence of these enzymes may be taken as the best indication that the four-way DNA junction, or a structure closely related to it, is likely to be functionally important in cells. These proteins carry out an exquisite feat of molecular recognition of a folded DNA structure.

Some indication of the manner of interaction of resolving enzymes and DNA junctions can be derived from a comparison of their cleavage sites. The cleavage positions for T4 endonuclease VII, yeast Endo X1, and a calf thymus resolving enzyme are shown in Fig. 7-17. Each enzyme cuts two of the four strands, consistent with the twofold structural symmetry of the junction. But the three enzymes cleave the junction in different ways, and the cleavage sites appear to have little in common at first sight, apart from the shared symmetry element. However, when these sites are placed on the stacked X-structure a clear relationship emerges. All the sets of cleavages are located on the minor groove side of the junction, suggesting that all three enzymes bind this face of the junction; this is consistent with hydroxyl radical footprinting of T4 endonuclease VII (Bhattacharyya et al., 1991; Parsons et al., 1990) and is also consistent with experiments showing a requirement for about one helical turn of each of the cleaved arms (Mueller et al., 1990). Thus it seems probable that the three enzymes have rather similar mechanisms for binding their substrates, despite their wide evolutionary separation. However, since the cleavages are introduced into different positions on the minor groove side, the binding and catalytic functions of these proteins could be separable, perhaps as different domains. Other resolving enzymes may well bind in different ways to four-way DNA junctions; there are a number of structurally distinct faces to the stacked X-structure that might be recognized by different proteins.

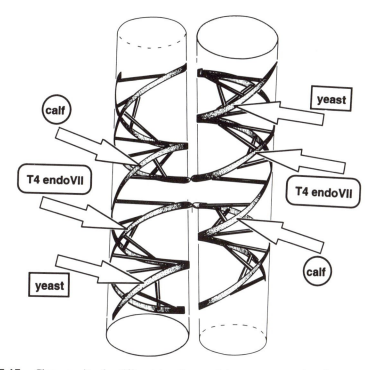

Figure 7-17. Cleavage sites for different junction-resolving enzymes on the minor groove side of the four-way DNA junction. The positions of cleavage by resolving enzymes obtained from phage T4, yeast, and calf thymus are indicated by arrows. All three enzymes cleave on the minor groove side of the junction, introducing two cleavages related by the central dyad axis.

Enzymes of the T4 endonuclease VII class are DNA structure-specific, with relatively low sequence preferences. Thus, if the sequence at the point of strand exchange in a given junction is changed such that the alternative stacking isomer becomes more stable, the cleavage pattern changes correspondingly (Duckett et al., 1988). If a junction of constant sequence is constrained to exist in one or other stacking isomer by means of tethering (Kimball et al., 1990), the resulting cleavage sites are determined by the isomeric form (Bhattacharyya et al., 1991). This demonstrates that the primary feature recognized by the enzyme is the structure of the four-way junction, and it also shows that sequence preferences are only of secondary importance.

How does an enzyme like T4 endonuclease VII achieve this feat of recognition of the structure of the four-way junction? Two possible models come to mind. First, the enzymes might recognize some local feature of the conformation of the junction. It has been demonstrated that the junction generates a high-affinity binding site for a number of intercalators (Guo et al., 1989, 1990; Lu et al., 1990a,b), and thus an aromatic side chain might probe the DNA structure in some way. Alternatively, the global structure of the junction could be recognized. The stacked X-structure provides a number of faces at which there is a relative inclination of two DNA helices; for example, two double-stranded DNA helices subtend an angle of approximately 120° on the minor groove side of the junction. If two subunits of the enzyme make precise contacts with these two helices, protein–protein interactions might allow the angle of inclination of the DNA to be measured (Bhattacharyya et al., 1991).

We have tended to favor the latter model. T4 endonuclease VII cleaves a variety of branched DNA structures, including three-way junctions (Jensch & Kemper, 1986) and base bulges (Kleff & Kemper, 1988), the common feature of which is the presentation of mutually inclined DNA helices. In addition, T4 endonuclease VII cleaves other DNA species containing bends or kinks, including A-tract curved DNA (Bhattacharyya et al., 1991) and DNA containing a single GpG or ApG site modified by *cis*-dichlorodiammineplatinum (II) (Murchie & Lilley, 1993). However, it is likely that the recognition processes are complex, with components of local and global recognition. Moreover, recent data indicate that interaction with some resolving enzymes may distort the global structure of the four-way junction (Duckett et al., 1995).

The kind of structure-specific interactions exhibited by T4 endonuclease VII may be generally exploited by enzymes involved in recombination and repair. We suspect that many such proteins will turn out to be modular, such that part of the protein contains the structure-recognition ability, while another part carries out a chemical reaction on the substrate. In the case of T4 endonuclease VII, the catalytic activity is of course nucleolysis. In *Escherichia coli* two proteins mediate the latter stages of homologous genetic recombination, at which a four-way junction appears to be involved. In promoting branch migration, RuvAB and RuvC appear to be chiefly responsible for the manipulation of the four-way junction (Iwasaki et al., 1992; Tsaneva et al., 1992) and resolution (Connolly et al., 1991; Iwasaki et al., 1991; Sharples & Lloyd, 1991), respectively. Both proteins are capable of the recognition and selective binding of the four-way junction (Parsons et al., 1992; Shiba et al., 1993), but then catalyze different reactions. In the case of RuvC the catalytic function is a nuclease, while in the case of RuvAB the catalytic function is a helicase (Tsaneva et al., 1993), which disrupts the helical structure and thereby promotes branch migration. The recognition and catalytic functions may coexist in the same protein molecule (as in T4 endonuclease VII or RuvC), or they may be in separate molecules (as in RuvAB). It is likely that proteins from higher organisms will function in a related way, and the resolving enzymes from yeast and calf thymus probably fall into this class. In each case the protein must recognize the structure of the junction, and this is likely to be in a manner that is related to that of T4 endonuclease VII.

Summary

The standard structure of DNA is that of a right-handed, fully base-paired, double-stranded helix with a straight axis. While this may be a good description of most sequences under normal circumstances, none of the aspects mentioned is sacrosanct; all can be altered under certain sequences and conditions. Base-pairing can be perturbed to include non-Watson–Crick mismatches or unpaired bases, or it can be rearranged more radically to form cruciform or triplex structures. DNA structures may be based on the association of two, three, or even four strands. The axis of double-helical DNA may be curved or kinked in a number of ways. Certain sequences, most famously those comprising oligoadenine tracts, lead to curvature of the trajectory of duplex DNA, while base bulges introduce a local kinking, the magnitude of which depends on the size and sequence of the bulge. The helical periodicity of duplex DNA is also dependent on sequence and conditions. Within the context of right-handed A and B structures the helical repeat can vary over the range of 10 to 11.5, but the handedness itself may change, exemplified by the formation of left-handed Z-DNA by alternating $(CG)_n$ and related sequences.

The conditions that lead to the formation of structural variants in DNA differ for each structure. Thus the sequence requirements for the formation of cruciform structures (inverted repeat symmetry), H-triplex structures (oligopurine sequences with mirror repeat symmetry), Z-DNA (alternating purine-pyrimidine sequences), and guanine tetraplex structures (oligoguanine sequences) are all particular to the individual structures. However, some aspects are common. Many rearranged DNA structures involve closer phosphate–phosphate approximations and are favored by elevated concentrations of certain ions. For example, magnesium ions are required to stabilize the four-way junction in DNA, while sodium or potassium ions stabilize guanine tetraplex structures. Most perturbed DNA structures generate a net underwinding of DNA and are therefore stabilized by negative supercoiling.

The four-way DNA junction is biologically significant as an intermediate in a number of DNA rearrangements. The structure of this junction has been extensively studied by a variety of techniques, leading to the proposal of the stacked X-structure. In this conformation the helices are associated in coaxial pairs, for which two isomeric forms exist. This structure is promoted by the presence of metal cations, and a binding site has been located at the center, the point of strand exchange. The four-way junction is the substrate for a class of ubiquitous nucleases, the junction-resolving enzymes. These enzymes exhibit a strong structure-selectivity in their action. Thus understanding the structure of the four-way junction and its recognition and manipulation by proteins is fundamental to the study of recombination mechanisms.

Acknowledgments

I thank colleagues and collaborators whose work I have discussed, including D. R. Duckett, A. I. H. Murchie, R. M. Clegg, B. Luisi, R. Bowater, F. Aboul-ela, M.-J. Giraud-Panis, A. Bhattacharyya, E. von Kitzing, N.E. Møllegaard, R. Pöhler, E. Palecek, G. Gough, K. M. Sullivan, and J. McClellan. I thank S. Diekmann and F. Azorin for comments, A. Rich for kind permission to reproduce Fig. 7-2, and the Cancer Research Campaign for financial support.

8

Reporter Groups for the Analysis of Nucleic Acid Structure

Maryanne J. O'Donnell and Larry W. McLaughlin

The ability to attach reporter groups to nucleic acids covalently permits the introduction of functionality or properties not normally present in the native biomolecule. These properties can include unique spectroscopic characteristics or oxidative/reductive capacity. Alternatively, the reporter groups may be small molecules that confer ligand/receptor binding capabilities, may result in further alkylation reactions, or may themselves be substrates for enzymatic activity; they may even be the enzyme itself. The presence of such reporter groups permits the study of nucleic acid structure and the processes in which they participate as a function of the new property or characteristic introduced into the DNA or RNA molecule. In this respect, the presence of a tethered reporter group can serve a number of functions:

1. It can report a hybridization event resulting from the binding of an appropriate complementary target RNA or DNA by a designed sequence probe.
2. It can assist in providing structural information regarding the nature of a particular nucleic acid complex.
3. It can provide information about the nature of protein–DNA(RNA) or ligand–DNA(RNA) complexes.
4. It can be used to report on new properties.
5. It can be employed to alter solubility, charge, or spectroscopic properties.

The initial hurdle to overcome in employing a reporter group of interest to study a particular structural or biological phenomenon is that involving its covalent introduction into the nucleic acid. Although it is possible to introduce reporter groups nonspecifically into nucleic acids, in most applications there is a decided advantage in knowing the location of the reporter group within the nucleic acid precisely.

Covalent Attachment of the Reporter Group

The introduction of specific labels into nucleic acids has been reviewed by Goodchild (1990) and by Manoharan (1993), and in the present chapter only selected examples of typical labeling approaches are examined. The first strategy for DNA conjugation with a reporter group employed nucleic acid polymerizing enzymes, such as DNA polymerase I, to introduce a nucleoside residue modified with the desired reporter group. This enzymatic approach can be valuable if large numbers of fluorophores or other agents are to be incorporated into the sequence of interest, but the use of DNA polymerases is less desirable when a single reporter group is to be incorporated at a specific site in the sequence. Fluorophores and other labels can be introduced directly into the nucleic acid sequence during chemical synthesis procedures; in these cases the label and the linkage tethering it to the oligonucleotide must withstand the acidic, basic, and oxidizing conditions necessary for the assembly and deprotection of the oligonucleotide. An alternative approach introduces a tether or other labeling site during the synthetic procedures which subsequently, in a postsynthetic manner, is modified with the agent of interest. The ideal labeling technique should be simple to employ and versatile in its application at a wide variety of positions within the sequence of interest. The site of labeling should minimize the disruption of the native nucleic acid structure, unless such disruption is the goal of the procedure. Three basic synthetic techniques have been employed to introduce new functionality such as fluorophores, intercalators, and crosslinking agents into chemically synthesized nucleic acids, namely, (1) end-labeling, (2) base/carbohydrate labeling, and (3) phosphate backbone labeling.

End-Labeling. One of the most common labeling techniques involves the introduction of a reporter group at either the 5'- or 3'-terminus of the oligonucleotide. Some procedures permit the *chemical* modification of a terminal phosphate for introduction of the agent of interest, while enzymatic reactions—for example, those employing T4 RNA ligase—can be employed with the appropriate substrate to introduce a modified 3'-terminal phosphate tethering the label. However, it has been more common to introduce a short linker molecule at one of the nucleic acid termini which carries a nucleophilic group such as a sulfhydryl or aliphatic amino group, which can subsequently be allowed to react with the desired functionality. End-labeling techniques are generally simple to employ and have been used successfully in a number of studies, but usually lack versatility; placement of the label at the terminus of an RNA or DNA fragment can be limiting for some applications.

Base/Carbohydrate Labeling. A variety of modified nucleoside residues have been prepared for site-specific incorporation into nucleic acids. Linkers have been attached most commonly to the C5 position of dT (or dU), the N4 amino group of dC, the N6 amino group of dA, the N2 amino group of dG, or the 2'-hydroxyl group of ribonucleosides. These linkers are generally attached during the synthesis of the modified nucleoside (ultimately as a phosphoramidite or other derivative), but more recently "convertible" nucleosides have been developed. Convertible nucleosides are introduced with some relatively labile functionality, such as the 6-phenoxy substituent of a hypoxanthine base, which can be displaced by a linker or a reporter group after assembly of the oligonucleotide. Tethering the label of interest to a base or sugar residue of a specific nucleoside enhances the versatility of the labeling procedure by permitting site-specific labeling at terminal as well as internal positions of an oligonucleotide sequence. The synthetic commitment necessary to prepare the modified nucleoside can be a drawback to this approach, and a number of reports indicate that the presence of linkers, particularly at sites involved in Watson–Crick hydrogen bonding, can result in significant destabilization of duplex DNA structure. Nonetheless, a wide variety of base/carbohydrate labeling procedures have been employed successfully in studies of structure or biological processes involving nucleic acids.

Backbone Labeling. Incorporation of a label into the backbone of a DNA or RNA molecule exploits the internucleotide phosphorus residue as the site for attachment of a tether or appropriate functional group. For example, a phosphorothioate diester can be introduced into a DNA sequence, and the sulfur atom is susceptible to alkylation reactions that permit covalent labeling specifically at this site. Linkers can also be incorporated into the backbone during oxidation of the trivalent phosphite triester to the pentavalent phosphoramidates as described originally by Todd and coworkers and extended more recently by a variety of research groups. As with terminal labeling, the linker can carry a masked amine or thiol functionality as a site for labeling. Backbone labeling is advantageous in that it provides versatility in the location of the label within the sequence and requires minimal commitment to the synthetic preparation of modified nucleosides. For duplex nucleic acids, the backbone is also located on the outer surface of the biomolecule where the presence of a reporter group is less likely to disrupt structure. However, backbone labeling procedures suffer from one potential disadvantage: Introduction of a linker to the internucleotide phosphorus results in two diastereomers. In many cases the diastereomeric character of the labeled sequence may be unimportant for the application of interest, and in some applications the use of individual diastereomers may be advantageous, but additional procedures are necessary to obtain a single diastereomer.

Applications. A very general categorization of applications for nucleic acids tethering reporter groups is one based simply on the number of labels present. One category would involve nucleic acids carrying multiple reporter groups. In these cases the property of interest is often that of detection, or localization, of the probe nucleic acid fragment. It is advantageous in these applications to incorporate multiple reporter groups into the probe sequence in order to enhance the sensitivity of detection. An early, but continuing application for multiply labeled probes involves their use for nonisotopic detection, typically in hybridization procedures. Similar studies have been described for end-labeled, base-labeled, and backbone-labeled DNAs containing multiple reporter groups. Hybridization has been a fundamental technique upon which gene-cloning studies have relied, and it has been important for the commercialization of this technology. Hybridization has also found important applications in genetic screening, including the in vitro detection of viruses and the in vivo location of genetic material. Laboratory techniques in widespread use such as DNA sequencing can also rely on hybridization techniques.

DNA sequences containing one or two strategically placed reporter groups have been used for biophysical studies, primarily to solve structural or mechanistic problems. Changes in the quantum yield or emission spectra for tethered fluorophores can often be employed to interpret the environment of specific sequences or individual base pairs. Fluorescence resonance energy transfer is used to determine the distance between donor and acceptor fluorophores. By preselecting the sites for attachment of the fluorophores, this technique can provide important information about the distance separating two sites within a nucleic acid complex. Spin labels or redox active reporter groups can also be employed to obtain important structural information.

Reporter groups have also been employed to alter the properties of nucleic acids. For example, the biological stability and efficacy of potential DNA-based therapeutic agents (e.g., antisense oligonucleotides) is dependent upon altering the properties of the probe to enhance cellular uptake and stabilize the phosphodiester linkages against nuclease degradation. The tethering of specific agents, such as cholesterol, can significantly alter properties such as cellular uptake. Studies of this type have been reviewed recently (Manohoram, 1993), and thus they are not detailed in the present chapter.

This chapter focuses on the techniques available to introduce one or more covalently tethered reporter groups into nucleic acids, and it describes some of the studies in which the derived conjugate is used to solve specific structural or mechanistic problems.

Covalent Attachment of Reporter Groups to Nucleic Acids

Labeling reactions are available for both the 5′- and 3′-termini of nucleic acids by using enzymatic or chemical synthesis procedures. A variety of modified nucleosides have been incorporated into nucleic acids, and some strategies for introduction of reporter groups of this type are illustrated in this chapter. Also described are methods that permit the attachment of reporter groups to the nucleic acid backbone.

Labeling Nucleic Acids with Reporter Groups at Either the 5′- or 3′-Terminus. Incorporation of a reporter group at the 3′-terminus of an oligodeoxynucleotide can be achieved enzymatically or synthetically, while 5′-labeling has only been achieved by synthetic procedures to date. It is possible to employ a terminal phosphomonoester or hydroxyl group to attach the reporter group of interest, but more commonly a linker molecule is incorporated at one of the nucleic acid termini during standard automated assembly of the oligonucleotide (cf. Chapter 2). At the 5′-terminus this may be accomplished by replacing the terminal nucleoside residue with the corresponding 5′-deoxy-5′-amino or 5′-deoxy-5′-mercapto derivative or by the incorporation of a non-nucleosidic phosphoramidite which possesses a linker incorporating an amino or sulfhydryl functional group. Introduction of an appropriate nucleophile for postsynthetic labeling at the 3′-terminus has been less popular because this terminus is bound to the synthesis support during standard solid-phase-based assembly procedures (Chapter 2). However, a modified nucleoside can be attached to the support such that it occupies the 3′-terminal position, and recent reports describe the modification of the synthesis support prior to assembly of the nucleic acid, which permits the incorporation of a masked reactive functionality at the 3′-terminus. Other 3′-terminal modifications have employed solution-phase synthesis or enzymatic methods to incorporate either (1) a functional group amenable to labeling or (2) the reporter group itself.

Labeling of the 5′-terminus of an oligonucleotide can generally be divided into two categories: (1) the functionalization of a nucleoside building block in order to accommodate a reporter group, along with its incorporation onto the 5′-terminus of the DNA/RNA fragment, or (2) the use of non-nucleosidic linkers coupled to the 5′-terminus of the oligonucleotide.

Since the 5′-hydroxyl group of the 5′-terminal nucleoside in a nucleic acid sequence remains unesterified, one strategy to provide a site for labeling involves the functionalization of one of the common nucleosides by replacement of the 5′-hydroxyl group with either a protected amine (Smith et al., 1985, 1986; Sproat et al., 1987a) or a sulfhydryl residue (Sproat et al., 1987b). The 5'-modified nucleoside phosphoramidites are incorporated into the support-bound oligonucleotide in the final coupling step of assembly of the oligonucleotide. Following deprotection and isolation, the reactive nucleophile is unmasked and can then be used for labeling reactions, as illustrated for the terminal 5'-amino-5'-deoxythymidine residue in Fig. 8-1.

As an alternative to the incorporation of a modified nucleoside phosphoramidite at the 5'-terminus of the oligonucleotide, it is possible to functionalize the 5'-hydroxyl group of the terminal nucleoside. For example, the 5'-hydroxyl group of a support-bound oligomer can be phosphorylated using a variety of reagents. In the presence of a sulfonyl triazolide activating agent, this phosphoryl group can be condensed with an alcohol possessing a reporter group, as has been demonstrated for biotin (Kempe et al., 1985). It has also been reported that the 5'-terminal hydroxyl group can simply be activated with 1,1'-carbonyldiimidazole followed by reaction with an amine to generate a carbamate linkage tethering the reporter group. In this latter work, the aliphatic amine was a non-nucleoside tether terminating in an amino group (De Vos et al., 1990; Wachter et al., 1986).

The use of non-nucleosidic linkers to incorporate a functional group for subsequent conjugation with a reporter group is probably the most popular approach for the 5'-terminal labeling of chemically synthesized nucleic acid oligonucleotides. A number of such linkers are commercially available and similar derivatives are easily prepared. Two common linkers are illustrated in Fig. 8-2. The linkers are designed and prepared as phosphoramidite derivatives (Agrawal et al., 1986; Connolly, 1987; Coull et al., 1986) so that they can be incorporated into the 5'-terminus of the sequence as a last coupling step. The corresponding H-phosphonate (Sinha & Cook, 1988) and phosphotriester derivatives (Ansorge et al., 1986; Francois et al., 1989) are also known. The linker arm contains a masked reactive functional group: Amino (Agrawal et al., 1986; Connolly, 1987; Coull et al., 1986) and sulfhydryl reagents (Connolly & Rider, 1985; Sinha & Cook, 1988) are most common, although the carboxyl derivative has also been employed (Kremsky et al., 1987). Both the sulfhydryl and amino linkers can be protected as triphenylmethyl (trityl) derivatives (Connolly & Rider, 1985), and the hydrophobic trityl protecting group can also be used for high-performance liquid chromatography (HPLC) isolation of the nucleic acid oligonucleotide in much the same way as the terminal dimethoxytrityl (DMT) group has been used for the isolation of unmodified oligonucleotides (cf. Chapter 2) (see, e.g., McLaughlin & Piel, 1984).

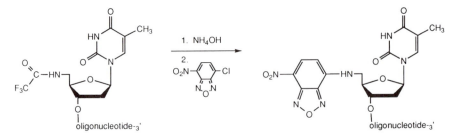

Figure 8-1. Unmasking and labeling a terminal 5'-amino-5'-deoxythymidine residue in a synthetic DNA oligonucleotide.

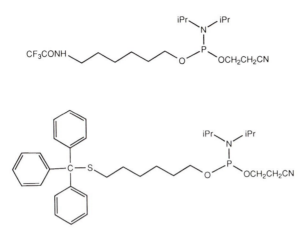

Figure 8-2. Two non-nucleoside linkers containing a masked amino or sulfhydryl reactive center.

A number of variations on this general approach have been reported. In addition to the trityl-based groups, base-sensitive protecting groups [e.g., trifluoroacetyl (Coull et al., 1986) or Fmoc (Agrawal et al., 1986)] have been used to mask the reactive amine center (see Fig. 8-2); these latter derivatives are removed during the standard ammonia deprotection conditions. The length of the tether between the masked functional group and the reactive phosphorus species can be varied to alter the distance between the 5'-terminal nucleoside residue and the tethered reporter group. In some cases the linker is derivatized directly with the reporter group of interest prior to its incorporation into the 5'-terminal position during synthesis (Asseline et al., 1992). Provided that the reporter group is stable to the conditions of oligonucleotide assembly and deprotection, this last approach simplifies the labeling procedure by eliminating the post-synthetic conjugation step.

A further extension of the terminal linker concept includes the use of multifunctional linkers that permit the attachment of multiple reporter groups to a single oligodeoxynucleotide. In one approach, a single terminal linker contains multiple nucleophiles; each of these can, in principle, be functionalized with a reporter group (Haralambidis et al., 1990). A second strategy employs a building block containing protected hydroxyl and amino groups (Nelson et al., 1989; Tang & Agrawal, 1990). The presence of the hydroxyl group permits the extension of the linker in a series of steps, each step resulting in a new site for introduction of the reporter group (Fig. 8-3).

A final method utilized in the 5'-functionalization of non-nucleotide derivatives is the direct incorporation of a reporter group which has been covalently attached to a phosphoramidite. This method involves the synthesis of a specifically functionalized phosphoramidite, as has been described for a number of fluorophores and other agents that can withstand the conditions of oligonucleotide synthesis and deprotection. Phosphoramidite derivatives of biotin (Alves et al., 1989), fluorescein (Cocuzza, 1989), acridine (Asseline et al., 1992; Stein et al., 1988), and anthraquinone (Mori et al., 1989) have all been synthesized and incorporated onto the 5'-terminus of oligodeoxynucleotides.

Functionalized nucleosides can, in principle, be incorporated onto the 3'-terminal position to provide a site amenable to 3'-terminal labeling. In this approach, the func-

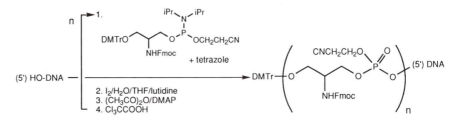

Figure 8-3. Generation of a multifunctional linker from a protected dihydroxyamino building block.

tionalization could occur on either the base or the sugar residue, and both of these approaches are described below. During solid-phase-based oligonucleotide synthesis, the 3'-hydroxyl group of the 3'-terminal nucleoside is attached to the solid support via an ester linkage, thus precluding its availability for functionalization during oligonucleotide assembly by procedures similar to those described for 5'-terminal labeling. Two general approaches have been developed to permit the attachment of reporter groups at the 3'-terminus: (1) the CPG (controlled-pore glass) support is suitably modified with the tether of interest prior to oligonucleotide assembly, or (2) postsynthetic labeling mediated by enzymes is employed.

One procedure that permits incorporation of a non-nucleosidic linker onto the 3'-terminal hydroxyl group involves the reaction of 2,3'-anhydrothymidine with 1,3-propanedithiol or some similar disulfide reagent (Zuckermann et al., 1987). The resulting nucleoside can be allowed to react with succinic anhydride and then attached to the long-chain alkyl amine CPG support, using essentially standard procedures (Fig. 8-4). After assembly of the oligonucleotide, ammonia deprotection removes the oligonucleotide containing the disulfide linker from the support. Reduction of the disulfide unmasks the reactive sulfhydryl functionality, and postsynthetic labeling can then be carried out.

The chemistry described to incorporate a sulfhydryl functional group has been extended to non-nucleosidic linkers, thus eliminating the need to have a thymidine residue present at the 3'-terminus of the oligonucleotide. In one study, an alkanethiol linker (Fig. 8-5) was protected with a DMT group and subsequently tethered to the DNA synthesis support through a simple disulfide linkage (Gupta et al., 1991). As shown, this permits the linker to be coupled to the support, after which the synthesis of the desired sequence can proceed; following ammonia deprotection and disulfide reduction, the thiol-containing linker can be generated, tethered to the 3'-terminal phosphate. This method can be employed to produce a free sulfhydryl residue, a protected thiol, or a 3'-terminal phosphate depending on the deprotection conditions employed and the choice of the spacer arm (Gupta et al., 1991).

Support modification can also permit the preparation of a linker containing an attached 3'-terminal amine. An aliphatic amino-alcohol can be immobilized on a solid support derivatized with 2,2'-dithiodiethanol via a carbamate linkage (Asseline & Thuong, 1989, 1990). Following deprotection with ammonia and cleavage of the disulfide, successive elimination of β-mercaptoethanol and decarboxylation produces a linker bound to the 3'-terminal hydroxyl group which contains a reactive amino functional group.

Similarly, it is possible to condense a solid support derivatized with 2,2'-dithiodiethanol with an appropriately protected nucleoside phosphoramidite, and then oxidize the intermediate phosphite triester with elemental sulfur. After completion of the synthesis, then treatment with aqueous ammonia and dithiothreitol, a 3'-terminal phosphorothioate results (Thuong & Asseline, 1991). Although less reactive than the sulfhydryl group, phosphorothioate esters are amenable to labeling reactions by alkylation of the sulfur residue.

Reporter groups can be attached to the 3'-terminus of oligodeoxynucleotides by utilizing enzymes such as T4 RNA ligase or terminal deoxynucleotidyltransferase. For short, chemically synthesized nucleic acids, the enzymatic procedures may not offer any advantages over the chemical ones, but enzymatic approaches can be invaluable for the labeling of DNAs and RNAs derived from naturally occurring nucleic acids.

T4 RNA ligase will utilize derivatives of the general type A(5')pp-R as donor substrates and will transfer p-R to a free 3'-hydroxyl group of the acceptor nucleic acid sequence. In this manner, fluorescein, biotin, and tetramethylrhodamine derivatives have been conjugated to adenosine pyrophosphate and have been utilized to label the 3'-terminus of tRNA in the presence of T4 RNA ligase (Richardson & Gumport, 1983). In a related study, T4 RNA ligase was used to transfer pCp(s) (containing a 3'-phosphorothioate) to the 3'-terminus of a tRNA. This terminal phosphorothioate could then be alkylated by a variety of reporter groups (Cosstick et al., 1984).

Multiple labeling of a naturally occurring DNA can be achieved by using terminal deoxynucleotidyltransferase. This enzyme utilizes dNTPs as substrates and catalyzes the addition of dNMP residues to the 3'-terminus of DNA fragments. Typically, about 200 nucleoside additions occur. In one case, the dNTP substrate was the 4-thio derivative of 2'-dUTP. After enzymatic polymerization, the thio group could be labeled with fluorescent probes (Eshaghpour et al., 1979).

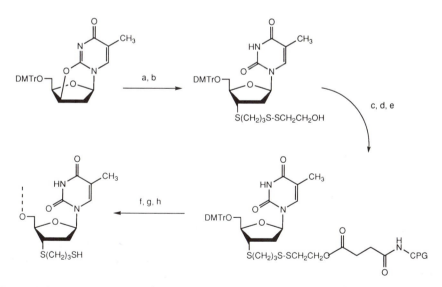

Figure 8-4. Incorporation of a 3'-terminal linker into a support-bound oligonucleotide. Conditions: a, 1,3-propanethiol; b, S-(2-thiolpyridyl)-2-mercaptoethanol; c, succinic anhydride/pyridine; d, p-nitrophenol/DCC; e, aminopropyl-CPG; f, DNA elongation; g, ammonia deprotection; h, dithiothreitol (DTT).

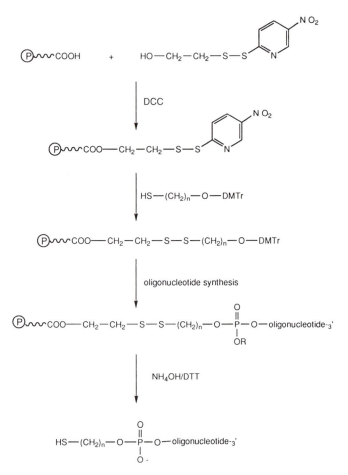

Figure 8-5. Modification of the DNA synthesis support (P) in order to incorporate a thiol linker into the 3'-terminus of the oligonucleotide (when $n = 2$, intramolecular nucleophilic attack by the thiol occurs generating ethylene sulfide and the 3'-terminal phosphate).

Labeling Nucleic Acids with Reporter Groups Attached to the Base or Carbohydrate Residue. For some studies it is preferable that the reporter group be located at a position within the interior of a given sequence. In these cases, tethering the label to either the 5'- or 3'-terminus is insufficient. Labeling within the interior of an oligonucleotide in a sequence-specific manner can be accomplished using a suitably modified nucleoside residue. Although, in principle, an appropriate tether can be incorporated into a nucleoside residue at virtually any site, the most common sites for nucleoside labeling include (1) the C5 and C4 positions of the pyrimidines, (2) the exocyclic amino groups and the C8 positions of the purines, and (3) the 2'-hydroxyl group of the carbohydrate residue. In most cases a nucleoside phosphoramidite is prepared that either carries the reporter group, has been functionalized with an appropriate linker tethering a masked functional group, or is appropriately functionalized such that a reporter group or linker can be attached postsynthetically. With this approach, both the position and the number of reporter groups present are determined simply by the position(s) and number of times the modified nucleoside is incorporated into the sequence.

In addition to the chemical synthesis of oligonucleotides containing modified nucleosides designed to contain attached reporter groups, enzymatic preparations have also been described. The most common approach is that in which the labeled nucleoside is prepared as the triphosphate derivative, and one of the DNA polymerases is employed to incorporate the nucleoside and attached label. While it is generally difficult to use this procedure for the incorporation of a single reporter group, the multiple labeling of nucleic acids using enzyme-based synthesis can be very effective (Langer et al., 1981; Leary et al., 1983).

The derivatization of the C5 position of pyrimidines has involved uridine nucleosides almost exclusively; the products are modified thymidines in which the C5 methyl group is attached to the reporter group of interest via a suitable linker. This functionalization is one of the most common of all the nucleobase modifications. The popularity of this site for the introduction of a reporter group undoubtedly reflects the lack of involvement of the C5 position in Watson–Crick hydrogen bonding; the reporter group at this site also extends into the major groove of duplex DNA such that little steric hindrance results. Derivatization of the C5 position is generally performed on deoxyuridine by mercuration followed by treatment with HCl . A palladium-catalyzed coupling is then employed to attach the linker to the base covalently (Bergstrom & Ogawa, 1978; Bergstrom & Ruth, 1976), and the nucleoside is subsequently converted to the protected phosphoramidite. A representative synthetic scheme is shown in Fig. 8-6, but a number of related syntheses have been reported in which the length of the alkylamine linker arm has simply been varied (Jablonski et al., 1986; Ruth, 1984). A variety of similarly designed linkers incorporated by the palladium-mediated coupling and terminating in an amino group have been reported (Dreyer & Dervan, 1985; Telser et al., 1989a,b,c).

A second route to a C5-modified thymidine employs the 5-halogenated nucleoside, such as 5-iodo-2'-deoxyuridine (Gibson & Benkovic, 1987). A similar palladium-mediated coupling generates the C5 linker tethering a masked amino group amenable to labeling reactions (Haralambidis et al., 1987; Meyer et al., 1989; Spaltenstein et al., 1988).

Derivatization of the C4 position of a pyrimidine (typically cytidine, thymidine, or uridine) suffers from the possible drawback that this site is involved in Watson–Crick base-pairing. Attachment of a linker to the C4 position of pyrimidines typically occurs by nucleophilic displacement of a leaving group introduced at this position. The first

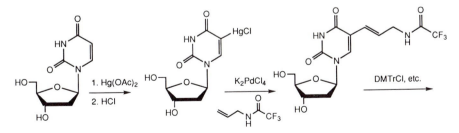

Figure 8-6. Incorporation of a linker containing a masked amino group at C5 of 2'-deoxyuridine by mercuration and palladium-catalyzed coupling.

reports of this approach employed a pyrimidine nucleoside functionalized at the C4 position with 1,2,4-triazole; this heterocycle could be displaced easily by an amine (Maggio et al., 1984; Sung, 1981, 1982; Urdea et al., 1988; Webb & Matteucci, 1986). In addition to the triazole derivative, pyrimidines functionalized as the *p*-toluenesulfonyl (Kierzek & Markiewicz, 1987), *o*-nitrophenyl (Sproat et al., 1989), and 2,4,6-trimethylphenyl (MacMillan & Verdine, 1990, 1991) derivatives undergo similar nucleophilic displacements. Similarly, sulfur may be displaced from 4-thiopyrimidine by a primary amine (Nikiforov & Connolly, 1992; Roget et al., 1989) and, in a related approach, it is possible to employ bisulfite-catalyzed transamination to displace the amino group of cytidine from the C4 position (Draper, 1984; Gillam & Tener, 1986; Telser et al., 1989a).

Amines are the most common nucleophiles employed for these functionalizations. It is possible to attach the reporter group directly to the C4 position provided that it carries a suitable amino group or other nucleophile, but more commonly a short linker such as 1,3-diaminopropane or a masked thiol is employed in this reaction. Suitable protection and conversion to the phosphoramidite affords a modified nucleoside bearing a functional group amenable to labeling reactions.

A recent advance in the use of C4-labeling is "convertible nucleosides," which are functionalized with a trimethylphenoxy group (MacMillan & Verdine, 1990, 1991) or with some other leaving group that will withstand the conditions of DNA assembly and deprotection. Postsynthetic reaction with a nucleophilic reagent permits the incorporation of a reporter group or appropriate linker (see Fig. 8-7). One advantage of this latter approach is that after a single DNA synthesis and isolation procedure, the sample can be divided for a series of postsynthetic modification reactions in order to attach a variety of reporter groups or to generate a series of fragments with a single reporter group tethered to a series of linkers of varying lengths.

Derivatization of purines has received far less attention than labeling of pyrimidines, but recent advances have simplified procedures for the introduction of appropriate linker molecules. One site for modification is that of the exocyclic amino groups. The pro-

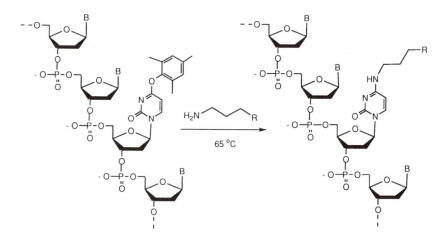

Figure 8-7. Incorporation of various tethers at the N4 position of cytidine by postsynthetic displacement of a trimethylphenoxy leaving group (R = NH$_2$, SH, COOH, etc.).

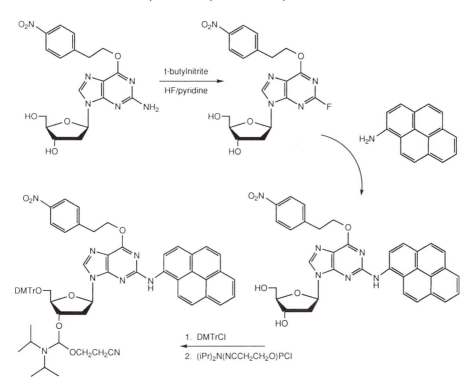

Figure 8-8. Derivatization of the exocyclic amino group of guanosine by functionalization of C2.

cedures for the introduction of a reporter group, or appropriate tether, generally parallel those described above for the C4 position of pyrimidines. Such derivatization has been accomplished by displacing a leaving group from the C2 (guanosine) (Erlanson et al., 1993; Harris et al., 1991; Kouchakdjian et al., 1989; Lee et al., 1990; Zajc et al., 1992) or C6 (adenosine) positions (Kazimierczuk et al., 1984; Lakshman et al., 1991, 1992; Lee et al., 1990). The leaving group is often chloride or fluoride, but phenoxy has also been employed for adenosine (Fereutz & Verdine, 1992).

For functionalization of the 2-position of dG, the 2-fluoro derivative is commonly used as a reactive intermediate (Erlanson et al., 1993; Harris et al., 1991). After preparation of the nucleoside analogue with a labile leaving group, functionalization can occur as part of an overall scheme (see Fig. 8-8) to generate a phosphoramidite building block (Lee et al., 1990) or in a postsynthetic manner (Erlanson et al., 1993; Harris et al., 1991) similar to that described above for the pyrimidines.

An N6-functionalized adenosine may be synthesized in a conceptually similar manner. Nucleosides containing 6-chloropurine (Kazimierczuk et al., 1984; Lee et al., 1990), 6-fluoropurine (Lakshman et al., 1991, 1992; MacMillan & Verdine, 1991), or 6-phenoxypurine (Ferentz & Verdine, 1992) nucleobases have been used for functionalization at the N6 position of adenosine.

The synthesis of C8-functionalized purines proceeds in a manner similar to that described above for the exocyclic amino groups. In this case the most common precursor is 8-bromoadenosine, which will react with protected cystamine derivatives to produce a linker tethering an amino group at the 8-position (Roduit et al., 1987). This approach to functionalization at the C8 position has been extended by others using such species as a dansyl fluorophore (Singh et al., 1990) or a psoralen crosslinking agent (Pieles et al., 1990).

The 3'- and 5'-hydroxyl groups have been employed to introduce reporter groups to the respective 3'- and 5'-termini of oligonucleotides, as described above in the section on terminal labeling. For labeling at interior sites within a given DNA or RNA oligonucleotide, only one position in the sugar portion of the nucleoside structure, the C2' carbon, has been examined for its ability to function as a site for the introduction of reporter groups. Two similar procedures have been described. In one case the 2'-hydroxyl group of a ribonucleotide is alkylated directly with the reporter group of interest (Yamana et al., 1991), or an appropriate linker tethering a masked amino (Guinosso et al., 1991; Manoharan et al., 1991) or thiol functional group (Manoharan et al., 1993) is employed. Although the initial 2'-O-alkylation product is contaminated by some 3'-O-alkyl isomer in these reactions, at least in the case of adenosine the two regioisomers can be readily resolved. In the second case, the related amino group of a 2'-amino-2'-deoxynucleoside can be alkylated with fluorophores or other agents (Aurup et al., 1994).

Functionalization of the Nucleic Acid Backbone. The functionalization of an internucleotide phosphodiester group at one of the nonbridging oxygens permits the introduction of a reporter group into an internal site of a nucleic acid sequence. The attachment of reporter groups to the DNA backbone offers a number of advantages over the modification of a nucleoside residue at an internal sequence position. For example, by using internal phosphodiester groups virtually any site within the sequence is amenable to introduction of the desired functionality, without the need for synthesis of a specifically modified nucleoside. The phosphate residues are not involved in interstrand Watson–Crick base-pairing, so the attachment of a linker or label at such sites should not drastically alter the stability of the nucleic acid complex, unless the agent of interest is itself designed to react/interact with the complex.

Modification of the prochiral phosphodiester residue with a single residue creates a chiral site and two phosphorus diastereomers (Rp and Sp). This diastereomeric character may be a disadvantageous in some applications, but offers some advantage for agents designed to bind, crosslink, or react at specific sites within the nucleic acid duplex. For double-stranded DNA(RNA) structures, one phosphorus stereoisomer directs the covalently attached derivative more toward the major groove of an A- or B-form helix, while the second diastereomer directs the agent more toward the minor groove. If the desired agent binds preferentially in one of the helix grooves, or reacts selectively with a functional group located in a specific groove, labeling of the backbone in a stereocontrolled fashion could assist in enhancing the desired reactivity. A number of modified DNA/RNA backbones have been reported; this functionalization can take the form of a phosphorothioate, phosphorodithioate, phosphoramidate, phosphotriester, methylphosphonate, or methylphosphonothioate. Of these possibilities, three have been employed for the introduction of reporter groups, namely, (1) phosphotriesters, (2) phosphorothioates, and (3) phosphoramidates.

Relative to the other two existing methods, labeling of phosphodiester groups with reporter groups that involve phosphotriester formation has been quite limited. It has not been possible to alkylate a specific phosphodiester moiety effectively in a postsynthetic manner; in order to select a specific site, it has been necessary to incorporate the reporter group or linker during the assembly of the oligonucleotide. In one study (Asseline et al., 1984) an acridine derivative was tethered to a phosphite diester in which the second substituent was the protected nucleoside monomer. Following conversion to the phosphoramidite, this modified nucleoside was coupled to the 5'-hydroxyl group of a growing oligonucleotide chain. After the completion of the oligonucleotide coupling procedures, deprotection and purification yielded a diastereomeric mixture of a DNA fragment containing the acridine derivative covalently bonded through a phosphotriester linkage.

A phosphorothioate-containing oligodeoxynucleotide results when one of the non-bridging oxygens of an internucleotide phosphodiester linkage is replaced with a sulfur atom. This is a conservative modification because the nucleic acid retains both its negative charge and its overall geometry; the subtle electronic differences between oxygen and sulfur permit the specific labeling of the phosphorothioate diester in the presence of normal phosphodiester groups.

Oligodeoxynucleotides containing multiple phosphorothioate internucleotide linkages can be synthesized by enzymatic means because the α-thionucleoside triphosphates are substrates for a number of polymerases (Vosberg & Eckstein, 1977). Chemical synthesis of phosphorothioate-containing oligodeoxynucleotides is the preferred technique for the introduction of a single labeling site into an oligodeoxynucleotide sequence. Phosphorothioate triesters can be introduced during the assembly of an oligodeoxynucleotide by oxidation of the intermediate phosphite triester at a specific site with a sulfur oxidation reagent. The phosphorothioate triester exists as a thione and is stable to the subsequent oxidation steps necessary to produce phosphotriester linkages in the remainder of the oligodeoxynucleotide. After assembly, deprotection of the oligodeoxynucleotide affords the phosphorothioate diester. Phosphorothioates can also be formed from the reaction of an internucleotidic *H*-phosphonate with elemental sulfur; however, in this approach the *H*-phosphonate is oxidized directly to the phosphorothioate *diester*. This species is not compatible with subsequent oxidations employing $H_2O/I_2/THF$/lutidine; the phosphorothioate diesters react quickly with iodine, resulting in desulfurization and conversion to the phosphate diester. The *H*-phosphonate method has been employed to introduce phosphorothioates efficiently at all positions within a sequence, but is less useful for the introduction of a single internal phosphorothioate diester.

Phosphorothioate esters are not as nucleophilic as the corresponding alkyl thiols, but reporter groups such as monobromobimane or those containing haloacetamide, aziridinylsulfonamide, or γ-bromo-α,β-unsaturated carbonyl linkers can be used effectively to modify the thioester functionality covalently (Fig. 8-9) (Fidanza & McLaughlin, 1989; Fidanza et al., 1992). Alkylation affords the corresponding phosphorothioate triester, and these triesters exhibit some base lability. Under neutral and slightly acidic conditions the phosphorothioate triesters are quite stable, but at higher pH hydrolysis of the triester proceeds, almost exclusively by P—S bond cleavage.

As noted above, the substitution of sulfur for a nonbridging oxygen creates a chiral center about the tetrahedral phosphorus, giving rise to Rp and Sp diastereomers. In some cases the two diastereomers can be resolved by HPLC, either before or after attachment of the reporter group, but this approach is unlikely to be successful in all cases. An individual phosphorothioate diastereomer can be obtained reliably by the synthesis of an isomerically pure phosphorothioate dinucleotide phospho-

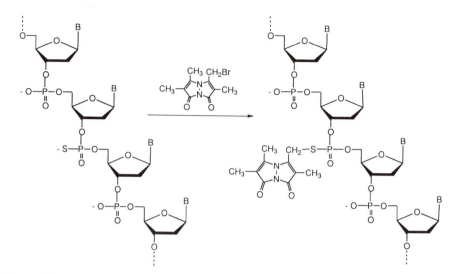

Figure 8-9. Labeling of a single internucleotidic phosphorothioate diester with monobromobimane.

ramidite building block (of either the Sp or Rp configuration) and incorporation of this dimeric unit into an oligodeoxynucleotide during its stepwise synthesis (Cosstick & Eckstein, 1985).

The substitution of one of the nonbridging oxygens of an internucleotidic phosphodiester linkage by an amine results in a phosphoramidate diester capable of tethering an appropriate reporter group. The amine may be attached to a reporter group, or an alkylamine linker terminating in a masked reactive center may be employed; both amine and thiol reactive sites have been used in this latter approach. The P—N bond of the labeled phosphoramidate is more stable at elevated pH values than the P—S bond in the phosphorothioate, but it has the same characteristic in producing two diastereomers about phosphorus. As with the other linkers described earlier, the length of the tether may be varied to extend an attached reporter group away from the oligodeoxynucleotide backbone as needed.

Phosphoramidates have been incorporated into oligomers using both phosphite triester-based (Letsinger & Heavner, 1975; Letsinger & Schott, 1981) and hydrogen phosphonate-based (Froehler, 1986) chemistries. An intermediate phosphite triester can be reacted with an aqueous solution of ethyl azidoacetate and followed by treatment with a 1,4-diaminobutane to introduce a linker tethering a reactive primary amine. This reaction has only been reported in solution for a dimer, but would likely be successful for longer sequences in solution or bound to a solid support. In a related report, a 1,5-diaminopentane linker was incorporated as a phosphoramidate and subsequently labeled with an acridine derivative (Jäger et al., 1988).

The more usual method for preparing internucleotide phosphoramidates involves the oxidation of a hydrogen phosphonate in the presence of a primary or secondary amine, as described originally by Todd and coworkers (Atherton et al., 1945; Atherton & Todd, 1947). In a recent development of this reaction, a nucleoside H-phosphonate is incorporated into a growing oligonucleotide chain at the site of labeling and the intermediate H-phosphonate linkage is immediately oxidized with carbon tetrachloride/pyridine in the presence of the amine linker or reporter group (Fig. 8-10). The use of di-

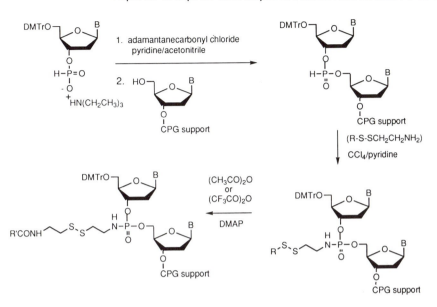

Figure 8-10. Incorporation of an internucleotidic cystamine linker to provide a masked sulfhydryl residue for covalent attachment of reporter groups [R = CH$_2$CH$_2$NH$_2$ or CH$_2$CH$_2$NHCOC(C$_6$H$_5$)$_3$. R′ = CH$_3$, CF$_3$, or C(C$_6$H$_5$)$_3$.].

aminoalkanes in this reaction results in the incorporation of a linker terminating in a primary amine, while the use of cystamine provides the corresponding terminal sulfhydryl residue.

As noted above, two phosphoramidate diastereomers result after oxidation of the intermediate phosphite triester or *H*-phosphonate. In principle, the preparation of an individual diastereomer could be accomplished by synthesis of the corresponding nucleoside phosphoramidate dimer, as accomplished in the case of phosphorothioates. However, a simpler approach is also available, at least for the cystamine-based phosphoramidate. Diastereomeric resolution (by HPLC using reversed-phase columns) is often enhanced after incorporation of the reporter group, particularly one that is large and hydrophobic. The cystamine derivative contains a terminal amino group that is lost after reduction of the disulfide to unmask the sulfhydryl reactive site. We have taken advantage of this observation and prepared the *N*-triphenyl acetyl protected cystamine linker. Upon incorporation into oligodeoxynucleotides as an internucleotidic phosphoramidate, the bulky triphenyl acetyl group permits resolution of the two diastereomeric sequences (up to sequence lengths of 20) (Fig. 8-11). Reduction of the disulfide prior to labeling unmasks the reactive thiol and releases that portion of the linker containing the triphenylacetamide (Fig. 8-11) (O'Donnell et al., 1993).

Selected Applications for Nucleic Acids Reporter Groups

The development of methods for the attachment of reporter groups to nucleic acids are numerous and the applications for these conjugates are just as varied. In the following

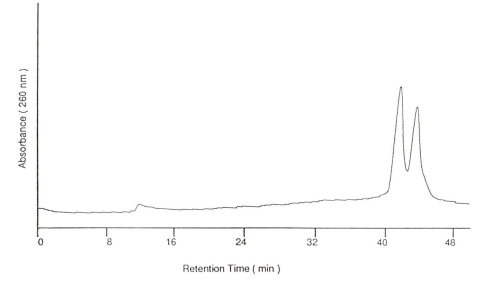

Figure 8-11. Resolution of the two cystamine-based phosphoramidate diastereomers.

section a series of applications are described in some detail; this is not a comprehensive list of such applications, but rather a group of selected applications that show the value and variety of studies that can be accomplished with suitably labeled nucleic acid fragments. We have chosen applications that encompass (1) the detection of nucleic acids, (2) the details of macromolecular complex formation, (3) structural analysis by the use of fluorescence energy transfer, (4) the use of redox chemistry to report on complex formation, (5) the study of specific complexes using photoaffinity agents, and (6) the study of new DNA properties.

Detection of Nucleic Acids. The two areas of study in which it is important to be able to detect and localize the presence of nucleic acids of specific sequences are those of DNA/RNA hybridization and DNA/RNA sequencing. In many cases where detection of a DNA sequence is necessary, terminal labeling, base labeling, and phosphorus backbone labeling procedures have been used successfully to monitor small quantities of nucleic acid. The incorporation of multiple labels has been very effective for the location of hybridization products in the absence of sophisticated detection equipment, while terminally labeled DNA fragments in conjunction with appropriate electronic detection can provide an effective means of DNA sequencing.

In the simplest approach to sequence location by hybridization techniques (Southern hybridization), the target DNA/RNA sequence is degraded into variously sized fragments that are resolved by electrophoresis, denatured, and transferred to a membrane such as nitrocellulose. The probe sequence is commonly radioisotopically labeled and added to the membrane. After incubation, the excess probe is removed and autoradiography can be employed to locate the hybridization product. An early example of multiple labeling to enhance detection sensitivity in hybridization processes involved the use of uridine triphosphate (UTP) or the corresponding 2′-deoxy derivative (dUTP) containing a linker attached to the C5 carbon atom. The terminal amine of the linker was conjugated with a biotin derivative to covalently label the base residue (Langer et al., 1981; Leary et al., 1983). The biotin-labeled nucleoside triphosphates (Bio-UTP or Bio-dUTP) are substrates for DNA and RNA polymerases and can be incorporated into DNA/RNA sequences by standard nick translation labeling procedures. Under the conditions used, the biotinylated nucleosides were incorporated into each site that normally required a uridine (or 2'-deoxythymidine) residue, such that the DNA/RNA fragment was multiply labeled with biotin residues. The biotin-labeled hybridization probes responded similarly to the native probes in hybridization experiments and resulted in unique hybridization products with little background effect.

Biotin itself is not fluorescent, but can result in the generation of a colored signal under appropriate conditions. For example, alkaline phosphatase can be crosslinked into a polymeric form and subsequently biotinylated and complexed with the protein avidin. The avidin complexed polymer can be further complexed with additional biotin ligands such as those present in the hybridization complex. After hybridization with a biotinylated probe, and complexation with the alkaline phosphatase crosslinked polymer, the enzyme complex is effectively located only at the site(s) of hybridization. The location of each hybridization complex is then determined by the addition of a colorless alkaline phosphatase substrate, which upon dephosphorylation generates a colored product. One such substrate is nitro blue tetrazolium, which generates a blue, insoluble product on the hybridization membrane. Although the number of biotinylated uracil bases actually complexed with the alkaline phosphatase polymer was not determined, the beauty of this early nonradioactive labeling approach was that the enzyme was used to effectively amplify the signal. For a single biotin ligand attached to the hybridization product, a polymeric enzyme complex could be attached, and each enzyme component would result in the generation of multiple molecules of the insoluble blue product. Picogram quantities of target sequence could be visualized using this method. Biotin has been employed in similar fashion to generate fluorescence (Broker et al., 1978) or chemiluminescence (Schaap et al., 1987) for the detection of hybridization products or DNA sequencing ladders.

The Sanger method of DNA sequencing relies upon the enzymatic synthesis of a series of DNA sequences complementary to the target DNA/RNA fragment and randomly terminating at a specific base residue (A, G, C, or T) (Sanger et al., 1977). Each of four reaction mixtures contains the template sequence, a primer, *one* of four possible 2',3'-dideoxynucleoside triphosphate (ddNTP) chain terminators, and the appropriate polymerase. Each reaction results in the desired series of fragments randomly terminating with one of the four dideoxynucleoside analogs (A's, G's, etc). Autoradiography of the polyacrylamide gel resulting from these four reaction mixtures results in a series of oligonucleotide ladders from which the nucleic acid sequence can be deduced easily. Radioisotopic labeling is commonly employed in order to detect the positions of the various bands in the sequencing ladders generated. However, more recent developments permit the use of fluorescence labeling as the means of band detection. Two similar procedures have been developed; in both cases four fluorophores are employed, one for each of the standard sequencing reactions (Prober et al., 1987; Smith et al., 1986). The fluorophores differ somewhat in their emission characteristics. The advantage of using four different fluorophores is that the emission characteristics of each fluorophore are related to a particular ddNTP chain terminator. Therefore, rather than using four different gel sequencing lanes, all four reactions can be resolved in a single lane by gel electrophoresis. The relative position in the gel of each band gives sequence location, and the emission characteristics of the fluorophore identify the base residue. The two procedures differ only in the manner in which the DNA sequences are labeled with the fluorophores. In one case the fluorophores are attached to the 5'-terminus of the primer sequence using either a nucleoside derivative or a non-nucleosidic linker (Smith et al., 1986). In the second approach the fluorophores are attached to the base residues of the ddNTP chain terminators (Fig. 8-12) (Prober et al., 1987).

Although both procedures are conceptually very similar, the latter approach may offer some advantages. By labeling the terminating ddNTP derivatives, the set of four labeled ddNTPs can be employed for a variety of sequencing reactions, while primer labeling requires the generation of four labeled primers for each sequencing reaction. Additionally, polymerases tend to exhibit some nonspecific dissociation from the primer template complex. While reassociation and continued DNA elongation can occur in these cases, small quantities of nonspecific termination products may be present in the sequencing reactions. When the 5'-terminus is labeled with the fluorophore, the nonspecific termination products would result in minor quantities of fluorescent bands at incorrect sequence locations. However, with the fluorophore present in the terminating ddNTP, any nonspecific termination product would not be labeled by the fluorophore; only those products terminated by incorporation of the dideoxynucleotide would contain the 3'-terminal fluorophore.

Each fragment contains only a single fluorophore, but with laser excitation and appropriate filters and photomultipliers, the detection of the DNA fragments containing a single fluorophore is reasonably efficient, and automated DNA sequencing technology has been developed to exploit these labeling procedures.

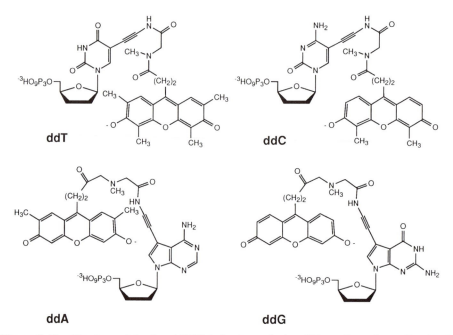

Figure 8-12. Structures of the four ddNTP derivatives tethering different fluorophores. Reprinted from Prober et al., 1987, with permission.

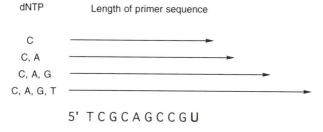

5' T C G C A G C C G U

3' A G C G T C G G C A G G T T C C C A A A

dNTP added	Relative Fluorescence	Peak max (nm)	Probe Position	Anisotropy
-	2.6	524	+1	0.2199
C	1.7	530	+2	0.1980
C, A	1.7	533	+4	0.1995
C, A, G	1.0	539	+7	0.1659
C, A, G, T	1.0	539	+10	0.1623
duplex, no pol.	1.0	538	+1	0.0776

Figure 8-13. (**Top**) The mansyl-labeled dU-containing (U) 11-mer/20-mer primer/template complex with the length of the elongated primer indicated in the presence of various dNTPs. (**Bottom**) The fluorescence characteristics of the variously elongated primer/template complexes. Reprinted from Allen et al., 1989, with permission.

Macromolecular Complex Formation. It can be very difficult to observe and characterize noncovalent binding between macromolecules such as nucleic acids and proteins. One approach used to observe DNA–protein binding involves the introduction of a group into the nucleic acid that can report on complex formation. Fluorophores tethered to nucleic acids can be valuable reporter groups in this respect. The emission characteristics of fluorophores are frequently dependent upon their environment, and a change in environment often results in either a shift in the emission maximum, a change in quantum yield, or both. Anisotropy effects also provide some indication of the rotational freedom available to a given fluorophore.

An example of the use of this approach to monitor complex formation is illustrated by the study of the DNA polymerase I/DNA complex (Allen et al., 1989). In this study, a mansyl fluorophore was tethered to the 5-position of a 2′-deoxyuridine nucleoside phosphoramidite using an amino linker. A primer/substrate complex was prepared in which the fluorophore was placed near the 3′-terminus of an 11-nucleotide primer sequence (Fig. 8-13). The 11-mer primer/20-mer template complex was designed such that the addition of the appropriate dNTPs would elongate the 11-mer in discrete steps. Using this technique, the fluorophore could be located at positions +1, +2, +4, +7, and +10 (position 0 was the 3′-terminus of the primer). As the 11-mer primer was lengthened by discrete amounts (using one or more dNTPs), the quantum yield for the fluorophore decreased and there was an associated red shift of the emission maximum.

These two properties are consistent with the fluorophore moving to a more hydrophilic environment as the polymerase translocates along the primer/template complex. These observations suggest the presence of strong protein–DNA interactions near the primer terminus. These interactions are still significant at positions $+2$ and $+4$ but are largely absent as the active site of the polymerase moves farther away from the site of the fluorophore (positions $+7$ and $+10$).

The anisotropy increases dramatically when the enzyme is added to the primer/template complex (Fig. 8-13). As the primer is elongated and the fluorescent probe is moved farther away from the 3′-terminus, the anisotropy values decrease, suggesting that the fluorophore regains some of its mobility as the enzyme complex passes over and beyond the site of the bound fluorophore.

The use of fluorescent probes as described in this study does not provide detailed structural information about the complex, but the ability to monitor binding and catalytic processes in solution under native conditions as reported by the fluorophore provides valuable information concerning the interaction of this enzyme with its DNA substrate.

Structural Analysis Using Fluorescence Resonance Energy Transfer. Spectroscopic analyses, particularly nuclear magnetic resonance (NMR), have become very important in the analysis of nucleic acid structure and dynamics. One of the most powerful NMR techniques available is the nuclear Overhauser enhancement (NOE) experiment in which magnetization is transferred between spin systems through space (rather than through bonds). One disadvantage of this particular technique is that only short distances (<4 Å) can be measured.

A spectroscopic ruler to provide longer distance measurements is available with fluorescence resonance energy transfer techniques. Electronic excitation energy can be transferred between a donor and an acceptor fluorophore over distances as great as 70 Å . The use of this method for distance estimation is based upon the theories of Förster (1965), who postulated that resonant energy could be transferred between fluorophores with overlapping emission and excitation spectra and that the efficiency and the rate of energy transfer would depend upon the inverse sixth power of the distance between the two chromophores ($1/r^6$). This postulate was subsequently verified by studies involving donor and acceptor molecules separated by known distances with the conclusion that fluorescence resonance energy transfer could be used effectively in the range of 10–60 Å. The challenge of this technique remains "to specifically label the macromolecule of interest with a suitable donor and acceptor" (Stryer & Haugland, 1967). Only recently, with the development of a variety of labeling procedures, has it been possible to place two reporter groups (the donor and acceptor fluorophores) effectively in specific locations within the complex of interest.

The distance between a donor and an acceptor reporter group is calculated according to the equation

$$R = R_0(\frac{1}{E} - 1)^{1/6} \qquad (8\text{-}1)$$

where E is the efficiency of energy transfer and R_0, the distance for 50% efficiency of resonance energy transfer, is calculated according to

$$R_0 = (9.79 \times 10^3)(J\kappa^2 n^{-4} \ \phi_D)^{1/6} \ \text{Å} \qquad (8\text{-}2)$$

where κ^2 is an orientation factor for the time scale of the fluorescence lifetime and a value of $^2/_3$ is often used for κ^2. The refractive index of the medium is n, and it is taken to be 1.4 in aqueous solution. ϕ is the quantum yield of the donor in the absence of the acceptor. J is the spectral overlap integral and can be calculated from the emission and excitation envelopes of the donor and acceptor fluorophores, respectively. The efficiency of energy transfer from the donor to the acceptor fluorophore can be determined from the quenching of the donor fluorescence in the presence of the acceptor, or from the enhancement of the acceptor fluorescence in the presence of the donor.

The four-way junction, a central intermediate in DNA recombination catalyzed by resolvase enzymes, is a covalent structure composed of two double-stranded DNA helices. A number of orientations of this structure are possible (see Fig. 8-14) and are differentiated by the nature of the DNA strands shared by both helices (crossed or noncrossed) and whether the orientation of the two helices is right-handed or left-handed. To determine which of the four possible structures was adopted by a four-way junction, a series of model junctions were prepared and labeled by tethering rhodamine and fluorescein reporter groups to two of the 5′-termini of the complexes using non-nucleoside linkers (Murchie et al., 1989). Various distances between the donor and acceptor fluorophores were generated by altering the length of the helices present in the complex. "This has the effect . . . of 'walking' the rhodamine dye around the H-arm helix" (Murchie et al., 1989). Resonance energy transfer between the donor and acceptor fluorophores was at a maximum when the rhodamine was attached to phosphates 15 and 16, and it was decreased for longer or shorter helix arms. From these relative distance data and modeling building experiments, the authors could conclude that the junction was present as a right-handed noncrossed junction (Fig. 8-14).

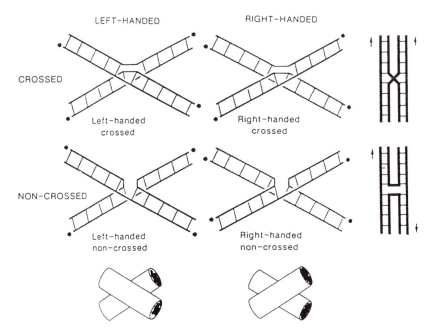

Figure 8-14. Possible topology of the DNA strands of the four-way junction. Reprinted from Murchie et al., 1989, with permission.

Fluorescence resonance energy transfer has the potential to provide important distance information in a wide variety of nucleic acid complexes, but suffers from the inaccuracy of the distance measurements (Ozaki & McLaughlin, 1992). While relative distances can be obtained effectively, as described in the study of the four-way junction, precise angstrom measurement remains elusive. The inability to obtain precise distance measurement stems largely from anisotropy effects that impact the orientation factor κ^2. With more information regarding the rotational and positional parameters for the fluorophores, the κ^2 parameter can be determined more precisely and distances using fluorescence resonance energy transfer can then be determined more accurately.

Use of Oxidation–Reduction Chemistry to Report on Complex Formation. Oxidation–reduction processes involving specifically tethered reporter groups can, in principle, be monitored directly using appropriate instrumentation. However, one of the most valuable applications of redox chemistry associated with tethered reporter groups is that resulting from reaction with the DNA itself (Chu & Orgel, 1985; Dreyer & Dervan, 1985; Hertzberg & Dervan, 1982; Schultz et al., 1982; Taylor et al., 1984). In these studies, a redox active complex is employed to generate a reactive species such as hydroxyl radical, which in the presence of DNA will result in strand cleavage (the details of this reaction are discussed in Chapter 5). Because the hydroxyl radical is short-lived, strand cleavage occurs only within a relatively short distance from the reporter group that produces the reactive species.

The use of covalently bound redox active groups has been used very effectively to report on the formation of triple helix complexes (Horne & Dervan, 1990; Maher et al., 1989; Moser & Dervan, 1987; Strobel et al., 1988). In these studies, an esterified ethylenediaminetetraacetic acid (EDTA) complex is tethered to a linker at the 5-position of a dT phosphoramidite and the pyrimidine-rich triple-helix strand is prepared with the EDTA-dT residue at one of the terminal positions of the sequence. After incubation with the target duplex, ferric ion is then added to generate the Fe–EDTA complex and redox activity is initiated by the addition of dithiothreitol in the presence of molecular oxygen. Under these conditions, oxidation–reduction cycling of Fe^{2+} and Fe^{3+} occurs and molecular oxygen is converted to hydroxyl radicals (Fig. 8-15). This radical species is nonspecific in its interactions with the duplex DNA; cleavage is not limited to a single site, but to a series of neighboring sites on either side of the Fe–EDTA-labeled nucleoside residue. The apparent unsymmetrical nature of the cleavage pattern (Fig. 8-15b) is a function of the two-dimensional representation of the cleavage sites. The Fe–EDTA complex tethered to the C5 carbon of dT is present in the major groove, and the released hydroxyl radicals diffuse to the phosphate sugar backbone on either side of the groove. The symmetrical diffusion of hydroxyl radicals in the three-dimensional structure of the major groove results in the apparent unsymmetrical cleavage patterns when viewed in two dimensions.

By placing the Fe–EDTA reporter group at the terminus of the DNA fragment and radioactively labeling one of the target strands, gel electrophoresis can be used to resolve the product bands, and the relative migration distance can be used to assign the location of cleavage. Using this assay, triplex formation at more than a single binding site will be readily apparent. Moving the Fe–EDTA complex to the 3'-terminus of the oligonucleotide and reassaying the cleavage pattern upon triplex formation can confirm the orientation of binding.

Hydroxyl radical cleavage is a very effective, if indirect, method for determining

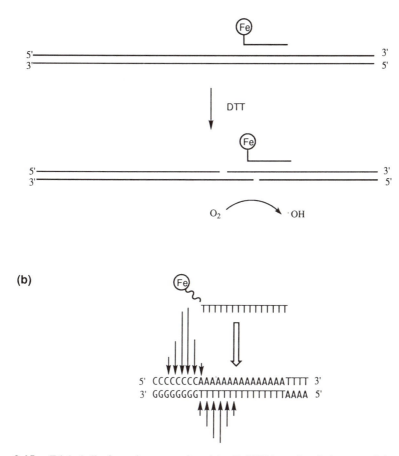

Figure 8-15. Triple helix formation as monitored by Fe-EDTA-mediated cleavage of the target DNA. (a) Formation of the triple helix with subsequent activation and generation of hydroxyl radical. (b) Typical cleavage pattern for a poly-dA-dT sequence targeted with a poly-dT triplex-forming oligonucleotide. Reprinted from Moser & Dervan 1987, with permission.

the location of the tethered redox-active reporter group. In addition to triplex formation, similar results have been obtained for cleavage of single-stranded DNA (Chu & Orgel, 1985; Dreyer & Dervan, 1985; Hertzberg & Dervan, 1982; Schultz et al., 1982; Taylor et al., 1984); when employed in protein studies, some cleavage of peptide linkages has been reported (Ermácora et al., 1992).

Photoaffinity Agents. Photoaffinity reagents are also employed to report upon the location of specific residues in macromolecular complexes. Although other types of affinity agents such as simple alkylating agents can be used to target, for example, a complementary DNA sequence, the photoactivation process present in photoaffinity agents provides an attractive trigger mechanism. In a study of DNA polymerase–substrate interaction described below, the polymerase complex could be formed and the enzyme could be translocated a specific distance from the affinity label; only after both processes had taken place was the affinity label activated (triggered). It would be difficult to accomplish an analysis of this type without the trigger mechanism provided by photoaffinity labeling.

Photoaffinity reagents have been employed to target DNA sequences in hybridization complexes (Praseuth et al., 1988). DNA sequences can be effectively photolabeled in protein–DNA complexes by attachment of the photoaffinity probe to an appropriate amino acid residue (Pendergrast et al., 1992). The complementary approach, involving (1) introduction of the affinity label at a specific site in the DNA and (2) probing the protein sequence for amino acid residues that may be important in recognition processes, has been an effective technique in the study of restriction endonuclease–DNA and the corresponding methylase–DNA complexes (Nikiforov & Connolly, 1992), as well as in the study of a variety of DNA (Catalano et al., 1990; Gibson & Benkovic, 1987; Mitina et al., 1990) and RNA (Grachev et al., 1987) polymerases. Two commonly used affinity labels are aromatic azides and benzophenone derivatives. Upon photolysis, the aromatic azides generate nitrenes that are effective in C—H insertion reactions, but many of these aromatic azides suffer from side reactions, most notably insertion of the nitrene into the adjacent aromatic C-H bond. This competing *intramolecular* insertion reaction can be largely eliminated, thus increasing the yield of nitrene available for *intermolecular* insertion reactions, by use of the fluorinated aromatic azides (Soundararajan & Platz, 1990). At least with a number of affinity-labeled proteins, benzophenones have been reported to result in higher yields of intermolecular insertion products, and their use may be advantageous in some cases.

The study of *Escherichia coli* DNA polymerase I again provides an intriguing illustration of the use of affinity labels, since the position of a sequence-specific DNA-tethered reporter group will change relative to the enzyme as the polymerase performs its DNA elongation function. In one study, *p*-nitrophenylazide was tethered to the C5 position of a 2'-deoxyuridine residue at the 3'-terminal position of a 10-mer/20-mer primer/template complex (Catalano et al., 1990). The relative positions of the polymerase and the affinity reporter group could be altered by extension of the primer sequence in discrete steps (Fig. 8-16). Locating the affinity label at the 3'-terminus of the 10-mer primer should place it near the active site of the polymerase, and in this position approximately 40% of the complexed protein could be modified upon photolysis. As the active site moved downstream from the affinity label during elongation, the ability to crosslink the protein to the primer strand decreased, as would be expected. As the azido label translocated out of the DNA binding domain, in the 17-mer and 20-mer primers, relatively little affinity labeling was observed.

The modification of the protein occurs by formation of a covalent bond between the aromatic azide and a specific amino acid residue, and the identification of the labeled peptide fragment and amino acid residue could be accomplished by isolation of the complex, digestion of the protein and nucleic acid, and then resolution and isolation of the labeled material. In the study cited, the 10-mer/20-mer primer/template complex yielded a tryptic peptide fragment spanning amino acid residues Ala-759 to Arg-775. Amino acid sequencing of the fragment identified the DNA-labeled residue as Tyr-766. This results indicates that Tyr-766 is in, or very near, the polymerization site of *E. coli* DNA polymerase I centered on the 3'-primer terminus in the primer/template complex.

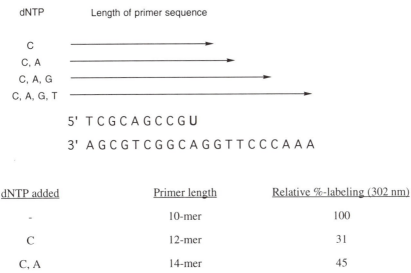

dNTP	Length of primer sequence
C	
C, A	
C, A, G	
C, A, G, T	

5' T C G C A G C C G U

3' A G C G T C G G C A G G T T C C C A A A

dNTP added	Primer length	Relative %-labeling (302 nm)
-	10-mer	100
C	12-mer	31
C, A	14-mer	45
C, A, G	17-mer	11
C, A, G, T	20-mer	9
no hv	11-mer	3

Figure 8-16. (**Top**) The *p*-nitrophenylazide-labeled dU-containing (**U**) 10-mer/20-mer primer/-template complex with the length of the elongated primer indicated in the presence of various dNTPs. (**Bottom**) The extent of photolabeled protein related to the length of the primer and position of the affinity label. Reprinted from Catalano et al., 1990, with permission.

New DNA Properties—Electron Transfer. A recent paper has described the phenomenon of photoinduced electron transfer through a DNA helix (Murphy et al., 1993). Electron transfer rates over long distances between two metal centers have been measured as a function of the intervening medium in a variety of systems (McLendon, 1988). These studies have led to the suggestion that aromatic groups could serve as "π-ways" and promote electron transfer over large distances. A double-stranded fragment of DNA provides a medium composed of a series of stacked aromatic residues, and it could thus function as a π-way for electron transfer. In the recent study by Murphy et al. (1993), metal intercalators were covalently attached to the 5'-terminus of a DNA fragment of 15 residues. Each fragment was prepared with a 5'-terminal non-nucleosidic linker terminating in a aliphatic amino group that was labeled with a metal intercalator. The "donor" reporter group, bis(phenanthroline)(dipyridophenazine)ruthenium(II)[$Ru(phen)_2dppz^{2+}$] (Fig. 8-17), does not luminesce in solution upon photoexcitation, but does exhibit strong luminescence in the presence of double-stranded DNA because the complex intercalates into the stacked aromatic base residues where the dppz ligand protects the phenazine nitrogens from aqueous quenching. The "acceptor" complex was bis(9,10-phenanthrenequinonediimine) (phenanthroline) rhodium(III) [$Rh(phi)_2phen^{3+}$] (Fig. 8-17).

Upon photoexcitation, a metal-to-ligand charge transfer transition occurred centered on the dppz ligand in the donor $Ru(phen)_2dppz^{2+}$ complex and resulted in luminescence. The excited state can be quenched simply by water or by appropriate electron acceptors. In the present case, the acceptor complex Rh(phi)$_2$phen^{3+} also contains ligand-to-metal charge transfer properties. Upon hybridization of two complementary oligonucleotides, one labeled at the 5′-terminus with the donor Ru complex and the second labeled at the 5′-terminus with the Rh acceptor, photoexcitation directed electron transfer from the Ru(II) donor to the dppz ligand, then through the stacked base residues of the double-stranded DNA to the π-acceptor ligand, and finally to the Rh(III) acceptor metal.

The steady-state emission spectrum (λ_{ex} 480 nm) of the Ru-modified oligonucleotide complexed to its unmodified complement was intense (λ_{em} 610 nm). Annealing of the two metalated strands resulted in complete quenching of the Ru luminescence. Time-resolved luminescence decay experiments could be used to set an upper limit of 2.5 nsec for the excited-state lifetime of the Ru-metalated DNA duplex. Picosecond single-photon counting experiments established a lower limit of 10^9 sec^{-1} for the quenching rate of the luminescence by the Rh acceptor.

Model building experiments indicated that the tethered metal complexes can only intercalate within one or two bases pairs from the ends of the duplex structures, and this property fixes the distances between the metal centers at no more than 48 Å and no less than 41 Å (Fig. 8-18). It is possible that the quenching observed was a function of electron transfer mechanisms or of simple Förster-type energy transfer mechanisms similar to those described earlier for fluorescent resonance energy transfer. However, Förster energy transfer is unlikely to occur in this case because there is no spectral overlap between the photoexcited donor Ru(II) complex and the acceptor Rh(III) complex. The use of a third complex, tris(phenanthroline) ruthenium(II) [Ru(phen)$_3$$^{2+}$], assisted in confirming the nature of the transfer process. A tethered Ru(phen)$_3$$^{2+}$ complex does not intercalate deeply into the double-stranded DNA, and thus the metal complex is not coupled to the π-stack. DNA tethering a Ru(phen)$_3$$^{2+}$ complex results in luminescence, but there is no quenching of the Ru center upon hybridization of the Rh-modified oligonucleotide. The uncoupling of the phenanthroline ligand in Ru(phen)$_3$$^{2+}$ from the π-stack of the DNA bases should not inhibit a resonate Förster-type quenching of the excited state, but would interrupt the metal–ligand–DNA–ligand–metal electron transfer process and prevent quenching of the excited state of the metal center.

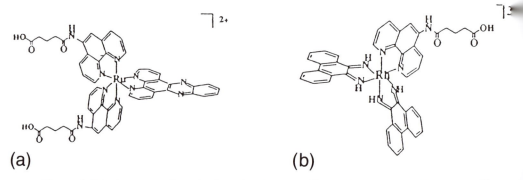

(a)   (b)

Figure 8-17. Structures of (a) the donor Ru(phen)$_2$dppz^{2+} and (b) the acceptor Rh(phi)$_2$phen^{+3} metal complexes. Reprinted from Murphy et al., 1993, with permission.

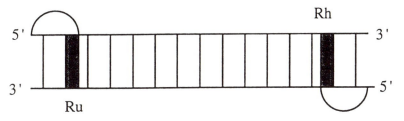

Figure 8-18. One of the possible complexes in which a tethered donor $Ru(phen)_2dppz^{2+}$ metal complex is quenched by a tethered acceptor $Rh(phi)_2phen^{3+}$ metal complex.

Summary

With the recent developments in nucleic acid conjugation chemistry, it is relatively simple to tether a reporter group of interest to one of the nucleic acid termini by employing a base or sugar functional group or by the incorporation of an appropriate non-nucleoside tether. When it is more advantageous to have the reporter group located internally, one can employ a suitable base functional group, the 2'-hydroxyl of the sugar residue, or an appropriately modified internucleotide phosphate as a site for conjugation. A number of examples illustrate the ability of various groups to report upon detection, structural, or mechanistic details, but a wide variety of applications for these materials remain to be developed.

Acknowledgments

Work in the laboratory of L. W. McLaughlin on conjugation chemistry has been supported by NIH research grant GM37065.

Nucleic Acid Binding and Catalysis by Metal Ions

Sergei A. Kazakov

Nucleic acids, which play critical roles in living organisms and viruses in the storage, transmission, and expression of genetic information, are actually salts (or complexes) of metal ions. Moreover, metal ions could have participated in the synthesis and evolution of the genetic molecules (Orgel & Lohrmann, 1974; Swaminathan & Sundaralingam, 1979).

It is not unusual for biochemists and molecular biologists dealing with nucleic acids to believe that the chemistry of metal ions is primitive in comparison with the complex properties of nucleic acids, presumably due to the small size of the metal ions; conversely, many inorganic chemists regard nucleic acids as just a large organic ligand. In reality, both nucleic acids and metal ions exhibit considerable complexity in their interactions.

At present, the involvement of metal-containing species in several facets of nucleic acid studies is well documented, including genetic information transfer and control of gene expression (Eichhorn, 1980; Grunberg-Manago et al., 1981; Mildvan & Loeb, 1981; O'Halloran, 1989, 1993), mutagenesis (Downey & So, 1989; Sirover & Loeb, 1976), chromosomal abnormalities (Christie & Costa, 1983; Sissoëff et al., 1976), and carcinogenesis (Andronikashvili & Mosulishvili, 1980; Flessel et al., 1980; Issaq, 1980; Sirover & Loeb, 1976). Unfortunately, it is often difficult to determine the actual molecular mechanism responsible for the observed effects, since many intracellular species, especially specific proteins, can act as mediators between the metal ion and nucleic acid in such processes in vivo (O'Halloran, 1989; Sissoëff et al., 1976).

Today, four major areas of research are providing the greatest impetus for characterization of metal–nucleic acid interactions. The first is the discovery of the antitumor and antiviral activity of complexes of platinum and several other metals, which are believed to have nucleic acids as a critical target (Cleare & Hydes, 1980; Köpf-Mayer & Köpf, 1987; Lippard, 1991; Reedijk, 1991). The second is the successful characterization of the three-dimensional crystal structures of metal–tRNA complexes, which permit a detailed analysis of specific metal–polynucleotide binding (Jack et al., 1977; Teeter et al., 1980). The third area involves the discovery of "ribozymes" ("catalytic" RNAs, see Chapter 13) which employ metal ions for both structural and catalytic functions. The fourth area involves the unique properties of some metal compounds as probes of nucleic acid structure and function (Barton & Lippard, 1980; Basile & Barton, 1989; Lippard, 1991; Tullius, 1989).

Because some "nonbiological" metal ions (e.g., $Pt^{2+/4+}$, Pd^{2+}, $Ru^{2+/3+}$, Au^{3+}, Hg^{2+}, and Ag^+) form inert complexes with nucleic acids that are convenient for isolation and investigation, numerous structural studies of such complexes have been described (Daune, 1974; Eichhorn, 1973; Gellert & Bau, 1979; Hodgson, 1977; Izatt et al., 1971; Kazakov & Hecht, 1994; Lippert et al., 1988; Martin, 1985; Martin & Mariam, 1979; Marzilli, 1977, 1981; Marzilli et al., 1980; Saenger, 1984; Sigel, 1989, 1993; Swaminathan & Sundaralingam, 1979; Tu & Heller, 1974; Weser, 1968). Apparently, only a few of the derived structures bear relevance to potential metal–nucleic acid binding modes in vivo. In contrast, the interactions of nucleic acids with so-called "biological" metal ions [i.e., K^+, Na^+, Mg^{2+}, Ca^{2+}, Mn^{2+}, $Fe^{2+/3+}$, $Co^{2+/3+}$, $Cu^{1+/2+}$, and Zn^{2+} (Williams, 1971)], result in unstable, short-lived complexes with nucleic acids. Usually, more than one such complex can exist in solution because of the complicated relationship between (1) the kinetic preferences of various sites in the same nucleic acid molecule toward interactions with metal ions and (2) the stability of the bonds so formed.

The best method for definition of metal–nucleic acid binding modes is X-ray crystallography (Gao et al., 1993; Geierstanger et al., 1991; Marzilli, 1981), but there is no guarantee that it bears relevance to the structure as it exists in solution. In some cases, the conformations of nucleic acids in crystals and in solution are not identical because there are more intermolecular interactions in a solid phase than in dilute solution (Sarma & Dhingra, 1981; Berman & Shieh, 1981). In crystals, even the alkali metal ions (e.g., Na^+ and Li^+) can exhibit direct coordination with nucleobases and ribose residues (Swaminathan & Sundaralingam, 1979); such interactions have never been detected in solution.

Coordination Chemistry of Metal–Nucleic Acid Interactions

Metal Binding Sites in Nucleic Acids. Because of the negative charge of internucleotide and terminal phosphate oxygens (Saenger, 1984), nucleic acids are polyanions in aqueous solutions at physiological pH. Therefore, their primary mode of binding with positively charged metal ions involves fast, electrostatic interaction that is nonspecific. The negative charge density, which attracts cations, is greater for multistranded regions of nucleic acids than for denatured or single-stranded ones (Gladchenko et al., 1986).

For many metal ions, initial nonspecific binding can be followed by more specific binding with electron-donor atoms of nucleic acids. At neutral pH, the principal metal-binding sites in nucleosides are believed to be (with pK_as at ionic strength $I = 0.1$ in parentheses) termini phosphate oxygens (1.0 and 6.6), internucleotide phosphate oxygens (1.0), and endocyclic atoms in guanosine (N7, 2.0), adenosine (N7 and N1, 3.8), and cytidine (N3, 4.2) (Martin & Mariam, 1979; Marzilli, 1977). The purine N3 atom is sterically hindered by the sugar, but changes in the glycosidic torsion angle could make this site available to bind metal ions. The O2 atom of cytosine is well established as the most important exocyclic metal-binding site at neutral pH, while the carbonyl groups of other nucleobases are relatively weak donors and usually can bind a metal ion only if the interaction is reinforced by chelation of other groups.

Once deprotonated, some endocyclic nitrogen atoms, like guanosine N1 (9.3), uridine N3 (9.5), and thymidine N3 (9.9) can become principal binding sites as well (Martin & Mariam, 1979; Marzilli, 1977). Exocyclic NH_2 groups of adenosine and cytosine, as well as ribose $2'$-OH, are poor ligands even when deprotonated ($pK_a > 12$). Especially where chelation is possible, metal binding at some specific sites can be so favorable that the binding will take place even at a pH considerably lower than the pK_a through proton substitution by the metal ion. In addition, the electric field produced by a nucleic acid polyanion in conjugation with compensating counterions increases the pK_a of the bases slightly and polarizes the hydrating water molecules bound to both the nucleic acid and metal ions.

The unsubstuted C5 atoms in cytidine and uridine can also bind Hg^{2+} via the only known metal–nucleic acid "covalent" (carbon–metal) bond (Jack, 1979). In contrast, most metal ions (electron pair "acceptors") can bind to nucleic acids only by "coordinative" bonds via electron pair "donor" atoms (e.g., N and O) to form a coordination complex.

A variety of minor bases found in tRNAs (Nishimura, 1979) also have modified metal-binding properties. Some of these modifications—for example, in N7-methylguanosine—restrict the metal-binding capability of the bases. In contrast, some functional groups having an intrinsically high affinity for metal ions, such as exocyclic thio (e.g., in 4-thiouridine or 2-thiocytidine) and amino acid functionalities (e.g., in 5-carboxymethylaminomethyluridine), enhance the number of potential metal-binding sites of the bases. Inosine (pK_a 8.9 for N1 H) is rare in nature but has proven very useful as a research tool in that it differs from guanosine only in lacking the 2-NH_2 group, but retains similar base-pairing (Leonard et al., 1992) and metal-binding properties (Marzilli, 1981).

Monofunctional attachment of a metal species to some particular nucleic acid donor atom can lead to multifunctional binding, where one metal ion is bound to several ligands; this may occur both intramolecularly as chelation (a term derived from the Greek word for claw, *chele*) and intermolecularly as crosslinking (Lepre & Lippard, 1990; Sherman & Lippard, 1987). Usually metal ions are able to react with more than one site on a nucleic acid and to favor different sites in a somewhat predictable manner (Eichhorn et al., 1981). In some cases, donor atoms alone may have weak metal-binding capabilities, but they can still be bound in a chelate as a consequence of favorable geometry created by metal ion binding with a strong binding site.

The reactions of metal species with nucleic acids that involve the formation of relatively stable bond(s) with specific group(s) on the nucleoside, nucleotide or polynucleotide molecule is considered to be specific binding. There are different kinds of metal-binding selectivity: (1) "atom-specific" (oxygen or sulfur in thiophosphates; donor nitrogen and oxygen atoms in nucleobases), (2) "group-specific" (phosphate, sugar, and nucleobases or some combination thereof), (3) base-specific in polynucleotides, (4) "sequence-specific", (5) "secondary structure-specific" (binding to single-stranded or double-stranded polynucleotide only), and (6) "tertiary structure-specific" (which is usual for RNA molecules and nucleoprotein macromolecular complexes).

The specificity of metal ion binding to the nucleic acids, which depends on both the nature and the structure of the nucleic acid as well as on the metal complex, can be affected significantly by variations in reaction conditions (e.g., metal concentration, pH, and temperature) as well as by ligand environment in the metal-complex species due to possible interligands interactions (Jack et al., 1977; Krotz et al., 1993; Ling et al., 1993; Lippert et al., 1988; Marzilli & Kistenmacher, 1977; Zou et al., 1993).

Nonspecific electrostatic binding of metal cations with nucleic acid phosphates is an *anticooperative* process, because bound cations reduce the net charge of a polyanion (Daune, 1974). In contrast, some specific binding of metal ions with a polynucleotide can favor binding of additional cations (a *cooperative* process) by changing the polynucleotide structure to create or make available an additional metal-binding site(s) (Danchin & Guéron, 1970; Labuda et al., 1985; Rialdi et al., 1972).

Coordination Properties of Metal Ions. In pure water, metal ions exist as the cationic aquo-complexes $[M(OH_2)_x]^{n+}$, which tend to form aqua–hydroxyl complexes $[M(OH_2)_{x-1}(OH)]^{(n-1)+}$ (see corresponding pK_as in Table 9-1) because the influence of the positive charge on the metal ion facilitates the loss of a proton from the coordinated water molecule. In addition to the water molecule and the hydroxyl anion, other single or multiatom ligands carrying negative charge (or which are neutral) can be bound to the same central metal ion. The total charge carried by the coordination complex, being the sum of the charges possessed by a metal cation and its ligand, depends on the mode of binding (i.e., with or without proton substitution) and pH, because of the ability of coordinated ligands to undergo deprotonation.

While the maximum number of ligands, including nucleobases, bound by single bonds to the metal ion is given by its normal coordination number (Table 9-1), the number of metal ions bound to the nucleic acid molecule in multinuclear complexes is limited by the quantity of separate available donor atoms. In spite of the fact that metal cation binding to a neutral nucleobase usually reduces its overall basicity (proton affinity), this effect is not sufficient to prevent the attachment of more than one metal species per base in the presence of excess metal ions (e.g., via both N7 and N1 atoms of adenine or via N7, N3, and deprotonated N1 atoms of guanine residues) (Lippert et al., 1988).

There are some correlations between coordination features of metal ions (in complexes with a nucleic acid) and metal-binding sites in nucleobases. Unhindered donor atoms (e.g., N7 of purines) can be involved in octahedral complexes, while atoms that are in sterically hindered environments (e.g., N3 of cytidine flanked by 4-NH$_2$ and O2 groups, or N1 of adenosine sterically influenced by the 6-NH$_2$ group) prefer to form complexes with metals having coordination numbers four and two (Gellert & Bau, 1979).

Stable binding between metal ions and ligands prohibits their exchange and permits the resolution of geometric isomers of the coordination compound, which can react differently with nucleic acids. In buffered, aqueous solutions the behavior of particular metal ions depends mainly on the kinetic lability and thermodynamic stability of the corresponding metal–ligand bonds, as well as on potential interligand interactions in both reactive metal species and the final products. In addition to inner-sphere complexes, in which the ligands are bound directly to the metal ion, there are so-called outer-sphere complexes in which a ligand is localized near the complexed metal ion via interaction with an inner-sphere ligand—for example, by hydrogen bonds with H$_2$O molecules coordinated to a metal ion.

Mechanism of Metal–Nucleic Acid Binding. Metal–nucleic acid complexes can both form (forward reaction) and dissociate (reverse reaction) as a result of ligand substitution (exchange) reactions, wherein inert ligands are retained while labile ones are replaced. Some anions, such as ClO$_4^-$ and NO$_3^-$, have no tendency to coordinate to metal ions even in rather concentrated aqueous solutions, but most others (e.g., Cl$^-$, CH$_3$COO$^-$, or phosphates, as well as the basic components of virtually all buffers) can compete with other coordinated or potential ligands for the metal ion (Cotton & Wilkinson, 1988).

Three basic classes of mechanisms for ligand substitution (replacement) reactions are generally considered: (1) *dissociative*, (2) *associative* (i.e., going through a transition state involving coordination of the new ligand in an inner-sphere complex) and (3) *interchange–exchange* of two ligands between inner and outer spheres of the metal ion (Cotton & Wilkinson, 1988).

In contrast to second-order, bimolecular *associative* reactions, the rate of a reaction by a *dissociative* mechanism does not depend on the concentration of nucleic acid because of the first-order rate law for unimolecular limited stage. Ligand substitution reactions in aqueous solutions rarely proceed directly. Instead, the leaving ligand is first replaced by H_2O, a nucleophile present in aqueous solution at a concentration of ~55 M; the entering ligand then attacks the aqua complex in the so-called anation reaction. Therefore, the rate constant of the forward reactions, which could be considered to be the *reactivity* of aquo-metal species, will parallel the rate of water substitution in the inner coordination sphere of the corresponding metal complexes (Frey & Stuehr, 1974) (Fig. 9-1). Generally, the reactions between aquo-metal complexes and nucleic acids under mild conditions are much faster ($t_{1/2}$ in the range 10^{-9}–10^{-3} sec) than most reactions to form covalent bonds (Dale et al., 1975; Frey & Stuehr, 1974; Williams & Crothers, 1975). However, if the limiting step of a substitution reaction is dissociation of a relatively stable coordination bond—for example, Pt—Cl in the reaction with AMP (Bose et al., 1985) or DNA (Bancroft et al., 1990)—the rate of this reaction could be quite slow ($t_{1/2}$ up to several hours).

The relative reactivities at neutral pH of the nucleobases with respect to substitutionally inert transition metal ions being under kinetic control are parallel to the order

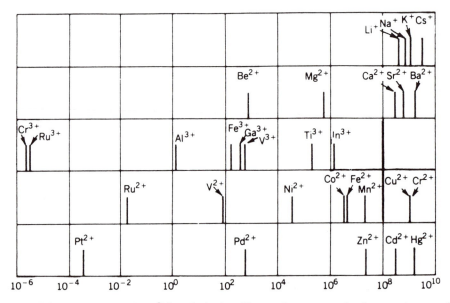

Figure 9-1. Rate constants (sec^{-1}) for substitution of inner-sphere water molecules on various metal ions. Reprinted from Cotton & Wilkinson, 1988, with permission.

of nucleophility of the nucleobases: G > A > C >> U or T (Barton & Lippard, 1980). This order parallels the relative rates of reaction for *cis*-$[(NH_3)_2Pt(OH_2)_2]^{2+}$, while the equilibrium constants for the same reactions are very similar (Barton & Lippard, 1980). Thus, the relative rates of such reactions can lead to a differences in initial product formation. Further transformations of the initially formed metal–nucleic acid complexes are under both kinetic and thermodynamic control (see below).

The metal ion concentration in the vicinity of the nucleic acid phosphate groups is far greater than that in the bulk solution, due to the electrostatic interactions between cations and polyanions; this "polyelectrolyte effect" can be expressed as a ratio of concentrations of ions proximal to and distant from the DNA. Divalent cations are attracted more strongly than monovalent ones. For example, a rough estimation (Daune, 1974) for a low ionic strength solution of DNA indicated the ratios of these concentrations for Na^+ and Mg^{2+} to be about 3×10^3 and 9×10^6, respectively. The high local concentration of positively charged reactive metal species would accelerate reactions with nucleic acids that proceed via associative mechanisms.

Stability and Lability of Metal–Nucleic Acid Bonds. Using the well-known classification of metal ions according to the thermodynamic stabilities of their complexes with different ligand donor atoms (Pearson, 1966), the corresponding metal ions would be classified similarly with respect to the usual donor atoms of nucleic acids (Angelici, 1973). According to that classification, "hard" (or class A) metal ions prefer O-donor atoms (phosphate oxygens) while "soft" (or class B) metal ions prefer N-donor atoms of the nucleobases, as well as sulfur atoms in thio derivatives of nucleic acids. Intermediate (or "borderline") metal ions have ambivalent properties (i.e., an ability to bind both kinds of donor atoms) (Fig. 9-2). Generally, for a fixed value of any ("covalent" or "ionic") index, the stability of metal ion complexes with most simple ligands increases with the magnitude of the other index. However, it is important to realize that biomolecules (including nucleic acids) are complex ligands and provide the donor atoms with metal ion interactions that are not completely consistent with the "hard–soft" classification (Martin, 1986).

"Hard" monovalent metal ions usually can bind nucleic acid phosphates only in a diffuse "*ion atmosphere*" manner, whereas for hard divalent (as well as polyvalent) metal ions, two more types of binding have been characterized: *inner-sphere* (i.e., directly between the metal ion and phosphate oxygen) and *outer-sphere* (i.e., through a water molecule coordinated to the metal ion) (Granot & Kearns, 1982). The increasing phosphate content (and hence negative charge) in the series N (nucleoside) < NMP < NDP < NTP determines the increase in affinity toward alkaline earth metal ions, which bind predominantly to the phosphate groups of nucleotides (Chen, 1978). Because the ionic radii of M^{2+} (Table 9-1) are about an order of magnitude less than the distance between adjacent phosphates in a stacked polynucleotide chain (~ 7 Å), divalent metal ions cannot chelate two such phosphates via inner-sphere binding (Pörschke, 1986). However, as was demonstrated in the interaction between Mg^{2+} and oligoriboadenylates (Pörschke, 1986), short flexible oligomers can bend to a loop structure with a sufficiently close approach of two internucleotide phosphates to chelate a divalent metal ion.

Nucleic acids such as tRNAs, which contain appropriate teritary structure, act as EDTA-like chelating agents capable of forming stable complexes even with "hard" divalent metal ions. Depending on the tRNAs, there are three to five strong and "intermediate"-strength site-specific magnesium-binding sites having stability constants

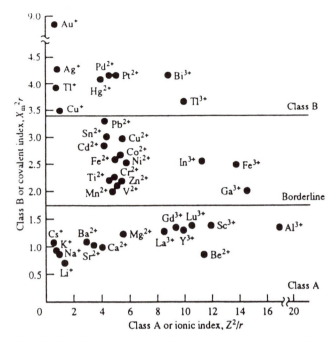

Figure 9-2. Classification of metal ions according to their preference for donor atom binding. Reprinted (with modifications) from Nieboer & Sanford, 1985, with permission. The "ionic index," being a function of charge-to-size ratio (Z^2/r), is a suitable measure of the ability of metal ions to participate in ionic interactions. The covalent index, which is a function of electronegativity and size ($X_m^2 r$), reflects the tendency for covalent bonding.

between 10^5 and 5×10^6 M^{-1}, as well as about 20 weak ones (5×10^2 to 10^4 M^{-1}) (Labuda et al., 1985; Reid & Cowan, 1990; Schimmel & Redfield, 1980). In addition to multiple outer-sphere bonds, there is inner-sphere coordination of metal ions with phosphate alone [e.g., for Mg^{2+} (Fig. 9-3A) (Teeter et al., 1980)] or with phosphate and a nucleobase residue [e.g., for Mn^{2+} (Fig. 9-3B) (Jack et al., 1977)] in the site-specific metal–tRNA complexes. It is interesting that some site-specific metal–RNA complexes can survive even in the presence of such strong chelating agents such as EDTA, presumably because of the electrostatic repulsion between the negatively charged EDTA and the polyanionic nucleic acid, as well as the inaccessibility to EDTA of the metal ions bound in internal regions of the RNA tertiary structure (Rialdi et al., 1972; Vary & Vournakis, 1984a; Wacker & Vallee, 1959; Wacker et al., 1963).

The order of relative stabilities of nucleobase–metal ion complexes is frequently given as G > A ~ C >> U or T (Eichhorn, 1973). However, because of the variety of sites to which metal ions can bind on the nucleobases, there are a number of exceptions. For example, the relative base affinities of Hg^{2+} and of Pb^{2+} appears to be T > C > A > G and C > G > A, respectively (Eichhorn, 1973). There have been fewer studies of metal complexes with pyrimidine nucleotides than for purine nucleotides because of the weaker coordination properties of the pyrimidine N3 atom relative to the N7 and N1 atoms of purines. However, the principal structures of such complexes have been determined (Gellert & Bau, 1979; Marzilli et al., 1980).

Metal ions that have borderline "hard–soft" properties can form mixed ternary complexes (composed of phosphate O and nucleobase N) (Martin & Mariam, 1979). For example, metal chelate structures can be formed with purine nucleoside-5′-phosphates (Sigel, 1989) in which the metal ion coordinates (1) inner-sphere to both the N7 and phosphate, (2) inner-sphere to N7 and outer-sphere (i.e. via water) to the phosphate

Table 9-1   Basic Features of Hydrated Metal Ions Involved in Metal–Nucleic Acid Interactions

Ion	Ionic Radius	Principal Coordination Number	Typical Geometry of Inner-Sphere Complex	First pK_a of $[M(OH_2)_x]^{n+}$ [a]
Ag^+	1.26	2 > 4	Linear	9.8–11.7
Al^{3+}	0.51	6 > 4	Octahedral	4.3–5.0
Ba^{2+}	1.34	8	Square antiprism	13.4–14.0
Be^{2+}	0.35	4	Tetrahedral	5.7–6.7
Bi^{3+}	0.96	6	Octahedral	Unstable
Ca^{2+}	0.99	8	Square antiprism	12.6–13.4
Cd^{2+}	0.97	6 > 4	Octahedral	7.6–10.2
Co^{2+}	0.72	6	Octahedral	7.6–9.9
Co^{3+}	0.63	6	Octahedral	0.9–2.0
Cr^{2+}	0.89	6	Octahedral	8.7–11
Cr^{3+}	0.63	6	Octahedral	3.8–4.4
Cu^+	0.96	4	Tetrahedral	Unstable
Cu^{2+}	0.72	4 > 6	Square planar	6.8–8.5
Eu^{3+}	0.95	9	Variable	4.8–8.5
Fe^{2+}	0.74	6	Octahedral	5.9–6.7
Fe^{3+}	0.64	6	Octahedral	~2.5
Hg^{2+}	1.10	2 > 4.6	Linear	2.4–3.7
In^{3+}	0.81	6	Octahedral	~3.7
Mg^{2+}	0.66	6	Octahedral	11.4–12.8
Mn^{2+}	0.80	6	Octahedral	10.6–10.9
Ni^{2+}	0.69	6	Octahedral	6.5–10.2
Pb^{2+}	1.20	6 > 9	Octahedral	6.5–8.4
Sn^{2+}	0.93	3 > 6	Pyramidal	3.7–6.8
Sr^{2+}	1.12	6	Octahedral	13.2 > 13.8
Th^{4+}	0.99	8–12	Variable	2.4–5.0
UO_2^{2+}	0.80 (U^{6+})	6 > 8	Octahedral	~5.7
VO^{2+}	0.63(V^{4+})	6	Octrahedral	~5.4
Zn^{2+}	0.74	4 > 6	Tetrahedral	8.2–9.8

[a]Dependent on ionic strength.

Source: Kazakov & Hecht, 1994.

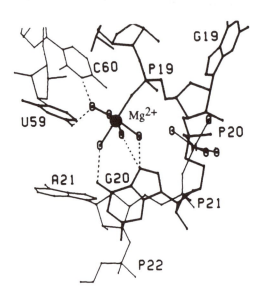

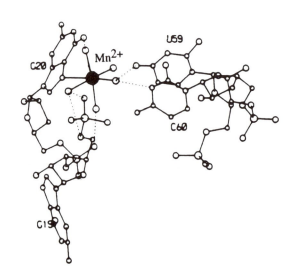

Figure 9-3. The binding modes of metal ions with the D loop of yeast tRNA^{Phe}. (**A**) [Mg(H$_2$O)$_5$]$^{2+}$ is bound directly to the phosphate-19 oxygen, whereas metal-coordinated H$_2$O molecules are involved in a number hydrogen bonds. Reprinted from Teeter et al., 1980, with permission. (**B**) [Mn(H$_2$O)$_5$]$^{2+}$ is bound directly to the N7 atom of guanine$_{20}$, whereas five H$_2$O ligands form hydrogen bonds with neighboring (in the tertiary structure of the tRNA) atoms. Reprinted from Jack et al., 1977, with permission.

group, (3) outer-sphere to N7, but inner-sphere to the phosphate group, and (4) only outer-sphere to both binding sites. The ratios of complexes having these structures will differ for different purine residues and metal ions. In contrast to the relatively simple structures (cases 1–3 above), the polymeric complex $[Cu_3(5'-GMP)_3(H_2O)_8]$ has three distinct coordination environments about different Cu^{2+} ions (Sletten & Lie, 1976). There are also many polymeric metal complexes of GMP or IMP involving inner-sphere binding of both purine N7 and phosphate O atoms, in the same nucleotide residue, to different metal ions (Marzilli et al., 1980).

Numerous combinations of inner- and outer-sphere bonds are possible in NDP and NTP complexes with ambivalent metal ions; these can involve participation of the nucleobase donor atoms or the phosphate groups alone (Phillips, 1966; Rawlings et al., 1993; Sigel, 1989, 1993; Weser, 1968). The release of atom N7 from the inner coordination sphere of the complexes between Zn^{2+} or Cd^{2+} and $(ATP)^{2-}$ upon binding of a second inert ligand, L (e.g., imidazole or NH_3) to form $[M(ATP)(L)]^{2-}$ suggests that N7 coordination is the "weak" point in such macrochelates (Sigel, 1989).

"Soft" metal species having two or more labile ligands (reactive coordinates) favor macrochelating or crosslinking between nucleobases via formation of a single coordinative bond between the metal ion and any of several bases. A less usual binding mode is intrabase-chelating—that is, the binding of a single metal ion with two donor atoms of the same nucleobase. Examples of this binding mode include chelates of A (via atoms N1 and 6-NH_2) or G (via atoms N1 and O6) (Lippert et al, 1988). However, an (N7,O6) chelate of G with cis-$[(NH_3)_2Pt^{2+}]$, which has been considered for many years to be a key DNA adduct in the mechanism of the antitumor action of cis-diclorodiammino-

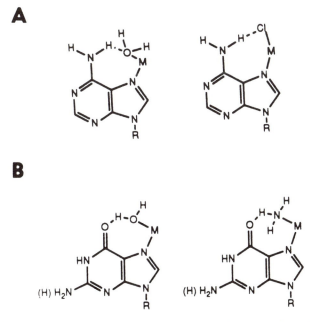

Figure 9-4. Principal types of outer-sphere metal chelates with adenine (**A**) and guanine (or inosine) (**B**) residues. Metal ions are shown coordinated directly to the bases by single inner-sphere bonds.

platinum(II), has not been proven essential (Lippert et al, 1988). Instead of the inner-sphere M—O bond, outer-sphere complexes via metal ion-coordinated H_2O, NH_3 or Cl^- ligands (Fig. 9-4) have been suggested (Chu et al., 1978; Hodgson, 1977).

The formation constant for metal–ligand bonds may be considered to be equal to the ratio of rate constants of the forward (combination of metal ion with the ligand) and reverse reaction (complex dissociation) at equilibrium (Martin, 1986). Therefore, despite the fact that the forward and reverse reactions of the same ligand with two different metal ions may differ greatly, the corresponding formation constants may be similar. For example, Ca^{2+} reacts with ATP about 100 times faster than does Mg^{2+}, but the equilibrium formation constants are nearly identical (Martin, 1964). The stability constants of the complexes of some metal ions with nucleosides and nucleotides (Izatt et al., 1971; Kim & Martin, 1984; Khan et al., 1988; Lönnberg & Apralahti, 1981; Lönnberg & Vihanto, 1981; Massoud & Sigel, 1988, 1989; Scheller & Sigel, 1983; Sigel, 1989; Sigel et al., 1987; Smith et al., 1991; Tu & Heller, 1974; Weser, 1968) are summarized in Table 9-2.

Despite the ability of metal ions to compete with protons for the same binding sites, the relative affinities of donor atoms toward metal ions are not a function of their basicity. For example, the N1 atom of A is intrinsically 320 times more basic than N7, but the ratios (in parentheses) of stability constants, {M-N1}/{M-N7}, for protons and selected metal ions vary as follows: H^+ (320) $>>$ [(dien)Pd^{2+}] (4) $>$ Ni^{2+} (3) $\geq$ Cu^{2+} (2.5) $>$ Zn^{2+} (1) (Martin, 1985).

Table 9-2 Logarithms of the Stability Constants of Metal–Nucleic Acid Complexes in Aqueous Solutions at 25°C

Nucleos(t)ide	Metal Ion							
	Mg^{2+}	Ca^{2+}	Mn^{2+}	Zn^{2+}	Co^{2+}	Cd^{2+}	Ni^{2+}	Cu^{2+}
Adenosine								0.16
5'-AMP	1.60	1.46	2.23	2.38	2.23	2.68	2.49	3.14
3'-AMP	1.49	1.36	2.06	1.98	1.80	2.32	1.89	2.75
5'-ADP	3.17	2.86	4.24	4.30	3.80		4.50	5.90
5'-ATP	4.29	3.91	5.01	5.16	4.97	5.34	4.86	6.34
Guanosine				0.80	1.00		1.40	1.90
5'-GMP	1.67	1.50	2.33	2.67	2.71	2.95	3.22	3.62
5'-GTP	4.02	3.58	4.73		4.63			
Cytidine				0.56		0.95	2.04	
5'-CMP	1.54	1.40	2.10	2.06	1.86	2.40	1.94	2.84
5'-CTP	4.20	3.85	4.90	5.03	4.78	5.05	4.52	6.03
5'-UMP	1.56	1.44	2.11	2.02	1.87	2.38	1.97	2.77
5'-UTP	4.27	3.94	4.91	5.01	4.73	5.10	4.47	5.87
5'-TMP	1.55	1.40	2.11	2.10	1.89	2.42	1.92	2.87
5'-TTP	4.23	3.85	5.01	5.03	4.78	5.09	4.52	5.83

Source: Kazakov & Hecht (1994), adapted with modifications.

The high affinity (thermodynamic stability) of a particular metal–ligand bond does not necessarily mean that the metal species will bind this ligand first and the corresponding metal complex will persist in solution under ordinary conditions for an indefinite length of time. This would be the case only for metal complexes having both high thermodynamic stability and kinetic inertness such as Pt^{2+} complexes with nucleobases, in which unforced dissociation does not occur within a reasonable length of time—for example, 100 hr at 37°C (Lippert et al., 1988). In comparison, the average lifetime of an Mg^{2+}—DNA bond is 4 msec at room temperature (Berggren et al., 1992).

In [(dien)Pd^{2+}]–nucleoside 5'-phosphate complexes, log K (stability constant) values of the Pd—N bonds are as follows (shown in parentheses): guanosine N7 (8.1) and N1 (7.9); adenosine N7 (4.1) and N1 (4.9); cytidine N3 (5.5); and thymidine N3 (8.7) (Martin, 1983). Pd^{2+} complexes, with stabilities close to those for Pt^{2+}, undergo ligand substitutions about 10^5 times faster than the corresponding Pt^{2+} complexes (Basolo et al., 1960). As result of kinetic lability, metal–base linkage isomerizations, involving cleavage of the N–metal bond formed initially under kinetic control, might occur to form more thermodynamically stable bonds. For example, the migrations of [(dien)Pd^{2+}] from the N7 atom to N1 of adenosine and from the N3 atom to N4 of cytidine have been reported (Lippert et al., 1988).

Metal Ions Affect the Structure of Nucleic Acids. The sites of binding of metal ions to nonstructured polynucleotides are similar to those of mononucleotides (Eichhorn, 1973). Generally, the affinity of a metal ion for multistranded polynucleotides is weaker than for denatured multistranded or single-stranded nucleic acids, because some potential metal-binding sites are rendered inaccessible by interbase H-bonds (Fig. 9-5). For example, for Mg^{2+} the relative affinities for individual types of nucleic acids are in order tRNA > poly(I) >> poly(A) ~ poly(A)·poly(U) ~ poly(I)·poly(C) ~ DNA $\geq$ denatured DNA > poly(C) > poly(U) >> mononucleotides ~ phosphate (Sander & Ts'o, 1971). In contrast to sites rendered inaccessible via H-bonding, a favorable steric arrangement of functional groups in the tertiary structure of a polymer could also result in a binding mode that is not favorable for the corresponding monomers.

In "cation-free" solutions (e.g., in deionized water) the secondary structure of a double-stranded DNA (Eichhorn, 1973) and the tertiary structure of tRNA molecules (Lindahl et al., 1966) denature even at 0°C. Monovalent cations at concentrations > 10^{-3} M prevent unwinding of a polynucleotide double helix in solution due to neutralization of the negatively charged phosphate groups, which would otherwise repel each other (Eichhorn, 1973).

Although most metal ions can stabilize double helixes, this property is not universal. Generally, an interaction of metal ions with phosphates leads to stabilization of a DNA double helix, whereas base binding (see Fig. 9-5A), or simultaneous base and phosphate binding by the same metal ion, results first in destabilization of the duplex and then in separation of the two polynucleotide strands (Eichhorn & Shin, 1968; Shin et al., 1972; Yamada et al., 1976). The ratios of stabilization/denaturing effects on DNA double helix for divalent metal ions are in the order $Mg^{2+} < Co^{2+} \sim Ni^{2+} < Mn^{2+} < Zn^{2+} < Cd^{2+} < Cu^{2+} < Hg^{2+}$ (Eichhorn & Shin, 1968). Generally the first three metal ions in this list provide only a stabilizing effect, whereas the others give increasing thermostability of DNA upon initial addition of metal ion, but then destabilize the double helix at higher concentrations (Eichhorn, 1973). However, even Mg^{2+}, having a

A

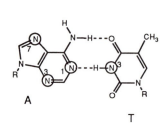
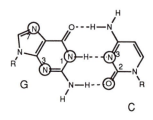

B

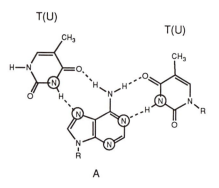

C

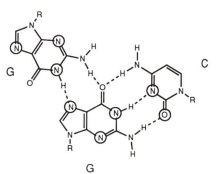

D

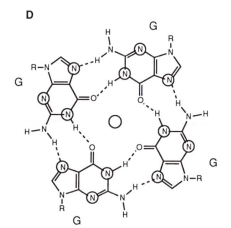

Figure 9-5. Potential metal ion-binding sites (circled) that participate in specific hydrogen bonds in Watson–Crick duplexes (**A**), triplexes (**B** and **C**) and a quadruplex (**D**).

low affinity toward nucleobases, can selectively bind adenine residues (via N7), both in DNA (Prakash et al., 1990) and RNA (Pörschke, 1979).

The destabilization is usually negligible if the concentrations of monovalent and divalent cations capable of binding the nucleobases are ≥0.1 M and ≤0.1 mM, respectively (Eichhorn, 1962). In such systems, the role of the monovalent cations (e.g., Na$^+$

or K^+) is to protect DNA against the divalent metal ion binding by creating an electrostatic barrier (Shin et al., 1972). Moreover, increasing the concentration of monovalent cations (to 0.1–1 M) can result in rewinding of DNA unwound by the divalent metal ions (e.g., by Cu^{2+}) via displacement of these divalent metal ions from DNA (Eichhorn, 1973). A similar effect is produced by chelating agents such as ethylenediaminetetraacetic acid (EDTA), or by dialysis against a solution lacking divalent metal ions.

The denaturing effect on DNA of high (5–15 M) concentrations of Li, Na, and K salts is considered to result from the action of the anions (rather than metal cations)— for example, I^-, ClO_4^-, CNS^-, Br^- and Cl^-—which are presumably capable of modifying the structure of water around the DNA helix and interacting with the nucleobases, thereby disturbing interbase stacking interactions (Felsenfeld & Miles, 1967).

The structural differences between ordered (folded) and denatured (unfolded) forms of RNA molecules are not quite as clear as in the case of intermolecular DNA duplexes, because of complex intramolecular RNA tertiary structures (Tinoco et al., 1987). Stabilization and denaturation (unfolding) of the "ordered" conformations of RNA by metal ions are based on principles similar to those for DNA but apparently involve more complex processes (Bujalowski et al., 1986; Flanagan & Jacobson, 1987, 1988; Reeves et al., 1970; Reid & Cowan, 1990; Rialdi et al., 1972; Stein & Crothers, 1976; Thomas et al., 1984; Vary & Vournakis, 1984a). It is believed, for example, that the specific tertiary structure of tRNA molecules can exist at high concentrations of Na^+, but that the tertiary structure is very unstable in the absence of divalent metal ions (Stein & Crothers, 1976). In turn, the tRNA tertiary structure provides the specific binding sites for divalent metal ions (Fig. 9-6) , which disappear with the liberation of the bound metal ions as a result of tertiary structure unfolding under denaturing conditions (Rialdi et al., 1972). However, if all endogenous Mg^{2+} ions were removed from a tRNA (e.g., by heating the solution in the presence of excess of EDTA followed by dialysis to remove EDTA at low temperature) and then added back to solution in exactly the original amount, the native tertiary structure could be restored, but not for each tRNA molecule (Reeves et al., 1970; Thomas et al., 1984), presumably because Mg^{2+} ions can stabilize both the "correct" secondary/tertiary structure and some other relatively stable structures. Because the folding and stabilization of RNA molecules during transcription in vivo can be directed and supported not only by metal ions but also cellular factors including polyamines and specific proteins (Mildvan & Loeb, 1981), the biologically active structure of RNAs need not be the same as the most stable one formed in vitro.

Complexes between nucleic acids (both DNA and RNA) and metal trivalent ions, such as Fe^{3+}, Co^{3+}, Cr^{3+} and Al^{3+}, are believed to exist in nature (Andronikashvili & Mosulishvili, 1980; Wacker & Vallee, 1959; Wacker et al., 1963), but it is not clear whether any of them are of importance biologically. It has been shown that these polyvalent metal ions are capable of forming inner-sphere bonds with nucleic acid phosphates and bases (Ascoli et al., 1973; Danchin, 1972, 1975; Gladchenko et al., 1982; Karlik et al., 1980; Pett et al., 1985) producing an effect on nucleic acid structure similar to that of divalent metal ions, but more pronounced due to the greater positive charge density of polyvalent metal ions. Tb^{3+} and Eu^{3+} ions, as well as some other lanthanides, which have been used widely as fluorescent probes of of nucleic acid structure (Gross & Simpkins, 1981; Rosental & Nelson, 1986), actually are not ideal for this purpose because they destabilize the double-helical structure and prevent complementary interactions between the single strands, despite their higher affinity toward

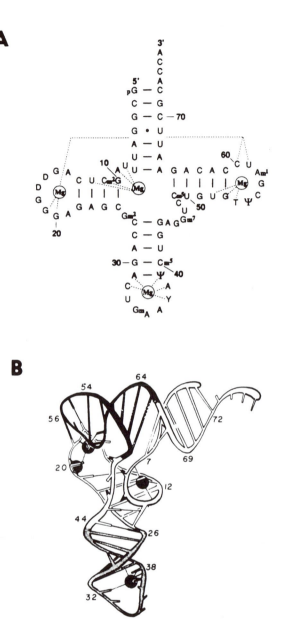

Figure 9-6. Sites of specific binding of Mg^{2+} ions in yeast tRNA^{Phe}. (**A**) Secondary structure. Reprinted (with modifications) from Hüttenhofer et al., 1992, with permission. (**B**) Tertiary structure. Reprinted (with modifications) from Holbrook et al., 1977, with permission.

double-stranded regions of polynucleotides relative to single-stranded ones (Gross & Simpkins, 1981).

Another kind of highly positively charged metal species is the so-called compact "inert" complex—for example, $[Co(NH_3)_6]^{3+}$. Such metal species, which can form only outer-sphere bonds with nucleic acids because the metal ion inner spheres are already occupied by slow-exchanging ligands, are unusually effective in stabilizing the secondary and tertiary structure of both DNA and RNA (Cowan, 1993; Karpel et al., 1975, 1980). Several complexes of this type bind tightly and selectively to double-helix motifs of nucleic acids, in comparison to single-strand ones, at low (5–10 mM) concentrations of monovalent cations (Karpel et al., 1980).

In addition to the simple effects on DNA duplex stability, the interaction of some divalent (e.g., Zn^{2+} and Ni^{2+}) and trivalent (e.g., Cr^{3+} and Tb^{3+}) metal ions, as well as polycationic inert metal species (e.g., $[Co(NH_3)_6]^{3+}$ and $[Ru(NH_3)_6]^{3+}$), can lead to dramatic changes in conformation of the double-helical structure, namely, transition from the right-handed B- to A- or C-forms and even to the left-handed Z-form of DNA (Chatterji, 1988; Eichhorn, 1981; Floro & Wetterhahn, 1984; Loprete & Hartmann, 1993; Narasimhan & Bryan, 1984; Nejedly et al., 1989; Schoenknecht & Diebler, 1993; Thomas & Messner, 1988). Metal ions are also known to play important roles in the formation of non-B DNA structures, the exact configurations of which are difficult to define at present. Such a delicate structural change of DNA—for example, that adopted by a single 5′ AGC 3′/5′ GCT 3′ trinucleotide under physiological conditions in the presence Zn^{2+} or Co^{2+} ions—may contribute to the some genetic diseases (Kohwi et al., 1993).

Some divalent metal ions, such as Zn^{2+} , Co^{2+} , Ba^{2+}, and Mn^{2+} (Laundon & Griffith, 1987) as well as cis-$[(NH_3)_2PtCl_2]$ (Leng, 1990), can promote sequence-directed bending of normally straight DNA duplexes that could be related to the control of gene expression . The efficacy of divalent metal ions in promoting the correct folding of four-way DNA cruciform junctions is in the order $Mg^{2+} > Mn^{2+} > Co^{2+} > Ni^{2+} \sim Cu^{2+}$ (Sullivan & Lilley, 1987). Recently, it was found that Al^{3+} ions can relax supercoiled DNA (Rao et al., 1993).

Most trivalent metal cations would be anticipated to cause DNA collapse (aggregation) into condensed toroidal structures, which are considered to be a model for DNA packaging in viruses and chromosomes (Eichhorn et al., 1981; Thomas & Bloomfield, 1988). $[Co(NH_3)_6]^{3+}$ induces this effect at five times lower concentration than spermidine (which also has a net charge of "3+" at physiological pH) under identical buffer conditions, whereas divalent metal ions and putrescine (net charge "2+" at physiological pH) are incapable of causing DNA condensation in aqueous solution (Thomas & Bloomfield, 1988; Widom & Baldwin, 1983).

The mechanisms of these structural interconversions are presumably based on the stabilization (by metal ions) of DNA conformations in which the phosphate groups can come closer together than in B-form DNA (Eichhorn et al., 1981).

Triple-stranded polynucleotide structures (Fig. 9-5B,C) are usually less stable than the related duplexes, at least in part because of the additional electrostatic repulsion of the third chain. However, when used at appropriate concentrations, metal cations can favor the formation of triplex structures by neutralization of the repulsive phosphate negative charges. For example, at Na^+ concentrations ≤ 0.2 M, poly(A)·poly(U) exists predominantly as a duplex, but at higher concentrations of monovalent cations, or in the presence of divalent metal cations (e.g., Mg^{2+} and Mn^{2+}), the duplex disproportionates to form triplex poly(A)·2 poly(U) and a free poly(A) strand (Felsenfeld & Miles, 1967; Eichhorn, 1973). Similar intramolecular processes are known for homopurine–homopyrimidine tracts in circular covalently closed DNA. Under superhelical stress, half the pyrimidine strand can be unpaired and form Hoogsteen base pairs with the remaining polypurine–polypyrimidine duplex, while the "released" purine strand remains single-stranded (H-DNA). A purine strand can also participate in intramolecular triplex formation (*H-DNA). Such unusual structural properties depend on the level of supercoiling, pH, and the presence of divalent metal ions. For example, Mg^{2+} was required at neutral pH to support the H-type, $(dA)_n \cdot 2(dT)_n$ structure in a recombinant plasmid (Fox, 1990).

In supercoiled plasmids, $(dG)_n \cdot (dC)_n$ sequences can fold into halves from the center of this tract, forming *H-type, $(dG)_n \cdot (dG)_n \cdot (dC)_n$ triplexes (Fig. 9-5C) in the presence of Mg^{2+}, Ca^{2+}, or Mn^{2+}, whereas Zn^{2+} ions cannot support this particular triplex structure (Kohwi, 1989). Moreover, Zn^{2+} can inhibit this structural conversion if Zn^{2+} and Mg^{2+} are both present, by interfering with magnesium binding to the $(dG)_n \cdot (dC)_n$ tract.

For heterogeneous polypurine–polypyrimidine tracts, the effect of various divalent metal ions on stability of the *H-DNA triplexes can be very different, depending on the nucleotide sequence (Bernués et al., 1990; Collier & Wells, 1990; Kang & Wells, 1992; Lyamichev et al., 1991; Malkov et al., 1993). For example, Cd^{2+}, Co^{2+}, Mn^{2+}, Ca^{2+}, and Zn^{2+} ions can stabilize the intramolecular triplex $(dTC)_n \cdot (dGA)_n \cdot (dAG)_n$, whereas Mg^{2+}, Ca^{2+}, Ba^{2+}, and Hg^{2+} cause destabilization (Malkov et al., 1993). For the sequence $(dCT)_{22} \cdot (dAG)_{22}$, the efficiency of metal ions in promoting the corresponding duplex–triplex transition was found to be $Zn^{2+} > Cd^{2+} >> Mn^{2+}$, whereas Mg^{2+} was inactive in this system (Bernués et al., 1990). Recently, the formation of an intermolecular Pu–Pu–Py triplex between the oligonucleotide dGAGAAAGGGG and a fragment of human papilloma virus DNA has been shown, specifically in the presence of Mn^{2+} but not in the presence of Mg^{2+} or Zn^{2+} (Malkov et al., 1993). Generally, the types of unusual structures adopted by long Pu–Py repeats in DNA are complex and heterogeneous with multiple different conformational isomers (Collier & Wells, 1990). Some monovalent cations, notably K^+ and Rb^+, were found to inhibit intermolecular Pu–Pu–Py triple-helical DNA conformation, presumably via promotion of other multistrand structures (Cheng & Van Dyke, 1993).

Several mono- and divalent metal ions can promote the transition from Watson–Crick duplexes to quadruplex structures containing the G-quartet assembly in a very unusual way (Fig. 9-5D) (Hardin et al., 1992). The relative efficiency of the metal ions $(K^+ > Ca^{2+} > Na^+ > Mg^{2+} > Li^+)$ indicates that these metal ions, especially K^+, are capable of stabilizing the $(G)_4$-DNA structure via putative accommodation inside the "cavity" of such tetrads (Hardin et al., 1992; Nagesh et al., 1992; Xu et al., 1993).

Catalytic Properties of Metal Ions Toward Nucleic Acids

Metal Ions Can Catalyze the Synthesis of Nucleic Acids. Because metal ions were present on earth before nucleic acids, it seems reasonable to believe that some reactions involving participation by metal ions may have played a role in the chemical synthesis, replication, and evolution of nucleic acids (Ochiai, 1988; Orgel, 1986; Orgel & Lohrmann, 1974).

First, adenosine, guanosine, and inosine are formed when the corresponding bases are heated as solids with ribose at about 100°C in the presence of inorganic salts such as $MgCl_2$, $MgSO_4$, $CaSO_4$, $MnSO_4$, or $MnCl_2$. The pyrimidine nucleosides cannot be formed under the same conditions (Orgel & Lohrmann, 1974).

The heating of nucleosides with inorganic phosphates and urea results in a mixture of the nucleoside 5'-phosphate and smaller amounts of the nucleoside 2'-, nucleoside

3'-, and cyclic 2',3'-phosphates as well as a variety of nucleoside polyphosphates and dinucleoside phosphates (Orgel & Lohrmann, 1974). In the presence of Mg^{2+} ions, the phosphorylation of nucleoside hydroxyl groups is suppressed, whereas pyrophosphate formation is greatly accelerated. For example, the major products of the solid-phase reaction between nucleoside 5'-phosphate and inorganic phosphate in the presence and absence of Mg^{2+} are nucleoside 5'-diphosphate and nucleoside 2'(3'),5'-bisphosphates, respectively. The mechanisms of these coupling reactions have been studied, but are not yet well understood (Joyce, 1989; Orgel & Lohrmann, 1974).

After direct combination of the monomers to form short oligonucleotides, these "primary" templates could direct the metal ion-promoted solution synthesis of nucleic acids having Watson–Crick complementary. Such template-directed reactions are possible only under conditions that are consistent with the formation and stability of the polynucleotide–mononucleotide complexes. For reasons discussed by Orgel, these reactions are restricted to polypyrimidine–monopurine systems where a mononucleotide substrate has an activated 5'-phosphate and free 2'(3') hydroxyl groups (Orgel, 1986). For example, Pb^{2+} ions promote polymerization of the activated nucleotide derivative guanosine 5'-phosphorimidazolide (ImpG) on a poly(C) template much more efficiently than Zn^{2+}, yielding products about 30 nucleotides in length; these have predominantly non-natural $2' \rightarrow 5'$ as well as natural $3' \rightarrow 5'$ internucleotide linkages (Lohrmann et al., 1980). In the poly(U)–ImpA system, Pb^{2+} ions preferentially produce $3' \rightarrow 5'$ oligoadenylates (Sleeper et al., 1979) whereas UO_2^{2+} ions promote $2' \rightarrow 5'$ phosphodiester bond formation (Sawai et al., 1990). In contrast to the double-helix poly(C)–ImpG system, the condensation of ImpA occurs in the triple-helical complex 2 poly(U)· mono(A) (Orgel, 1986). In addition to the metal ions, the nature of the phosphate activating group also affects the $2' \rightarrow 5' : 3' \rightarrow 5'$ ratios in the polymerization products (Orgel, 1986; Sawai et al., 1990). In contrast to the corresponding enzymatic processes, the metal ion-catalyzed polymerization is very sensitive to the structure of the templates and terminates once the template is filled by the complementary copy (Joyce, 1987; Orgel, 1986).

The ability of the metal ions to promote formation of the $3' \rightarrow 5'$ products is especially interesting because the sugar 2'-OH group is more reactive than the 3'-OH group in the absence of a template (Lohrmann & Orgel, 1978). It is assumed that some kind of specific interaction of these metal ions with the template, substrate(s), or both provides an orientation of reactive substrate groups that is favorable for formation of the $3' \rightarrow 5'$ phosphodiester bond (Joyce, 1987; Kanaya & Yanagawa, 1986). Recently a mechanism by which the ability of UO_2^{2+} ions to control the polymerization of adenosine 5'-thiophosphoroimidazolide with remarkably high regio- and stereoselectivity can be understood has been proposed (Shimazu et al., 1993).

The possibility of template-assisted ligation of unmodified oligoribonucleotides on polynucleotide templates by metal ion-dependent mechanisms was demonstrated by condensation of oligo(A) on a poly(U) template in the presence of cyanogen bromide and imidazole (Kanaya & Yanagawa, 1986). The relative efficiency of individual metal ions in promoting this transformation was $Mn^{2+} > Co^{2+} > Ni^{2+} > Fe^{2+} > Zn^{2+} > Mg^{2+} \geq Cu^{2+}$; the overall yield of the metal-promoted coupling reaction was 10% to 40%, whereas without metal ions the yield was only 0.5%. Mixtures of $2' \rightarrow 5'$ and $3' \rightarrow 5'$ phosphodiesters as well as 5'–5' pyrophosphate bonds between coupled oligoadenylates were formed. Ni^{2+}, Co^{2+}, and Zn^{2+} gave a greater proportion of the $2' \rightarrow 5'$ isomer, while Mn^{2+} and Cu^{2+} favored the formation of the 5'–5' isomer. In comparison with other metal ions, Ni^{2+} gave the highest proportion of the natural $3' \rightarrow 5'$ isomer in the derived oligoadenylates—that is, about 30% of the total products. In contrast to the reaction at 4°C, where the products of the ligation are stable, the coupling yields were maximal after incubation at 25°C for 18 hr and then began to diminish, presumably as a result of hydrolytic degradation of the phosphate bonds catalyzed by the same metal ions.

Strikingly, the template-directed polymerization of ribonucleotides was much more efficient than deoxynucleotides or nucleotide derivatives with sugar residues other than ribose (Orgel, 1986). This observation provides one reason to believe that RNA may have emerged before DNA on the earth, and it supports the "RNA world" hypothesis (Joyce, 1989; Pace, 1991).

The prebiotic hydrosphere contained a substantial concentration of many multivalent ions capable of depolymerizing RNA (Ochiai, 1988; Pace, 1991); this would be especially true at high temperature. It seems possible that RNAs exposed to such an environment would have been at a selective disadvantage relative to DNAs. Most likely this was one of the evolutionary factors guiding the appearance of DNA, which is more stable hydrolytically than RNA. Therefore, RNA prebiotic synthesis could have occurred by surface reactions under moderately dry conditions, which would have minimized hydrolytic processes but still supported macromolecular chemistry (Ferris & Ertem, 1993; Pace, 1991).

The discovery of ribozymes (see Chapter 13 of this book) in recent years has led to speculation about the role of "RNA catalysis" in the origin of life. RNA "self-splicing," phosphotransferase activity, template-directed ligation, primer elongation, and "self-replication" (Been & Cech, 1988; Doudna et al., 1991, 1993a; Green & Szostak, 1992; Joyce, 1987; Orgel, 1986; Zaug & Cech, 1986) are actually metal ion-dependent processes and, therefore, may involve metal ions in the catalytic step (see below). There is much evidence that ligation occurs through a simple reversal of the metal-promoted RNA "self-cleavage" mechanism. Recently it has been shown that the metal ion requirements for the ligation reaction catalyzed by the "hairpin" ribozyme are the same as those for the cleavage reaction (Chowrira et al., 1993).

Metal Ions Catalyze Nucleic Acid Cleavage. Despite the poor metal-binding properties of the 2'-OH group of RNA, RNA and DNA can exhibit very different behavior in reactions with metal ions. One process that proceeds exclusively with RNA is the cleavage of phosphodiester bonds via transesterification catalyzed by metal ions and metal hydroxides (Fig. 9-7A) (Eichhorn, 1973; Kochetkov & Budovsky, 1972). There are only a few examples of metal ion-catalyzed hydrolysis of phosphodiester bonds in DNA (Basile et al., 1987; Chin, 1991; Chin & Zou, 1987; Divakar et al., 1987; Kesicki et al., 1993; Pratviel et al., 1993; Schnaith et al., 1994; Tamm et al., 1952).

In comparison, most reagents that cleave polynucleotides via metal-dependent redox mechanisms exhibit little distinction between degradation of DNA and RNA, and some also fail to discriminate between single- and double-stranded structures (Celander & Cech, 1990; Holmes et al., 1993; Holmes & Hecht, 1993; Pope & Sigman, 1984; Sigel, 1969). However, in at least one study the peroxidative activity leading to degradation of the nucleotide bases was clearly different for the derived RNA and DNA metallo-complexes (Sigel, 1969, 1993).

A large variety of metal ions catalyze the cleavage of internucleotide phosphodiester bonds in RNA, and some of them also catalyze hydrolysis of nucleoside phos-

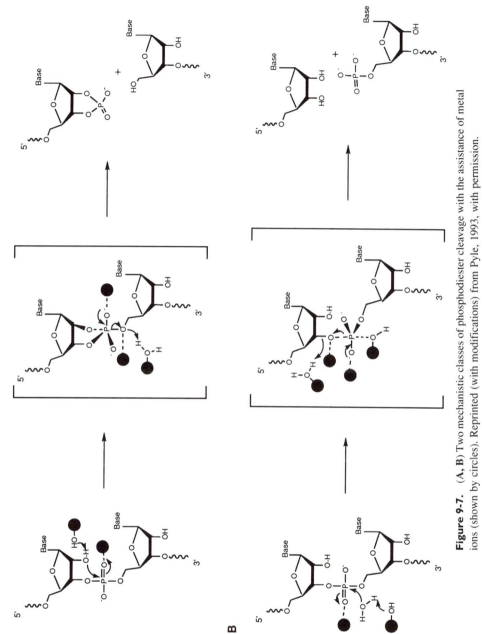

Figure 9-7. (A, B) Two mechanistic classes of phosphodiester cleavage with the assistance of metal ions (shown by circles). Reprinted (with modifications) from Pyle, 1993, with permission.

phomonoesters (e.g., Pb^{2+}, La^{3+}, and $Th^{3+)}$ or cyclic phosphates (e.g., Zn^{2+} and Cd^{2+}) (Kochetkov & Budovsky, 1972; Sumaoka et al., 1992). Promotion of the hydrolysis of both inorganic and nucleoside polyphosphates by metal ions and metal complexes is also well documented (Bose et al., 1985; Cooperman, 1976; Haight, 1987; Milburn et al., 1985; Ramirez & Marecek, 1980; Sigel et al., 1984; Tafesse et al., 1985; Utyanskaya, 1989). Recently, the unique capability of Zr^{4+}, Th^{4+}, and Cu^{2+} ions to provide selective hydrolysis of the pyrophosphate bridge without cleavage of phosphodiester bonds in 5'-capped RNAs (Baker, 1993; Visscher & Schwartz, 1992) has been discovered. Metal-promoted cleavage of RNA is found to result initially in an oligomer having a 2',3'-cyclic phosphate at the 3'-terminus, which can be subsequently hydrolyzed to a mixture of 2'- and 3'-phosphates (e.g., by Zn^{2+} ions (Breslow & Huang, 1991; Butzow & Eichhorn, 1971)) or selectively to the 3'-phosphates [e.g., by Pb^{2+} ions (Farkas, 1968)].

The rate and specificity of polyribonucleotide cleavage varies markedly with (1) the nature of the metal ion complexes and RNA substrates and (2) the conditions of the experiment (e.g., metal concentration, metal/nucleotide ratio, temperature and pH) (Hayashi et al., 1993; Kolasa et al., 1993; Morrow et al., 1992; Stern et al., 1990; Vijay-Kumar et al., 1984). The fact that metal-dependent cleavage of RNA may occur even under mild conditions must always be kept in mind when carrying out experiments with RNA.

High temperature enhances of the rate of RNA cleavage but decreases its specificity. For example, Mg^{2+} and Mn^{2+} ions at millimolar concentrations can be used to produce a sequencing ladder (i.e., random cleavage of a ^{32}P-end-labeled RNA oligomer) (Forster & Symons, 1987; Pieken et al., 1991).

The pH dependence of metal-dependent RNA cleavage is related to ionization state of both aqua–metal complexes and nucleic acid constituents. As a rule, RNA cleavage promoted by metal ions has a maximum rate in solution at pH values around the pK_as of first metal-bound H_2O molecules (see Table 9-1)—that is, when the metal species are still carrying positive charge (Brown et al., 1985; Butzow & Eichhorn, 1965b; Kochetkov & Budovsky, 1972). Surprisingly, insoluble metal hydroxides lacking a charge—for example, $Bi(OH)_3$ and $Al(OH)_3$—can also promote depolymerization of RNA at high temperature, presumably via a heterogeneous reaction (Cooperman, 1976; Eichhorn, 1973).

At neutral pH, the most rapid cleavage of RNA has been achieved by the action of Pb^{2+}, rare metal ions (e.g., Eu^{3+}, La^{3+}, Ce^{3+}, Tb^{3+}, and Lu^{3+}), and Zn^{2+} (Kochetkov & Budovsky, 1972). Zn^{2+} cleaves RNA only about 4% as well as lead (Farkas, 1968). Other catalytically active metal ions—for example, Al^{3+}, Cd^{2+}, Mn^{2+}, Cu^{2+}, Co^{2+}, Ni^{2+}, or Mg^{2+}—are one to two orders of magnitude less active than Zn^{2+} (Breslow & Huang, 1991; Butzow & Eichhorn, 1965b; Kochetkov & Budovsky, 1972; Stern et al., 1990).

It has been observed that the rates of phosphodiester bond cleavage promoted by metal ions in polynucleotides is higher than those in short oligonucleotides (Butzow & Eichhorn, 1971; Kochetkov & Budovsky, 1972). In RNA oligomers, the nature of the adjacent phosphate group can affect the rate and primary position of metal-dependent cleavage of internal phosphodiester bonds. For example, the relative rates (shown in parentheses) of Zn^{2+}-dependent cleavage (marked by "!") of short adenosine oligomers were found to decrease in the order Ap!A3'-p ($\sim$125) $>>$ Ap!ApA ($\sim$7) $>$ Ap!A2'-p ($\sim$3) $>$ ApAp!A ($\sim$2) $>$ Ap!A ($\sim$1) (Butzow & Eichhorn, 1971). The fact that the 3'-terminal phosphate group selectively enhances the rate of the cleavage of the neighboring internucleotide phosphate suggests a specific exonuclease-like activity of Zn^{2+} ions. A similar type of transformation, which affords cleavage in higher yield, was recently found for Pb^{2+} ions (Kazakov & Hecht, unpublished data).

The nucleotide composition and the nature of the nearest-neighbor base in RNA molecules also affect the rate of metal-promoted cleavage (Eichhorn, 1973). For example, the rates of metal ion-dependent cleavage of polyribonucleotides at pH 7 decrease in the following order (Butzow & Eichhorn, 1965a,b; Farkas, 1968):

Pb^{2+}:   poly(A) $\geq$ poly(U) > poly(C) > poly(I)
La^{3+}:   poly(C) > poly(U) > poly(A) > poly(I)
Zn^{2+}:   poly(U) ~ poly(A) > bulk yeast RNA ~ poly(C) ~ poly(I)

The reactivity of the dinucleotides toward Zn^{2+} decreases in the order UpUp > ApAp > CpCp > GpGp, which is exactly opposite to the order of the affinity of the base residues for this metal ion (Eichhorn et al., 1971; Ikenaga & Inoue, 1974). The differences in reactivity found for isomeric pairs of dinucleotides (i.e., ApGp > GpAp, CpAp > ApCp, CpGp > GpCp, ApUp $\geq$ UpAp and GpUp $\geq$ UpGp) indicates that RNA cleavage by Zn^{2+} depends on both the base composition and sequence of the dinucleotides (Ikenaga & Inoue, 1974). To explain this sequence dependence, it is necessary to consider the specific interactions (e.g., stacking, H-bonds) between adjacent bases which determine the local conformation of the ribose–phosphate backbone at the site of cleavage (Dock-Bregeon & Moras, 1987; Eichhorn et al., 1971; Ikenaga & Inoue, 1974; Kierzek, 1992). These interactions might be introduced, modified, or diminished by the binding of metal ion(s) (Eichhorn et al., 1971). The zinc-dependent cleavage of polyribonucleotides is inhibited dramatically in the presence of metal species having a high affinity for nucleobases, such as Ag^+ (Eichhorn et al., 1967) or cis-$[(NH_3)PtCl_2]$ (Wherland et al., 1973). Conversely, poly(I) and to a lesser extent inosine-5′-phosphate, but not poly(A) or AMP, suppress the degradation of poly(A) by Zn^{2+} ions in solution (Butzow & Eichhorn, 1965b). Therefore, the presence of guanosine (or its "analog" inosine) in a heteropolyribonucleotide might not decrease the rate of phosphodiester bond cleavage in proximity to its own position in the polymer, but rather affect the cleavage elsewhere on the polymer, either intra- or intermolecularly (Butzow & Eichhorn, 1971).

Polynucleotides possessing a more stable secondary structure are cleaved more slowly than single-stranded ones in the presence of the same metal ions (Kochetkov & Budovsky, 1972; Werner et al., 1976). For example, Pb^{2+} ions cleave poly(A) or poly(U) alone three times faster than the duplex poly(A)·poly(U); poly(I)·poly(C) shows no detectable cleavage under these conditions (Farkas, 1968). One of the principal reasons for this could be specific interbase interactions (stacking and H-bonds) in a helical conformation, either single- or double-stranded. These interactions, which can stabilize the conformation of a dinucleoside phosphate moiety such that the phosphorus atom is moved away from the sugar 2′-OH group and rotation about the C3′—O bond is restricted (Tinoco et al., 1987), can interfere with the conversion of the polynucleotide to a conformation suitable for the transesterification reaction that leads to cleavage (Ikenaga & Inoue, 1974; Kierzek, 1992; Kochetkov & Budovsky, 1972) . The requisite conformations can occur only at certain types of bends in the RNA chain (Brown et al., 1985; Dock-Bregeon & Moras, 1987; Werner et al., 1976).

Some metal ions, bound to suitable "pockets" within the RNA tertiary structure, can induce very specific cleavage of the RNA near the binding sites. Several metal ions— for example, Pb^{2+} (Brown et al., 1985; Ciesiolka et al., 1989, 1992; Deng & Termini, 1992; Gornicki et al., 1989; Pan & Uhlenbeck, 1992; Streicher et al., 1993; Sundaralingam et al.,1984; Werner et al., 1976), Eu^{3+} (Ciesiolka et al., 1989; Kazakov & Altman, 1991; Rordorf & Kearns, 1976), Zn^{2+} (Kazakov & Altman, 1991; Rordorf & Kearns, 1976), Fe^{2+} (Vary & Vournakis, 1984b), and Mn^{2+} (Holmes & Hecht, unpublished data)—can induce site-specific cleavage of different RNAs at neutral pH. Other metals, such as Mg^{2+} (Ciesiolka et al., 1989; Kazakov & Altman, 1991; Sanger et al., 1979; Wintermeyer & Zachau, 1973) as well as Cu^{2+}, Ca^{2+}, Ba^{2+}, and Sr^{2+} (Kazakov & Altman, 1991), require a higher pH ($\sim$9.5) to promote the specific cleavage of RNA molecules. The best-studied example of this phenomenon is the site-specific cleavage of $tRNA^{Phe}$ by Pb^{2+} (Behlen et al., 1990; Brown et al., 1985; Sundaralingam et al., 1984). Available X-ray diffraction data from crystals of yeast tRNA^{Phe}, soaked in a dilute Pb(II) acetate solution, provides considerable insight into the relationship between RNA structure and the mechanism of cleavage (Brown et al., 1985; Sundaralingam et al., 1984). Only one of the three tightly bound Pb^{2+} ions provides effective cleavage whereas the other two are inactive, presumably because of an unfavorable structure of these Pb^{2+}-binding sites (i.e., the distance and spatial orientation between the metal ion and the relevant internucleotide bond). The latter sites must also lack flexibility within the phosphodiester bond, restricting its ability to undergo conformational changes conducive to the cleavage reaction (Brown et al., 1985; Sundaralingam et al.,1984). A study of several different tRNAs, including yeast tRNA^{Phe} and structurally related species, indicates that the specificity and efficiency of lead-induced cleavage depend more on the conformation of the substrate polynucleotide than on its sequence (Behlen et al., 1990; Ciesiolka et al., 1989; Ciesiolka et al., 1992; Pan et al., 1991; Werner et al., 1976). Therefore, metal ions capable of cleaving RNA could, in principle, be used to probe both specific metal-binding sites within an RNA and conformational changes in RNA structure.

Catalytic RNAs, or "ribozymes" (see also Chapter 13), perform the most specific metal-dependent cleavages of RNA, (Altman et al., 1993; Cech & Bass, 1986; Long & Uhlenbeck, 1993; Pyle, 1993; Symons, 1992). There are two kinds of ribozyme cleavage reaction: those that occur in *cis* (or self-cleavage), when both "enzyme" and "substrate" domains are parts of the same RNA molecule; and those that occur in *trans*, when these domains are parts of separate RNAs. Because metal ions can play a crucial role in the folding and stabilization of an RNA, including both the active conformation of the ribozyme and the formation of a productive complex between the ribozyme and substrate, which in turn determines or affects function, it is difficult to separate the structural and catalytic role of metal ions (Brown et al., 1985; Dahm & Uhlenbeck, 1991; Grosshans & Cech, 1989; Guerrier-Takada et al., 1986; Kazakov & Altman, 1991; Piccirilli et al., 1993; Pyle, 1993 Smith et al., 1992b; Surratt et al., 1990). However, there is strong evidence for the direct involvement of metal ions in the active site chemistry of some catalytic RNAs (Dahm et al., 1993; Dahm & Uhlenbeck, 1991; Piccirilli et al., 1993; Sawata et al., 1993; Smith et al., 1992b), which may, therefore, be regarded as true metalloenzymes. Studies of ribozymes have suggested that weak but specific bonds with metal ions are required to support RNA catalysis (Grosshans & Cech, 1989; Smith & Pace, 1993; Sugimoto et al., 1988). Definition of the mechanism of metal ion catalysis is presently considered to be a central problem in the chemistry of ribozymes (Pyle, 1993).

RNA cleavage is a result of phosphoryl transfer reactions of metaphosphate, PO_3^-, from one nucleophilic atom to another, presumably through a pentacoordinate trigonal bipyramidal transition state (Figs. 9-7 and 9-8) (Cooperman, 1976; Haydock & Allen, 1985; Yarus, 1993). In this transition state structure, there are three sites at which metal ions can interact, namely, the phosphate moiety and the leaving and entering groups. Although it is not known how metal ions catalyze RNA phosphoryl transfer reactions in every case, there are putative mechanisms that can accommodate the experimental observations. Generally, catalysis takes place when metal ion(s) binds more tightly to the presumed intermediate/transition state than to the ground state of reactants (Bender et al., 1984; Jencks, 1969). No cleavage will take place when the metal ion binds less tightly to the transition state than to the ground state, whereas ligation may occur when the metal ion binds more tightly to the transition state than to the products of the cleavage reaction (Cooperman, 1976).

Other possible catalytic roles of metal ions in enzymatic and nonenzymatic cleavages of phosphodiester bonds has been reviewed (Brown et al., 1985; Cech, 1987; Cedergren et al., 1987; Cooperman, 1976; Haydock & Allen, 1985; Jencks, 1969; Pyle, 1993; Toh et al., 1987; Yarus, 1993), providing precedent for or evidence against the possible mechanisms listed below .

The first mechanism could involve bidentate binding of two oxygen atoms of a single phosphate group to the metal ion (Alexander et al., 1990; Hendry & Sargeson, 1989) which could compress the tetragonal O—P—O angle to a trigonal geometry. This would create an apical position susceptible to nucleophilic attack and thus reduce the activation energy necessary for transition to the bipyramidal intermediate (Fig. 9-8A).

A second possible mechanism involves electrostatic or coordinative binding of the leaving group in the transition state by a metal ion (Fig. 9-7), which would result in an increase in the acid strength of this group. This should stabilize the development of electron density on the leaving group in the transition state and weaken the bond between the phosphorus atom and the oxygen of leaving group, thus accelerating the cleavage reaction. Metal ion catalysis by interaction with the anionic leaving group is an example of general acid catalysis.

Another mechanism would result from direct (inner-sphere) binding of the metal ion to the phosphate oxygen, thus withdrawing electron density and enhancing the electrophilicity of the phosphorus atom. This would make the phosphorous more susceptible to nucleophilic attack by the neighboring 2'-OH group (transesterification) (Fig. 9-7A) or by an external nucleophile—for example, a water molecule (hydrolysis) (Fig. 9-7B). But such polarization could have no or little catalytic effect both because the increasing positive character of the phosphorus atom could be offset by a corresponding inhibition of leaving-group expulsion and because the withdrawal of electron density from the oxygen bound to metal ion can be compensated by increased π-bonding between the phosphate O and P atoms (Cooperman, 1976). Moreover, the transition state might be destabilized by the electrophilic metal interaction with the phosphoryl oxygen atom(s) (Herschlag & Jencks, 1987).

Because of strong polarization of metal-coordinated water molecules (see Table 9-1), an aquo-metal ion bound to RNA can be a source of highly reactive metal-bound hydroxyl anion (M^{2+}—OH^- or M^+—OH) (Chin & Zou, 1987; Hendry & Sargenson,

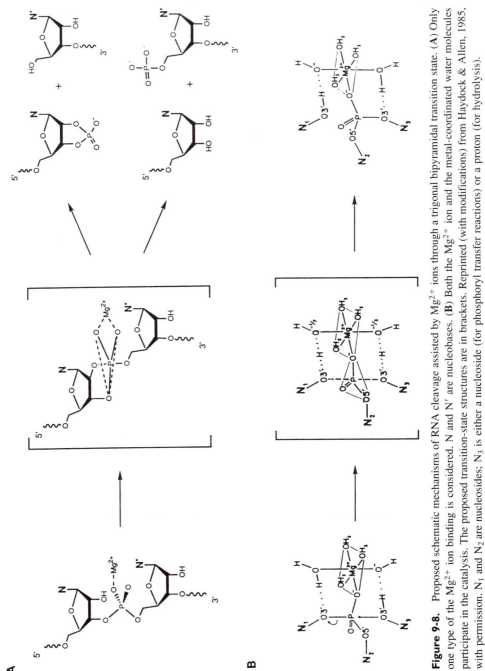

Figure 9-8. Proposed schematic mechanisms of RNA cleavage assisted by Mg²⁺ ions through a trigonal bipyramidal transition state. (**A**) Only one type of the Mg²⁺ ion binding is considered. N and N′ are nucleobases. (**B**) Both the Mg²⁺ ion and the metal-coordinated water molecules participate in the catalysis. The proposed transition-state structures are in brackets. Reprinted (with modifications) from Haydock & Allen, 1985, with permission. N₁ and N₂ are nucleosides; N₃ is either a nucleoside (for phosphoryl transfer reactions) or a proton (for hydrolysis).

1989; Morrow & Trogler, 1988). The nucleophilicity of the bound hydroxide is expected to be lower than that of free hydroxide, but might be greater than that of free water. Large perturbations of the pK_as of aqua–metal complexes (e.g., Mg^{2+}, Ca^{2+}, or Mn^{2+}) bound to RNA (e.g., into the active center of a ribozyme) seem quite possible and would be consistent with the frequent occurrence of a corresponding effect in protein enzymes (Guerrier-Takada et al.,1986; Kao & Crothers, 1980). A metal ion-bound hydroxyl anion, oriented via appropriate H-bonding at the proper distance and orientation relative to the sugar $2'$-OH group or outer-sphere H_2O molecule, would enhance the nucleophilicity of these two species via proton abstraction (Fig. 9-7A) (Brown et al., 1985; Haydock & Allen, 1985). These three types of activated hydroxyl groups would be better able to perform a nucleophilic attack on the neighboring internucleotide P atom, forming either $2',3'$-cyclic phosphate and $5'$-OH termini (Fig. 9-7A) or $3'$-OH and $5'$-phosphate ends at the cleavage site (Fig. 9-7B). Alkali, acids, and free metal ions, as well as most ribozymes, perform transesterification reactions with the formation of $2',3'$-cyclic phosphate and $5'$-OH termini (Kochetkov & Budovsky, 1972; Yarus, 1993). Only one kind of catalytic RNA, namely RNA subunits of eubacterial RNases P, actually promote hydrolysis (or solvolysis), affording products containing $3'$-OH and $5'$-phosphate termini (Guerrier-Takada et al., 1986).

The aqua and hydroxyl ligands of the metal ion can transfer protons between the incoming nucleophile and leaving group attached to the internucleotide phosphate undergoing cleavage (Fig. 9-8B). A related mechanism could involve some other inner-sphere ligands—for example, amines (Chin & Zou, 1988; Hendry & Sargeson, 1989) or imidazole (Atwood & Haake, 1976; Breslow et al., 1989; Kazakov & Doronina, unpublished data)—which are capable of accepting or donating protons and can function similarly to metal ion-coordinated H_2O molecules.

By shielding the negative charges of the RNA backbone phosphates, metal cations could allow the "anionic" nucleophile to overcome charge repulsion from the target phosphate. The divalent metal ion, as a center of positive charge, could also favor the rearrangement of the negatively charged oxygens of the phosphate group from their normal tetrahedral geometry toward the bipyramidal transition state, where the distance between the oxygen atoms is shorter and, therefore, repulsive charge interactions are stronger. Additionally, charge shielding by metal ions would facilitate both intra- and intermolecular specific RNA interactions (e.g., between a ribozyme and its RNA substrate) that might otherwise be restricted by electrostatic repulsion (Reich et al., 1988).

Tertiary complexes, consisting of either (1) phosphate diester, metal ion(s), and nucleophile or (2) ribozyme, metal ion(s), and substrate, could be formed via the specific chemical crosslinking of multiple RNA sites by both inner- and outer-sphere coordination with metal ion(s) in the tertiary structure of tRNA molecules (see Fig. 9-3A). The proper orientation of groups in the reactive complex is clearly very important, because nonspecific binding could lead to numerous nonproductive (i.e., catalytically inactive) complexes.

Some RNA-bound transition metal ions, possessing the ability to exist in solution in more than one oxidation state, could be available for single-electron "redox" reactions. This has been documented for tRNA complexes with $Co^{2+/3+}$ (Danchin, 1972), $Cr^{2+/3+}$ (Danchin, 1975), and $Fe^{2+/3+}$ (Vary & Vournakis, 1984b). Metal ions with variable valence can catalyze redox processes via bridging and electron transfer between oxidizing and reducing atoms, with or without reversible changes in the oxidation state of the metal ion (Bender et al., 1984). Metal-dependent redox reactions can cause the site-specific modification of nucleotides (Holmes et al., 1993; Sigel,1969) or selective cleavage of internucleotide bonds (Kazakov & Altman, 1991; Kazakov et al., 1988; Lickl et al., 1988; Samuni et al., 1983; Vary & Vournakis, 1984b; Wang & Van Ness, 1989; Zhou et al., 1988) near the sites of metal binding. One example of a phosphoryl transfer reaction that proceeds via a putative metal-dependent redox mechanism is catalyzed by the phosphatase from porcine uterus where two iron ions, which can exist in both "2+" and "3+" oxidation states, have been found in the enzyme catalytic center (David & Que, 1990). The hydrolytic cleavage of DNA molecules by a simple complex via a specific redox process, modeling the di-iron oxo-protein, was recently described (Schnaith et al., 1994). The reported redox activity of some RNA nucleoside residues (Yanagawa et al., 1992) provides an additional argument for the potential importance of such processes.

A unique feature of metal catalysis is the potential for performing complex catalytic transformations by combining the individual types of catalysis discussed above by a single multifunctional metal ion (Cooperman, 1976; Haydock and Allen, 1985). However, there is much evidence suggesting that catalysis of phosphoryl transfer reactions by two or even three metal ions could be rather general and applicable to both protein metalloenzymes (Beese & Steitz, 1991; Bratty et al., 1993; David & Que, 1990; Davies et al., 1991; Hough et al., 1989; Karlin, 1993; Kim & Wykoff, 1991; Vallee & Auld, 1993; Vincent et al., 1992; Volbeda et al., 1991) and ribozymes (Chowrira et al., 1993; Haydock & Allen, 1985; Kazakov & Altman, 1992; Smith & Pace, 1993; Steitz & Steitz, 1993; Uchimaru et al., 1993; Yarus, 1993), as well as to simpler model systems (Chin & Banaszcyk, 1989; Cooperman, 1976; Rawji et al., 1983; Sumaoka et al., 1992). There are three principal roles of metal ions in multimetal active sites: *catalytic*, participating directly in the bond-breaking or -making steps; *cocatalytic*, operating in concert both to perform catalysis and to stabilize active-site conformation; and *structural*—that is, direct or indirect stabilization of a catalytically active conformation of a biopolymer (Vallee & Auld, 1993).

Specific Cleavage of RNAs at GAAA Sites Catalyzed by Mn^{2+} Ions

Studies of the minimal structures that can function as metal-dependent ribozymes probably provide the best demonstration that nucleic acids and metal ions can actually be equal partners in the complexity and specificity of their interactions required for catalysis. Ribozyme size reduction is highly desirable also for structural investigation by X-ray crystallography (Doudna et al., 1993b) and NMR (Limmer et al., 1993), as well as for computational analysis of structural models (Denman,1993; Karlin, 1993).

Recently, the smallest example of a metal-dependent self-cleaving RNA motif has been reported; this motif employs Mn^{2+} and Cd^{2+} for RNA cleavage (Dange et al., 1990; Kazakov & Altman, 1992; Pyle, 1993; Van Atta & Hecht, 1994). In contrast to other known catalytic RNA systems, in which both Mn^{2+} and Mg^{2+} can promote specific cleavage of the RNAs (Chowrira et al., 1993; Dahm & Uhlenbeck, 1991; Gardiner et al., 1985; Grosshans & Cech, 1989; Guerrier-Takada et al., 1986; Kazakov & Altman, 1991; Lehman & Joyce, 1993; Morales et al., 1989; Piccirilli et al., 1993; Pieken et al., 1991; Smith et al., 1992b; Uhlenbeck, 1987; Yang et al., 1990), Mg^{2+} alone cannot substitute for Mn^{2+} in this case (Figs. 9-9 and 9-10) (Dange et al., 1990; Kazakov & Altman, 1992).

Among the biologically important divalent cations, both Mn^{2+} and Mg^{2+} are essential components of many enzymatic reactions involving DNA and RNA (Luck & Zimmer, 1972; Martin, 1990). Most reactions of this type can proceed in the presence of either Mg^{2+} or Mn^{2+}, but dramatic changes in the kinetics and specificity of some of these reactions take place when Mn^{2+} is substituted for Mg^{2+}. Both Mn^{2+} and Mg^{2+} have been found in nucleic acids isolated from a variety of biological species as firmly bound metal ions (Sissoëff et al., 1976; Wacker & Vallee, 1959).

Recently, Dange et al. (1990) have reported the Mn^{2+}-dependent, site-specific cleavage of a 31-nt RNA (Fig. 9-9A) incorporating the 15-nt hairpin excised from the 5'-end of the *Tetrahymena* intron I catalytic RNA during autocyclization (Zaug et al., 1985). The products of this remarkable transformation include fragments with 5'-OH and 2',3'-cyclic phosphate termini; as described below, investigation of this system revealed that the core structure capable of undergoing this transformation constitutes the smallest ribozyme motif identified to date.

Conditions Required for the Mn-Dependent, Site-Specific Cleavage. The optimum conditions for the specific RNA cleavage were found to be close to physiological conditions, as regards ionic strength, pH, and temperature. No RNA cleavage was detected below pH 6.6, and only random cleavage was found at pH $\geq$ 8.4; the maximum ratio of specific to random cleavage was observed at pH 7.5. The specific cleavage was detectable at Mn^{2+} concentrations as low as 0.25 mM, but was maximal at 20 mM Mn^{2+}. No RNA cleavage was detected below 15°C even after 24-hr incubation. The maximum ratio of the rates of specific to random cleavage of the RNA in the presence of 10–20 mM Mn^{2+} ions was found to occur at 45°C. Both the specific and random RNA cleavage increased with increasing temperature. The conditions which were found to be as optimal were 10 mM $MnCl_2$, 100 mM NaCl, and 50 mM Tris-HCl (pH 7.5) at 37–45°C; the usual reaction time was 1–2 hr.

Which RNA Conformation is Required for Cleavage? The ability of RNA molecules to form a number of alternative conformations in solution is well known (Favre & Thomas, 1981; Fedor & Uhlenbeck, 1990; Forster & Symons, 1987; Le et al., 1993; Prasad et al., 1992; Rigler & Wintermeyer, 1983; Walstrum & Uhlenbeck, 1990; Woodson & Cech, 1991; Wu & Lai, 1990). The location of the cleavage site at the junction between the putative single-stranded region and double-stranded stem, which is usual for both the enzymatic and self-cleavage of RNA (Campbell et al., 1987; Dock-Bregeon & Moras, 1987), suggested that the hairpin structure could be a key element of this self-cleaving system. Examination of the nucleotide sequence of RNA **1** (Fig. 9-9A) reveals that several alternative secondary structures which contain one (Fig. 9-9B,C) or more than one molecule of RNA **1** (Fig. 9-9D–F) can, in principle, be formed via Watson–Crick base-pairing. The general feature of all these structures is the location of a potential metal ion-binding "pocket" within hairpin loops or at the intersection of the three duplex stems near the cleavage site. Similar structural features

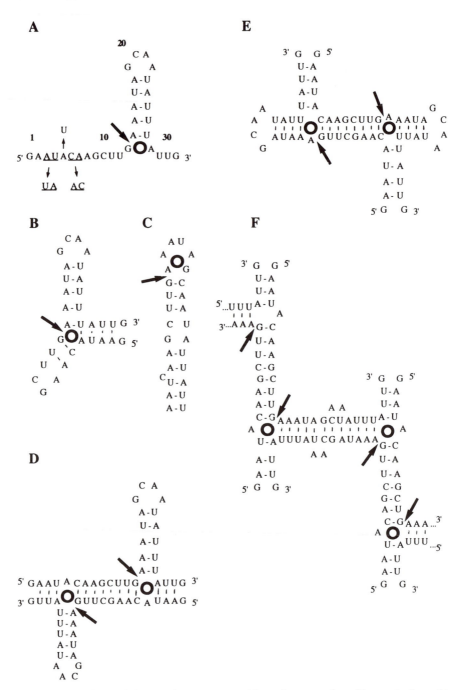

Figure 9-9. (A–F) Possible secondary structures with maximum number of base pairs formed by RNA **1**. Reprinted (with modifications) from Kazakov & Altman, 1992, with permission. Arrowheads show the sites of specific cleavage catalyzed by Mn^{2+} ion(s). Small arrows show the nucleotide substitutions in the derivatives of RNA **1**. Bold circles indicate putative sites of specific binding of Mn^{2+} ions near the cleavage sites.

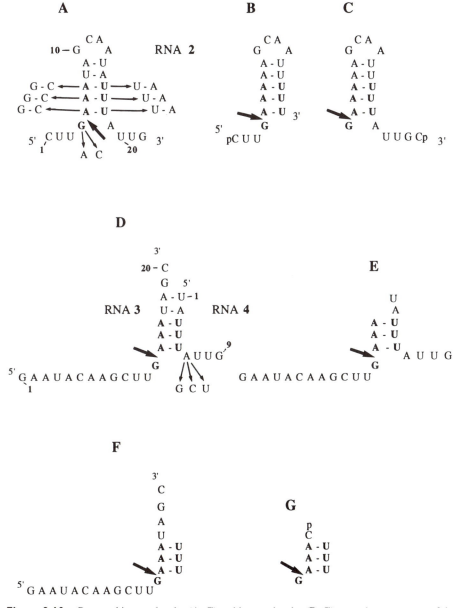

Figure 9-10. Proposed intramolecular (**A–C**) and intermolecular (**D–G**) secondary structures of the synthetic oligoribonucleotides used to determine the sequence and structural requirement for the Mn^{2+}-induced specific cleavage of RNA. Reprinted from Kazakov & Altman, 1992, with permission. Arrowheads show the sites of specific cleavage. Small arrows indicate the nucleotide substitutions in the derivatives of RNA **2** and RNA **4**. Nucleotides shown in boldface are the smallest active structure found.

have been found earlier in "hammerhead" and "hairpin" "ribozymes" (Long & Uhlenbeck, 1993).

Probing of RNA secondary structure by single-stranded specific nucleases S_1 and T_2, as well as double-stranded specific nuclease V_1, indicated that RNA 1 could undergo a conformational rearrangement in the presence of Mn^{2+}, suggesting the possibility of affecting the conformational equilibrium via appropriate metal ion(s) binding (Van Atta & Hecht, 1994). Because the structure of the metal ion–RNA complex necessary for catalytic activity probably does not require high thermodynamic stability, but rather a specific orientation of key functional groups and conformational flexibility of selected bonds, it seems impossible to predict the catalytically active structure simply on the basis of conformational analysis of RNA structure, or even based on data derived from enzymatic, chemical, or physicochemical probing. Size minimization of the catalytic domain and the use of functional tests for potentially important RNA–RNA and RNA–metal ion(s) interactions were used to derive information about the conformation actually required to obtain Mn^{2+}-dependent cleavage.

To determine whether active structures derived from RNA 1 contain more than one molecule of RNA, by analogy with the "hammerhead" ribozyme (Forster et al., 1988), the rate of self-cleavage as a function of the concentration of RNA 1 analog was investigated. The initial rate of Mn^{2+}-dependent self-cleavage of the RNA at a concentration of 100 nM at 45°C has been found found to be about 10-fold greater than that at a concentration of 10 nM (Kazakov & Altman, 1992). The second-order equation for a bimolecular reaction, however, predicts a 100-fold difference in rate under these conditions, and higher-order reactions would exhibit even greater dependency on concentration. Therefore, despite the fact that some intermolecular interaction(s) affect the rate of the specific cleavage of RNA 1, there is no compelling evidence for a dimer or more complex structure.

Minimum Size and Structure Requirements. The introduction of a few nucleotide substitutions at positions 2–6 of RNA 1 (Fig. 9-9A), which would be expected to destroy certain base pairs that are critically important for the stability of the proposed alternative secondary structures (Fig. 9-9B–F), did not affect the self-cleavage reaction. Therefore we proposed that at least part of the 5′-sequence of RNA 1 was not essential for the self-cleavage reaction. In fact, the structure of RNA 2 (Fig. 9-10A), whose sequence corresponds to nucleotides 10–31 of RNA 1, exhibited the same self-cleavage activity as RNA 1.

To identify the minimal size requirements for self-cleavage of the hairpin-like RNAs, RNA 2 was ^{32}P-end-labeled at either its 5′- or 3′-terminus, and then treated under mild conditions with snake venom phosphodiesterase or with alkali, in the absence of Mn^{2+} ions, to generate fragments that had an intact and radiolabeled 5′- or 3′-terminus. The truncated RNAs were then separated by gel electrophoresis, isolated from polyacrylamide gels, and then incubated separately in the presence of Mn^{2+} ions for induction of specific cleavage. The shortest derivatives of RNA 2 that underwent cleavage are presented in Fig. 9-10B,C.

To determine the possible requirement for a hairpin structure, the loop structure was eliminated by division of RNA 1 into two fragments—that is, RNA 3 and RNA 4 (Fig. 9-10D)—and then monitoring reconstitution of the active structure from the separate fragments. It was shown that RNA 3, as well as derivatives of RNA 3 that lack ≤4 nt from the 3′-end, could be site-specifically cleaved in the presence of RNA 4 (Fig.

9-10E). Therefore, neither the loop sequence nor the hairpin shape of RNA **1** are necessary to support the specific cleavage of RNA.

The shorter RNA fragments, in which five or more nucleotides are deleted from the 3'-end of RNA **3**, cannot be cleaved under the same conditions. On the other hand, the trinucleotide UUU can promote the specific cleavage of RNA **3** (Fig. 9-10F). Therefore, just the three complementary base pairs between the AAA sequence in substrate RNA and the UUU sequence of the "external guide sequence" RNAs appear to be necessary for promotion of the cleavage in *trans*. It seems logical to believe that the structure of the active complex is the same for the specific cleavage both in *cis* and in *trans*.

Mn^{2+}-Dependent "Ribozyme". Using the optimal concentrations of the RNAs [RNA **4** (50 nM); RNA **3** (75 nM)] that provided the maximum yield of specific cleavage products in the buffer utilized for RNA **1**, 1.14 ± 0.02 moles (determined in seven trials) of cleaved RNA **3** were formed per mole of RNA **4** during an 8-hr incubation at 37°C. Therefore, RNA **3** can be regarded as a "substrate" and RNA **4** can be considered a "catalyst" as part of the ternary complex with Mn^{2+}.

When RNA **3** was incubated in the absence of any other RNA under standard conditions for >8 hr at 37°C, oligonucleotides corresponding to the specific self-cleavage between G_{13} and A_{14} were somewhat more prominent than the products of random degradation induced by Mn^{2+} ions. It may be possible that this internucleotide bond can be made more labile by the folding of RNA **3** itself, as in the case of site-specific hydrolysis of the *Tetrahymena* IVS RNA (Zaug et al., 1985).

Sequence Requirement. Both nucleotide substitutions at G_4 and the replacement of any one of three neighboring A·U base pairs in the putative hairpin structure of RNA **2** by a G·C or a U·A base pair as shown in Fig. 9-10A resulted in RNAs that would not undergo cleavage in the presence of Mn^{2+}. These results indicate the strict GAAA/UUU sequence requirement near the site of specific cleavage. Additional nucleotides flanking the minimal sequence are not necessary for the cleavage reaction to take place, but can affect this process by promoting (or suppressing) the formation of the RNA conformation required for cleavage.

The results of both the structure minimization and nucleotide substitution experiments described above have provided evidence that the trinucleotide UUU sequence might be sufficient to promote metal-dependent specific cleavage (between G and A) in *trans* of an RNA containing the GAAA sequence. In fact, it was shown that both UUU and poly(U) can promote the Mn^{2+}-dependent specific cleavage of RNA **3** at the same site as does RNA **4**. Poly(I), poly(A), and poly(C) exhibited no activity under the same conditions. Higher nucleotide concentrations of both poly(U) and UUU, as compared with RNA **4**, were needed to promote the cleavage of RNA **3**, presumably because the complementary complex (**GAAAUA**)·(**UAUUU**) formed between RNA **3** and RNA **4** (Fig. 9-10D) is more stable than the (**GAAA**)·(**UUU**) complex. RNA **5** and RNA **6** (see Fig. 9-11) form even longer and, therefore, more stable complementary complexes with RNA **3** than does RNA **4** [(**GAAAUAGC**)·(**GCUAUUU**) and (**CAAGCUUGAAAUA**)·(**UAUUUCAAGCUUG**), respectively] and promote specific cleavage of RNA **3** at 37°C more effectively than does RNA **4** at low oligonucleotide concentrations (see Figs. 9-10D and Fig. 9-11).

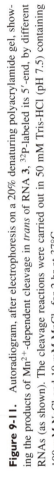

Figure 9-11. Autoradiogram, after electrophoresis on a 20% denaturing polyacrylamide gel, showing the products of Mn^{2+}-dependent cleavage in *trans* of RNA 3, ^{32}P-labeled its 5′-end, by different RNAs (as shown). The cleavage reactions were carried out in 50 mM Tris-HCl (pH 7.5) containing 100 mM NaCl and 10 mM $MnCl_2$ for 2 hr at 37°C.

Furthermore, the ability of UUU, poly(U), and RNA **4** to promote the specific cleavage of the pentamer GAAACp (the 3'-terminal C was added as a consequence of 3'-end labeling GAAA with [5'-^{32}P]pCp) provides additional evidence for the proposed minimal sequence requirement (Fig. 9-10G).

Both UUU and GAAACp function less efficiently as participants in the formation of a scissile complex than do longer RNA fragments containing the same sequences, even if the extra sequences do not provide extra base-pairing between RNA "substrate" and "enzyme." This indicates that the phosphates (or nucleotides) flanking these sequences are not essential for the cleavage reaction per se, but can affect metal ion–RNA complex formation.

Double or Triple Helix? Because local triple helices of the type A·2U can be formed between GAAA and UUU motifs under conditions used in the presence of 10 mM Mn^{2+} (Felsenfeld & Miles, 1967; Han & Dervan, 1993; Usher & McHale, 1976), it seems possible that the Mn^{2+}-dependent specific cleavage, especially in the presence of an excess of the catalytic RNA, could occur in a triple-helical structure. There are a few facts, described above, which make the functional importance of triplex formation seem less likely, including the fact that poly(U) is a less effective catalytic than RNA **3**, in contrast to the greater propensity of poly(U) to form a triple helix with the AAA sequence. In addition, the maximum cleavage of RNA **3** (containing the GAAA motif) by RNA **4** (containing UUU) has been found to obtain at a 1.5:1 molar ratio of these RNAs, and the same specific cleavage occurs at low RNA concentrations both in the putative double-stranded stem of the hairpin RNAs and in the completely complementary duplex (RNA **3**)·(RNA **6**) (Fig. 9-11), in which nucleotide sequences are not conducive to triple-helix formation.

High Temperature Does Not Abrogate the Specific Cleavage Reaction. Because the formation of the complex GAAA·UUU appears to be essential for the promotion of the specific cleavage between G and A, the temperature dependence of the RNA cleavage has been investigated both in *cis* and in *trans*. Surprisingly, we found that rate of the specific cleavage both in *cis* and in *trans* increased with increasing temperature up to 95°C, at which point it was about 10-fold faster than at 37°C. Only insignificant amounts of the Watson–Crick duplex GAAAUA·UAUUU could persist at 95°C because of its low anticipated melting temperature (Felsenfeld & Miles, 1967; Freier et al., 1986; Kanaya & Yanagawa, 1986). In contrast, some specific RNA complexes with Mn^{2+} can survive at high temperature, as in the case of tRNA molecules where multiple manganese-binding sites have been found even at 100°C (Dancin & Guéron, 1970). Moreover, it was shown that even at ~80°C, the site-specific complex between Pb^{2+} and yeast tRNA^{Phe} survived and could induce the selective cleavage of this RNA (Krzyzosiak et al., 1988). It seems reasonable to expect that the dissociation of the specific metal–RNA complex anticipated at higher temperatures may be offset by a greater rate of metal-promoted RNA cleavage.

As suggested for some other self-cleaving RNAs (Forster & Symons, 1987; Rosenstein & Been, 1990; Ruffner et al., 1989; Smith et al., 1992a; Wu & Lai, 1990), conditions that destabilize an RNA structure may actually enhance the cleavage reaction, either by promoting the interconversion between various RNA conformations, assuming that only some of these forms are active, or by facilitating dissociation of the products of the cleavage reaction.

When RNA **2**, or a mixture of RNAs **3** and **4**, were initially mixed with Mn^{2+} at room temperature, subsequent heating at 95°C produced specific cleavage at the usual site with increasing yields of cleavage products during the course of the 5-min incubation at 95°C. In contrast, preincubation of the same RNAs for 2 min at 95°C before the addition of Mn^{2+}, followed by incubation with Mn^{2+} at 95°C for an additional 3 min, resulted in about 50% less cleavage in *cis* and no specific cleavage in *trans*. Preincubation of RNA **2**, or the mixture of RNAs **3** and **4**, in the presence of 10 mM Mn^{2+} for 3 hr at 22°C did not result in specific cleavage, but following incubation for 3 min at 95°C, cleavage of both RNAs occurred at rates about twice those of controls carried out without preincubation. The requirement for a "prebinding" phase seems consistent with the accumulation of some preformed Mn^{2+}–RNA complexes from which the catalytically competent conformation was accessible with higher probability.

Preincubation of RNA **3** with 10 mM Mn^{2+} at 37°C for ≥ 30 min before the addition of RNA **4** also resulted in a much lower rate of specific cleavage of RNA **3**. This diminution of cleavage may be explained by aggregation of single-stranded oligonucleotides by Mn^{2+} crosslinking via guanine moieties (Yurgaitis & Lazurkin, 1981) and by metal binding to adenine residues (Sorokin et al., 1982) because spatial and conformational restrictions inside the aggregates can preclude both inter- and intramolecular interactions. It was shown that metal–base interactions in the complexes of both ApG and GpA with Mn^{2+} were an order of magnitude more abundant than with any other dinucleoside monophosphate (Anderson et al., 1971; Bean et al., 1977; Chan & Nelson, 1969).

There are two more observations in the present case that may result from the ability of Mn^{2+} to induce aggregation of nucleic acids. First, the rate of the specific cleavage both in *cis* and in *trans* diminishes dramatically after the cleavage reaction is about 50% complete. Also, we found that RNA cleavage in the presence of 10 mM Mn^{2+} could be inhibited completely by the addition of any oligo- or polynucleotide to the reaction mixture at concentrations ≥ 50 ng/μl, including the same ribozyme oligomers. Under these conditions, fast aggregation with subsequent coprecipitation of nucleic acids is known to obtain (Eisinger et al., 1965). This phenomenon constitutes a real barrier to the application of nuclear magnetic resonance (NMR) and X-ray crystallographic methods for the structural study of this Mn^{2+}-dependent ribozyme.

Metal Ion Coenzymes and Inhibitors. A survey of numerous metal ions showed that, aside from Mn^{2+}, only Cd^{2+} could promote the same site-specific cleavage of RNA both in *cis* and in *trans* when present at the same concentrations as those at which Mn^{2+} ions are effective.

That the rate of this reaction is actually enhanced with increasing concentrations of Na^+ ions suggests the need for the screening of phosphate negative charge repulsion to allow RNA folding into the structure that leads to cleavage.

The competition between Mn^{2+} and different divalent metal ions has been studied in solutions having moderate ionic strength (100 mM NaCl plus 50 mM Tris-HCl). It was found that the specific cleavage of RNAs induced by 10 mM Mn^{2+} was unaffected by Mg^{2+} up to 25 mM. However, at lower concentrations of manganese ions, 1 mM Mg^{2+} significantly enhanced the catalytic properties of 1–3 mM Mn^{2+}, while

Mg^{2+} ions at 10 mM decreased the ability of Mn^{2+} ions (at ≤ 10 mM concentration) to promote the self-cleavage of RNA. Because both Mg^{2+} and Mn^{2+} are similar in their ability to chelate or crosslink phosphate oxygens in nucleic acids (Granot & Kearns, 1982; Granot et al., 1982; Mildvan & Grisham, 1974; Yamada et al., 1976), this mode of binding of divalent metal ions must be important in the catalytic mechanism.

Furthermore, since Mg^{2+} ions alone are inactive, some additional kind(s) of essential Mn^{2+}-binding site(s) must exist. This role could be played by heterocyclic N atoms of the bases, which have a much greater affinity toward Mn^{2+} than toward Mg^{2+} (Anderson et al., 1971; Clement et al., 1973; Cohn et al., 1969; Danchin & Guéron, 1970; Kasyanenko et al., 1987; Martin, 1990; Murray & Flessel, 1976; Pörschke, 1979; Shin, 1973; Shin et al., 1972). Co^{2+}, which can support the GAAA-specific cleavage reactions, albeit to much lesser extent than Mn^{2+}, as well as Ni^{2+} (which does not support these reactions by itself) at concentrations of 10 mM did not affect the self-cleavage reaction promoted by 10 mM Mn^{2+}. At the same concentration, Cu^{2+} diminished the activity of Mn^{2+} ions by about 50%. Zn^{2+} ions at 0.1 mM concentration completely inhibited RNA cleavage in the presence of 10 mM Mn^{2+}, without causing random degradation of the RNA. At higher concentrations, Zn^{2+} induced nonspecific cleavage of the RNAs. Because the relative phosphate/base binding ability of the metal ions tested is $Mg^{2+} < Co^{2+} \leq Ni^{2+} < Mn^{2+} < Zn^{2+} < Cd^{2+} < Cu^{2+}$ (Eichhorn, 1973), these results support the thesis that the ability of the active metal ions to bind both to phosphates and to nucleobase(s) is essential for promoting specific RNA cleavage. Zn^{2+} ions might be expected to compete with Mn^{2+} ions by binding near the catalytic site(s), but Zn^{2+} could not promote GAAA-specific RNA cleavage.

Features of "Metal–Nucleic Acid Binding" Required for Specific Cleavage. Apparently, the ability of these metal ions to promote or to inhibit the specific cleavage of the same RNAs derives from their coordination properties in complexes with nucleic acids, which actually is much more complicated than just the physicochemical properties of the metal ions, or their affinity toward phosphate and heterocyclic bases.

There is no correlation between the catalytic activity of the metal ions and their ionic radii, first pK_a of coordinated water molecules, geometry of corresponding coordinating inner spheres (see Table 9-1), and the relative rates of substitution of H_2O molecules bound with the metal ions (see Fig. 9-1).

It has been shown that as a rule, the interaction of divalent metal ions with polynucleotides leads to the stabilization of helical structures as result of phosphate binding (e.g., by Mg^{2+}) or to destabilization via predominant purine base binding (e.g., for Cu^{2+} and Cd^{2+}) (Eichhorn & Shin, 1968; Murray & Flessel, 1976; Shin et al., 1972; Shin, 1973). Mn^{2+}, Co^{2+}, and Zn^{2+} do not behave in a simple manner; they can both specifically stabilize some helical structures and destabilize others to induce a transition from a multistrand helix to a single helix or to a random coil, depending on the conditions, the nature of the nucleotide bases and metal–base binding mode, and the preferred type of intra- or intermolecular interactions in the corresponding polynucleotide structure (Eichhorn, 1973; Eichhorn & Shin, 1968; Shin, 1973; Shin et al., 1972). But there is no rule without exceptions. Although uracil has the lowest affinity among the RNA bases toward these metal ions (Cohn et al., 1969; Shin, 1973), Cu^{2+}, Cd^{2+} and Mn^{2+} ions can destabilize double-helical poly(U), whereas Mg^{2+} and Zn^{2+} ions promoted the formation of this duplex (Shin, 1973). In contrast, Mn^{2+} and Cd^{2+} ions, but not Mg^{2+} or Zn^{2+} ions, can stabilize noncanonical base-pairing (e.g., between inosine and uridine or adenosine residues in polynucleotide complexes) in the polynucleotide complexes (Murray & Flessel, 1976).

Determination of the Atoms and Groups Essential for the Promotion of Specific RNA Cleavage. Studies of the effects of sequence changes within the "active core" of ribozymes can indicate the nucleotide residues that are critical for the cleavage activity (Heidenreich et al., 1993; Usman & Cedergren, 1992). These results can reflect the disturbance, direct or indirect, of essential interactions between functional groups, the RNA binding characteristics, and position of the metal ion cofactor. The next logical step after determination of the conserved nucleotide sequences is mutagenesis at the level of single atoms. A "substitution/interference" assay can be used to determine the functionally essential atoms whose substitution (or modification) inhibits specific RNA cleavage (Bratty et al., 1993; Heidenreich et al., 1993). For this purpose, the introduction of deoxyribose in place of ribose, phosphorothioates in place of phosphates, and derivatives of the heterocyclic bases into the NGAAAN/NUUUN sequence by chemical or enzymatic synthesis of corresponding modified oligonucleotides could provide much useful mechanistic information. This approach has been used previously with the GAAA sequence in the "hammerhead" ribozyme motif (Bratty et al., 1993; Dahm & Uhlenbeck, 1991; Fu et al., 1993; Fu & McLaughlin, 1992; Perreault et al., 1991; Ruffner & Uhlenbeck, 1990; Slim & Gait, 1992; Tuschl et al., 1993; Usman & Cedergren, 1992; Williams et al., 1992).

There are many ways in which the 2'-OH group of ribose, being both a proton donor and an acceptor, can interact specifically with other RNA donor or acceptor atoms (via hydrogen bonds) as well as with metal ions (via outer- or inner-sphere complexes). Therefore, by substitution of 2'-OH with 2'-H in individual sugar residues, it should be possible to identify which 2'-OH groups are necessary for catalysis, even if their roles are not understood at a molecular level.

In contrast to phosphate oxygen atoms, the negative charge in the phosphorothioate moiety is largely localized on the more polarizable sulfur atom (Frey & Sammons, 1985; Liang & Allen, 1987). Hard metal ions (e.g., Mg^{2+}) prefer to bind only the phosphorothioate oxygen, whereas softer metal ions (e.g., Co^{2+}, Mn^{2+} and Zn^{2+}) can chelate both oxygen and sulfur. Cd^{2+} ions, in comparison, have a stronger affinity toward sulfur (Burgers & Eckstein, 1979; Herschlag et al., 1991; Pecoraro et al., 1984). Using RNA oligomers having a phosphorothioate with a bridging S atom at the cleavage site, or the individual Sp or Rp stereoisomers with nonbridging sulfur atoms, provides the opportunity to determine whether direct interaction of a particular phosphate oxygen(s) with the metal ion(s) is essential for promotion of the specific RNA cleavage; the stereochemical course of the cleavage reaction can also be determined (Heidenreich et al., 1993; Herschlag et al., 1991; Piccirilli et al., 1993; Ruffner & Uhlenbeck, 1990; Slim & Gait, 1991).

The introduction of modified heterocyclic bases at selected positions of the RNA is the most recent approach in the chemical mutagenesis of RNA. In fact, the bases play a key role in nucleic acid structure and function. Many modified nucleobases exist in nature [e.g., in tRNAs (Nishimura, 1979)], and an enormous variety of structures can be synthesized in the laboratory. Among the several examples of base modifications described, perhaps the most interesting for the study of the Mn^{2+}-dependent ribozymes would be alterations in the GAAA sequence involving the purine exocyclic amino groups of adenine and guanine, which could be replaced by purine and hypoxanthine bases, respectively (Slim & Gait, 1992). The deletion of the exocyclic O6 group of guanine via substitution with 2-aminopurine (Chowrira et al., 1991) would also be interesting, as would the deletion of the heterocyclic N7 of adenine by substitution with 7-deazadenosine (Fu & McLaughlin, 1992). Besides the well known interbase and metal ion–base interactions, specific H-bonds between the sugar 2'-OH group and base atoms [e.g., with N3 of purine or O2 of pyrimidine bases (Dock-Bregeon & Moras, 1987; Ts'o, 1974)] can be formed in some RNA conformations. The base–ribose interactions may be of special importance because, for example, it is proposed that both the exocyclic 6-amino group of adenosine and the 4-amino group of cytidine participate in the promotion of RNA cleavage at UA and UC sequences, respectively (Dock-Bregeon & Moras, 1987; Kierzek, 1992).

For the shielding of specific metal-binding sites, direct partial modifications (i.e., less than one atom per polynucleotide molecule) of selected nucleic acid residues such as ethylation of phosphate oxygen by ethylnitrosourea or methylation of atoms N7/N1 in purines by dimethylsulfate (Ehresmann et al., 1987) might also be useful.

Studies of the types described above are in progress, but we know already that the 2'-OH group of the guanosine moiety in the GAAA sequence of Mn^{2+}-dependent ribozymes is essential for formation of the 2',3'-cyclic phosphate at the end of the cleavage product, whereas the 2'-OH groups of the uridine moiety in the UUU sequence are not, because poly(dU) promotes the specific cleavage of RNA **3** in the same fashion as poly(U), albeit only when present at threefold higher concentration.

Mechanism of Catalysis. This study is far from complete because there is as yet no direct evidence for the mechanism of the Mn^{2+}-dependent cleavage of RNA in the GAAA/UUU motif. Nonetheless, a working hypothesis can help to move the investigation forward.

The computer-simulated model of the complementary complex of GAAA with UUU indicates that there could be specific binding sites for up to three Mn^{2+} ions (Fig. 9-12); these could be involved both in the folding of the RNAs and in the cleavage of the phosphodiester bond between G and A in the GAAA motif. The hypothetical mechanism based on this model (Kazakov & Altman, 1992) suggests that the function of the UUU strand could consist of orienting complementary adenine bases and facilitating binding of the metal ions in a catalytically productive fashion. The key interaction in this mechanism could be the bond between the ribose 2'-oxygen atom and the divalent metal ion. Such an unusual metal-binding mode has been found for Cd^{2+} in the complex with inosine 5'-phosphate (Goodgame et al., 1975).

Recently, it was shown for an A-form RNA double helix that the predominant metal-binding sites, namely the N7 and N1 atoms of purines, were buried in the deep and narrow major groove (Gao et al., 1993). Therefore, a substantial conformational distortion of the A-form duplex would be necessary to allow metal ion binding to the N7

A

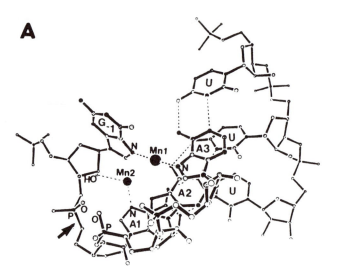

B

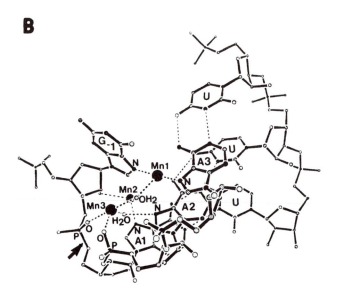

Figure 9-12. Hypothetical models of the binding of two (**A**) or three (**B**) Mn^{2+} ions to the double-helical complex formed by GAAA and UUU to promote specific cleavage (the site of cleavage is shown by an arrow), generating guanosine $2',3'$-cyclic phosphate and $5'$-$_{HO}$AAA. Reprinted (with modifications) from Kazakov & Altman, 1992, with permission.

atom of the adenine residues, as shown in Fig. 9-12. In contrast, the N3 atoms of the purines in the exposed minor groove can become the principal metal-binding sites in a double-stranded RNA region (Gao et al., 1993).

Space-filling models indicate that any reasonable complex of Mn^{2+} ion(s) with GAAA, occurring in an A-form duplex like RNA **6** / RNA **3** (Fig. 9-11) where the guanine residue is base-paired, cannot provide specific activation and cleavage of the phosphodiester bond between G and A. Nevertheless, cleavage in the RNA **6** / RNA **3** system occurs as fast as in the partially complementary complexes RNA **4** / RNA **3** or RNA **5** / RNA **3** (Fig. 9-11). This could indicate that cleavage takes place in oligonucleotide substrates that are not (fully) annealed.

Finally, a "four-step-mechanism" of Mn–ribozyme action is proposed. The first step involves the formation of a complementary complex between GAAA and UUU sequences. The second step is specific Mn^{2+} binding to the N3 atom of an adenine residue(s) of the formed duplex. The sterically inaccessible adenine N7 and N1 atoms, whose uncomplexed nature is important for the next step, do not bind Mn^{2+}. The third step involves dissociation of the Mn^{2+}-bound GAAA·UUU duplex, followed by the formation of a GAAA tetraloop-like structure to produce the catalytically active tetranucleotide complex containing prebound Mn^{2+} ion (see below). The fourth step is the actual chemical transformation that produces the specifically cleaved fragments.

The GAAA tetraloop structure, determined by NMR (Heus & Pardi, 1991; Orita et al., 1993; SantaLucia et al., 1992), suggests the presence of a complex hydrogen-bonding network (Fig. 9-13) which provides extraordinary stability to the mini-hairpins (Antao et al., 1991; Franzen et al., 1993; Hirao et al., 1989, 1992). It seems conceivable that the GAAA motif itself (without flanking hairpin base pairs) could form a similar loop-like structure in an RNA molecule, at least as one of a number of conformational isomers.

The study of molecular models of $pGpA_1pA_2pA_3$ hairpin loop shows that a single Mn^{2+} ion could bind specifically both to the N3 atom of A_1 and to the oxygen anion of the phosphate connecting G and A, as well as to an additional oxygen anion of the internucleotide phosphate at the 5′ end of G. Such a metal binding mode is chemically and sterically allowed (Yamada et al., 1976). The structure of the GAAA tetraloop provides the irregularity of base stacking for A_1, which according to Yamada et al. (1976) could favor the binding of Mn^{2+} simultaneously to the two phosphates and the adenine ring (Fig. 9-14). The freedom of movement of the base and a sharp change in the direction of the ribose-phosphate backbone would differentiate the ribonucleotide at the cleavage site from all others in self-cleaving RNAs (Mei et al., 1989).

There are two key bonds which could provide specific activation of the two chemical groups in the complex between the loop $pGpA_1pA_2pA_3$ motif and Mn^{2+} ion (Fig. 9-13). First, the H-bond between the N7 atom of A_2 and the 2′-OH group of the G sugar should increase the nucleophilicity of the 2′-OH group. This suggests an important role for the adenine base in catalysis. Indeed, the imidazole ring of the adenine nucleoside was shown to be able to catalyze the hydrolysis of the ester bond in the model compound *p*-nitrophenyl acetate (Maurel & Décout, 1992). Second, Mn^{2+} could perform a structural and catalytic function by coordination of this ion to the oxygen atom of the GpA_1 internucleotide phosphate, thus enhancing both the stability of the catalytically active structure and the electrophilicity of the corresponding phosphorus atom. In addition to the activation of the 2-′OH group and internucleotide phosphate, the transesterification reaction can occur only if the proper orientation of the activated

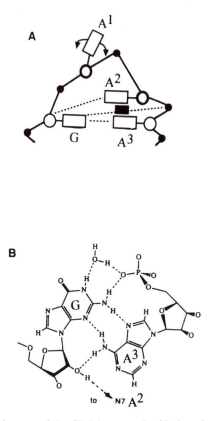

Figure 9-13. Structural features of the GAAA tetranucleotide loop in the hairpins. (**A**) General scheme illustrating specific contacts between nucleobases (boxes), sugars (circles), and phosphates (boldface circles). Hydrogen bonds and stacking interactions are shown by dotted lines and by the black box, respectively. (**B**) Putative hydrogen-bonding network for the guanosine residue in the GAAA tetraloop structure. The N7 atom of the A_2 residue can form a hydrogen bond with the 2'-OH group of G. Reprinted (with modifications) from Heus & Pardi, 1991, and SantaLucia et al., 1992, with permission.

reactive groups is achieved. It seems likely that a specific conformational change of the GAAA-loop structure (promoted e.g., by binding of this complex to an additional Mn^{2+} ion(s) may be needed to achieve the requisite transition state geometry (Fig. 9-8A). We found that the GAAA tetraloop could not be self-cleaved by Mn^{2+} ions in the hairpin RNAs having no internal UUU sequence. This could be the result of either (1) inappropriate binding of Mn^{2+} ions to the adenine base(s) (e.g., predominantly to the atoms N1 and N7) or (2) the very high stability of the hairpin GAAA tetraloop, which would restrict its conformational flexibility.

A New Kind of RNA Restriction Enzyme. The reaction in *trans* can be utilized to cleave large RNAs at specific sites. For example, RNA **4**, when hybridized to M1

RNA, the catalytic RNA subunit of *E. coli* RNase P, induces cleavage at three accessible GAAA sequences in the presence of Mn^{2+} ions. Furthermore, *E. coli* tRNA^{Lys}, which has the anticodon ^{mnm5s2}UUU (where the first anticodon base is 5-methylaminomethyl-2-thiouridine), was used as an "external guide sequence" (EGS) to induce specific cleavage, in the presence of Mn^{2+}, in the anticodon loop GAAA^{mtp} (where the anticodon-adjacent base is N6-(Δ^2-isopentenyl)-2-methylthioadenosine) of *E. coli* tRNA^{Phe}.

Moreover, an RNA molecule having several GAAA sites could potentially be cleaved specifically at only one site by using the difference in primary sequence adjacent to these sites. As shown in Fig. 9-11 for one case, the higher the complementarity between an EGS and its substrate RNA, the lower the concentration of the EGS that was needed for efficient cleavage of the RNA substrate. Therefore, it may be possible for a specific GAAA site to be cleaved at a concentration of the appropriate EGS RNA that leaves the other GAAA sites intact.

The Mystery of the GAAA Motif. Beyond the obvious in vitro "restrictase" and in vivo therapeutic applications of this sequence-directed cleavage of an RNA in *trans*, study of the mechanism of the Mn^{2+}-dependent specific cleavage of GAAA might help to clarify the mysterious role(s) of this tetranucleotide sequence in the biochemistry of catalytic RNAs.

A　　　　　　　　　　　　**B**

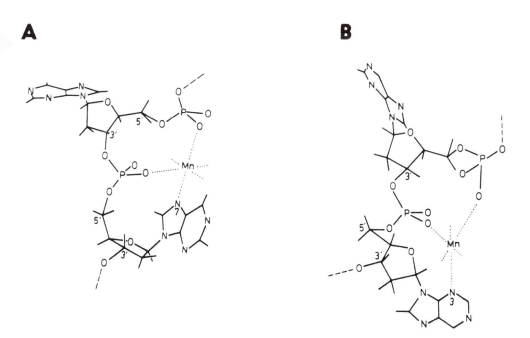

Figure 9-14. Schematic representation of two sterically allowed structures for Mn^{2+}–nucleic acid complexes, where Mn^{2+} is bound to two phosphates and to N7 (**A**) or N3 (**B**) atoms of the adenine residue. Reprinted from Yamada et al., 1976, with permission.

The only conserved sequence motif in bacterial RNase P RNAs is an 11-nt segment, GAGGAAAGUC (Darr et al., 1992). GAAA also has been found in the consensus sequence of self-splicing group I and group II introns from different sources, including viroids (circular pathogenic plant RNAs) and virusoids (plant satellite RNAs) (Dinter-Gottlieb, 1986). The conserved GAAA sequence is an essential element of both "hammerhead" and "hairpin" ribozymes (Long & Uhlenbeck, 1993). It is interesting that the conjugate of *Tetrahymena* and "hammerhead" "ribozymes" can undergo Mg^{2+}-dependent self-cleavage between G and A at the "hammerhead" consensus GAAA site (Grosshans & Cech, 1991). Recently, the involvement of the conserved GAAA motif in the specific interactions which provide a conformational change of the *Tetrahymena* ribozyme and promote the cleavage of an RNA substrate in *trans* has been described (Wang et al., 1993).

RNA editing in trypanosomes also represents a kind of RNA catalysis and involves the addition and deletion of $(U)_n$ within the coding region in several mitochondrial mRNA transcripts (Blum et al., 1991). According to the transesterification mechanism proposed, the transfer of uridine residues occurs from the 3′ oligo(U) tail of a small (40–60 nt) gRNA to an editing site, with the number of uridine residues determined by their base-pairing with A and G "guide" nucleotides (including two GAAA sequences) in the gRNA .

The precursor tRNA[Leu] having a UAA anticodon from diverse cyanobacteria and major groups of chloroplasts contains a single group I intron with two conserved GAAA sequences; one of them is the splicing site with the self-cleaving phosphodiester bond between G and A (Xu et al., 1990). The intron homology across chloroplasts and cyanobacteria implies that the consensus nucleotide sequence has been maintained in their genomes for at least 1 billion years (Kuhsel et al., 1990).

Summary

Interest in metal–nucleic acid interactions is based on the involvement of metal ions in nucleic acid biosynthesis, processing, and degradation, as well as genetic information transfer and expression, mutagenesis, chromosomal abnormalities, and carcinogenesis. The antitumor and antiviral activities of some metal compounds are also of special interest. It may also be noted that some metal-containing reagents are used as heavy atom labels in X-ray crystallography of nucleic acids and as chemical probes of nucleic acid structures in solution. There are a number of metal-binding sites in nucleic acids having different affinity for different metal ions. In spite of the fact that metal ions are much smaller in size than nucleic acids, they exhibit considerable complexity in their interactions with nucleic acids due to three features. First, chemical and stereochemical properties allow metal ions to form both ionic and coordination bonds with nucleic acid constituents, either directly or through water molecules. The second relevant feature is the variety of thermodynamic stability and kinetic liability of the formed metal–nucleic acid bonds. Finally, the third feature is the ability of metal ions to stabilize or destabilize certain nucleic acid structures. Metal ions and their aqua–hydroxy complexes, capable of carrying both electrophilic and nucleophilic functions, can catalyze both the synthesis and hydrolysis of nucleic acids. The joining of the catalytic properties of RNA molecules and metal ions in ribozymes makes them functionally similar to protein metalloenzymes. This chapter has attempted to describe the variety and complexity of elementary modes of metal–nucleic acid interactions and to illustrate the problems that exist in determining the structures and functions of the complexes formed between biologically relevant metal ions and nucleic acids.

10

Small Molecule–DNA Interaction

John A. Mountzouris and Laurence H. Hurley

DNA is the presumed intracellular target for a variety of quite structurally diverse low molecular weight ligands. A number of DNA-reactive synthetic and natural products or their derivatives have achieved success as clinically important antitumor agents, including bleomycin, doxorubicin, cisplatin, actinomycin D, mitomycin, and cyclophosphamide. The field of nucleic acid targeted drug design has been highlighted in a series of recent articles and texts (Hurley, 1989, 1992; Propst & Perun, 1992; Pullman & Jortner, 1990). It draws on powerful structural techniques such as X-ray crystallography, multidimensional nuclear magnetic resonance (NMR), and computational chemistry, which have begun to disclose how DNA structure bears the imprint of its sequence. DNA-binding proteins such as restriction endonucleases and transcriptional factors have evolved so that their structures are refined to recognize specific DNA sequences in order to carry out the precise genetic manipulations ongoing in a living system. Low molecular weight ligands, while exploiting more limited means, also recognize and interact with DNA in ingenious ways. Increasingly, these molecules are of chemical, biological, and medical significance as potential artifical gene regulators or cancer chemotherapeutic agents.

This chapter is intended to provide an introduction to relevant and important examples in this area. The first part is a survey of the principal strategies used by specific ligands to interact with DNA. In the second part, we follow the development of work primarily from our laboratory and our collaborators at The Upjohn Company on the antitumor antibiotic (+)-CC-1065 and its analogues, providing an overview of how specific structural elements of the (+)-CC-1065 molecule recognize DNA structure and modulate the interactions of DNA-reactive proteins with DNA. In specific cases, attempts are made to relate *structural* modification of DNA by (+)-CC-1065 to the *biochemical* and *biological* effects it produces.

The main categories for interaction of small ligands with DNA include (1) DNA strand cleavage, (2) noncovalent association with the minor groove of DNA, (3) intercalation between DNA base pairs, (4) alkylation of a component nucleotide, and (5) a combination of (3) and (4).

DNA Cleavage

The neocarzinostatin and kericidin chromophores, esperamicins, dynemicins, and calicheamicins are members of the enediyne class of antitumor antibiotics (Fig. 10-1).

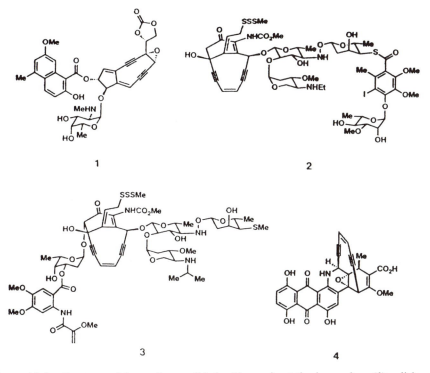

Figure 10-1. Structures of the enediyne antibiotics. Neocarzinostatin chromophore (**1**), calicheam-icin γ_1^I (**2**), esperamicin A1 (**3**), and dynemicin A (**4**).

Their intriguing molecular architecture and unique molecular mechanism for interaction with, and cleavage of, DNA has spurred a wealth of characterization and synthetic efforts, as summarized in several reviews (Lee et al., 1991b; Nicolaou & Dai, 1991; Tanaka et al., 1991). The remarkable biological potencies of these natural products are thought to be a consequence of their ability to bind to cellular DNA and produce single- or double-stranded breaks with great efficiency. The molecular mechanism of single- and double-stranded breakage by the bleomycins is distinct from that utilized by the enediyne class (Stubbe & Kozarich, 1987). While the bleomycins are clinically important, the enediynes produce a toxicity intolerable for clinical use, fueling the search for analogs with improved therapeutic potential. While the details of DNA binding and strand breakage differ from one molecule to another, for present purposes calicheamicin γ_1^I is taken as an example.

Calicheamicin γ_1^I (**2**) was isolated from *Micromonospora echinospora* ssp. *calichensis* at Lederle Laboratories. With 10^3 times the cytotoxic potency of adriamycin against murine cell tumors, a brief listing of its biological activities would include 50% reduction in HeLa cell DNA synthesis at low drug concentrations (50 pg/ml), mutagenesis in *Escherichia coli*, and chromosome aberrations in human diploid lung fibroblasts (Lee et al., 1991a,b; Zein et al., 1988, 1989b). The calicheamicin γ_1^I structure consists of three moieties: the glycosylated hydroxylamino sugar, the labile trisulfide fuctionality, and the bridgehead aglycone containing the conjugated diacetylenic ring system (Fig. 10-1). Analysis using high-resolution polyacrylamide gel electrophoresis (PAGE) of a 5'-^{32}P-end-labeled DNA cleaved by calicheamicin γ_1^{I} identified the primary cleavage sites as involving transformation of the 5'-cytidine in 5'T<u>C</u>CT or 5' thymidine in 5'C<u>T</u>CT sequences. Scission on the opposing strand was offset by two nucleotides toward the 3'-end of the 5'AGGA complement, due to the right-handed twist of DNA (Zein et al., 1988). The 3'-offset asymmetric DNA cleavage pattern by calicheamicin γ_1^{I} is regarded as evidence for modification of the DNA by the drug positioned within the *minor* groove, in accordance with Van Dyke and Dervan's model for asymmetric cleavage by minor groove ligands (Van Dyke & Dervan, 1984). This information can be used to determine the DNA sequence-selectivity of the drug (Fig. 10-2). The pathway leading to DNA cleavage is proposed to begin with sequence-specific delivery of the drug to the minor groove by binding of the oligosaccharide to the 3'-end of the cleavage site. ^{1}H-NMR studies show that the carbohydrate moiety of calicheamicin γ_1^{I} adopts an extended and fairly rigid conformation in polar and nonpolar solvents, enforced by a high barrier of rotation about the N—O bond (Walker et al., 1990). In addition to imparting a curvature complementary to the minor groove (Walker et al., 1991; Yang et al., 1991), this structural domain contributes hydrogen bonding functionality, electrostatic interaction, hydrophobic effects (Ding & Ellestad, 1991), and polarizability of the halogen group (Hawley et al., 1989), which have all been proposed to contribute to the positioning of this portion of the drug in the minor groove poised for its molecular interaction with DNA. From the information presently available, the proposal for the cascade of molecular events leading to DNA cleavage is depicted in Fig. 10-3 (Golik et al., 1987a,b; Lee et al., 1987a,b). A nucleophile (e.g., glutathione) is thought to reductively cleave the allylic trisulfide to the corresponding thiolate anion. The resulting thiolate undergoes a conjugate addition to the α,β-unsaturated ketone to afford the dihydrothiophene intermediate. Strain due to sp^3 hybridization at C7 is released by a cycloaromatization via a Bergman cyclization, leading to the benzenoid diradical. The reactive diradical is then well positioned to abstract two hydrogen atoms, one from the C5' position of deoxycytidine and the other from the C4' hydrogen of the nucleotide two base pairs toward the 3'-side of the complementary 5'AGGA-containing strand, leading to a double-stranded break (DeVoss et al., 1990; Zein et al., 1989).

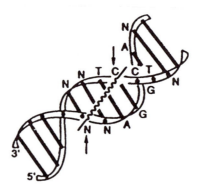

Figure 10-2. DNA cleavage in minor groove by calicheamicin γ_1^{I}. Arrows mark sites of hydrogen abstraction by the diradical intermediate. Reprinted from Lee et al., 1991b, with permission.

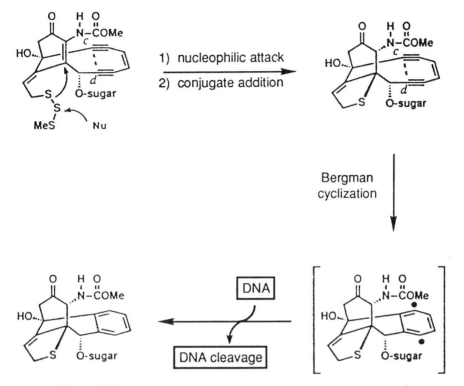

Figure 10-3. Proposed mechanism for DNA cleavage by calicheamicin γ_1^I.

The idiosyncrasies of neocarzinostatin, esperamycin, dynemicin, and calicheamicin DNA chemistry are elaborated extensively in the primary literature and in reviews; the key underlying theme is that the extraordinary potency of these molecules lies in their common ability to (1) bind to specific sequences on DNA and (2) unleash sp^2 carbon-centered radicals upon appropriate chemical triggering, which abstract protons from the deoxyribose of DNA, leading to strand fragmentation. In recent work, the surprising discovery of the importance of the apoprotein of kericidin in cleaving histone molecules to unmask naked DNA has been illustrated (Zein et al., 1993).

Minor Groove Binding

A host of biophysical methods such as ultraviolet/visible spectroscopy, circular dichroism, NMR, and X-ray crystallography have disclosed that many ligands that interact noncovalently with the minor groove of double-helical DNA bind predominantly to AT-rich sequences. A series of crystallographic studies of the complexes of a number of these agents such as netropsin (**5**) (Coll et al., 1989; Kopka et al., 1985a,b), distamycin A (**6**), and Hoechst 33258 (**7**) (Pjura et al., 1987; Quintana et al., 1991; Teng et al., 1988) (Fig. 10-4) have illuminated most powerfully the particular hydrogen bonds and steric contacts resulting in a close match of drug molecular architecture to the shape of the AT minor groove. In the majority of cases, a 1:1 ratio of drug to binding site

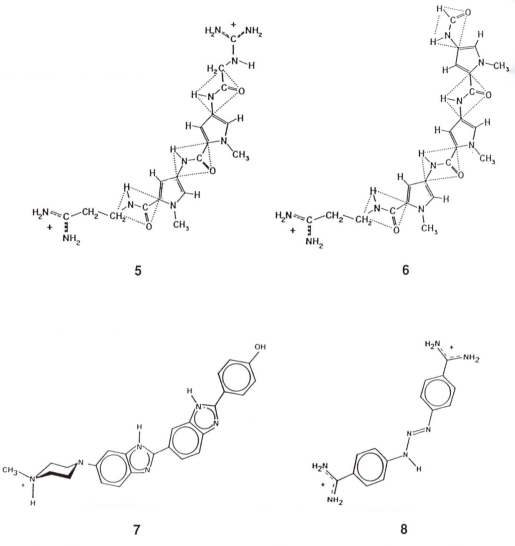

Figure 10-4. Structures of netropsin (**5**), distamycin A (**6**), Hoechst 33258 (**7**), and berenil (**8**).

has been found. However, in the NMR investigation of a distamycin A·d(CGCAA-ATTGGC)·d(GGCAATTTGCG) complex, a 2:1 ratio was indicated such that two drug molecules were bound side by side in the same region of the minor groove, with significant expansion of the minor groove required to accommodate stacking of the aromatic chromophores (Pelton & Wemmer, 1989). In addition to steric interference of the bulky 2-amino group of guanine in the minor groove, the greater negative electrostatic potential of AT-rich regions in comparison to GC-rich regions has also been proposed to account for the AT-selectivity of a number of these ligands bearing positive charge. A recent crystallographic structural analysis of the diarylamidine berenil (**8**)

(Fig. 10-4) complexed with the self-complementary DNA duplex d(CGCGAATT-CGCG)$_2$ provides a good example of minor groove noncovalent association (Brown et al., 1990). NMR characterizations of berenil–duplex DNA complexes have also been reported (Lane et al., 1991; Yoshida et al., 1990).

Berenil, which possesses mild cytotoxic and antiviral properties, consists of two phenylamidine rings linked by a triazene bridge (De Clercq & Dann, 1980). The crystal structure of the free ligand discloses that the molecule has a slight curvature, with a right-handed twist of 9.6 Å between the phenyl rings (Pearl et al., 1987). In the structure of the complex with d(CGCGAATTCGCG)$_2$ determined by X-ray crystallography, the drug spans the three-base-pair site 5′AAT (Fig. 10-5) (Brown et al., 1990). Situated asymmetrically with respect to the dyad axis of the self-complementary oligonucleotide, the primary intermolecular interactions occur with adenosine$_5$ and adenosine$_{18}$ on opposing strands. The observed position of the berenil molecule has the hydrophobic phenyl rings oriented between the more hydrophobic sugar and C5′ atoms of the two backbone strands, while the more polar triazene group of the drug is positioned between the two phosphodiester groups of the backbone. This places the terminal amidinium groups in a position suitable for hydrogen bonding to donor ni-

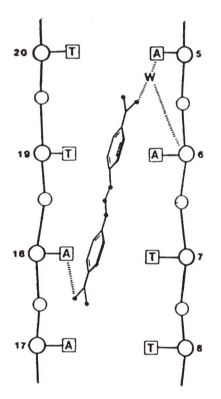

Figure 10-5. Berenil complex with d(CGCGAATTCGCG)$_2$. Dotted lines indicate hydrogen bonding from berenil amino group of amidine to N3 of adenosine$_{18}$ and through water (W) to N3 of adenosine$_5$. The water molecule also bridges O4′ of adenosine$_6$. Reprinted from Brown et al., 1990, with permission.

trogens on the edge of the heterocyclic bases. The drug is anchored by hydrogen bonds from each amidinium tail to opposing strands on the DNA. One amidinium tail bonds directly through one amino group to N3 on the edge of adenosine$_{18}$ at a distance of 3.11 Å and with a somewhat longer bond of 3.4 Å to O$_2$ on thymidine$_8$. The other amidinium tail, positioned near adenosine$_5$, does not participate in any hydrogen bonds directly to this base but bonds instead to a water molecule (Fig. 10-5). This water molecule participates in hydrogen bonding with N3 of adenosine$_5$, one NH2 of the amidinium tail, and the O4′ sugar atom of adenosine$_6$ and is shifted relative to its position in the native DNA structure.

To optimize the binding of berenil, the intrinsically narrow AT-rich minor groove of d(CGCGAATTCGCG)$_2$ widens by approximately 1.5 Å, in line with the effect produced by netropsin and distamycin. Yet while the latter drugs also significantly alter other base-pair parameters relative to the crystal structure of the oligonucleotide in the absence of drug, the influence of berenil on DNA structure appears modest and localized. At the 3′-end AT base pair, there is an increase of 4° in the propeller twist, necessary to permit more optimal hydrogen bonding geometry between N3 of adenosine$_{18}$ and the berenil amidinium proton. This is accompanied by small changes in base-pair buckle in the AAT-binding region and roll at the first AT step, thereby reducing steric clashes between the bases and the inward-facing hydrogen atoms of the berenil phenyl rings, as well as facilitating base–berenil hydrogen bonding.

A conclusion to be drawn from the berenil d(CGCGAATTCGCG)$_2$ structure is that the triazene group and the phenyl rings position berenil in the AT minor groove, while the amidinium moieties select particular bases. The water molecule at the 5′-end is situated for coordination to the adenine base through N3. In the case of proteins, the indirect association with DNA through bridging water molecules is perhaps best established for the phage 434 operator-repressor complex, where Arg43 is bridged via three water molecules to base pairs in the minor groove. Indeed, Neidle and coworkers speculate that water may serve to mimic and act as a relay for amino acid electron donors and acceptors, "filling in the gaps not occupied by respective side chains" (Brown et al., 1990).

Intercalation

DNA intercalating agents possess planar aromatic ring systems that are able to insert (intercalate) between adjacent base pairs in DNA, forming reversible complexes; members of this class of compounds are important for cancer chemotherapy (Denny, 1989). Classical (mono)intercalators such as ethidium bromide and acridine orange "sandwich" between two successive base pairs, parallel to each. Since the distance between normal base pairs in DNA is approximately 3.4 Å, separating these base pairs to create the cavity necessary to accommodate a monointercalator requires unwinding the DNA substantially. Actinomycin D, adriamycin, and daunomycin are intercalating agents that have had widespread clinical use (Fig. 10-6) (Blum & Carter, 1974; Pratt et al., 1994). In the case of adriamycin and some of the other clinically useful intercalators, topoisomerases are implicated in the mechanism of action (Liu, 1989). Protein (topoisomerase)-associated DNA strand breaks are produced as a consequence of treatment of cells with these drugs, as the drugs trap intermediates in topoisomerase-mediated strand exchange.

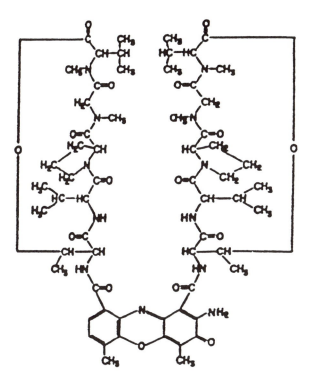

actinomycin D

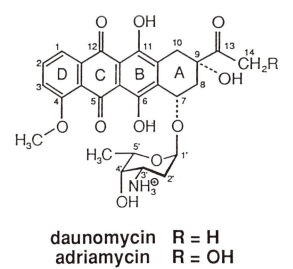

daunomycin   **R = H**
adriamycin   **R = OH**

Figure 10-6.   Structures of actinomycin D (**9**), daunomycin (**10**) and adriamycin (**11**).

The potent antitumor and antimicrobial agent echinomycin is a cyclic octadep-sipeptide antibiotic bearing two quinoxaline rings that bisintercalate into DNA (Fig. 10-7). Chemical and enzymatic cleavage inhibition patterns with DNases I and II (Low et al., 1984) and methidiumpropyl-EDTA-Fe(II) (Van Dyke & Dervan, 1984) indicate a strong preference for binding sites centered at 5'CpG steps, with a binding site size of around four base pairs. AT base pairs next to drug-binding sites show hypersensi-tivity to cleavage relative to free DNA, indicating an alteration of DNA conformation. A summary from polyacrylamide gel analysis showing the sequence selectivity is pre-sented in Fig. 10-8 (Van Dyke & Dervan, 1984). From crystal structures of echino-mycin with the DNA hexamer d(CGTACG)$_2$, the unexpected finding is that echino-mycin, along with select analogs, causes a transition from Watson–Crick to Hoogsteen base-pairing in the AT partners adjacent to the intercalation site (Fig. 10-9) (Wang et al., 1984). The results of NMR studies that demonstrate the sequence-dependence and stability of Hoogsteen bisintercalated complexes of echinomycin having Hoogsteen base-pairing are presented below.

Bisintercalation of Echinomycin. Echinomycin complexes with the self-complemen-tary tetranucleotides d(ACGT)$_2$ and d(TCGA)$_2$ have been characterized by NMR (Gao & Patel, 1988). From an earlier structure of the echinomycin–d(CGTACG)$_2$ complex determined by X-ray crystallography (Wang et al., 1984), it was predicted that the quinoxaline rings would intercalate at (A$_1$-C$_2$)·(G$_3$-T$_4$) steps in the d(ACGT)$_2$ complex and at (T$_1$-C$_2$)·(G$_3$-A$_4$) steps in the d(TCGA)$_2$ complex (Fig. 10-10) in solution, which turned out to be true. Evidence for bisintercalation in solution is twofold. Studies on monointercalators such as actinomycin D show a downfield shift of phosphate reso-nances at the intercalation steps (Patel, 1974; Petersheim et al., 1984) and an upfield

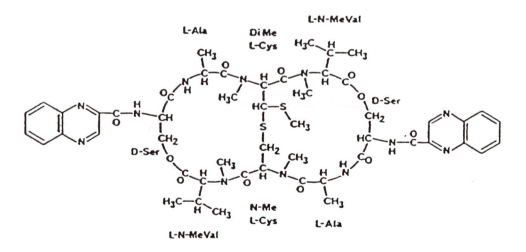

echinomycin

Figure 10-7. Structure of echinomycin.

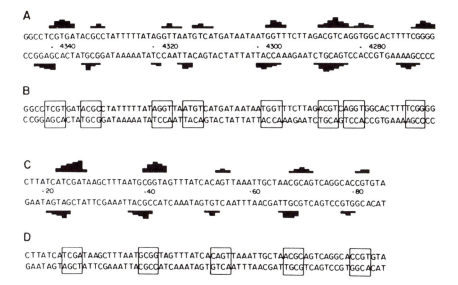

A

```
      GGCC TCGTGATACGCC TATTTTTATAGGT TAATG TCATGATAATAATGGT T TCT TAGACGTC AGGTGGC ACTTTTCGGGG
        · 4340              · 4320              · 4300              · 4280
      CCGGAGC ACT ATGCGGAT AAAAATATCC AAT TACAGTACT AT TAT TACC AAAGAATCTGCAGTCC ACCGT GAAAAGCCCC
```

B

```
      GGCC TCGT GATACGC CTATTTTTATAGGTTAATGT CATGAT AATAATGGT TTCTTAGACGTCAGG TGGCACTTTTCGGGG
      CCGG AGCACTATGCGGATAAAAAATATCCAATTACAGTACTATTATTACCAAAGAATCTGCAGTCCACCGTGAAAAGCCCC
```

C

```
      C T TATCATCGATAAGCT T TAATGCGGTAGTT TATCACAGT TAAATTGCT AACGCAGTCAGGCACCGTGTA
        · 20              · 40              · 60              · 80
      GAATAGTAGCTAT TCGAAATTACGCCATC AAATAGTGTC AATTTAACGATTGCGTCAGTCCGTGGC ACAT
```

D

```
      C T TATCATCGATAAGCT T TAATGCGGTAGTT TATCACAGT TAAATTGCTAACGCAGTCAGGCACCGTGTA
      GAATAGTAGCTATTCGAAATTACGCCATCAAATAGTGTCAATTTAACGATTGCGTCAGTCCGTGGCACAT
```

Figure 10-8. MPE·Fe(II) footprints of echinomycin at 200 μM concentration on 80-base-pair (**A, B**) and 70-base-pair (**C, D**) restriction fragments, determined by densitometry of the autoradiograms. The MPE·Fe(II) footprints are shown as histograms; the height is proportional to the reduction in cleavage at each nucleotide compared with the MPE·Fe(II) cleavage of unprotected DNA. The top strand patterns are for 5′-end-labeled DNA, while the bottom strand patterns are for 3′-end-labeled DNA. Boxes indicate the echinomycin-binding sites. Reprinted from Van Dyke & Dervan, 1984, with permission.

shift of the resonances corresponding to imino protons of base pairs next to intercalation sites (Feigon et al., 1984; Patel, 1974), presumably due to electron density shifts from unwinding and other structural changes in the DNA. Two of the three phosphates in both $d(ACGT)_2$ and $d(TCGA)_2$ shift downfield upon admixture of echinomycin (Fig. 10-11). For the $d(ACGT)_2$ complex, the guanosine$_3$ and thymidine$_4$ imino protons shift upfield significantly (Fig. 10-11). These results position intercalation of the quinoxaline chromophores of echinomycin between A_1-T_4 and C_2-G_3 base pairs, corresponding to the $(A_1$-$C_2) \cdot (G_3$-$T_4)$ steps in the $d(ACGT)_2$ duplex. A parallel situation exists for the $d(TGCA)_2$ duplex.

Minor Groove Binding and Intermolecular Interactions of Echinomycin. In the two-dimensional NOESY experiment, through-space interactions are observed only between protons separated by less than ~5 Å. Minor groove binding of the cyclic octapeptide is established by NOEs to H1′ and H4′ protons of DNA that reside in the minor groove (Table 10-1). These include medium-to-strong NOEs for the methyl groups of L-Ala and L-NMe-Val to H1′ and H4′ protons of residue cytidine$_2$ and for the CH_3 group of L-Ala to H1′ and H4′ of guanosine$_3$.

In the X-ray structure of the echinomycin–$d(CGTACG)_2$ complex, hydrogen bonding of the amide proton of L-Ala to guanosine N3 contributed to the sequence preference for intercalation. In the NMR study, large downfield shifts of the amide protons of L-Ala in both oligomer complexes are consistent with that observation. Also, inter-

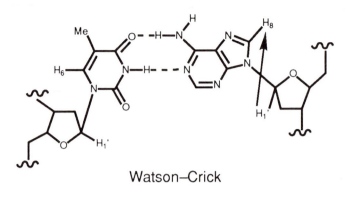

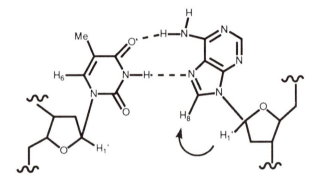

Watson–Crick

Hoogsteen

Figure 10-9. Schematic illustration of Watson–Crick and Hoogsteen A·T base-pairing. Arrows denote H8 to H1′ through-space connectivity used to distinguish these two types of base-pairing in two-dimensional NMR experiments.

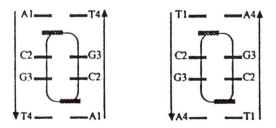

Figure 10-10. Intercalation sites for echinomycin in the d(ACGT)$_2$ and (TCGA)$_2$ complexes. Reprinted from Gao & Patel, 1988, with permission.

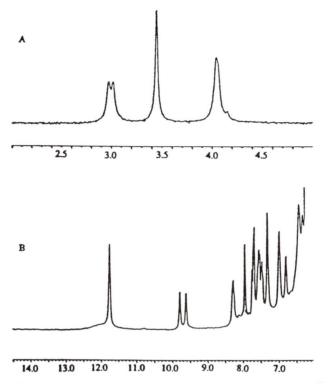

Figure 10-11. Echinomycin–d(ACGT)$_2$ complex. (A) Proton-decoupled 121-MHz ^{31}P-NMR spectrum (referenced to external trimethyl phosphate) in D$_2$O buffer, pH 7.0, at 25°C. (B) 400-MHz ^{1}H-NMR spectrum of imino protons in H$_2$O buffer, pH 6.0, at 26°C. Reprinted from Gao & Patel, 1988, with permission.

molecular NOEs indicate that in the d(ACGT)$_2$ complex the quinoxaline ring stacks primarily with adenosine$_1$ and cytidine$_2$, not with guanosine$_3$ and thymidine$_4$ (Table 10-1 and Fig. 10-10). The NOE connectivities in the d(TCGA)$_2$ complex also demonstrate that the ring stacks with pyrimidines thymine$_1$ and cytidine$_2$, not with purines guanosine$_3$ and adenosine$_4$.

Base-Pairing. The glycosidic bond of adenosine is in the *anti* conformation in the Watson–Crick base pair and in the *syn* conformation in the Hoogsteen base pair (Fig. 10-9). A Watson–Crick-to-Hoogsteen transition requires rotation about the glycosidic bond, permitting thymidine NH1 to hydrogen-bond to adenosine N7. Distance-dependent NOEs from H8 to H1′ are predicted to be weak or strong for the Watson–Crick or Hoogsteen base pairs, respectively (Patel et al., 1982). In the echinomycin–d(ACGT)$_2$ complex, there is a strong NOE between the H8 and H1′ protons for adenosine$_1$, indicating Hoogsteen base-pairing, while the base-to-sugar contact is weak for all other residues, which are, therefore, identified as Watson–Crick in conformation (Fig. 10-12). By contrast, all base-to-sugar through-space interactions are weak in the d(TCGA)$_2$ complex, indicating uniform Watson–Crick base-pairing (Fig. 10-12).

Table 10-1 Intermolecular NOEs Between Echinomycin and Oligonucleotide Protons

Residue	H8, H6	H5/H2	Oligonucleotide Proton[b] H1'	H3'	H2',2"	H4'
A. In the Echinomycin-d(TCGA) Complex in D₂O at 22°C[a]						
T_1	Q6, Q7 (w)		Q7, Q8 (w) Val-CH₃ (m)[c]	Q7 (w)	Q6, Q7, Q8 (w) Val-CH₃ (m)[c]	
C_2	Q6, Q7 (w) Ala-CH₃ (w)	Q5 (w) Q6 (m) Q7 (w)	Q7, Q8 (w) Ala-Hα (w) Ala-CH₃ (s) Val-CH₃ (m)[c]	Q6 (w) Q7 (m)	Ala-CH₃ (w)	Cys-SCH₃ (w)[d] Ala-CH₃ (w) Val-CH₃ (m)[c]
G_3	Q3 (w) Ala-CH₃ (w)		Q3 (w) Ala-CH₃ (s)		Q3 (w) Ala-CH₃ (w)	Ala-CH₃ (m)
A_4		Ala-CH₃ (w)	e		e	
B. In the Echinomycin-d(ACGT) Complex in D₂O at 22°C[f]						
A_1	Q7, Q8 (m) Val-CH₃ (m)		Q7, Q8 (s) Val-CH₃ (s)	Q6 (w) Q7, Q8 (m)	Q7, Q8 (s) Q6 (w)	Q7 (m)

C$_2$	Q6 (w) Q7, Q8 (m) Ala-CH$_3$ (m) Cys-SCH$_3$ (m)	Q5 (m) Q6 (s) Q7 (m)	Q7, Q8 (s) Ala-CH$_3$ (s) Val-CH$_3$ (m) Cys-SCH$_3$ (w)	Q6 (w) Q7 (m) Ala-CH$_3$ (w) Val-CH$_3$ (w) Cys-SCH$_3$ (w)	Q6 (w) Q7 (m) Ala-CH$_3$ (w) Val-CH$_3$ (w) Cys-SCH$_3$ (w)	Ala-CH$_3$ (s)	Q7 (m) Val-CH$_3$ (m) Cys-SCH$_3$ (w) Ala-CH$_3$ (m)
G$_3$	Q3 (w) Ala-CH$_3$ (m)		Q3 (w) Ser-Hα (w) Ala-CH$_3$ (s)	Q6 (w) Q7 (m) Ala-CH$_3$ (w) Val-CH$_3$ (w) Cys-SCH$_3$ (w)		Q3 (w) Ser-Hα (w) Ala-CH$_3$ (m)	Ala-CH$_3$ (s)
T$_4$		Ser-Hα (s) Ser-Hβ (m)	Ser-Hα (s) Ser-Hβ (m)			Ser-Hα (w)	

^{a}0.1 M NaCl, 10 mM phosphate in D$_2$O, pH 6.5.

^{b}The NOEs are designated strong (s), medium (m), and weak (w).

^{c}Val-CH$_3$ overlaps with C2(H2'), preventing definitive NOE assignment.

^{d}Cys-SCH$_3$ overlaps with C2(H2'), preventing definitive NOE assignment.

^{e}Ser-Hα protons resonate at the same position as HOD resonance, and predicted NOEs to sugar protons of A$_4$ cannot be detected.

^{f}0.1 M NaCl, 10 mM phosphate in D$_2$O, pH 7.0.

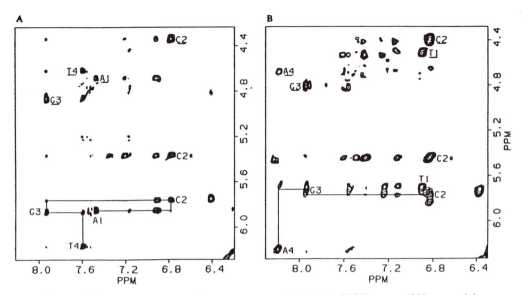

Figure 10-12. Contour plots of the phase-sensitive 400-MHz NOESY spectra (250-msec mixing time) demonstrating through-space connectivities between the base and quinoxaline protons (6.2–8.2 ppm) and the sugar H1', H3', and H5 protons (4.2–6.4 ppm). (**A**) The echinomycin–d(ATGC)$_2$ complex and (**B**) the echinomycin–d(TCGA)$_2$ complex in D$_2$O buffer at 22°C. NOEs between each base and its own sugar H1' proton (5.6–6.4 ppm), as well as its own sugar H3' proton (4.3–5.0 ppm), are labeled in the spectra, with the base–H3' connectivities underlined. Reprinted from Gao & Patel, 1988, with permission.

Interestingly, NMR studies on an echinomycin–d(ACGTACGT)$_2$ complex indicate that while at low temperatures the AT base pairs on either side of the intercalation site adopt the Hoogsteen conformation, as the temperature is raised the Hoogsteen base pairs in the duplex interior are destabilized and exchange between the Hoogsteen base pair and either an open or Watson–Crick base-paired state, even at 20°C (Gilbert et al., 1989). The terminal AT base pairs remained stably Hoogsteen base-paired up to at least 45°C. Because virtually all AT base pairs in cellular DNA are in the duplex interior and physiological temperatures are above 20°C, this invites further work to determine the biological relevance of the structural accommodation of the duplex to the drug. Stable internal Hoogsteen base pairings between the third and second strands in triplex DNA, which are of considerable interest as a tool for probing gene regulation, are known to occur. Of course, the Watson–Crick base pairs of the first and second strands are not required to be disrupted for this event. On an aesthetic level, these studies afford an appreciation of yet one more way in which the flexible polymer DNA may surprise the observer with unexpected contortions. Progress will continue if we can correlate these changes with biological activity and decode the DNA and drug features responsible for such action.

Alkylation

A number of potent cancer chemotherapeutic drugs and mutagens such as mitomycin, aflatoxin B$_1$, and the nitrogen mustards interact covalently with DNA and produce intriguing structural and biological effects. For a review of the molecular mechanism for sequence recognition of DNA by covalent interaction, see Warpehoski & Hurley (1988). Our focus in this section is on anthramycin (Fig. 10-13), whose sequence-selectivity is correlated with the ability to distort DNA structurally, most likely by bending. Anthramycin is a member of the pyrrolo[1,4]benzodiazepine family of antibiotics, potent antitumor agents thought to mediate their cytotoxicity and antitumor activity through covalent bonding to guanine in DNA (Graves et al., 1984; Kohn et al., 1974; Petrusek et al., 1981), resulting in inhibition of nucleic acid synthesis (Kohn, 1975). Significantly, these low molecular weight ligands show sequence-selective reaction with DNA (Hertzberg et al., 1986; Hurley et al., 1988). Footprinting techniques reveal the preferred sequence for pyrrolo[1,4]benzodiazepine bonding to be 5'-PuGPu-3', while 5'-PyGPy-3' sites are least preferred. 5'-PyGPu-3' and 5'-PuGPy-3' sites show intermediate bonding preference. Anthramycin adducts with DNA sequences of differing selectivity have been investigated by two-dimensional NMR, fluorescence, and molecular modeling [for a review, see Hurley (1992)]. From these studies and others, it is proposed that the reaction proceeds via nucleophilic attack of N2 on C11 of the pyrrolo[1,4]benzodiazepine imine form to produce the covalent adduct (Barkley et al., 1986), with the drug bound within the minor groove (Fig. 10-14). The stereochemistry at the covalent linkage site and the orientation of the drug in the minor groove are both sequence- and drug-dependent.

A theoretical investigation of the DNA sequence-selectivity of anthramycin bonding suggested a potentially important role for sequence-dependent nucleic acid flexibility in directing the reaction of pyrrolo[1,4]benzodiazepines with DNA (Zakrzewska & Pullman, 1986). Using the successive infinitesimal rotations methodology, Zakrzewska and Pullman calculated that for anthramycin adducts 5'-PuGPu-3' sequences are more flexible and require the least energy (about 20 kcal/mol) to deform to the adduct conformation. 5'-PyGPy-3' sequences are most resistant to distortion, requiring about 30 kcal/mol to achieve their final adduct conformation (Table 10-2). The rank order of bonding preference 5'-PuGPu-3' >> 5'-PyGPy-3' hinted at the role of sequence flexibility in determining adduct stability and is in close agreement with the experimental results using footprinting experiments. Calculations predict that 5'-PyGPy-3' sequences are slightly favored over 5'-PuGPu-3', although the experimental observations indicate the opposite. However, the energy difference is a minor one, most likely attributed to approximations in the methodology.

The suggestion from modeling that sequence flexibility is a determinant of sequence recognition inspired an experimental comparison of drug-induced distortion of DNA involving different sequences (Kizu et al., 1993). To establish whether sequence-

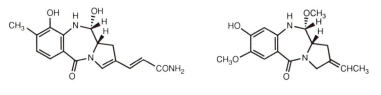

anthramycin   tomaymycin

Figure 10-13.   Structures of anthramycin and tomaymycin.

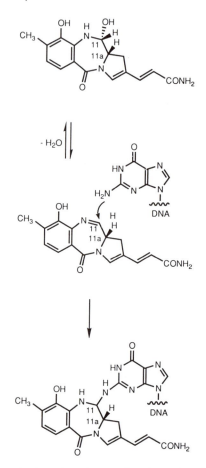

Figure 10-14. Proposed mechanism of covalent reaction of pyrrolo[1,4]benzodiazepines with DNA through N2 of guanine. Reprinted from Barkley et al., 1986, with permission.

dependent distortion is manifested as DNA bending, DNA electrophoretic mobility was monitored. Five different 21-base-pair oligonucleotides (Fig. 10-15) were modified with anthramycin and then ligated into multimers using T4 ligase; the mobility of each was then analyzed on a 6% nondenaturing polyacrylamide gel (Fig. 10-16). If the drug bends the 21-base-pair DNA, the ligated fragments should amplify this bending, resulting in fragments that assume an overall curvature and, therefore, multimers that migrate more slowly through the gel relative to control DNA. While some drug-modified ligation products show measurable retardation of electrophoretic mobility (see lanes 3, 6, 9, and 12, Fig. 10-16) when compared to unmodified controls, other ligation products show little or no retardation (Kizu et al., 1993). The bending angles were calculated for each of the five sequences modified with anthramycin using a calibration equation for gel mobility anomalies derived by Koo et al. (1986) and by Koo and Crothers (1988) (Table 10-3). Table 10-4 displays kinetic rate constants for the reaction of anthramycin with the same DNA sequences, determined by time-course high-performance liquid chromatography (HPLC) studies. As can be seen by comparison of the data in these tables, those sequences that anthramycin most prefers react most

Table 10-2 Consensus Sequence Analysis of Footprinting Data for Anthramycin, Tomamycin, and Sibiromycin[a]

	Anthramycin[b]		Tomaymycin[b]		Sibiromycin[b]	
	+[c] (%)	−[c]	+[c] (%)	−[c]	+[c] (%)	−[c]
5'-PuGPu-3'	54 (83)	11 (17)	22 (45)	27 (55)	16 (67)	8 (33)
5'-PyGPy-3'	11 (26)	32 (74)	3 (10)	28 (90)	8 (36)	14 (64)
5'-PyGPu-3'	21 (38)	34 (62)	9 (20)	35 (80)	12 (40)	18 (60)
5'-PuGPy-3'	16 (26)	46 (74)	15 (32)	32 (68)	11 (41)	16 (59)
5'-NGN-3'	102 (45)	123 (55)	49 (29)	122 (71)	47 (45)	56 (55)

[a]All guanines were scored for the presence or absence of drug binding on the basis of their ability to inhibit MPE·Fe(II) cleavage significantly. There was no significant preference for a base or base type one base pair on either side of the covalently modified guanine. Likewise, individual base analysis (G, C, A, T) at positions adjacent to the covalently modified guanine did not reveal any significant preference.

[b]A total of approximately 515, 363, and 194 bp, respectively, were analyzed for anthramycin-, tomaymycin-, or sibiromycin-binding sites.

[c]Plus (+) and minus (−) refer to the presence or absence of a footprint as determined by inhibition of MPE·Fe(II) cleavage.

Source: Zakrzewska & Pullman, 1986, adapted with permission.

rapidly and exhibit the most bending upon adduct formation (Kizu et al., 1993). It should be stressed that the bending anthramycin induces in DNA is very modest compared to that introduced by other drugs that interact covalently with DNA, such as (+)-CC-1065, which is described below.

The significant finding from this study is that the hierarchy of covalent bonding

```
AGA:   5'  ATATIATCIAGAACTCICITT   3'
           3'  ACTAICTCTTIAICICAATAT   5'


CGA:   5'  ATATIATCICGAACTCICITT   3'
           3'  ACTAICICTTIAICICAATAT   5'


AGC:   5'  ATATIATCIAGCACTCICITT   3'
           3'  ACTAICTCITIAICICAATAT   5'


CGC:   5'  ATATIATCICGCACTCICITT   3'
           3'  ACTAICICITIAICICAATAT   5'


TGC:   5'  ATATIATCITGCACTCICITT   3'
           3'  ACTAICACITIAICICAATAT   5'
```

Figure 10-15. Synthetic oligonucleotides used in anthramycin bending study. Sequences were designed to contain a single anthramycin-binding site covering the range of drug bonding specificities. All but one guanosine were replaced by inosine to direct anthramycin binding to the desired nucleotide. Reprinted from Kizu et al., 1993, with permission.

sites based on sequence parallels the hierarchy for magnitude of drug-induced DNA bending by anthramycin and for its rate of reaction with DNA. Assuming the right-handed twist of anthramycin does not exactly match the twist of the minor groove and, as a consequence, the DNA has to distort around the noncovalently bound drug to produce an energetically favorable transition state for covalent reaction, then those sequences that are more flexible should react preferentially and may also result in a more distorted DNA structure relative to the unmodified duplex. Conversely, less pliable sequences will react more slowly, be more resistant to distortion, and result in less distorted drug–DNA adduct structures. Once again, it appears that the inherent properties of the DNA sequence might act to set the terms for the molecular partnership with a low molecular weight ligand.

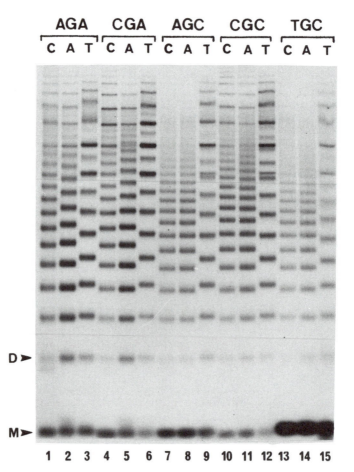

Figure 10-16. Electrophoretic behavior of the ligation products of AGA, CGA, AGC, CGC, and TGC oligonucleotides (Fig. 10-15) saturated with anthramycin and tomaymycin on a 6% nondena-turing polyacrylamide gel. Lanes labeled C represent ligated multimers of control oligomers (lanes 1, 4, 7, 10, and 13). Lanes labeled A and T represent oligomers modified with anthramycin (lanes 2, 5, 8, 11, and 14) and tomaymycin (lanes 3, 6, 9, 12, and 15), respectively. M and D represent the position of monomers (21-base-pair oligonucleotide) and ligated dimers (42-base-pair oligonucleotide), respectively. Reprinted from Kizu et al., 1993, with permission.

Table 10-3 Calculation of DNA Bending Angles Produced by Anthramycin and Tomaymycin in 21-Base-Pair Oligonucleotides[a]

Sequence	Drug	Bending Angle (Degrees)
5'-AGA-3'	Anthramycin	6.3–8.4
5'-CGA-3'	Anthramycin	5.6–7.4
5'-AGC-3'	Anthramycin	0
5'-CGC-3'	Anthramycin	0
5'-TGC-3'	Anthramycin	0
5'-AGA-3'	Tomaymycin	10.5–14.0
5'-CGA-3'	Tomaymycin	10.9–14.4
5'-AGC-3'	Tomaymycin	10.3–13.6
5'-CGC-3'	Tomaymycin	10.0–13.2
5'-TGC-3'	Tomaymycin	8.3–10.9

[a]The estimate of bending angle of each sequence was calculated using a calibration equation for gel mobility anomalies derived by Koo & Crothers (1988).

Table 10-4 Rate Constants for the Binding of Anthramycin to 21-Base-Pair Oligonucleotides

Sequences	k (min^{-1})
5'-AGA-3'	1.12×10^{-1}
5'-CGA-3'	2.89×10^{-2}
5'-AGC-3'	2.10×10^{-2}
5'-CGC-3'	8.77×10^{-3}
5'-TGC-3'	5.05×10^{-3}

Bifunctional Interaction: Intercalation and Covalent Reaction

Altromycin B is an antitumor antibiotic recently isolated from an actinomycete (strain AB 1246E-26) at Abbott Laboratories, Chicago (Brill et al., 1990; Jackson et al., 1990). It belongs to the pluramycin family of antitumor antibiotics, which also includes kidamycin, hedamycin, pluramycin, neopluramycin, DC92-B, and rubiflavin A (Fig. 10-17) [for a review, see Sequin (1986)].

A synopsis of recent work defining the structure of the altromycin B–DNA adduct and a proposed prototype DNA adduct structure for the pluramycin antitumor antibiotics is of interest (Sun et al., 1993). Experiments using supercoiled DNA have demonstrated that altromycin B and related drugs intercalate into DNA. However, the presence of both the planar 4*H*-anthra[1,2-*b*]pyran 4,7,12-trione moiety and an epoxide in altromycin B suggested the possibility that alkylation may occur in combination with intercalation. An experiment using high-resolution gel electrophoresis demonstrated that while the altromycins and hedamycin both intercalate and covalently modify gua-

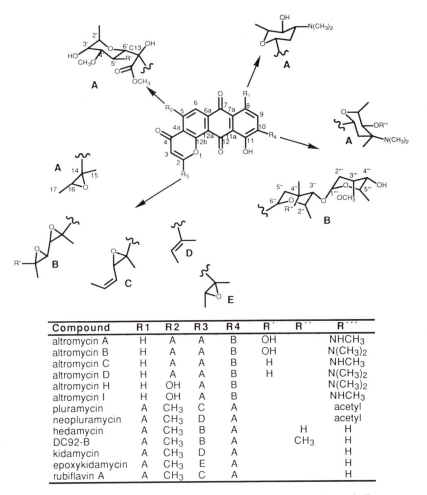

Compound	R1	R2	R3	R4	R′	R″	R‴
altromycin A	H	A	A	B	OH		NHCH₃
altromycin B	H	A	A	B	OH		N(CH₃)₂
altromycin C	H	A	A	B	H		NHCH₃
altromycin D	H	A	A	B	H		N(CH₃)₂
altromycin H	H	OH	A	B			N(CH₃)₂
altromycin I	H	OH	A	B			NHCH₃
pluramycin	A	CH₃	C	A			acetyl
neopluramycin	A	CH₃	D	A			acetyl
hedamycin	A	CH₃	B	A		H	H
DC92-B	A	CH₃	B	A		CH₃	H
kidamycin	A	CH₃	D	A			H
epoxykidamycin	A	CH₃	E	A			H
rubiflavin A	A	CH₃	C	A			H

Figure 10-17.   Structures of the pluramycins, including altromycin B.

nine in DNA, kidamycin and pluramycin, which lack analogous epoxides, are only able to intercalate into DNA. It was observed that only the former induce strand breaks in DNA upon thermal treatment in the presence of 1 M piperidine. The results of this experiment support the proposal that the epoxide common to the altromycins and hedamycin is essential for DNA alkylation. The fact that altromycin B induces strand breakage narrows the possible sites of alkylation of guanine to N3 or N7. Implication for N7 as the alkylation site is further strengthened by the fact that N7-deazaguanine incorporated into DNA is not a substrate for alkylation by altromycin B. Mass spectrometry of the altromycin B–base adduct gave a parent ion (M + 1) at m/z 1077.429 (calc. 1077.430), which is consistent with the molecular formula $C_{52}H_{65}N_6O_{19}$. This supports the structure of the altromycin B–guanine adduct shown in Fig. 10-18. A downfield chemical shift of 2.25 ppm for H16 and 16.3 ppm for C14, along with an upfield shift of 8.5 ppm for C16 of altromycin B upon covalent attachment, is also consistent with attachment of nitrogen (N7 of guanine) at C16. Two-dimensional NMR through-bond connectivities such as ^1H–^{15}N long-range coupling between the 17-methyl group of al-tromycin B and N7 of guanine, as well as through-space connectivities between the 17-methyl group and H8 of guanine, also indicate N7 alkylation.

The altromycin B molecule is comprised of four components: the planar aglycone moiety, which consists of a 4H-anthra[1,2-b]pyran ring system; the disaccharide at C10; the glycosidically bound neutral sugar at C13; and the epoxide at C2 (Fig. 10-16). The

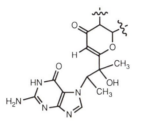

Figure 10-18. Proposed chemical structure of the altromycin B–DNA adduct. Reprinted from Sun et al., 1993, with permission.

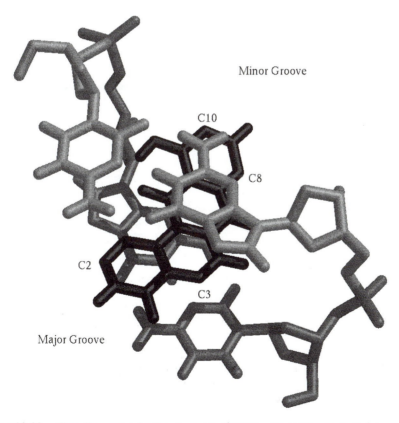

Figure 10-19. Threading model for the altromycin B–DNA adduct. Altromycin B is in gray and DNA is in black. Reprinted from Sun et al., 1993, with permission.

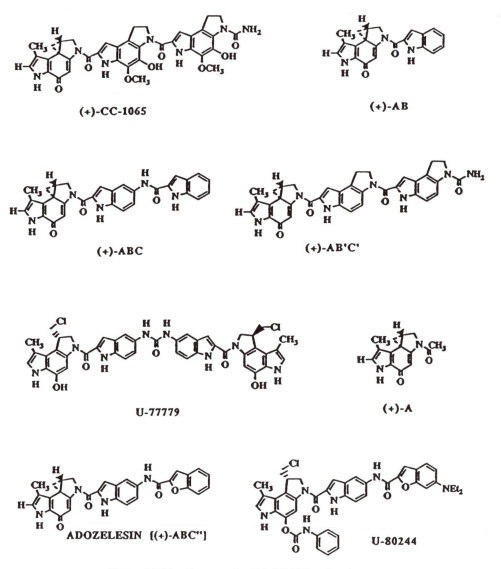

Figure 10-20. Structures for (+)-CC-1065 and analogues.

aglycone moiety is similar in structure to the intercalators nogalamycin and the synthetic anthracene 9,10-dione group of antitumor antibiotics, which interact with DNA through a threading mechanism (Islam et al., 1985). This threading type of mechanism is differentiated from classical intercalation because, in the case of threading, one each of two side chains or rings occupy the major and minor DNA grooves, rather than just one groove, as observed for classic intercalators. Based on the known structure of the nogalamycin–DNA complex and the knowledge that altromycin B alkylates N7 of guanine in the major groove, a structure of the threaded altromycin B (N7-guanine)–DNA adduct has been proposed (Fig. 10-19). In this model the aromatic chromophore inter-

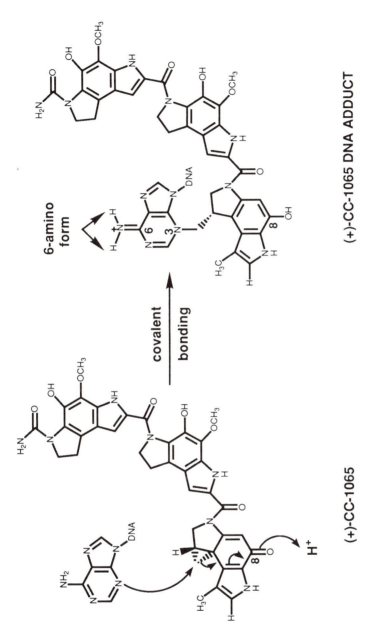

6-amino form

covalent bonding

(+)-CC-1065

(+)-CC-1065 DNA ADDUCT

Figure 10-21. Reaction of (+)-CC-1065 with double-stranded DNA to form the (+)-CC-1065 (N3-adenine)-DNA adduct. Reprinted from Hurley et al., 1984, with permission.

calates between the base pairs and positions the side chains at C2 (epoxide) and C5 (C-glycoside) in the major groove and the disaccharide at C10 in the minor groove of DNA. The disruption of DNA structure to achieve this would most easily be accommodated if the threading were to take place from the major to the minor groove. The intercalative binding of the aromatic ring would then position the epoxide in the major groove in proximity to N7 of guanine to facilitate nucleophilic attack at C16 and acid-catalyzed opening of the epoxide ring. In the proposed model, the intercalation site is placed arbitrarily between the covalently modified GC base pair and the base pair to the 5'-side. Structural characterization of an altromycin B–DNA adduct is under way (see note added in proof, p. 323).

This model, which can be generally extended to other epoxide-containing pluramycins, provides a novel motif for a conjoint intercalation–alkylation with DNA that also incorporates a threading mechanism; it should spur investigation into additional compounds that select multiple options from the menu of DNA interactions available, perhaps with valuable biological consequences.

In the foregoing sections, one drug was presented to exemplify each of the general categories of drug–nucleic acid interactions. Featured were specific DNA molecular responses for each drug—for example, calicheamicin γ_1^I strand breakage. In this section, our own laboratory's work on (+)-CC-1065 and its analogues is described (Fig. 10-20). (+)-CC-1065, a potent antitumor agent that covalently interacts with DNA, was isolated from *Streptomyces zelensis* by scientists at The Upjohn Company in 1979 (Hanka et al., 1978). Three synthetic analogues are now in Phase I and II clinical trials (DeKoning et al., 1990; Li et al., 1990, 1992). Previous studies have demonstrated that this antibiotic is extraordinary for both its base and DNA sequence-selectivity. Structurally, (+)-CC-1065 consists of three repeated pyrroloindole subunits attached via amide linkages that are approximately 15° out of plane, providing the drug molecule with a right-handed crescent shape. Subunit A contains the DNA-reactive cyclopropane ring that alkylates N3 of adenine when it binds within certain reactive sequences (Fig. 10-21) (Reynolds et al., 1985; Scahill et al., 1990). The predominant tautomeric species of the covalently modified adenine is the doubly protonated 6-amino form, with the positive charge delocalized over the entire adenine molecule (Lin & Hurley, 1990).

Structure of (+)-CC-1065–DNA Adducts and Determination of Sequence Specificity

It has been shown previously that when (+)-CC-1065–DNA duplex adducts undergo thermal treatment (100°C, 30 min), depurination and β-elimination take place at the covalently modified adenine (Hurley et al., 1984; Reynolds et al., 1985). Subsequently, under basic conditions a second β-elimination occurs, releasing a modified sugar moiety that was originally part of the covalently modified adenosine (Fig. 10-22). This provides a convenient method to determine the location of (+)-CC-1065 bonding sites on DNA. The (+)-CC-1065 bonding sequences on DNA are determined on Maxam–Gilbert sequence gels by locating (+)-CC-1065 thermally induced DNA strand breaks on singly 5'- or 3'-^{32}P-end-labeled DNA restriction enzyme fragments isolated from T7 DNA, as well as the early promoter region of SV40 DNA. A detailed analysis of the derived data led to the identification of two subsets of DNA sequences, 5'-

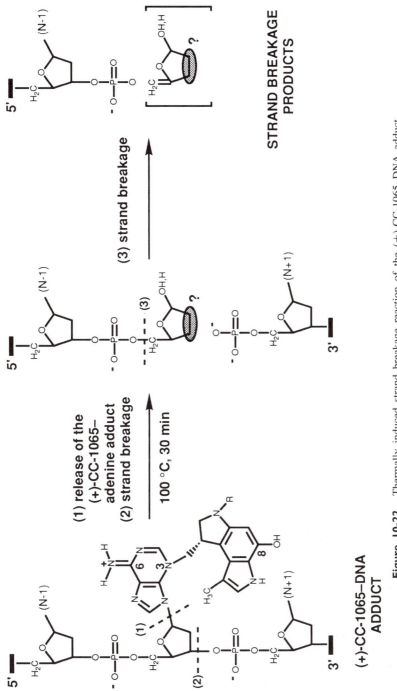

Figure 10-22. Thermally induced strand breakage reaction of the (+)-CC-1065–DNA adduct. Reprinted from Reynolds et al., 1985, with permission.

313

PuNTTA-3'* and 5'-AAAAA-3'* (Pu = purine, N = any base, and * denotes the co-valent attachment site), which are highly preferred for (+)-CC-1065 bonding. Figure 10-23 illustrates one of the autoradiograms from these experiments.

Molecular Basis for the DNA Sequence-Specificity of (+)-CC-1065

Surprisingly, the alkylating subunit alone (subunit A; Fig. 10-20) contains sufficient structural information to encode the primary molecular basis for sequence-selectivity, and this subunit is also essential for antitumor activity. However, the noncovalent bind-ing interactions of the B and C subunits with DNA can modulate or fine-tune this sequence-selectivity and, in the case of (+)-CC-1065, produce winding of DNA (Hurley et al., 1990; Warpehoski & Hurley, 1988).

The covalent bonding reaction of (+)-CC-1065 with DNA has been proposed to in-volve catalytic activation of the covalent reaction between (+)-CC-1065 and DNA and to be at least partially responsible for the molecular basis for sequence-selective recog-nition of DNA by the alkylating subunit of (+)-CC-1065. In addition to catalytic ac-tivation, we have also proposed that DNA conformational flexibility is an important component of sequence recognition in the (+)-CC-1065 bonding reaction (Hurley et al., 1990; Warpehoski & Hurley, 1988).

Importance of Ternary Complexes in Understanding the Molecular Basis for the Biological Effects of (+)-CC-1065

For several DNA-reactive drugs, the biological effects are considered to result from modulation of protein–DNA interactions. Examples of ligands whose role in DNA–drug–protein complexes has been examined include (1) the quinolone antibac-terials, which inhibit bacterial gyrase (Shen et al., 1989), (2) the topoisomerase in-hibitor amsacrine (Nelson et al., 1984), and (3) ditercalinium as a target for the UvrABC endonuclease (Lambert et al., 1989). In the case of (+)-CC-1065, the wide array of analogs has facilitated sharp definition of those structural elements that induce DNA bending, helix stiffening, and winding as a consequence of N3-adenine covalent adduct formation (Fig. 10-24). These effects have been matched to specific effects on the ac-tivity of DNA metabolizing enzymes such as polymerases, ligases, and helicases. These enzymes are involved in biochemical events crucial to DNA synthesis and repair.

DNA Polymerases. (+)-CC-1065 induces local bending (17° to 19° toward the minor groove), winding, and helix stiffening of DNA molecules as a consequence of N3-adenine covalent adduct formation (Lee et al., 1991a; Lin et al., 1991; Sun & Hurley, 1992a,b,c). In addition, previous work has shown that in L1210 cells, (+)-CC-1065 inhibited DNA synthesis much more than it inhibited RNA or protein synthesis (Bhuyan et al, 1982). Inhibition of DNA replication due to the strong irreversible binding of (+)-CC-1065 to double-stranded DNA has been proposed to be responsible for the strong cytotoxicity of (+)-CC-1065 (Li et al., 1982). The DNA replication process con-sists of three steps: recognition of replication origins, unwinding of duplexes, and syn-thesis of a new complementary DNA strand. In a recent study, the complementary

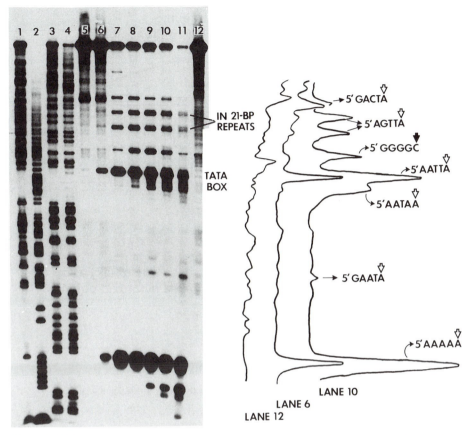

Figure 10-23. Effect of (+)-CC-1065 concentration on DNA strand breakage produced by thermal treatment of (+)-CC-1065–DNA adducts. The reaction of (+)-CC-1065 with an SV40 DNA fragment. Lanes 1–4, Maxam–Gilbert sequencing reactions: G, A + G, C, and T + C, respectively. Lanes 5–12, (+)-CC-1065 treatments. (+)-CC-1065–DNA adducts were prepared by incubating the DNA for 24 hr at 4°C with various dilutions of a parent stock solution of (+)-CC-1065: Lanes 5–11 are 1:10^6, 1:10^5, 1:10^4, 1:10^3, 1:10^2, 1:10, and undiluted stock solution of (+)-CC-1065, respectively. Lane 12 contained untreated DNA. The DNA was then precipitated in ethanol, resuspended in a sodium citrate buffer, heated (100°C, 30 min), again precipitated, and then analyzed by electrophoresis. Densitometric analysis of the autoradiograph is shown. The open arrows indicate the adenines that are covalently bound to (+)-CC-1065. Reprinted from Reynolds et al., 1985, with permission.

strand synthesis on a single-stranded template has been chosen as a model for the replication process (Sun & Hurley 1992a). For this purpose, a DNA template containing a site-specific (+)-CC-1065–DNA adduct was constructed and used to monitor the passage or termination of the DNA polymerase at the site of the drug lesion. At the site of the drug modification, two different consensus sequences were utilized, 5'-GATTA*-3' and 5'-AGTTA*-3' (* indicates drug-modified adenine), the latter being one of the most reactive bonding sites for (+)-CC-1065 and its analogues (Reynolds et al., 1986). A comparison of (+)-CC-1065–DNA adducts on slightly different sequences allowed

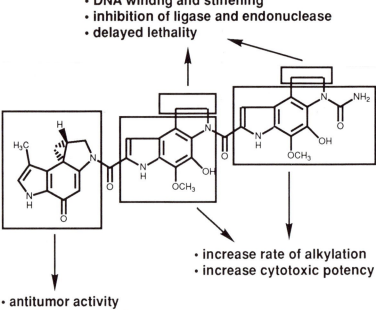

- **modulation of sequence selectivity**
- **DNA windng and stiffening**
- **inhibition of ligase and endonuclease**
- **delayed lethality**

- **increase rate of alkylation**
- **increase cytotoxic potency**

- **antitumor activity**
- **sequence selectivity**
- **DNA bending**
- **UVRABC nuclease recognition and incision**

Figure 10-24. Structure–activity relationships for (+)-CC-1065 and analogues.

the investigation of the sequence dependency of bypass or termination by the polymerase at the drug modification site. In these studies, four (+)-CC-1065 analogues [(+)-AB'C', (+)-ABC", (+)-ABC, and (+)-AB] were compared with (+)-CC-1065.

(+)-CC-1065 and its analogs were found to block DNA polymerase strongly in experiments employing drug-modified single-stranded DNA templates (Sun & Hurley, 1992a). In control studies using high concentrations of dNTPs and unmodified DNA molecules, the major products were fully elongated primers. In the presence of drug-modified templates, only small amounts of full-length molecules were synthesized (Fig. 10-25). The same two major termination sites were found for all five (+)-CC-1065 analog-modified templates. At the highest concentration of dNTP (lane 3), the termination step was one nucleotide farther than at lower dNTP concentrations. The termination sites were mapped precisely using a dideoxy chain termination procedure. The primary termination step that was found at 10 μM dNTP concentration (lane 2) was determined to be the second nucleotide prior to the drug-modified adenine, while at the highest dNTP concentration (100 mM, lane 3) this termination site disappeared and a new one appeared just one nucleotide prior to the drug-modified adenine. All the drug–DNA adducts caused the same pattern of termination in DNA synthesis mediated by the Klenow fragment, implying that each drug–DNA adduct changed the DNA structure in a manner such that the base-pairing step was impaired at the nucleotide prior to modified adenine, as well as at the drug-modified adenine.

DNA polymerase incorporates the nucleotide one base pair prior to the site of drug modification only with low efficiency at 10 μM dNTP concentrations. Therefore, we evaluated whether this step might also be carried out with a fidelity less than usual. For this experiment, four parallel incubations were prepared; each of these contained 10 μM dNTPs, but individual tubes contained one of the four dNTPs at 100 μM concentration. As expected, at 100 μM dGTP the primary pausing at the second nucleotide from the covalently modified adenine was overcome, since dGTP was the complementary nucleotide for incorporation at that site (Fig. 10-26, lane 3). However, while supplementation with 100 μM dTTP or dCTP did not cause elongation past the primary pausing site, supplementation with 100 μM dATP allowed considerable passage to the secondary pausing site, which is presumably due to misincorporation of adenine opposite cytosine. Further results showed that the (+)-CC-1065–DNA adduct was particularly prone to produce misincorporation of the nucleotide prior to the modified adenine when it was covalently attached to the sequence 5'-GATTA*-3', while the other analogues did not induce significant misincorporation. However, in the particular se-

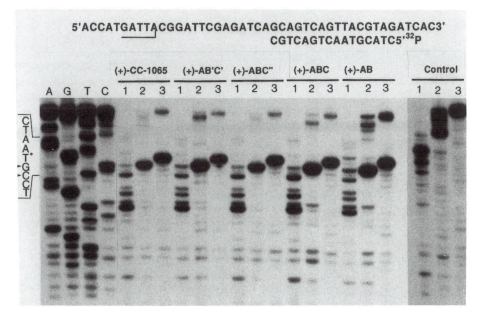

Figure 10-25. Effect of dNTP concentration on the termination site of in vitro DNA synthesis by DNA polymerase (Klenow fragment) on the single-stranded drug-modified template shown at the top of the figure. Lanes A, T, G, and C refer to the dideoxy sequencing reactions carried out with Klenow fragment on 5'-^{32}P-end-labeled 16-mer annealed to the 45-mer template. Lane headings refer to drug molecules used in template modification. Lanes 1, 2, and 3 are reactions containing 1,10, and 100 μM dNTPs, respectively. 5'-^{32}P-end-labeled 16-mers (50 nM) annealed to either the 45-mer template or the 45-mer template containing drug–DNA adduct were incubated with 0.1 unit/ml (180 nM) Klenow fragment at 30°C for 0.5 hr as a function of dNTP concentration. The sequence within brackets is that of the primer-extended strand at the end of the sequence, and lower and upper arrowheads indicate the predominant termination sites of DNA synthesis at 10 and 100 μM dNTPs, respectively. The thymidine nucleotide indicated by an asterisk is the nucleotide opposite the drug-modified adenine. Reprinted from Sun & Hurley, 1992a, with permission.

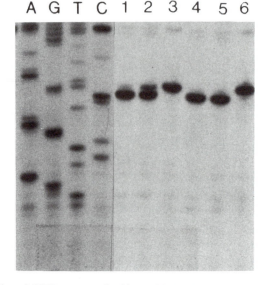

Figure 10-26. Effect of dNTP concentration bias on DNA polymerase termination sites in the presence of site-directed, (+)-CC-1065-modified, single-stranded templates. $5'$-^{32}P-end-labeled 16-mer (50 nM) annealed to 45-mer template that had been modified with (+)-CC-1065 was incubated with 0.1 unit/ml of the Klenow fragment of DNA polymerase for 30 min in the presence of the indicated amount of dNTPs. A, G, T, and C refer to dideoxy sequencing reactions with unmodified templates. Lanes 1–6 contained 10 μM dNTPs; 10 μM dNTPs and 100 dATP; 10 μM dNTPs and 100 μM dGTP; 10 μM dNTPs and 100 μM dTTP; 10 μM dNTPs and 100 μM dCTP; and 100 μM dNTPs in the reaction mixtures, respectively. Reprinted from Sun & Hurley, 1992a, with permission.

modified adenine. Overall, these results imply that the potent mutagenicity of (+)-CC-1065 may be caused by the misincorporation of nucleotides during DNA synthesis.

Ligases. T4 DNA ligase was used as a model system to evaluate how (+)-CC-1065 and related compounds can affect enzymatic ligation of DNA (Sun & Hurley, 1992b). This ligase uses DNA molecules as its substrate, and structural changes occurring in DNA may affect the ligation reaction. T4 DNA ligase can mediate the formation of a phosphodiester bond on one strand of duplex DNA, provided that the fragments to be ligated terminate in a $3'$-hydroxyl group and a $5'$-phosphoryl group, respectively. The enzyme uses ATP as an energy source for this purpose (Lehman, 1974; Shore et al., 1981). It seemed logical to expect that the introduction, at the ends of the duplex near the annealing site, of several drug modifications known to have different effects on DNA structure might produce differential effects on the ligation process. Specifically, different (+)-CC-1065 analogues produce variable extents of elevation in T_m, and (+)-CC-1065 and (+)-AB$'$C$'$ (Fig. 10-20) are unique in that they produce unusual DNA winding effects that can affect the relative orientation between the $3'$-hydroxyl group and the adjacent $5'$-phosphoryl group of an annealed intermediate between two pieces of DNA (Lee et al., 1991a). As expected, while (+)-ABC and (+)-ABC$''$ induced moderate inhibition of the ligation of an unmodified strand when ligation involved a site within 10 base pairs of the $5'$-end of the site of drug modification, (+)-CC-1065 and

(+)-AB′C′ induced more significant inhibition, implying that the DNA winding effect of these latter two compounds can change the relative orientation of the 3′-hydroxyl group and the adjacent 5′-phosphoryl group and, consequently, make it more difficult for nucleophilic attack on the activated 5′-phosphoryl group by the 3′-hydroxyl group. In contrast to the observed *inhibition* of ligation of unmodified DNA strands by these molecules, drug modification *enhanced* the ligation rate of the drug-modified strand. This was presumably due to drug-induced alteration of the position (and possibly the mobility) of the 5′-phosphoryl group, making it easier for the 3′-hydroxyl group to attack the activated 5′-phosphoryl group, as required for the formation of a phosphodiester bond. We speculate that this differential effect of (+)-CC-1065 on ligation on each strand results from the fact that the drug-induced winding and helix-stabilizing effects occur predominantly to the 5′-side of the covalently modified adenine (i.e., toward the drug overlap site). The overall conclusion is that the helix-stabilizing effect restricts the mobility of the 5′-phosphoryl group of the (+)-CC-1065-modified strand and the 3′-hydroxyl group on the unmodified strand. Furthermore, the winding effect of (+)-CC-1065 and (+)-AB′C′ can change the proximity between the 3′-hydroxyl group and the adjacent 5′-phosphoryl group, resulting in greater inhibition of the ligation of (+)-CC-1065- and (+)-AB′C′-modified DNA.

In addition to the proximal inhibitory effect of drug molecules on DNA ligation, a distal inhibitory effect of (+)-CC-1065, (+)-AB′C′, and (+)-ABC″ was also observed, indicating that these drug molecules can change the flexibility of DNA molecules, as well as have other effects on DNA, such as bending and winding (Sun & Hurley, 1992a,b,c). The molecular basis for distal ligation inhibition by drug molecules is still not completely clear, but two important points have been addressed from this study. First, it was discovered that T4 DNA ligase does not favor the formation of 180° out-of-phase bent DNA from individual bent oligomers. This conclusion may be rationalized as follows. T4 DNA ligase may require maximum contact with the two DNA substrates for successful ligation to occur. If bent DNA molecules are aligned in a 180° out-of-phase manner, T4 DNA ligase cannot attain the required amount of contact with the DNA substrates, compared to linear or in-phase bent DNA molecules. This might be analogous to the observation that an overall in-phase bent structure in the prokaryotic upstream promoter region can enhance the transcription rate by achieving maximum contact between the promoter region and RNA polymerase, which facilitates the formation of an open complex for transcription initiation (McAllister & Achberger, 1989). Second, the formation of out-of-phase bent DNA was only inhibited when (+)-CC-1065-, (+)-AB′C′-, and (+)-ABC″-induced bent oligomer DNA molecules were used as ligase substrates, whereas the bent oligomer DNA induced by (+)-ABC can form out-of-phase bent DNA just as efficiently after the ligation reaction as control DNA. To rationalize this observation, we propose that T4 ligase can still distort DNA structure sufficiently to ensure the maximum contact between enzyme and DNA substrates with (+)-ABC-modified bent DNA; however, DNA modified with (+)-CC-1065 cannot be so easily distorted by T4 DNA ligase, presumably because of its helix-stiffening effect.

Helicases. It seemed particularly interesting to measure the ability of DNA helicases to function in the presence of drug-adducted substrates for several reasons. First, replicative helicases are thought to be located in the vanguard of the multiprotein complexes that catalyze DNA synthesis and play a crucial role in providing the DNA polymerase with a single-stranded template that can be utilized efficiently (Alberts, 1984). Therefore, the helicase would be the first component of the replication complex to encounter the (+)-CC-1065-modified DNA. If the drug, which is known to stabilize a duplex, interferes with helicase-mediated unwinding, one would predict that this would have severe consequences for further replication fork progression. In addition, other major pathways of DNA metabolism are known to utilize DNA helicases, such as homologous recombination and DNA repair. Therefore, the ability of helicases to melt (+)-CC-1065-adducted substrates is relevant to the effect the drug may have on these processes as well.

Escherichia coli helicase II is a single-stranded DNA-dependent ATPase with helicase activity that translocates progressively along single-stranded DNA in a $3' \rightarrow 5'$ direction (Matson, 1986). In order to evaluate the effect of drug modification of DNA on the efficiency of helicase II, partial duplexes were constructed that had a $3'$ single-stranded tail and contained a unique drug bonding site (5'-GATTA*-3') on either the long or short strand. The rationale for using both substrates (i.e., drug modification on either the long or short strand) derived from an earlier study (Maine et al., 1992), which showed that the drug-induced winding and helix-stabilizing effects on DNA molecules occurred predominantly to the 5'-side of the covalently modified adenine (i.e., toward the drug overlap site). Thus, if drug molecules were positioned in both orientations relative to the direction of helicase translocation, it would be possible to test the differential effect of drug orientation on helicase II-catalyzed unwinding of DNA. In this study (Sun & Hurley, 1992c), several analogs of (+)-CC-1065 [(+)-ABC, (+)-AB, (+)-AB'C', and (+)-ABC''; see Fig. 10-20] were also included in an attempt to relate these results to the differential biological activities of these drug molecules. In order to measure both the helicase unwinding activity and the extent and selectivity of drug modification at the desired site in the same experiment, only the drug-modified strands were 5'-^{32}P-end-labeled. The results of this study, using the thermal strand breakage assay, clearly demonstrated that the extent of inhibition of DNA unwinding mediated by both helicase II and *E. coli* rep protein is partially correlated with the drug-induced DNA stabilizing effect. However, drug-induced stabilization of DNA is in itself insufficient to explain the full extent of helicase inhibition. For example, while the effect of (+)-AB'C' on the ΔT_m of DNA is almost the same as that of (+)-ABC'', (+)-AB'C' produced a dramatically increased inhibition of DNA unwinding compared to (+)-ABC''. The additional extent of helicase II inhibition common to (+)-CC-1065 and (+)-AB'C' appears to be correlated with their unique winding effects (Sun & Hurley, 1992c). Significantly, the increase in the amount of helicase inhibition is also dependent upon the orientation of the winding effect of drug molecules, which has to be opposite with respect to the direction of helicase II translocation. There is a striking correlation of the orientation-specificity of the winding effect of (+)-CC-1065 and a similar orientation-specificity of the ter-binding protein, which also produces contrahelicase activity (Lee et al., 1989). It is interesting to speculate whether the mechanisms for helicase inhibition in these two cases might be related. The (+)-CC-1065 molecule has its major interactions with DNA to the 5'-side of the covalently modified adenine, and the drug-induced winding and helix-stiffening effects occur predominantly to the 5'-side of the covalently modified adenine (i.e., toward the drug overlap site). Our results show that helicases have more difficulty in unwinding a duplex that has a drug adduct on the displaced strand, compared to that on the helicase-bound strand. We speculate that the presence of drug molecules may prevent helicases from entering the drug modification site by stabilizing or winding the duplex when drug molecules were positioned on the displaced strand, whereas helicases have difficulty in translocation through the covalently modified strand because of steric interaction due to the drug–adenine adduct when drug molecules are present on the helicase-bound strand.

The inclusion in this same study of a gapped duplex (GD1) (Fig. 10-27) as a substrate for helicase II and *E. coli* polymerase I provided important insights into understanding the effect of drug-induced inhibition of DNA unwinding mediated by helicase II and the possible further impact on DNA metabolism, such as the DNA repair process. It is well known that one pathway for repair of oligonucleotides containing DNA damage is through the combined actions of the UvrD protein (helicase II) and DNA polymerase after the UvrA, UvrB, and UvrC gene products of *E. coli* have identified the damaged site and produced 3' and 5' incisions (Husain et al., 1985; Lee at al., 1989). These proteins recognize DNA damaged by bulky adducts such as pyrimidine–pyrimidine cyclobutane dimers and produce incisions at the fourth or fifth phosphodiester bond on the 3'-side of the damage and at the eighth phosphodiester bond on the 5'-side (Sancar & Rupp, 1983). The results (Fig. 10-27) showed that the (+)-CC-1065- and (+)-AB'C'-damaged nucleotides are resistant to excision by the combined action of UvrD and DNA polymerase, which could result in persistent DNA strand breaks after incision by UvrAB and UvrC protein. The real possibility of generation of persistent DNA strand breaks during the repair process was suggested in a previous study using a eukaryotic system, in which it was shown that (+)-CC-1065 produced depletion of NAD levels in repair-proficient and -deficient (xeroderma pigmentosum) human cells, which appears to be related to poly(ADP)ribosylation and persistent DNA strand breakage (Jacobson et al., 1986). If persistent DNA strand breaks are generated as a result of the normal repair processes, it can be speculated that drug molecules such as (+)-CC-1065 and (+)-AB'C' are likely to have much more potent biological effects in vivo compared to other analogs lacking these DNA winding effects. This leads to the question of whether the unique winding effects of (+)-CC-1065 and (+)-AB'C' may be related to the delayed lethality produced in mice by these same compounds (Warpehoski & Bradford, 1988).

In conclusion, covalent modification of DNA molecules with (+)-CC-1065 and its analogues was found to be a strong hindrance to in vitro DNA synthesis mediated by *E. coli* DNA polymerase, Klenow fragment, and T4 polymerase and to cause pretermination of polymerization around the adduct site, which then inhibits the DNA replication process (Sun & Hurley, 1992a–c).

It has been shown that (+)-CC-1065 and its analogs cause the proximal inhibition of T4 DNA ligase through their helix-stabilizing effect and unusual winding effect. Furthermore, the helix-stiffening effect of (+)-CC-1065, (+)-AB'C', and (+)-ABC″ has been demonstrated by using circularization experiments; the distal ligation inhibition effect can be rationalized as being the consequence of the helix-stiffening effect of the drug molecules. Finally, it is proposed that the delayed lethality in rodents caused by (+)-CC-1065 and (+)-AB'C' that is associated with the inside-edge substituents (Fig. 10-24) might result from these unusual DNA-winding effects and strong DNA helix-stiffening effects of these drug molecules. These results clearly define some of the biochemical consequences of drug-induced winding and helix-stabilizing of DNA on the activity of unwinding enzymes such as helicase II and *E. coli* rep protein.

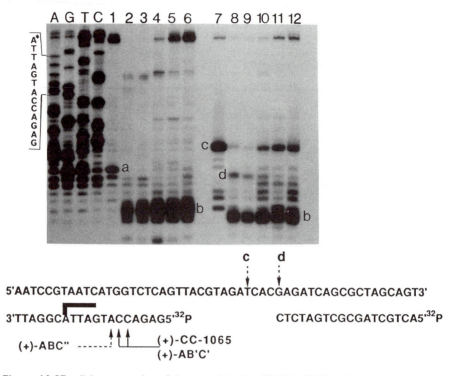

Figure 10-27. Primer extension of the gapped duplex (GD1) by DNA polymerase I alone (lanes 1–6) or by the combined action of helicase II and polymerase I (lanes 7–12). Reactions in lanes 1–12 contained 20 ng of DNA substrate, 50 μM of the four dNTPs, and 1 unit of DNA polymerase I in helicase II unwinding buffer. Reactions were carried out at 30°C for 30 min. For reactions in lanes 7–12, DNA substrates were unwound by 100 ng of helicase II for 15 min and then incubated for 30 min with the addition of 1 unit of DNA polymerase I and 50 μM of dNTPs. Reactions contain unmodified DNA in lanes 1 and 7, (+)-CC-1065-modified DNA in lanes 2 and 8, (+)-AB'C'-modified DNA in lanes 3 and 9, (+)ABC″-modified DNA in lanes 4 and 10, (+)-ABC-modified DNA in lanes 5 and 11, and (+)-AB-modified DNA in lanes 6 and 12. Band "a" represents single-stranded DNA displaced by DNA polymerase I coupled to DNA synthesis. Band "b" represents the strand breakage product of drug-modified strands. Bands "c" and "d" represent prematurely terminated DNA synthesis. The four lanes to the extreme left are dideoxy-DNA sequencing lanes. The sequence complementary to the covalent modification site is shown on the left. The sequence of GD1 is shown at the bottom of the figure, identifying the drug bonding site and the position of bands "c" and "d", as well as the pausing sites for DNA polymerase near the adduct sites for (+)-ABC″, (+)-CC-1065, and (+)-AB'C'. Reprinted from Sun & Hurley, 1992c, with permission.

Summary

The subject of how low molecular weight ligands interact with DNA is, on a microscopic scale, one of molecular recognition. This survey has sought to demonstrate both the wide array of interaction motifs that ligands exploit and the interdisciplinary approach required to unveil the coupling of structure and function inherent in these molecules. The characterization of a particular ligand complex with DNA by as many biochemical and biophysical methods as are available is critical to developing a rigorous model of the structure. This effort is a natural beneficiary of improvements and innovations in recent years in molecular biology techniques and hardware and software for X-ray crystallography, NMR, and computational modeling.

Note added in proof. More recent publications on the structural characterization of the pluramycins (Hansen et al., 1995; Hansen & Hurley, 1995) as well as their DNA sequence specificity (Sun & Hurley, 1995a,b) and unique reaction with the TBP-TATA box complex have appeared. The DNA sequence specificity (Sun & Hurley, 1993) and NMR-determined solution structures of Bizelesin crosslinked adducts (Thompson et al., 1995; Thompson & Hurley, 1995) have also appeared. Finally, the importance of (+)-CC-1065 bending in the early promoter region of SV40 has been determined (Sun & Hurley, 1994a,b).

Acknowledgments

This research was supported by grants from the U.S. Public Health Service (CA-49751), the Welch Foundation, The Upjohn Company, and the Burroughs Wellcome Scholars Program. John A. Mountzouris was supported by fellowships from The American Foundation for Pharmaceutical Education and the ACS Division of Medicinal Chemistry. We are grateful to David Bishop for editorial assistance and preparation of the manuscript.

11

The Chemistry of Protein–DNA Interactions

Christopher J. Larson and Gregory L. Verdine

Proteins regulate the expression of genes at all stages from transcription initiation to posttranslational modification. Among the many regulated events that culminate in the appearance of a functional gene product in the cell, one of the earliest is recruitment of the RNA biosynthetic machinery to the transcriptional startpoint of a gene. The role of recruiter is served by proteins that bind to specific sequences flanking a gene, from which stance they establish direct protein–protein contacts with components of the transcriptional apparatus (Dynan, 1989; Mitchell & Tjian, 1989; Tjian & Maniatis, 1994). In the 20 years since the first sequence-specific DNA-binding protein was isolated (Gilbert and Müller-Hill, 1966), much progress has been made toward understanding the chemical basis for molecular recognition in the interface between protein and DNA. These gains in understanding have been enabled in part through technical advances in chemical synthesis and sequencing of DNA, cloning and overproduction of proteins, and high-resolution structural analysis through X-ray crystallography and nuclear magnetic resonance (NMR) spectroscopy. As of this writing, the structures of more than 50 sequence-specific protein–DNA complexes have been reported (Table 11-1). Virtually all of these protein–DNA complexes have also been studied by biochemical experiments that test the effects of changes in the protein–DNA interface on formation of the protein–DNA complex. In the cases of most known DNA-binding proteins, for which no detailed structural information is yet available, such biochemical data have provided important insights into the functional groups that are involved in molecular recognition, if not their three-dimensional arrangement. Although the information gained through such indirect methods is moderate in resolution when compared to X-ray or NMR structures, the former provide an essential, independent check on the latter, and can be obtained on a routine basis without resorting to specialized equipment or expertise. In one case, revision of an X-ray structure was prompted by its inability to be reconciled with biochemical data (Heitman & Model, 1990; Kim et al., 1990); in another case, a surprising and unusual protein–DNA co-crystal structure has been strongly supported by extensive chemical modification studies (Joachimiak et al., 1994; Smith et al., 1994). Moreover, chemical probes of protein–DNA interactions can provide insight into the energetic contributions of interfacial contacts inferred from high-resolution structural studies (Lesser et al., 1990; Mazzarelli et al., 1992; McLaughlin et al., 1987); such energetic information cannot be obtained by X-ray or NMR.

This chapter discusses the current state of understanding in the area of protein–DNA

Table 11-1 High-Resolution Structures of Protein–DNA Complexes

Protein	Resolution	Function	Reference
E. coli EcoRI	3.0 Å	Restriction endonuclease	Kim et al., 1990 McClarin et al., 1986
434 repressor	2.5 Å 3.5–4.5 Å 2.5 Å 2.5 Å	Transcription factor	Aggarwal et al., 1988 Anderson et al., 1987 Rodgers & Harrison, 1993 Shimon & Harrison, 1993
E. coli trp repressor	2.4 Å NMR	Transcription factor	Lawson & Carey, 1993 Otwinowski et al., 1988 Zhang et al., 1994
λ repressor	1.8 Å 2.5 Å	Transcription factor	Beamer & Pabo, 1992 Jordan & Pabo, 1988
434 *cro* repressor	3.2–5.5 Å	Transcription factor	Wolberger et al., 1988
Antennapedia[a]	NMR	Transcription factor	Otting et al., 1990
Engrailed[a]	2.8 Å	Transcription factor	Kissinger et al., 1990
λ *cro* repressor	3.9 Å	Transcription factor	Brennan et al., 1990
E. coli CAP-cAMP	3.0 Å	Transcription factor	Schultz et al., 1991
Matα2[a]	2.7 Å	Transcription factor	Wolberger et al., 1991
Zif268[a]	2.1 Å	Transcription factor	Pavletich & Pabo, 1991
Glucocorticoid receptor[a]	2.9 Å	Transcription factor	Luisi et al., 1991
GAL4[a]	2.7 Å	Transcription factor	Marmorstein et al., 1992
E. coli met repressor	2.8 Å	Transcription factor	Somers & Phillips, 1992
Bovine papillomavirus-E2	11.7 Å	Transcription factor	Hegde et al., 1992
GCN4[a]	2.9 Å	Transcription factor	Ellenberger et al., 1992 Konig & Richmond, 1993
GATA-1[a]	NMR	Transcription factor	Omichinski et al., 1993
GLI[a]	2.6 Å	Transcription factor	Pavletich & Pabo, 1993
Estrogen receptor[a]	2.4 Å	Transcription factor	Schwabe et al., 1993
*Eco*RV endonuclase	2.5 Å 3.0 Å	Restriction endonuclase	Winkler et al., 1993 Winkler et al., 1993
TATA-binding protein	2.5 Å 2.3 Å	Transcription factor	Kim et al., 1993a Kim et al., 1993b
Max	2.9 Å	Transcription factor	Ferré-D'Amaré et al., 1993
HNF-3/Fork Head	2.5 Å	Transcription factor	Clark et al., 1993
Tramtrack	2.8 Å	Transcription factor	Fairall et al., 1993
*Hha*I methyltransferase	2.8 Å	DNA methyltransferase	Klimasauskas et al., 1994
Hin recombinase	2.3 Å	DNA recombinase	Feng et al., 1994
Arc repressor	2.6 Å	Transcription factor	Raumann et al., 1994
MyoD[a]	2.8 Å	Transcription factor	Ma et al., 1994
E47[a]	2.8 Å	Transcription factor	Ellenberger et al., 1994
Oct-1 POU domain[a]	3.0 Å	Transcription factor	Klemm et al., 1994

[a]These structures represent only the DNA-binding domain of the protein, rather than the full-length protein, bound to DNA. In mose cases, the strength and specificity of DNA binding by the fragment closely mirrors that of the full-length protein.

recognition, highlighting critical experiments that have played a particularly important role in shaping our view and also describing ongoing efforts to elucidate the principles governing protein–DNA recognition. We focus primarily on (1) noncovalent interactions between proteins and DNA and (2) the structural and energetic origins of such interactions. An exhaustive survey of protein–DNA interactions would fill several volumes, hence we have chosen to provide an overview of protein–DNA interactions from a chemical viewpoint, with particular emphasis on established and emerging biochemical techniques for probing sequence-specific contacts. Unfortunately, these boundaries exclude discussion of such worthy subjects as structural motifs used for recognition by proteins (Harrison, 1991; Pabo & Sauer, 1992), the biophysical methods used in high-resolution structural determination (Harrison, 1991; Russu, 1991), covalent transactions involving proteins and nucleic acids (Suck, 1992; Verdine, 1994; Winkler, 1992), interactions of proteins with single-stranded DNA (Kneale, 1992) and duplex RNA (Nagain, 1992), and recognition of noncanonical DNA structures by proteins (Churchill & Travers, 1991; Travers, 1989, 1992), as well as the coupling of protein folding to DNA binding (Spolar & Record, 1994). The reader is directed to the several excellent treatments of these subjects noted above (see also Travers, 1993).

Interactions and Structures

Noncovalent interactions between proteins and DNA are characterized most notably by their great stability and exquisite specificity: equilibrium dissociation constants (K_d's) on the order of 10^{-9}–10^{-12} M and ratios of specific to nonspecific binding of 10^3–10^7 are typical. A number of forces contribute to the exceptionally strong and specific binding; some of these forces act locally in distinct regions of the protein–DNA interface, whereas others exert a more global influence on complexation. Local forces are responsible for many of the direct protein–DNA contacts that are inferred from high-resolution structural and biochemical probing; these contacts are of several types: hydrogen bonds, ionic interactions, and nonpolar interactions. Global forces include long-range electrostatic steering (Kim et al., 1993a,b; Record et al., 1976), DNA-induced protein folding and conformational changes (Anthony-Cahill et al., 1992; Brennan et al., 1990; Klimasauskas et al., 1994; Spolar & Record, 1994;), protein-induced DNA distortion such as bending or twisting (Aggarwal et al., 1988; Kim et al., 1993a,b; Schultz et al., 1991; Steitz, 1990; Travers, 1993; Winkler et al., 1993), and cooperativity gained through simultaneous DNA recognition by multiple protein modules (Härd et al., 1990; Hochschild & Ptashne, 1986; Johnson et al., 1979; Jordan & Pabo, 1988; Mao et al., 1994; Metzger et al., 1993; Meyer et al., 1980; Sasse-Dwight & Gralla, 1988). These global forces, although significant, are still poorly understood and thus represent an important, underdeveloped research frontier. On the other hand, local forces have been studied extensively, particularly as manifest in protein–DNA contacts, and hence will form the principal focus of this review.

Proteins contact DNA through a combination of interactions with the deoxyribose-phosphate backbone and with the pendant heterocyclic bases (Fig. 11-1). Because the backbone is chemically uniform in DNA, contacts to it are generally considered not to be a dominant factor in determining the specificity of protein–DNA interactions. In most cases, contacts involving DNA bases are the principal determinant of specificity. In duplex DNA, the bases stack upon one another so as largely to block access to the

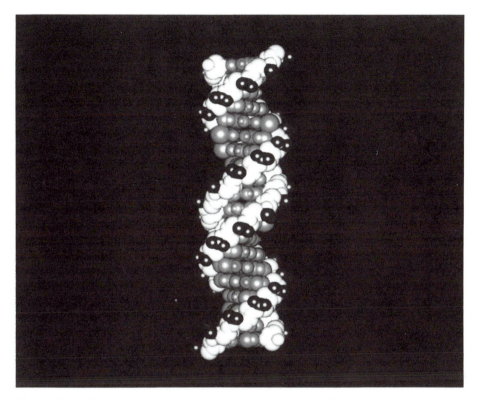

Figure 11-1. The structure of B-DNA illustrating the minor (small) and major (large) groove of DNA, the phosphate groups regularly spaced along the backbone, and the nucleobases H-bonded and oriented perpendicular to the helix axis.

π face of the heterocycles (Fig. 11-1); thus most base contacts involve edge-on interactions with the functional groups that project from the σ plane of the heterocycles.

The folding of duplex DNA leads to the generation of two distinct groove-like feature on its surface; these are termed the *major* and *minor* grooves (Fig. 11-1). The majority of known proteins interact with bases in the major groove. The observed propensity toward major groove interactions may arise from its greater accessibility in B-DNA: The major groove is wide and deep (12 Å and 8.5 Å, respectively, including van der Waals radii), whereas the minor groove is narrow and deep (6 Å and 7.5 Å, respectively). Although β-sheets and regions of extended polypeptide chain are sometimes utilized for DNA recognition (Ghosh et al., 1995; Kim et al., 1990, 1993a,b; Kissinger et al., 1990; Klemm et al., 1994; Liu et al., 1994; Müller et al., 1994; Somers and Phillips, 1992; Wolberger et al., 1991), α-helices are more commonly employed as DNA-recognition modules. An α-helix is able to fit into the major groove but cannot fit into the minor groove without causing gross DNA distortion; since, e.g., the diameter of a polyalanine α-helix, including van der Waals radii, is ~8.5 Å. Consistent with this notion, the known cases of minor groove binding involve either (1) radically distorting DNA so as to widen the minor groove or (2) binding by a flexible peptide-like extension (Kim et al., 1993a,b; Kissinger et al., 1990; Klemm et al., 1994; Otting et al., 1990; Wolberger et al., 1991). In addition, the edges of the bases present a greater variety of functionality into the major groove than that presented by the bases in the minor groove (Fig. 11-2). For example, the N^2 amino group in C·G and G·C base pairs occupies roughly the same position in the minor groove, and a similar degeneracy exists for the lone pairs on O2 of T and N3 of A in the A·T and T·A base pairs (Seeman et al., 1976). Thus the minor groove may possess a reduced potential for specificity-determining interactions compared with the major groove.

The patterns of locally interacting groups such as hydrogen bonding moieties on the contact surfaces of the protein and the DNA provide chemical complementarity of the protein to correct, but not to incorrect, sequences of DNA. Although contacts to the deoxyribose–phosphate backbone are generally considered "nonspecific," they have an important role in providing strength to the complex. One way to view the important role employed by nonspecific contacts is to consider that they can generate a complex with a K_d of 10^5–10^8, requiring only another 3–5 kcal/mole of specific binding energy in order to acquire biologically useful levels of strength and specificity.

Having considered the loci in DNA that participate in protein binding, we now turn our attention to the roles played by the various kinds of interactions.

Hydrogen Bonds. More emphasis has been placed on the role of hydrogen bonding in protein–DNA interactions than on any other contributing factor. This situation may have resulted from the fact that hydrogen bonds follow relatively simple rules. Namely, hydrogen bonds are formed between an electron-rich "acceptor" atom A and a proton attached to an electronegative "donor" atom D. Although bonding is stronger when all of the atoms in the A--- H—D array are arranged linearly, there is good reason to believe that hydrogen bonds with bond angles of 135° or less can make important contributions to protein–DNA interactions. A useful guide is that the H—D bond length is ~1 Å, and the hydrogen bond is twice that length; hence, the A—D bond distance of a typical hydrogen bond is generally 3 Å or slightly less.

Hydrogen bonds normally make up a large fraction of the total number of contacts in a protein–DNA complex. These involve both contacts to DNA bases and hydrogen bonds of neutral protein functionality with the charged phosphodiester oxygens of the backbone. Despite their prevalence and the importance historically accorded them, ev-

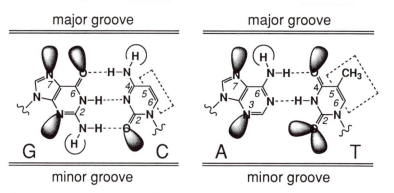

Figure 11-2. Potential contact functionally present on base pairs in DNA. Shaded lobes, hydrogen bond acceptors; circles, hydrogen bond donors; dashed rectangles, nonpolar surfaces.

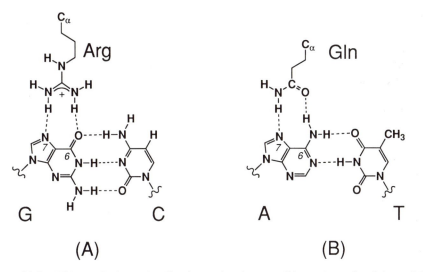

Figure 11-3. Bidentate hydrogen bonding interactions between (A) guanine and arginine and (B) adenine and glutamine.

idence supports that hydrogen bonds contribute only modestly to the overall energetics of specific complexation (Ha et al., 1989; Spolar & Record, 1994). Even so, hydrogen bonds, especially those involving DNA bases, appear to serve a critical role in determining the *specificity* of protein–DNA interactions. Formation of matched sets of hydrogen bonds and other types of contacts to the DNA bases of the optimal, cognate site allows the entire protein to come into intimate contact with the DNA surface, permitting the formation of a large number of energetically favorable nonspecific contacts. On a noncognate site, repulsive interactions between prospective hydrogen bonding or hydrophobic groups prevent the formation of an intimate complex. Although this view is undoubtedly oversimplified, it nevertheless represents a useful approximation that holds for most cases.

Several types of hydrogen bonding contacts to the edges of DNA bases are commonly observed, including (1) direct hydrogen bonds to amino acid side chains, (2) direct hydrogen bonds to the polypeptide backbone amide functionality, and (3) water-mediated contacts to side chains or the main chain of the polypeptide. Of these, the first type has thus far been the most commonly observed. Contrary to early predictions (Seeman et al., 1976), there does not seem to be any "recognition code" relating particular amino acid residues to particular bases: Some amino acids contact a variety of bases and vice versa. Within a series of closely related structures, some conservation of contact pairs is observed, but comparison of unrelated proteins yields no discernible pattern (Pabo & Sauer, 1992). Tabulation of contact partners, however, does reveal some interesting trends (Pabo & Sauer, 1992). For example, hydrogen bonds to guanine are particularly prevalent, perhaps because the N7 atom of G is the most electron-rich acceptor available in B-DNA. Furthermore, the N7 and O6 positions of guanine offer the potential for bidentate hydrogen bonding to arginine (Fig. 11-3A); because this interaction involves a neutral and a charged species, it is expected to be stronger than the corresponding bidentate interaction between adenine and glutamine (Fig. 11-

3B). If the greater prevalence of bidentate contacts to guanine residues indeed does reflect an energetically more favorable interaction than that with adenine, another source of this greater stability could be the "Jorgensen effect"—that is, the presence of favorable secondary interactions between neighboring H-bond donors and acceptors (Jorgensen & Pranata, 1990).

The backbone amide function possesses a substantial partial positive charge on H and partial negative charge on O and thus presents an excellent potential hydrogen bonding partner; however, the main chain of a protein is typically less accessible than the side chains. Therefore, it may come as no surprise that direct polypeptide main-chain contacts to DNA bases have only been observed infrequently in protein–DNA complexes thus far (Hayashibara & Verdine, 1992; Jordan & Pabo, 1988; Klimasauskas et al., 1994; MacMillan et al., 1993; Marmorstein et al., 1992), and only to the exocyclic—NH of cytosine. Hydrogen bonds to the backbone NH have not been observed; only contacts to the backbone amide carbonyl have been seen. It is interesting to note that if one disregards conformational preferences and the secondary amine of proline, main-chain contacts are not specific with regard to any particular amino acid.

Whereas the observation of water-mediated base contacts (Otwinowski et al., 1988) was initially greeted with skepticism (Staacke et al., 1990), it now appears that such contacts are both real and in some cases indispensable (Hegde et al., 1992; Joachimiak et al., 1994; Luisi et al., 1991; Ma et al., 1994; Schwabe et al., 1993; Smith et al., 1994; Steitz, 1990). Dunitz has estimated the energetic cost of "tying down" a water molecule to be 0–7 kcal/mole at 37°C (Dunitz, 1994). Considering that a water molecule can make up to four hydrogen bonds, each of which contributes $\sim$1.5 kcal/mol in enthalpy, it is apparent that tightly bound water molecules are not necessarily energetically deleterious and might even be energetically favorable in some cases. As with direct amino acid–DNA backbone contacts, water-mediated hydrogen bonding interactions may play a greater role in determining specificity than in conferring strength to the protein–DNA complex. Recent work by Sigler and coworkers (Joachimiak et al., 1994) has reaffirmed the importance of water-mediated contacts in the case of the *trp* repressor operator complex, in addition to demonstrating directly that the stereochemical relationship of bound water to its hydrogen bonding partners is critical.

Direct hydrogen bonds between DNA-binding proteins and the backbone phosphates account for about half of all direct hydrogen bonds observed in protein–DNA complexes (Pabo & Sauer, 1992). Again, there does not seem to be any discernible pattern of amino acid usage apart from the rules of hydrogen bonding: Direct and water-mediated hydrogen bonds between amino acid side chains or the polypeptide amide backbone and the phosphates of DNA seem to play significant roles in a number of complexes (Harrison, 1991; Pabo & Sauer, 1992; Steitz, 1990; Travers, 1993). Any functionality with the ability to donate a hydrogen bond to the phosphate oxygens should be able to serve in this capacity, and short polar side chains and the polypeptide backbone may actually have an entropic advantage over the long side chains of positively charged amino acids such as lysine and arginine in fulfilling this role.

Nonpolar Interactions. The second major type of contact that plays a role in protein–DNA interactions is the interaction of nonpolar groups on the protein and DNA. As with most ligand–receptor interactions in water, the dominant force driving the formation of protein–DNA complexes is the reduction of solvent-exposed nonpolar surface area (Ha et al., 1989). Although strident debate continues as to the origins of this phenomenon, called the *hydrophobic effect*, the most widely accepted explanation is that it reflects the favorable entropy change that accompanies the release of ordered water molecules surrounding a nonpolar surface into the bulk solvent (Dill, 1990). Thus, the driving force behind nonpolar interactions is a process leading to the establishment of a contact, rather than the contact itself. This has led to some confusion in the literature regarding nonpolar interactions. What can be observed in, for example, a typical X-ray structure of a protein–DNA complex is encounters of nonpolar groups to within van der Waals contact distance—that is, roughly 3.2 Å for the distance separating carbon atoms between hydrocarbons. The contact itself is thus most accurately termed a *van der Waals contact*—namely, the enthalpically favorable interactions that result from dispersion forces between nonpolar molecules. What cannot be observed in a crystal structure is the hydrophobic effect. The relative contributions of the hydrophobic effect and van der Waals interactions as individual forces driving the formation of nonpolar contacts remain a controversial issue, although the present weight of evidence favors the former as being more important (Ha et al., 1989).

Whereas hydrogen bonds require a fairly precise geometric alignment of the interacting partners, nonpolar contacts can take place at any relative orientation that is sterically accessible. This high degree of orientational freedom has presumably allowed for rapid evolution of nonpolar interactions in protein–DNA interfaces: Formation of a sequence-specific protein–DNA complex is typically accompanied by the burial of 2000–4000 Å^2 of nonpolar surface area (Ha et al., 1989). The hydrophobic effect is estimated to contribute 22–25 cal per Å^2 of nonpolar surface removed from exposure to solvent; the amino acid leucine possesses close to 100 Å^2 of nonpolar surface area (Richards, 1977).

With regard to the nonpolar interactions that play an important role in determining sequence-specificity, examination of the structures listed in Table 11-1 reveals that the most widely noted contact site is the C5 methyl group of thymine. Virtually all structurally characterized complexes involving a protein bound in the major groove utilize this type of contact. The amino acid partners vary, with Ala being the most common (Fig. 11-4) and bulky hydrophobic residues being less common; involvement of the hydrocarbon side chains of polar residues like Gln is frequently observed. The methyl group of T alone contributes ~15% of the solvent-exposed nonpolar surface area in typical mixed-sequence B-DNA, with a water-accessible surface area of ~26 Å^2.

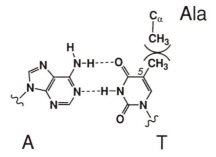

Figure 11-4. Schematic illustration of a nonpolar contact between the side chain of Ala and the vinyl-CH$_3$ of T. Half-circles represent van der Waals surfaces.

Nonpolar contacts to other bases are probably more widespread than is currently appreciated. In addition to the methyl group of T, the 5,6 C=C of T and C is hydrophobic and therefore represents a suitable target for a nonpolar residue on a protein; in at least one case, such a contact has been has been inferred to play an important energetic role (Hayashibara and Verdine, 1992).

The ~53-Å^2 exposed surface of the ribose sugar ring of any nucleotide is the single most prevalent solvent-exposed nonpolar surface in DNA, accounting for ~30% of the solvent-exposed nonpolar surface area in B-DNA (Ha et al., 1989). Examination of a model of B-DNA reveals that, although access to the major groove side of the ribose sugar is blocked by both the base and the phosphate group, the minor groove side of the ribose is accessible to potential ligands. However, as noted above, the dimensions of the minor groove preclude entry of the normal protein recognition modules, and thus contacts to the minor groove side of riboses in solved structures (Kim et al., 1993a,b) involve significant widening of the minor groove by the protein, so as to gain access to its potentially vast riches of nonpolar surface area.

Ionic Interactions. The third major type of local interaction in protein–DNA complexes is the ionic interaction or salt bridge, namely, the coulombic attraction of atoms having opposite formal charge. Charge can be readily dissipated by hydrogen bonding with water, hence salt bridges are probably important only when the interacting proteins are separated by no more than one or two shells of water. Like nonpolar contacts, the formation of salt bridges involving the phosphate backbone is driven in large part by an effect that cannot be observed in structures of protein–DNA complexes: the favorable entropy associated with the release of small molecule counterions from the DNA backbone, also known as the *polyelectrolyte effect*. Evidence suggests that this is second only to the hydrophobic effect as a driving force behind formation of protein–DNA complexes (Ha et al., 1989). Approximately half (~45%) of the total solvent-accessible surface area in B-DNA belongs to the phosphate groups of the backbone.

Chemical Methods for Determining Contacts in
Protein–DNA Complexes

Although X-ray and NMR provide the most conclusive determinations of intermolecular contacts in protein–DNA complexes, chemical methods of contact determination also serve an important role. In particular, these methods fulfill the pressing need for methods of contact analysis that (1) can be routinely carried out in the biochemical laboratory without resorting to specialized expertise or equipment; (2) provide insight into the energetic contribution of particular interactions to the whole of the protein–DNA association; and (3) provide an independent means of verifying structural models resulting from X-ray or NMR analysis.

In this section we discuss the various methods that have been developed for probing protein–DNA interactions through chemical modification of DNA. The structural and chemical basis of each will be discussed, as will their strengths and weaknesses.

DNA Footprinting: Protection Versus Interference. Several of the most informative and widely used techniques are referred to as *footprinting*, an anthropomorphism that alludes to the fact that binding of a protein to DNA distinguishes the bound regions of DNA chemically. Footprinting can be carried out in two distinct modes, termed *protection* and *interference*.

In protection footprinting (Fig. 11-5), the protein is bound to end-labeled DNA and the resulting complex is treated with a reagent. The reagent attacks the DNA so as to produce lesions that either directly or through subsequent chemical steps give rise to

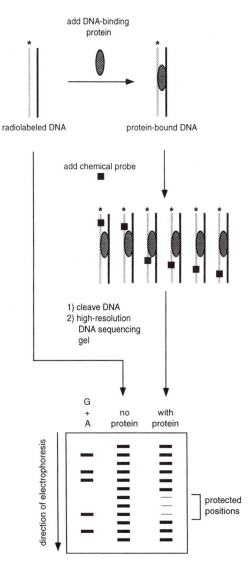

Figure 11-5. Schematic illustration of protection footprinting. The lane labeled G + A refers to Maxam–Gilbert sequence revealing G and A residues only, which serves to calibrate the footprint with the DNA sequence. Asterisk denotes radioactive end-label. Reprinted from Hayashibara, 1993, with permission.

strand cleavage. The conditions of the chemical treatment are adjusted such that each DNA molecule suffers, on average, one "hit" by the reagent, thereby producing a statistical mixture of singly modified DNA molecules. The conceptual basis of this experiment is that binding of a protein to DNA renders certain positions inaccessible to the reagent, thereby protecting that position of the DNA from attack by the reagent. Because the sites of attack by most reagents vary with the DNA sequence, it is important to carry out a control reaction on the same DNA fragment having no bound protein. Comparison of the cleavage patterns of the protein-bound and protein-unbound DNA reveals certain positions in the sequence, visualized as bands on a high-resolution DNA sequencing gel, that are reduced in relative intensity by binding of the protein; these positions are said to be protected. Just as sites in DNA can be protected by a protein, it is not uncommon to observe sites at which attack of a reagent is facilitated by a protein; such sites are said to be enhanced.

Interference footprinting, on the other hand, involves pretreatment of the end-labeled DNA with a reagent, so as to produce a probe of statistically modified DNA molecules (Fig. 11-6). This pool is then incubated with the DNA-binding protein of interest and the resulting mixture fractionated by native gel electrophoresis, which separates protein-bound from unbound DNA on the basis of their difference in size. The bound and unbound DNA fractions are recovered from the gel and are subjected to conditions that cause DNA backbone cleavage; the resulting fragments are separated on a high-resolution DNA sequencing gel and are visualized by autoradiography. A control lane is also included in which the entire pool of modified DNA is not exposed to the protein. The conceptual basis of the interference footprinting experiment is that those positions at which chemical modification occurs at a contact site will fail to bind the protein, and hence the corresponding bands will be enriched in the unbound fraction. Conversely, those positions at which chemical modification occurs at a noncontact position will bind the protein as if they were unmodified; hence the corresponding bands will be enriched in the bound fraction. Bands enriched in the bound lane of the gel thus represent noncontact positions, and bands enriched in the unbound lane represent contact positions.

There are important differences in the ways in which protection and interference experiments probe protein–DNA interfaces, and these must be kept in mind when interpreting the results from each. Protection footprinting is fundamentally a kinetic experiment in that it measures the effect of a protein on the *rate* of attack by a reagent. Any position that is occluded upon binding of a protein, whether or not the position is involved in a contact, will exhibit protection. For this reason, it is not safe to equate individual sites in a protection footprint with contact positions. Protection footprints, however, are comparatively simple and often the method of choice for gross localization of a protein-binding site within a long stretch of DNA. The intrinsically kinetic basis of protection footprints can be of particular value in gauging the presence of protein-induced DNA distortion, which is often manifested as enhancement of bands in the gel.

Interference footprints are fundamentally thermodynamic in nature because they detect the effect of DNA modification on the equilibrium binding strength of a protein. Bearing in mind that any perturbation experiment looks not at the natural system but instead a modified version of it, it is generally considered safe to interpret positions at which interference is evident as contact sites. Exceptions to this rule are known (Ellenberger et al., 1992).

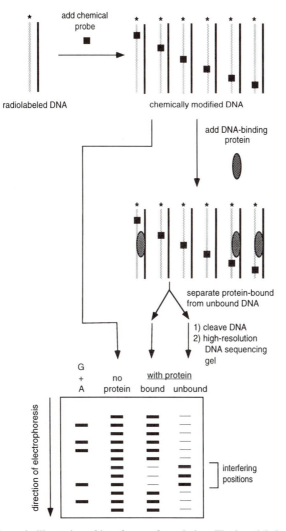

Figure 11-6. Schematic illustration of interference footprinting. The lane labeled G + A refers to Maxam–Gilbert sequence revealing G and A residues only, which serves to calibrate the footprint with the DNA sequence. Asterisk denotes radioactive end-label. Reprinted from Hayashibara, 1993, with permission.

Footprinting Methods That Detect Global Features of Protein–DNA Complexes. Several enzymatic and chemical footprinting techniques that are commonly utilized to assess protein occupancy and distortion of DNA are described below. The first foot-printing method to be developed (Galas & Schmitz, 1978), *DNase I footprinting*, remains one of the most valuable tools available for studying protein–DNA interactions. In most cases, a protein-binding site on DNA is localized to a stretch of several hundred base pairs through genetic experiments or electrophoretic mobility shift assay (Chodosh et al., 1986). DNase I footprinting is then used to localize the protein-binding site to within tens of base pairs. DNase I is an ~30-kD endonucleolytic glycoprotein that cleaves DNA in relatively sequence-independent fashion (Hogan et al., 1989). The protein requires divalent metal cations, usually Ca^{2+} or Mg^{2+}, to catalyze single-stranded nicks in DNA; in the presence of Mn^{2+}, double-stranded cleavage takes place (Suck et al., 1988). DNase I footprinting takes advantage of this enzymatic activity, assaying the ability of a bound protein to protect a region of DNA from attack by the protein. The ability to infer how a sequence-specific DNA-binding protein straddles its recognition site has been greatly facilitated by the availability of X-ray crystallographic structures that reveal how DNase I, the probe molecule in this footprinting method, interacts with DNA (Suck & Oefner, 1986;. Suck et al., 1988). DNase I binds in the minor groove of duplex DNA, widening the minor groove and bending the DNA toward the major groove. The requirement for induced distortion, in addition to the greater stiffness of protein-bound DNA, may explain in part why even major groove-binding proteins generally exhibit strong DNase I footprints. Because of the large size of DNase I, it generally yields footprints that are 4–5 base pairs greater in size on each end of a binding site than the actual region of DNA occupied by a protein. Proteins that bind DNA with a K_d on the order of 10^{-7} M or less can be footprinted using DNase I, and a method has been reported for extracting thermodynamic data from these footprints (Brenowitz et al., 1986). Examples of studies utilizing DNase I include delineation of cooperative binding by a λ repressor (Hochschild & Ptashne, 1986), detection of factors that interact with the human β-interferon regulatory region (Zinn & Maniatis, 1986), discovering interaction between the adenovirus E1a protein and the transcription factor ATF (Liu & Green, 1990), and identification of an octamer protein-binding site in the human kappa light-chain enhancer (Nelms & van Ness, 1990).

Exonuclease III footprinting is complementary in many respects to DNase I footprinting. Exonuclease III is a multifunctional bacterial protein that catalyzes processive cleavage of nucleotides from DNA, beginning at the 3'-end and moving toward the 5'-end ($3' \to 5'$ direction). The presence of a protein at a DNA site blocks procession of exonuclease III. Thus, the 3'-most band on an exonuclease III footprint represents the 3'-end of the binding site. One advantage of this technique over DNase I footprinting is the greater effectiveness of exonuclease III on short oligonucleotides and its ability to cleave certain stretches of DNA that are poor substrates for DNase I. One disadvantage is that the technique cannot provide information on the 5'-boundary of a protein binding for each DNA strand. Examples of studies that have used this technique include identification of protein-binding sites in chromatin (Wu, 1984) and occupation of heat-shock elements in crude nuclear extracts (Wu, 1985).

MPE·Fe(II) footprinting (Harshman & Dervan, 1985; Van Dyke & Dervan, 1983) involves a reagent having an intercalating ligand, methidium, linked to a generator of hydroxyl radicals, EDTA·Fe(II) (Fig. 11-7). The methidium moiety slides between base pairs, thereby localizing the EDTA·Fe(II) close to the minor groove surface of DNA. Through a series of well-studied reactions, the EDTA·Fe(II) reduces H_2O_2 to H_2O and ·OH, an exceedingly reactive species that abstracts hydrogen atoms from carbon at a practically diffusion-controlled rate. Generation of ·OH in the vicinity of DNA results in degradation of the 2'-deoxyribose moiety through initial attack at H1' or H4', both of which are located in the minor groove. Binding of a protein to either groove of DNA inhibits intercalation of MPE·Fe(II) in the minor groove, thereby protecting adjacent sequences from attack of ·OH.

Because ·OH is so reactive, it has little chance to diffuse from the site of its gen-

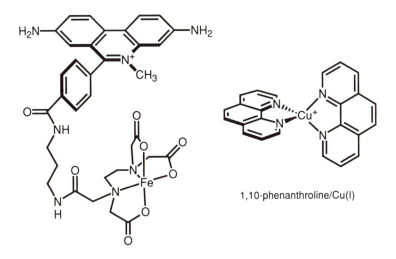

Methidiumpropyl-EDTA/Fe(II)

1,10-phenanthroline/Cu(I)

Figure 11-7. Structures of the DNA-cleaving agents methidiumpropyl-EDTA·Fe(II) and 1,10-phenanthroline·Cu(I).

eration; hence MPE·Fe(II) footprints correspond reasonably closely to the actual site bound by a protein. MPE·Fe(II) may be somewhat more sensitive than DNase I toward small molecules bound to DNA, and it has been widely used for footprinting such complexes (Van Dyke & Dervan, 1983). Despite the potential utility of MPE·Fe(II) in footprinting protein–DNA complexes, this method has not seen widespread use, probably because the reagent is not commercially available. Examples of its use include studies of open promoter complex formation with RNA polymerase (Bateman & Paule, 1988), interactions of a resolvase with a recombination crossover point (Hatfull et al., 1987), and binding of a transcription factor to an upstream enhancer sequence (Sawadogo & Roeder, 1985).

A close relative of MPE·Fe(II) is 1,10-phenanthroline·Cu(I), or $(OP)_2Cu^+$ (Spassky & Sigman, 1985). Like MPE·Fe(II), $(OP)_2Cu^+$ converts H_2O_2 to a species that abstracts H atoms in the minor groove, primarily at H1′. Like MPE·Fe(II), $(OP)_2Cu^+$ binds in the minor groove, but unlike MPE·Fe(II), it generates a reactive intermediate that is not freely diffusible in solution. $(OP)_2Cu^+$ is commonly available from several sources. 1,10-Phenanthroline·Cu(I) has been used extensively in the analysis of bacterial RNA polymerase–promoter complexes (Amouyal & Buc, 1987; Kuwabara & Sigman, 1987) and has also been applied to studies of eukaryotic transcription factors (Chen et al., 1995; Lowenthal et al., 1989).

One advantage of MPE·Fe(II) and $(OP)_2Cu^+$ over DNase I is the greater insensitivity of these nonenzymatic reagents to DNA sequence. Whereas DNase I footprints typically contain gaps due to poor cutting at certain sequences, especially AT-rich sequences, MPE·Fe(II) and $(OP)_2Cu^+$ yield footprints in which a band is observed at each position.

Hydroxyl radical footprinting (Tullius & Dombrowski, 1985) essentially disconnects the EDTA·Fe(II) moiety from MPE·Fe(II). In doing so, it allows one to generate ·OH in close proximity to DNA, without the requirement for prior intercalation.

Thus, whereas MPE·Fe(II) detects the ability of a protein to protect from intercalation into its binding site, EDTA·Fe(II) detects the ability of a protein to protect the carbohydrate portion of DNA against H atom extraction by hydroxyl radical. Because the reagent is given broad access to the protein–DNA complex in hydroxyl radical footprinting, it is able to detect subtle features of DNA structure, such as protein-induced distortion (O'Halloran et al., 1989), and alterations in intrinsic groove width. As with MPE·Fe(II), one limitation of the method is that the mechanistic basis of DNA nucleotide degradation by ·OH remains poorly understood. It is virtually certain that hydroxyl radical, a small diffusible species, abstracts predominantly the sugar $1'$ and $4'$ protons, which are located in the minor groove; therefore hydroxyl radical might reasonably be expected to serve solely as a minor groove probe. However, many proteins that bind exclusively in the major groove have been shown to protect DNA against hydroxyl radical cleavage, and as the origins of such effects become better understood, hydroxyl radical footprinting will become an even more powerful technique for investigation of protein–DNA interactions. Examples include studies of *trp* repressor (Carey, 1989) and nuclear hormone receptor (Chalepakis et al., 1988) complexes.

Diethyl pyrocarbonate (DEPC) footprinting involves the transfer of a carboxyethyl group ($-CO_2CH_2CH_3$) to N^7 of adenine (major product) and guanine (minor product) to form adducts that undergo spontaneous hydrolytic opening of the purine five-membered ring (Vincze et al., 1973). In DNA, the ring-opened products slowly undergo glycosidic bond cleavage leading to subsequent strand scission (Herr, 1985). However, the strand cleavage reaction can be accelerated by treatment of the adducted DNA with base, typically 1 M aqueous piperidine. The reaction of DEPC with canonical B-form DNA is so inefficient as to be of little practical value; however, Z-DNA (Herr, 1985), other noncanonical DNA structures (Lyamichev et al., 1989), and single-stranded regions of DNA (Furlong & Lilley, 1986) exhibit increased reactivity toward DEPC. Thus, DEPC can serve as an effective probe of protein-induced DNA distortion, and it has been used as such (Bateman & Paule, 1988). DEPC footprinting in the interference mode has been reported (Heuer & Hillen, 1988). In this case, a predenaturation step was employed to render the adenine residues in DNA susceptible to attack by DEPC, and then the DNA was renatured prior to incubation with the protein.

KMnO4 (Bayley & Jones, 1959) and *OsO4* footprinting (Beer et al., 1966) involve strong oxidizing agents that attack the 5,6 double bond of thymine and to a lesser extent cytosine. OsO_4 generates a 5,6-*cis*-diol, whereas the stronger oxidant $KMnO_4$ converts the initially formed diol to more highly oxidized products. Following treatment with $KMnO_4$ and OsO_4, the DNA is treated with base to effect strand scission at the modified positions. Much the same as with DEPC footprinting, B-form duplex DNA is not sufficiently reactive with $KMnO_4$ and OsO_4 to footprint most proteins; however, noncanonical structures such as Z-form or single-stranded regions in DNA will react effectively with the oxidants (Furlong et al., 1989). $KMnO_4$ has been used to carry out in vivo protection footprinting of the bacteriophage λ initiator protein (R. McMacken, personal communication) as well as in vitro studies of RNA polymerase complexes (Jeppesen & Nielsen, 1989).

$KMnO_4$ and OsO_4 have limited usefulness as interference footprinting reagents. In order to produce duplex DNA containing the products of these reagents, the duplex must be denatured, oxidized, and then renatured. Furthermore, the products themselves cause severe local disruption of duplex structure (Kao et al., 1993), suggesting that they can cause apparent interference at noncontact sites through a secondary mechanism.

Footprinting Methods That Probe Phosphate Contacts. As mentioned above, the establishment of electrostatic and hydrogen bonding contacts to the phosphate moieties in DNA is a major driving force for protein–DNA interactions. The determination of these contacts can provide a wealth of information on groove location of protein modules, groove crossovers, the compactness of a protein–DNA interface, and the actual size of a protein-binding site. The latter information can be especially useful in guiding the design of oligonucleotides for co-crystallization studies. Furthermore, patterns of backbone contacts can be useful in aligning a variety of DNA sites bound by the same protein, which in some cases is ambiguous on the basis of sequence information alone.

Ethylation interference footprinting is the technique used most widely to detect contacted phosphates (Siebenlist & Gilbert, 1980). The method derives its name from the fact that it measures the interference of protein binding caused by ethylation of the nonbridging phosphate oxygens. The ethylation reaction involves treating the DNA with ethylnitrosourea to produce a mixture of two diastereomeric ethyl phosphotriesters (Fig. 11-8). Following separation of protein-bound and unbound DNA, the positions of the phosphotriesters are determined by specific cleavage with base to yield a phosphodiester and alcohol. Any of the three phosphotriester substituents can undergo cleavage: Hydrolysis of the ethyl substituent yields native DNA with an intact backbone, whereas hydrolysis of the 5'- or 3'-substituents results in backbone cleavage. The products corresponding to the 5'- and 3'-OH's serving as leaving groups are formed as the major and minor products, respectively, and these possess different electrophoretic mobilities; hence at each position a major and minor cleavage band is observed.

Because phosphate contacts are often stereospecific with respect to the nonbridging oxygens, each of the two diastereomeric ethyl phosphotriesters presents a different functional group (either a P=O or a P—OCH$_2$CH$_3$ group) to the protein, and hence the protein is likely to interact with each isomer in a different way. Despite this complication, the technique appears to generate reliable phosphate contact information, as judged by comparisons of ethylation interference footprints with X-ray structures (Bushman et al., 1985; Ellenberger et al., 1992). The conditions of time and temperature necessary to bring about this reaction are incompatible with proteins; hence, it is not possible at present to carry out ethylation protection footprinting.

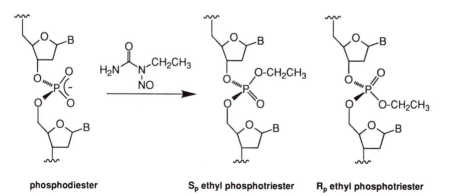

phosphodiester   S$_p$ ethyl phosphotriester   R$_p$ ethyl phosphotriester

Figure 11-8. Generation of the interference probe employed in ethylation interference footprinting.

Recent examples describing the use of ethylation interference to study phosphate contacts have involved Trp repressor (Carey, 1989), the mammalian transcription factor EBP1 (Clark et al., 1989), and *E. coli* DNA photolyase (Baer & Sancar, 1993). Despite the power of the ethylation interference footprinting experiment, it has not been as widely used as, for example, DNase I footprinting or DMS footprinting (see below). This situation may result from the technical difficulty of the technique, which arises primarily from the sometimes capricious nature of the ethylation reaction. This limitation may eventually be overcome by development of reagents that will modify the DNA phosphate oxygens selectively and under mild, easily controlled conditions.

Uranyl photofootprinting, a recently described procedure (Jeppesen & Nielsen, 1989), may offer a convenient alternative to ethylation interference footprinting for the determination of contact phosphates. Exposure of a protein–DNA complex to uranyl(IV) ion, UO_2^{2+}, results in binding of the ion to phosphates, except those that are in close contact with a protein. Irradiation with long-wavelength ultraviolet light (300–420 nm) converts the UO_2^{2+} to a strongly oxidizing excited-state species, which cleaves DNA through an unknown mechanism. The mechanism is believed to involve oxidation of the carbohydrate moiety that is proximal to the bound UO_2^{2+}. The one cautionary note is that this new technique has not yet seen use in analysis of protein–DNA complexes, although it has proven effective in studies of other ligands bound to DNA. In cases reported so far (Nielsen et al., 1988), uranyl photofootprinting and ethylation interference footprinting yield analogous results for the same ligand–DNA complex.

Although uranyl photofootptinting is procedurally characterized as a protection method, it is unlike all others in that the probe reagent, UO_2^{2+}, is allowed to equilibrate with the ligand–DNA complex, during which time the system presumably selects thermodynamically for those sites at which UO_2^{2+} interferes with binding of the ligand. Hence, in terms of interpretation, uranyl photofootprinting is more akin to an interference experiment.

Footprinting Methods That Detect Base Contacts. Even though all intermolecular contacts are arguably sequence-specific, those involving bases appear to provide the principal basis for sequence discrimination. For this reason, methods that detect contacted bases are the most widely practiced of all footprinting methods.

Methylation protection and methylation interference footprinting are extensions of chemical DNA sequencing technology, and they rely on the selective alkylation of DNA by dimethylsulfate; hence, they are sometimes referred to as *dimethylsulfate (DMS) footprinting* (Carey, 1989; Gronostajski et al., 1985; Heuer & Hillen, 1988). Treatment of DNA with S_N2-type alkylating agents such as DMS result in a variety of adducts, of which the predominant species are *N*7-methylguanosine (m^7G) and *N*3-methyladenosine (m^3A). The ratio of m^7G to m^3A is on the order of 5–10:1, depending on the sequence. Methylation of G at N7 blocks access to the major groove acceptor sites of guanine, N7 through bonding and O6 through steric congestion, and hence disrupts any possible major groove interactions with G (see Fig. 11-2). In the same way, interaction of a major groove protein with G is expected to protect it against attack by DMS. The methyl group of m^7G is located in a position analogous to that of thymine and therefore is not expected to affect DNA secondary structure; this hypothesis has been supported by NMR studies on m^7G-containing oligonucleotides (Ezaz-Nikpay and Verdine, 1994). The less prevalent adduct m^3A bears a methyl group blocking the principal minor groove acceptor site of adenine; thus, methylation is expected to prevent binding of a protein to A in the minor groove and vice versa. However, for reasons that are not well understood, some proteins that are known to bind exclusively in the major groove show interference by m^3A (Ellenberger et al., 1992). Thus, although m^3A interference is often used to argue in favor of minor groove binding, in the absence of other data it cannot be considered conclusive.

Both m^7G and m^3A are thermally labile, undergoing spontaneous hydrolysis of the glycosidic bond to yield a free base and an abasic site. The abasic site is also labile, undergoing β-elimination to effect strand scission. However, the time scale of these spontaneous events is too slow to be practical in a footprinting experiment, and so aqueous piperidine is used to effect the degradation of m^7G and m^3A and their attached sugars, leaving behind only the pendant phosphate groups attached to the 5' and 3' cleavage products.

Missing contact footprinting (Brunelle & Schleif, 1987) (Fig. 11-9) is an interference method that relies on Maxam–Gilbert chemistry to generate lesions at the positions occupied by any of the four bases in DNA. Guanosine and adenosine are depurinated by treatment with formic acid, and cytosine and thymine are degraded to the 1'-ureide using hydrazine. *Missing nucleoside footprinting* (Hayes & Tullius, 1989) (Fig. 11-9) is also an interference method, but in this case hydroxyl radical, generated in much the same way as in hydroxyl radical footprinting, is used to generate anucleosidic sites in DNA.

Both of these methods have the advantages of permitting analysis of all four bases in DNA and of being operationally straightforward. On the other hand, both methods

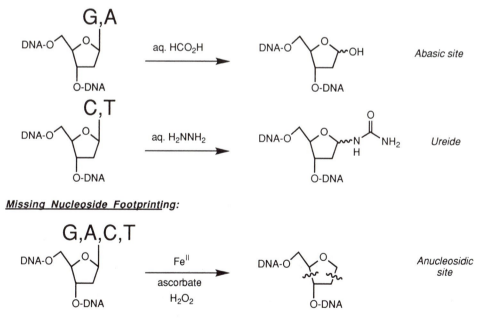

Figure 11-9. Generation of the interference probe employed in missing contact and missing nucleoside footprinting.

have the distinct disadvantage of generating lesions that substantially perturb duplex DNA structure (Kalnik et al., 1989). Such distortion is likely to cause interference outside the immediate position occupied by the lesion, thereby substantially diminishing the resolution of contact information. Indeed, comparison of results from missing contact (Brunelle and Schleif, 1987) and missing nucleoside (Hayes & Tullius, 1989) analysis of a lambda repressor–operator complex with the high-resolution X-ray structure (Jordan & Pabo, 1988) revealed the presence of substantial interference at sites not contacted by the protein. An added disadvantage is that loss of an entire base precludes one from assigning a contact locus to the major or minor groove—the base is missing from both grooves.

Template-Directed Interference (TDI) Footprinting

The optimal footprinting method for determining base contacts should possess a number of characteristics, including the following: (1) The method should be capable of assaying contacts to all four bases in both grooves, (2) the structure of the interfering probe moiety should be known, (3) the interference probe should minimally perturb DNA secondary structure relative to its natural counterpart, and (4) the method should be operationally simple. In one case or another, all of the methods described above fall short of these goals. The underlying cause of this deficiency stems in all cases from the need to rely on reagents that should be capable of selectively modifying contact functionality in DNA without affecting its conformation. To appreciate the difficulty inherent in such an approach, consider the case of the vinyl —CH_3 of T. Although this is one of the most important contact functions used by proteins, it is also one of the least reactive groups on the major groove surface. How could one expect to generate a reagent that would modify the vinyl —CH_3 of T without, for example, attacking the hydrocarbon portion of the DNA backbone? Several years ago, when we first began considering this problem, we concluded that footprinting approaches that rely on reagent control are fundamentally incapable of being generalized. TDI footprinting was therefore developed as a potential solution to this problem (Hayashibara & Verdine, 1991, 1992; Mascareñas et al., 1993). To circumvent the difficulty of achieving specific DNA modification through reagent-based approached, TDI footprinting relies upon the design and synthesis of modified monomers that are incorporated into DNA enzymatically.

TDI footprinting analogues (Fig. 11-10) are designed to (1) possess alterations in prospective contact functionality that will disrupt interactions with a protein, (2) cause the least possible perturbation of duplex DNA structure, and (3) permit selective DNA backbone cleavage so as to detect the presence of the analogue.

The technique is illustrated in Fig. 11-11. Asymmetric PCR, in which an excess of one primer is present, is used to generate a single-stranded DNA *template* containing a binding sequence for the protein of interest. A ^{32}P-end-labeled oligonucleotide *primer* is *annealed* through base-pairing to the template to generate a *primed template*. This primed template is mixed together with the four naturally occurring monomer units for enzymatic DNA synthesis: 2'-deoxyadenosine 5'-triphosphate (dATP), dCTP, dGTP, and dTTP. In addition, a TDI footprinting analogue, present as the corresponding 5'-triphosphate, is included in the reaction mixture. A ratio of the natural triphosphate (dNTP) to the TDI analogue (dN*TP) is used that has been predetermined empirically to result in one analogue incorporation event per extension to full-length duplex DNA;

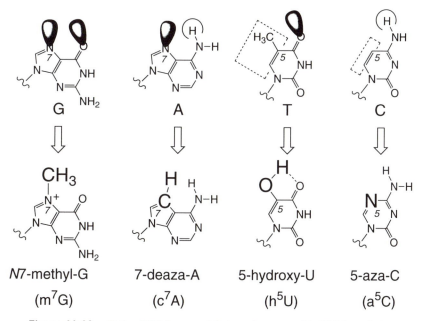

Figure 11-10. Native DNA bases and their analogues used in TDI footprinting.

this ratio varies among analogues according to their acceptability by the DNA polymerase enzyme. The DNA synthesis reaction is initiated by addition of a DNA polymerase, usually a mutant form of T7 DNA polymerase that is commercially available. To footprint all four bases in one experiment, four parallel extensions can be carried out, each with a different analogue. The product of the extension is a pool of duplex DNA in which one strand has a particular DNA base statistically substituted by the analogue. The remainder of the procedure is analogous to that described above for methylation interference footprinting: The pool of modified DNA is incubated with protein, the protein-bound DNA is then separated from unbound DNA, and the two DNA fractions are recovered from the gel and subjected to chemical cleavage. The cleaved fragments are then separated on a high-resolution DNA sequencing gel and visualized by autoradiography.

The analogue used for TDI footprinting of guanines, 7-methyl-2′-deoxyguanosine, is the same probe as that used in methylation interference footprinting. However, the pools of modified DNA produced in the two experiments are not identical: Whereas the pool of modified DNA generated by enzymatic incorporation of m^7dG into DNA contains only m^7dG, the corresponding pool generated by methylation of DNA contains m^7dG, m^3dA and a host of other minor products. Thus, TDI footprints of guanines (TDI-G footprints) tend to be "cleaner" than the corresponding methylation interference footprints (Hayashibara & Verdine, 1991).

The TDI-C footprinting analogue, 5-aza-2′-deoxycytidine (a^5dC), was the second reported (Hayashibara & Verdine, 1992). In this analogue, the hydrophobic 5-C of cytosine has been changed to a hydrophilic N. Evidence indicates that a^5dC can interfere not only with nonpolar contacts to the 5,6 double bond, but also with hydrogen bonding contacts to the 4-NH_2 (Hayashibara & Verdine, 1991). The close isosterism of a^5dC with C makes it unlikely that the analogue perturbs DNA secondary structure; this hy-

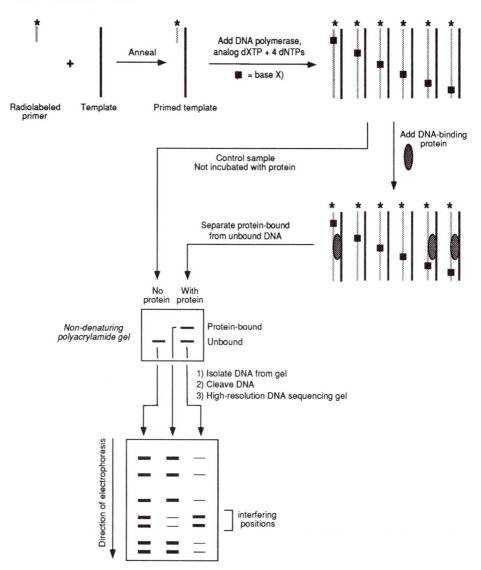

Figure 11-11. Schematic illustration of TDI footprinting. In this experiment, no G + A lane is required, since the footprinting shows a band at all positions occupied by one of the four bases (the base that is substituted by the analogue) and is, therefore, internally calibrated. Asterisk denotes radioactive end-label. Reprinted from Hayashibara, 1993, with permission.

pothesis is supported by the finding that noncontact a^5dC residues paired to contact G residues or situated on the 5'- or 3'-side of contact bases fail to cause interference (Hayashibara & Verdine, 1992; Larson et al., 1995).

The TDI-T analogue is 5-hydroxy-2'-deoxyuridine (h^5dU) (Mascareñas et al., 1993). In place of the hydrophobic vinyl —CH_3 of T, h^5dU possesses a hydrophilic —OH group, which is expected to disrupt nonpolar contacts ordinarily made to T. In addition, the —OH group may internally hydrogen bond to the C4 C=O, thereby competing for hydrogen bonding of the carbonyl to a intermolecular donor. Because h^{5}dU is isosteric with dT, the analogue is expected to cause little perturbation of DNA secondary structure. On the other hand, replacement of the vinyl —CH$_3$ with —OH profoundly affects the nucleophilicity of the 5,6 double bond: h^{5}dU residues in duplex DNA are at least 10^4-fold more reactive toward oxidation by KMnO$_4$ than dT residues. Hence, treatment with dilute KMnO$_4$ followed by aqueous piperidine can be used for specific detection of h^{5}dU residues in DNA.

As of this writing, a chemically cleavable analogue of dA suitable for use in TDI-A footprinting has not yet been developed. However, a reversed-sense TDI experiment has been devised for the footprinting of contact dA residues (Hayashibara, 1993). In this experiment, the analogue, 7-deaza-2'-deoxyadenosine (c^{7}dA), is *less* susceptible to cleavage by K$_2$PdCl$_4$ than is dA (Iverson & Dervan, 1987), because the mechanism of cleavage requires coordination to the N atom at position 7 that is missing in c^{7}dA. Hence, the sense of footprints using c^{7}dA is reversed relative to an ordinary TDI footprint using a cleavable analogue: Interference is evident as a band that is enriched in the bound fraction (enriched in dA versus c^{7}dA) and depleted in the unbound fraction (enriched in c^{7}dA versus dA). A significant drawback of the reversed-sense TDI experiment is that interference becomes difficult to detect when more than two interfering adenines are present in a strand, because the fractional decrease in band intensity at interfering positions becomes difficult to distinguish (Hayashibara, 1993). Notwithstanding this drawback, reversed-sense TDI-A footprinting has proven useful in footprinting several protein–DNA complexes (Hayashibara, 1993).

Future developments in TDI footprinting will include (1) extension of the method to include forward-sense footprinting of dA residues in the major groove and (2) preparation of analogues suitable for TDI footprinting of minor groove contacts.

Methods Other Than Footprinting for Determination of Base Contacts

Thus far we have focused our discussion on the structural basis for protein–DNA interactions and footprinting methods to detect these interactions. Although a comprehensive discussion of the other "chemical" methods for determining protein–DNA contacts is beyond the scope of this review, several approaches are particularly noteworthy.

Synthetic Nucleoside Analogues. One of the most powerful approaches involves site-specific incorporation of nucleoside analogues through solid-phase oligonucleotide synthesis; the effect of the modification on the equilibrium dissociation constant is then measured. The advantages of this method are that it allows very detailed contact analysis and provides a rigorous energetic value for the magnitude of the interference. The disadvantage of this method is that it is very labor-intensive, requiring an oligonucleotide to be synthesized for each site analyzed. For these reasons, the modified oligonucleotide approach is used most effectively as a secondary contact analysis for issues such as the energetic value of a particular interaction as inferred from the cost of perturbing that contact. Examples of base analogues used in these experiments include purine (Mazzarelli et al., 1992; Newman et al., 1990), 7-deazaadenine (Mazzarelli et al., 1992; Newman et al., 1990), 3-deazaadenine (Newman et al., 1990), 7-deazaguanine (Seela & Driller, 1986), 2-aminopurine (Aiken et al., 1991), 2,6-diaminopurine (Aiken et al., 1991), 5-methyl-2-pyrimidinone (Mazzarelli et al., 1992; Newman et al., 1990), 4-thiothymine (Newman et al., 1990), 2-thiothymine (Newman et al., 1990), uracil (Aiken et al., 1991; Mazzarelli et al., 1992; Newman et al., 1990), N6-methyladenine (Larson et al., 1995), N4-methyl-cytosine (Larson et al., 1995), 8-oxoguanine (Larson et al., 1995), and hypoxanthine (Larson et al., 1995; McLaughlin et al., 1987).

PCR Random Site Selection. Another method that deserves particular mention is a randomized site-selection scheme based on the polymerase chain reaction (PCR) (Blackwell & Weintraub, 1990; Blackwell et al., 1990; Thiesen & Bach, 1990). In this method, a random pool of DNA sequences is incubated with a protein, and the small subset of sequences that binds the protein is separated from incompetent sequences. This "enriched pool" is amplified by PCR and bound to the protein at higher stringency. After several iterations of this process, the pool is cloned into bacteria and individual members are sequenced. The analysis leads to the generation of a so-called consensus binding sequence that is determined by purely thermodynamic selection. The random site selection procedure reveals those positions that exhibit a strong sequence selection, but does not reveal whether the protein interacts *directly* with a particular base pair; in the same way the method does not allow one to distinguish which base in a base pair is contacted nor in which groove the contact takes place. Despite these limitations, site selection is an extremely powerful method for deducing the preferred DNA-binding site of a protein.

Photocrosslinking Methods for the Determination of Contact Amino Acids

Whereas much effort has been expended on development of chemical methods for identifying contacted DNA bases, much less progress has been made toward methods for determination of the amino acid contact partners. Several recent reports suggest that this situation is beginning to change. Matthews and coworkers were the first to report identification of contact amino acids through site-specific crosslinking to the photoactive nucleoside 5-bromo-2′-deoxyuridine (Allen et al., 1991; Wick & Matthews, 1991). Several subsequent reports have reinforced the utility of BrdU crosslinking (Blatter et al., 1992; Ebright, 1991). A significant drawback of this method, generally poor yields of the crosslinked product, has recently been overcome by use of laser photolysis in conjunction with 5-iodo-2′-deoxyuridine (IdU) (Willis et al., 1993). Chemistry for nitrene-based crosslinking to amino acid residues has been developed, through methodology that permits site-specific incorporation of 8-azido-2′-deoxyadenosine (N_3dA) into oligonucleotides (Liu & Verdine, unpublished results). Notwithstanding these advances, much progress remains to be made with regard to generalizing crosslinking approaches and making them routinely applicable in the biochemical laboratory.

Summary

In this review we have discussed the progress, problems, and promise of chemical methods for analysis of protein–DNA interactions. Although much progress has been made toward understanding molecular recognition in these macromolecular complexes, many significant questions remain. Gaining the answers to these questions represents a formidable challenge of immense significance for the synthetic, physical, and structural chemist.

Antisense/Antigene Oligonucleotides

Paul S. Miller

The genetic information encoded in double-stranded DNA and transcribed into single-stranded mRNA is an attractive target for biochemical manipulation and potential therapeutic intervention. A variety of chemical reagents, both natural and synthetic, are capable of reacting with DNA or RNA. Such reaction can lead to inactivation or destruction of these molecules. This mode of action forms the basis for many of the chemotherapeutic agents used today.

Reactions occurring with DNA can also lead to mutation of the genetic information. Whether such mutations are harmful or beneficial to the organism harboring them is not readily controllable. Recently it has become possible to introduce new genetic information into cells via the agency of retroviruses. This technique, which has been termed *gene therapy*, opens up the possibility of introducing normal, functioning genes to replace defective, nonfunctioning genes that arise from genetic mutations. The therapeutic potential of this technique is only now beginning to be explored.

In addition to replacing defective genes, it would be desirable to have techniques available that allow selective prevention of expression of unwanted genes. For example, the ability to suppress viral genes without simultaneously affecting the expression of host cellular genes could lead to the development of highly effective antiviral agents. Such selective inhibition is theoretically possible by taking advantage of the known selective hydrogen bonding interactions which take place between complementary bases of nucleic acids. Using these base–base interactions, it has been possible to design short nucleic acid molecules or oligonucleotides, which can selectively bind to either RNA or DNA molecules. Oligonucleotides that interact with single-stranded RNA have been given the name antisense oligonucleotides, whereas those which interact with double-stranded DNA are called *antigene oligonucleotides*.

A number of monographs (Cohen, 1989; Crooke & Lebleu, 1993; Erickson & Izant, 1992; Murray, 1992; Wickstrom, 1991) and reviews (Cohen, 1991; Cook, 1991; Crooke, 1992; Dolnick, 1990; Ghosh & Cohen, 1991; Hélène & Toulme, 1990; Malcolm et al., 1992; Miller & Ts'o, 1988; Milligan et al., 1993; Stein, 1992; Tidd, 1990; Uhlmann & Peyman, 1990) describing various aspects of antisense and antigene oligonucleotides have been published recently. This chapter describes the principles behind the design of antisense and antigene oligonucleotides. In addition, it explores the development of oligonucleotide analogues which have been designed with the goal of eventual therapeutic use in mind. Finally, the biological activities of some selected antisense and antigene oligonucleotides are discussed.

Antisense Oligonucleotides

An antisense oligonucleotide is a chemically synthesized oligonucleotide or an oligonucleotide analogue whose sequence is complementary to a specific nucleotide sequence within a target mRNA. The antisense oligomer is designed to bind to the mRNA sequence via normal Watson–Crick hydrogen bonding interactions (Chapter 1). As a consequence of binding, mRNA function or expression is disrupted. In this manner gene expression is controlled at the mRNA level by the antisense oligonucleotide.

RNA as a Target for Antisense Oligonucleotides. In order to understand the principles behind the design of antisense oligonucleotides, it is necessary to understand some of the basic elements of mRNA structure and biochemistry. Messenger RNA is a single-stranded ribonucleic acid containing the genetic information that codes for the amino acid sequence of its corresponding protein. Cellular messenger RNA is transcribed from DNA in the nucleus of eukaryotic cells. In addition to coding sequences called *exons*, most eukaryotic mRNA transcripts contain extra nucleotide sequences called *introns* which do not appear in the final cytoplasmic version of the mRNA. In the nucleus the introns are excised from the RNA transcripts and the exons are joined by a process called *splicing* to create, after additional processing steps, the mature mRNA. The mature mRNA, which is transported to the cytoplasm, usually contains the following elements: a 5'-cap region; an untranslated leader sequence; an AUG initiation codon which signals the ribosomes to start protein synthesis; the coding region consisting of triplet sequences of nucleotides or codons, each of which codes for a specific amino acid; a UAA, UGA or UAG termination codon, which signals the ribosomes to stop protein synthesis; and finally a 3'-untranslated sequence of nucleotides terminating with a polyadenylate tract. The processing steps and basic structure of eukayrotic mRNA are shown schematically in Fig. 12-1.

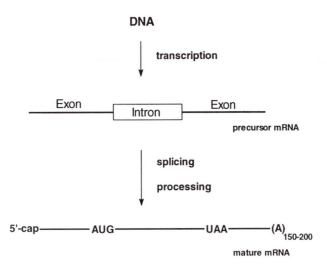

Figure 12-1. Structure and biosynthesis of messenger RNA.

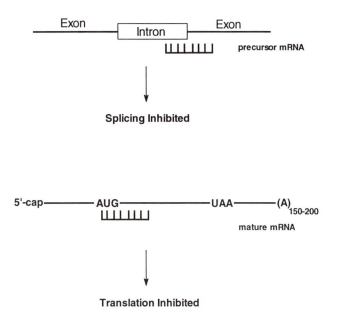

Figure 12-2. Inhibition of mRNA splicing or translation by antisense oligonucleotides.

In principle, mRNA function can be disrupted either during biosynthesis/ processing steps or by preventing the mRNA from being translated. This is shown schematically in Fig. 12-2. Antisense oligonucleotides designed to disrupt biosynthesis/processing are usually targeted to the intron/exon regions, or so-called splice junctions, of the precursor mRNA. The oligomers may be complementary to either the exon/intron interface (donor splice junction) or the intron/exon interface (acceptor splice junction). In either case the oligomers are designed to interfere with splicing of the message by blocking access of the spliceosome to the splice junction. Because heterogeneous nuclear RNA undergoes rapid turnover in the nucleus, it is believed that mRNA that is not properly spliced is degraded in the nucleus.

Most antisense oligonucleotides have been designed to prevent translation of mRNA. There are a number of sites in the mRNA that could logically be targeted by antisense oligomers. Oligomers that bind to the 5'-cap region of the mRNA would be expected to interfere with initial binding of the 40S subunit of the ribosome, the first step in ribosome assembly on the message. Oligomers complementary to the initiation codon region of the mRNA would be expected to interfere with assembly of the complete ribosome by preventing joining of the 60S subunit to the 40S subunit after this subunit has scanned down the 5'-untranslated region to the initiation codon (Kozak, 1989). Finally, oligomers complementary to the coding region of the mRNA would be expected to interfere with the elongation step of protein synthesis as the assembled ribosome translates the message.

Ribonuclease H. Antisense oligonucleotides were originally believed to inhibit mRNA function by physically blocking interactions between the ribosome and mRNA in a process called *hybridization arrest of translation* (Paterson et al., 1977). It was later realized, however, that ribosomes possess "unwinding" activities capable of disrupting mRNA secondary structure and could therefore destabilize duplexes formed between mRNA and complementary DNA during the elongation step (Liebhaber et al., 1984; Shakin & Liebhaber, 1986). Such unwinding activity would be expected to render ineffective antisense oligomers complementary to the mRNA coding region. Nevertheless, in certain systems such oligomers were found to be effective inhibitors of translation. Closer examination revealed that an enzymatic activity, RNase H, was responsible for the inhibitory activity (Minshull & Hunt, 1986; Walder & Walder, 1988). RNase H cleaves the RNA strand of duplex hybrids formed between the RNA and DNA. It appears to be found in many, if not most, eukaryotic and bacterial cells; in the case of eukaryotic cells it is believed to be involved in the degradation of RNA primers used during DNA replication.

When oligodeoxyribonucleotides are hybridized to single-stranded RNA, RNase H will selectively cleave the RNA strand at the oligonucleotide-binding site (Minshull & Hunt, 1992). The oligonucleotide itself is not cleaved and is free to bind to another RNA molecule. Thus an essentially catalytic cleavage reaction can obtain in which a few oligodeoxyribonucleotides can give rise to the cleavage of a large number of RNA molecules.

Targeting Strategies and the Role of Secondary Structure. Not all oligonucleotides can form hybrids with RNA that can be cleaved by RNase H. Those which contain a deoxyribose sugar phosphate backbone are effective. However, as discussed in a later section, oligonucleotides with modified backbones usually do not produce cleavable hybrids. Thus, it becomes quite important to carefully consider the targeting site in the mRNA sequence. It appears that in the absence of RNase H, the most effective target sites in mRNA are those in the 5'-cap region or those at or near the initiation codon region (Goodchild et al., 1988a,b; Liebhaber et al., 1984). In both cases, oligonucleotides bound to these regions are believed to be capable of preventing ribosome binding to the mRNA. Thus the oligomers are not readily displaced from the mRNA, nor is it necessary to rely on the activity of RNase H to cause inhibition. Less work has been done using oligomers directed to the splice sites of mRNA, although such oligomers have been found to be effective when targeted to these sites (Agrawal et al., 1991; Becker et al., 1989; Kulka et al., 1989; Morrison, 1991; Smith et al., 1986).

Although mRNA is a single-stranded nucleic acid, it is important to recognize that the molecule is not simply a linear sequence of nucleotides devoid of any secondary structure. Most mRNAs contain many internal sequences that are complementary or partially complementary to other internal regions of the molecule. This situation leads to the formation of a considerable amount of secondary structure within the molecule, which manifests itself in the form of double-stranded helical stem structures that terminate with a single-stranded loop region. Such structures are called *hairpins*. One also finds double-stranded helices that contain bulges and internal loops. Depending upon the sequence of the particular mRNA, a considerable portion of the nucleotides in the mRNA can be involved in intramolecular base-pairing interactions. Tertiary folding can also result from interactions between nucleotides in different stem/loop regions. In addition, eukaryotic mRNAs are bound by proteins. These structural features and interactions with proteins would be expected to affect the ability of antisense oligonucleotides to bind to their complementary target sequences.

Little is known about the secondary or tertiary structures of most mRNAs. It is possible to carry out structure mapping studies on RNA molecules using a variety of chemical and enzymatic reagents (Ehresmann et al., 1987). However, such studies are difficult and require substantial quantities of RNA. Algorithms have been developed which use thermodynamic parameters to predict minimum energy secondary structures based on the nucleotide sequence of the RNA molecule (Jaeger et al., 1989). Such computational methods have been extended to attempts to predict tertiary interactions in RNA as well (Hubbard & Hearst, 1991). Although these methods have been used to successfully predict ribosomal RNA structure, their general applicability to predict mRNA structure accurately remains largely untested.

It has been demonstrated that the structure of the binding site could have profound effects on the ability of an antisense oligonucleotide to interact with its target (Chastain & Tinoco, 1993; Ecker et al., 1992; Hjalt & Wagner, 1992; Lima et al., 1992). If the binding site for a particular oligomer is part of a stem in the mRNA, it is unlikely that the oligomer will be able to bind efficiently to this sequence. Oligomers which are targeted to loop regions of RNA hairpins show different binding affinities depending upon the side of the loop to which the oligomer is targeted. In at least one study, however, attempts to correlate predicted secondary structure with the inhibitory effectiveness of antisense oligomers was not successful (Stull et al., 1992). This may have been because the predicted secondary structures were not accurate representations of the real structures, or because of other factors such as the ability of the oligomer to successfully compete with secondary structures plays a role. There does appear to be a better correlation between the inhibitory effectiveness of an antisense oligomer and the calculated free energy of duplex formation between the oligomer and its complementary mRNA target sequence (Monia et al., 1992; Stull et al., 1992; Wagner et al., 1993). This calculation makes no assumption about the structure, or lack thereof, of the binding site for the oligomer.

In designing antisense oligonucleotides, a major goal is to achieve a high degree of selectivity in the interaction between the oligomer and its designated target mRNA. In other words, the antisense oligomer should be capable of binding only to its target mRNA and to no other mRNA. What length oligomer is required optimally to achieve such specificity? Based on statistical arguments it can be shown that, depending on its base composition, an oligonucleotide 15–19 nucleotides in length will be uniquely complementary to a single nucleotide sequence within the human genome which contains approximately 4×10^9 base pairs (Hélène & Toulme, 1989). If one takes into consideration the percentage of the genome, which is actually expressed as mRNA (0.5%), the required unique length decreases to 11–15 nucleotides. Thus, at least statistically, an oligonucleotide 15 nucleotides in length should be able to bind selectively to a single mRNA species within a human cell. Depending upon the nucleotide sequence, such selectivity may require that the oligomer be capable of discriminating a single mismatch when binding to the mRNA. Such single-base mismatch discrimination can be achieved under in vitro conditions, where it is possible to adjust hybridization conditions such as salt concentration and temperature to obtain maximum binding and selectivity. However, such manipulations are not possible in the cellular environment in which antisense oligonucleotides would be required to function therapeutically. Nevertheless, there exist several examples of antisense oligomers which, under physiological conditions, are apparently capable of selectively interacting with mRNA targets which differ by a single base (Chang et al., 1991; Monia et al., 1992; Saison-Behmoaras et al., 1991).

Antisense RNA and Ribozymes. Chemically synthesized oligonucleotides and many oligonucleotide analogues are readily prepared on automated DNA synthesizers and thus are attractive candidates for use as antisense agents. However, other forms of the antisense nucleic acids have also been developed. These include antisense RNA and antisense ribozymes.

Antisense RNA is prepared in situ by inserting a cloned form of the target mRNA gene into a plasmid in an orientation such that the coding strand of the gene is copied (Izant, 1992; Murray & Crockett, 1992). This produces an RNA transcript that is complementary to the coding strand, and thus is antisense to the mRNA. Antisense RNA can be produced from the plasmid by in vitro transcription. In cell culture experiments the antisense RNA can be introduced into the cells by microinjection. Alternatively, cells can be transfected with the plasmid carrying the "antisense gene." In this case the antisense RNA is produced by the cells in either a transient manner or, if the antisense gene has become incorporated into the host cell genome, as a permanently expressed RNA (Nellen et al, 1992; van Blokland et al., 1993).

Ribozymes are RNA molecules that are capable of catalyzing RNA cleavage and splicing reactions (Altman, 1990; Cech, 1990; Cech et al., 1992; Guerrier-Takada et al., 1983; Kruger et al., 1982; Pyle, 1993) (Chapter 13). Antisense ribozymes have been constructed that retain the ability to catalyze the cleavage reaction and which also contain binding sequences capable of interacting with complementary sequences in mRNA (Chowrira & Burke, 1992; Goodchild, 1992; Hendry et al., 1992; Kikuchi & Sasaki, 1991; Odai et al., 1990; Paolella et al., 1992; Parker et al., 1992; Uhlenbeck, 1993). Upon binding to the mRNA target, the catalytic core of the antisense ribozyme is positioned for cleavage of the mRNA. Antisense ribozymes are attractive from the standpoint of their potential ability to cleave a large number of target mRNA molecules; however, much work needs to be done to improve their catalytic efficiency. Nevertheless, a number of laboratories have recently reported the successful use of antisense ribozymes in cell culture (Heidenreich & Eckstein, 1992; Homann, et al., 1993; Kashni-Sabet et al., 1992; Koizumi et al., 1992; Ojwang et al., 1992; Scanlon et al., 1991; Sexena & Ackerman, 1990; Sioud et al., 1992).

Antigene Oligonucleotides

Antisense oligonucleotides are designed to interact with and to inhibit the expression or function of mRNA within cells. Because inhibition occurs at the level of the mRNA molecule, control of gene expression is indirect. Inhibition requires the continued presence of the oligomer, and the possibility exists that the cell can simply produce increased levels of the target mRNA in order to overcome the inhibitory effect of the antisense oligomer. In theory, expression of a gene could be controlled directly at the DNA level by oligonucleotides which are capable of binding to double-stranded DNA. Such oligomers are called *antigene oligonucleotides*.

Triplex Formation: pyr·pur·pyr Motif. Oligonucleotides can bind to homopurine sequences in double-stranded DNA to form triple-stranded structures or triplexes (Dervan, 1989). Two basic binding motifs, the pyrimidine·purine·pyrimidine (pyr·pur·pyr) motif and the purine·purine·pyrimidine (pur·pur·pyr) motif, have been described for triplex formation. In this notation, the first base is in the third strand and the next two bases are base pairs in the duplex target molecule.

The structure of a triplex with a pyr·pur·pyr motif is shown in Fig. 12-3. Two types of base triads, the T·A·T triad and the C^+·G·C triad, allow formation of the triplex. The sugar phosphate backbone of the oligomer which lies in the major groove of the duplex has the same polarity as that of the homopurine strand to which it binds. Thymine bases of the third strand can form two Hoogsteen-type hydrogen bonds with the adenine of the A·T base pair. Cytosine bases of the third strand, however, must be protonated in order to form two hydrogen bonds with the guanine of the G·C base pair of the duplex (Howard et al., 1964; Lipsett, 1963; Thiele & Guschlbauer, 1971). Thus triplex formation involving cytosines occurs more readily at acidic pH values, usually between 5 and 7. It has been found that substitution of cytosine by 5-methylcytosine allows triplex formation to occur at pH values which are closer to those found under physiological conditions (Lee et al., 1984; Povsic & Dervan, 1989). It is also important to note that the bases in the T·A·T and C$^+$·G·C triads are isomorphous. That is the bases in each triad have the same relative orientations and positions in the triplex, thus allowing formation of a regular helical structure.

Although pyr·pur·pyr triplexes are quite stable at acidic pH, the stability markedly decreases at pH 7 and above. This is particularly true when the triplex contains multiple deoxycytidines or 5-methyldeoxycytidines. The effect of pH on triplex stability is seen by examining the melting temperature, T_m, of the third strand of the triplex. The melting temperature is measured by observing the change in absorbance at 260 nm as a function of temperature as shown in Fig. 12-4A. As the complex denatures, the absorbance of the solution increases, due to loss of hydrogen bonding interactions and unstacking of the bases in the oligomer strands. In the case of triplexes, two tran-

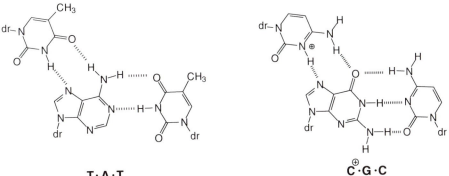

T·A·T   C̊$^{\oplus}$·G·C

Figure 12-3.   Base triads in the pyrimidine·purine·pyrimidine triplex motif. The strand orientation is shown in the schematic below the base triads. The homopurine strand is represented by the thickest line.

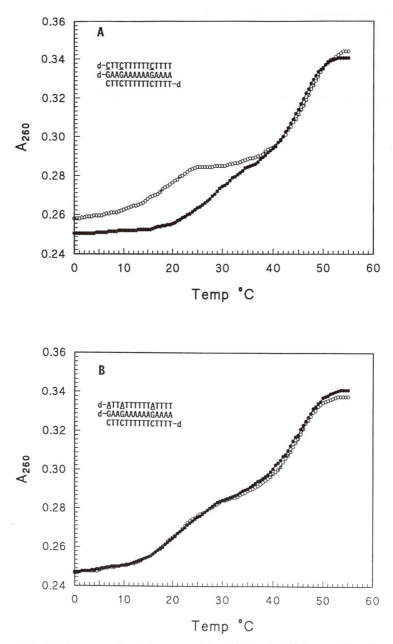

Figure 12-4. Melting curves for triplexes containing (**A**) 5-methyl-2'-deoxycytidine, C̲, or (**B**) 8-oxo-2'-deoxyadenosine, A̲, in the third strand. Solid circles and open circles were from measurements carried out at pH 7.0 8.0, respectively. Reprinted from Miller et al., 1992, with permission.

sitions are usually observed. The first transition corresponds to melting of the third strand, whereas the second transition corresponds to melting of the target duplex. The T_m of the third strand is defined as the midpoint in the first transition curve and represents the point where half of the third strand has dissociated from the triplex. As shown in Fig. 12-4A, the T_m for the third strand, d-CTTCTTTTTTCTTTT, is 28°C at pH 7.0 but decreases to 17°C at pH 8.0 (Miller et al., 1992).

Several base analogues have been developed that are capable of forming base triads with G·C base pairs without the necessity of base protonation. These analogues include pseudoisocytidine (Ono et al., 1991, 1992), 6-methyl-8-oxoadenine (Krawczyk et al., 1992), and 8-oxo-adenine (Davison & Johnsson, 1993; Jetter & Hobbs, 1993; Miller et al., 1992). The proposed hydrogen bonding schemes for triads formed by these bases are shown in Fig. 12-5A.

Pseudoisocytidine is a C-nucleoside which contains two hydrogen bond donor groups at the N2 and N3 positions. It can thus form two hydrogen bonds with the N7 and O6 of guanine. Unlike cytosine, base protonation is not required to form these hy-

A

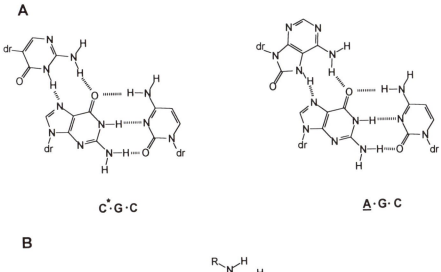

$$\overset{*}{C}\cdot G\cdot C \qquad\qquad \underline{A}\cdot G\cdot C$$

B

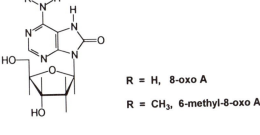

R = H, 8-oxo A

R = CH₃, 6-methyl-8-oxo A

1

Figure 12-5. (A) Base triad formation by pseudoisocytidine, C*, or 8-oxo-2'-deoxyadenosine, A, with G·C base pairs. (B) The structure and conformation of 8-oxo-2'-deoxyadenosine and its 6-methyl derivative.

drogen bonds. The bases 6-methyl-8-oxoadenine (**1**, R = CH$_3$) and 8-oxoadenine (**1**, R = H) exist in the keto tautomeric form. Furthermore in the nucleoside, the bases preferentially adopt the unusual *syn* conformation. In this conformation, which is shown in Fig. 12-5B, the hydrogens at the N6 and N7 positions can form hydrogen bonds with the O6 carbonyl oxygen and the N7 of the guanine base of the G·C base pair. As shown in Fig. 12-4B, when 8-oxoadenine replaces 5-methylcytosine in the third-strand oligomer, the T_m of the third strand is unchanged when the pH is increased from 7.0 to 8.0.

Triplex formation is not confined simply to third-strand oligodeoxyribonucleotides. Recent studies have shown that oligoribonucleotides and 2'-*O*-methylribonucleotides also participate in triplex formation. It appears that the backbone of the target duplex and the third-strand oligomer plays an important role in determining triplex stability. For the pyr·pur·pyr motif, the order of stability is D·*D*·D, R·*D*·D, R·*D*·R, D·*D*·R > R·*R*·D, R·*R*·R >> D·*R*·R, D·*R*·D (Han & Dervan, 1993; Roberts & Crothers, 1992). Here R indicates a ribose sugar backbone and D represents a 2'-deoxyribose backbone. The purine strand of the target duplex is shown in italics. Third-strand oligonucleotides which contain a 2'-*O*-methylribose sugar backbone have also been shown to form very stable triplexes with double-stranded DNA targets (Shimizu et al., 1992).

Triplex Formation: pur·pur·pyr Motif. The other common triplex motif is pur·pur·pyr, whose structure is shown in Fig. 12-6 (Beal & Dervan, 1991; Broitman et al., 1987; Chastain & Tinoco, 1992; Chen, 1991; Pilch et al., 1991; Radhakrishnan et al., 1991a,b). In this case, triplex formation involves A·A·T and G·G·C triads. Again, the sugar phosphate backbone of the oligomer lies in the major groove of the duplex. Unlike the pyr·pur·pyr-type triplex, however, the polarity of the sugar phosphate backbone appears to be antiparallel to that of the homopurine strand to which it binds. The adenine base of the third strand forms two hydrogen bonds with the adenine of the duplex. In a similar fashion, guanine binds via formation of two hydrogen bonds with guanine of the duplex. Because both adenine and guanine of the third strand have suitable hydrogen bond donor and acceptor groups, protonation of the bases is not required and thus triplex formation is relatively insensitive to pH conditions. Unlike the

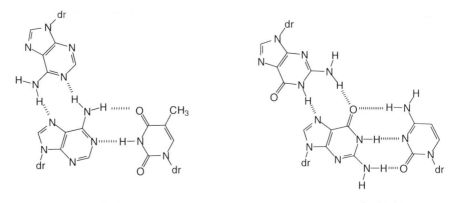

A·A·T **G·G·C**

Figure 12-6. Base triads in the purine·purine·pyrimidine triplex motif.

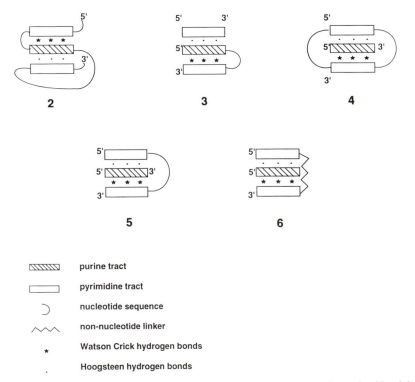

2   **3**   **4**

5   **6**

⬚⬚⬚⬚⬚	purine tract
▭	pyrimidine tract
⊃	nucleotide sequence
∿∿	non-nucleotide linker
*	Watson Crick hydrogen bonds
.	Hoogsteen hydrogen bonds

Figure 12-7. Schematic illustrations of intramolecular and bimolecular oligonucleotide triplexes.

pyr·pur·pyr-type triplex, however, the base triads of the pur·pur·pyr motif are not iso-morphous. A variation on this motif which utilizes T·A·T and G·G·G triads has also been described (Cooney et al., 1988).

Intramolecular and Bimolecular Triplexes. In addition to intermolecular triplexes that involve interaction between separate oligonucleotide strands, other types of triplexes have been observed. These are illustrated schematically in Fig. 12-7. Single-stranded oligonucleotides (**2**) which contain appropriately positioned tracts of pyrimidine and purine sequences can form intramolecular triplexes (Chen, 1991; Durand et al., 1992; Haner & Dervan, 1990; Jayasena & Johnston, 1992; Radhakrishnan & Patel, 1993; Radhakrishnan et al., 1992; Wang et al., 1992). Such intramolecular triplexes have proven particularly valuable in the study of triplex structure by nuclear magnetic res-onance spectroscopy.

Triplexes can be formed between two oligonucleotides. Oligopyrimidines can, for example, interact with the purine strand of the double-helical stems of hairpin oligonu-cleotides (**3**) (Manzini et al., 1990; Roberts & Crothers, 1992; Xodo et al., 1991). Bi-molecular triplex formation has also been observed in systems where the target strand consists of a homopurine tract and the binding strand contains two homopyrimidine tracts. In this case the homopyrimidine tracts of the binding strand interact through for-mation of both Watson–Crick and Hoogsteen hydrogen-bonding motifs. Examples of such binding strands include circular oligodeoxyribonucleotides (**4**) (Kool, 1991;

Prakash & Kool, 1991, 1992), stem-loop oligodeoxyribonucleotides (**5**) (Brossalina & Toulme, 1993; Souza & Kool, 1992; Xodo et al., 1990), and oligodeoxyribonucleotides in which the homopyrimidine tracts are joined by a non-nucleotide linker (**6**) (Giovannangeli et al., 1991; Salunkhe et al., 1992). Binding strands of this type may be valuable in targeting single-stranded regions of DNA or RNA molecules.

Extending the Range of Triplex Formation with Base Analogues. Both the pyr·pur·pyr and the pur·pur·pyr motifs are limited in the range of sequences that can be recognized, because the third-strand oligomer must interact with a homopurine sequence in the DNA target. Introduction of pyrimidines in the purine strand results in "mismatches" which tend to destabilize the triplex (Griffin & Dervan, 1989; Macaya et al., 1991; Mergny et al., 1991; Miller & Cushman, 1993; Radhakrishnan & Patel, 1993; Radhakrishnan et al., 1991a; Wang et al., 1992; Yoon et al., 1992). Triplex recognition could be much more universal if bases or base analogues were discovered which could interact with such pyrimidine interruptions. Natural bases can be used to a limited extent to interact with pyrimidine interruptions in the homopurine tract. The best-known example is the formation of a G·T·A triad, where G is in the third-strand oligomer (Griffin & Dervan, 1989; Mergny et al., 1991; Yoon et al., 1992). Based on nuclear magnetic resonance studies it appears that the N2 amino group of G forms a single hydrogen bond with the O4 carbonyl group of T of the T·A base pair (Radhakrishnan et al., 1991b; Radhakrishnan & Patel, 1993; Wang, et al., 1992). An additional hydrogen bond between the other N2 amino proton of G and the O4 carbonyl of T of a neighboring A·T base pair also contributes to stabilizing this triad.

Several base analogues have also been described that are capable of interacting with a pyrimidine interruption in a homopurine tract. These are shown in Fig. 12-8. The synthetic base 4-(3-benzamidophenyl)imidazole (**7**) binds selectively to C·G and T·A base pairs in a pur·pyr duplex (Griffin et al., 1992; Kiessling et al., 1992). Recent nuclear magnetic resonance (NMR) evidence suggests that this base interacts as a result

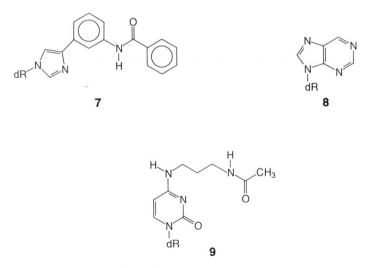

Figure 12-8. Triplex-forming base analogues.

of intercalation rather than direct hydrogen bonding with the base pairs of the target duplex (Koshlap et al., 1993). 2'-Deoxynebularine (**8**) binds to C·G base pairs and also interacts with A·T base pairs (Stilz & Dervan, 1993). Formation of a single hydrogen bond between the N1 of nebularine and the N4 amino group of C of the C·G base pair has been proposed. N^4-(3-Acetamidopropyl)deoxycytidine (**9**) has been found to selectively interact with a C·G base pair in a target duplex (Huang et al., 1993). In this case the side arm of the cytosine base may be able to reach across the major groove of the duplex and allow a hydrogen bond to form between the NH of the acetamido group and the O6 carbonyl oxygen of G. It appears likely that as new base analogues are developed and as further knowledge of the structures of triplexes becomes available, it will be possible to use third-strand oligomers to interact with an increasing variety of sequences in duplex DNA. This ability will increase the potential for using these oligomers to interact specifically with genes of interest.

In Vitro Activity of Antigene Oligonucleotides. Oligonucleotides capable of forming triple-stranded complexes with DNA have been investigated for their abilities to prevent protein binding to DNA and to prevent transcription of DNA in vitro. Oligodeoxyribopyrimidines containing C or 5-methyl C have been shown to be able to interfere with binding of the Sp1 transcription factor to the plasmid DNA (Maher et al., 1989). The oligomer can bind to a homopurine sequence inserted into and overlapping with an Sp1 transcription factor binding site in a mouse MT-1 promoter site cloned into a plasmid. This homopurine tract also overlapped a restriction site recognized by *Ava* I, *M. Taq* I and *Taq* restriction endonucleases. The oligopyrimidines were also effective at inhibiting cleavage at this site by the restriction endonucleases. Oligomers containing 5-methyl C were more effective inhibitors than those containing C, an observation which is consistent with the greater binding affinity of 5-methyl C-containing third-strand oligopyrimidines at physiological pH. Similar inhibition of restriction endonuclease cleavage of DNA by triplex-forming oligodeoxyribopyrimidines was observed by Francois et al. (1989).

The promoter regions of many eukaryotic genes contain homopurine tracts. These tracts provide sites for triplex formation by antigene oligonucleotides. Oligonucleotides targeted to these regions can prevent transcription of the genes in cell-free systems. For example, an oligomer targeted to the purine-rich tract located 115 base pairs upstream from the transcription origin of the human c-*myc* gene was shown by polyacrylamide mobility shift analysis to bind a DNA restriction fragment carrying this sequence. As a consequence of binding, transcription in a HeLa cell extract was inhibited by the oligonucleotide (Cooney et al., 1988). Similarly, oligodeoxyribopyrimidines have been targeted to a homopurine tract which overlapped the bacteriophage T7 RNA polymerase promoter region. These oligomers were shown to effectively inhibit transcription in a sequence-specific manner (Maher et al., 1992; Skoog & Maher, 1993).

Oligopyrimidines containing 8-oxo-6-methyl-2'-deoxyadenosine retard transcription elongation when targeted to a homopurine tract in a double-stranded DNA cassette, and as a consequence they produce truncated transcripts (Young et al., 1991). However, prolonged incubation with polymerase resulted in the synthesis of full-length transcripts. This result suggests that third-strand oligomers bound to regions of DNA which are undergoing transcription are incapable of inhibiting transcription permanently. Once the antigene oligomer dissociates from the DNA, transcription can proceed. Indeed the number of truncated transcripts increased when the antisense oligomer was equipped with a modified base which allowed the oligomer to form a covalent adduct with the homopurine target strand of the DNA.

As shown by Maher et al. (1989), antigene oligonucleotides can interfere with binding of transcription factors to DNA. Such oligomers are also effective inhibitors of in vitro transcription. Thus an oligodeoxypyrimidine 15 nucleotides in length targeted to a pur·pyr tract located upstream of the regulatory site of the interleukin-2 receptor α-chain gene inhibited NF-κB binding and was also shown to repress in vitro transcription of the gene (Grigoriev et al., 1992).

Oligonucleotide Analogues

Some of the first antisense experiments were performed using oligodeoxyribonucleotides. These were used to inhibit Rous sarcoma virus protein synthesis and replication both in vitro and in cell culture (Stephenson & Zamecnik, 1978; Zamecnik & Stephenson, 1978). Subsequently, numerous laboratories have demonstrated that antisense oligodeoxyribonucleotides can, under certain conditions, function successfully in cell culture experiments. Despite these successes, however, it was realized early on that the susceptibility of the phosphodiester internucleotide linkage of oligodeoxyribonucleotides would very likely be an impediment to the use of oligodeoxyribonucleotides as therapeutic agents. This prediction is based on the observation that the internucleotide phosphodiester linkage of these oligomers is susceptible to degradation by both exo- and endonucleases, particularly 3′-exonucleases (Elder et al., 1991; Shaw et al., 1991a; Tidd & Varenius, 1989). Nucleases are a ubiquitous component of serum and are found within the intracellular environment.

A number of strategies have been developed to circumvent this problem. Antisense oligodeoxyribonucleotides can be used in cell culture if the cells are grown in serum-free culture medium or in medium containing heat-inactivated serum. Heat inactivation reduces or eliminates nuclease activity. The oligomers can be further protected from exonuclease degradation by modifying the ends of the molecule with, for example, alkylamine groups (Orson et al., 1991). Such modification is successful because endonuclease activity is apparently very low or nonexistent in many cells. Although these methods permit experiments to be carried out in cell culture, it is unlikely such modified oligomers would survive for significant lengths of time in animals or humans.

Analogues Containing Modified Phosphodiester Linkages. Antisense oligonucleotide analogues have been designed whose internucleotide linkages are resistant to nuclease hydrolysis. Two types of analogues have been developed, those with phosphorous containing internucleotide linkages and those whose linkages lack phosphorous. Examples of the former type of analogue are shown in Fig. 12-9. These include oligonucleotide phosphorothioates (**10**), α-anomeric oligodeoxyribonucleotides (**11**), and oligonucleoside methylphosphonates (**12**).

Oligonucleotide Phosphorothioates. Oligodeoxyribonucleotide phosphorothioates contain internucleotide linkages in which one of the nonesterified oxygens of the phosphate group is replaced by sulfur. This substitution results in the creation of a new chiral center in the molecule. The phosphorothioate group can thus exist in either the Rp or Sp configuration. Oligonucleotide phosphorothioates are easily prepared on auto-

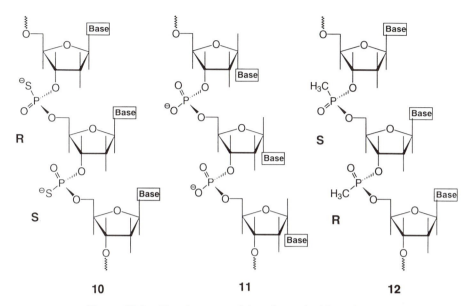

Figure 12-9. Phosphorus-containing oligonucleotide analogues.

mated DNA synthesizers by treatment of the support-bound, protected oligonucleotide phosphite or *H*-phosphonate intermediates with solutions containing elemental sulfur (Koziolkiewicz et al., 1986; Stec et al., 1984) or with 3*H*-1,2-benzodithiol-3-one 1,1-dioxide, the so-called Beaucage reagent (Iyer et al., 1990). Because the stereochemistry of the sulfurization cannot be controlled, the resulting oligomer contains 2^n diastereoisomers, where n is the number of phosphorothioate linkages in the oligomer. Thus if the oligomer is 15 nucleotides long and contains 14 internucleotide phosphorothioate linkages, it will consist of a mixture of 16,384 diastereoisomers. The two phosphorothioate configurations are shown in Fig. 12-9. Recently, oligonucleotide phosphorodithioates were prepared in which both nonesterified oxygens of the phosphodiester linkage are replaced by sulfur (Bjergärde & Dahl, 1991; Brill et al., 1989; Piotto et al., 1991). The phosphorous atom of the phosphorodithioate linkage is achiral, thus eliminating the problem of diastereoisomers.

Although the phosphorothioate linkages of **10** are resistant to hydrolysis by both exo- and endonucleases, they can be cleaved by these enzymes slowly (Campbell et al., 1990; Stec et al., 1984; Stein et al., 1988b). Oligonucleotide phosphorothioates hybridize with complementary single-stranded nucleic acids. It appears that the configuration of the phosphorothioate linkage affects the stability of the duplex. In a self-complementary deoxyribonucleotide containing a single phosphorothioate linkage, the Rp isomer melted approximately 2°C lower than the oligomer with normal phosphodiester linkages (LaPlanche et al., 1986). The stability of duplexes formed between oligodeoxyribonucleotide phosphorothioates and complementary RNA targets is reduced relative to that for the unmodified oligodeoxyribonucleotides (Stein et al., 1988b). In this case, GC content affects the stability; oligomers with chain lengths greater than 15 and with GC content greater than 50% generally appear to have melting temperatures above 37°C.

The RNA strand of hybrids formed between oligodeoxyribonucleotide phosphorothioates and RNA are substrates for RNase H (Cazenave et al., 1987; Stein et al., 1988b). Depending upon the conditions, the susceptibility of the phosphorothioate-containing substrates may be somewhat greater than that associated with an unmodified oligodeoxyribonucleotide (Ghosh et al., 1992). The phosphorothioate oligomers are also capable of binding to a variety of proteins, including DNA polymerases and reverse transcriptase (Gao et al., 1989; Majumdar et al., 1989). It thus appears that phosphorothioate oligomers may function by an antisense mechanism as well as by acting as inhibitors of polymerases and possibly other enzymes.

These properties may account for the particularly high inhibitory effects that phosphorothioate oligomers show against HIV in HIV-infected T cells. In one series of experiments both antisense oligomers and dSC_n—that is, oligomers containing phosphorothioate linkages and consisting of various lengths of deoxycytidine residues—were all found to be effective in protecting T cells from the cytopathic effects of the HIV virus (Matsukura et al., 1987). The most effective inhibitor was dSC_{28}, which most likely acts as an inhibitor of the viral reverse transcriptase (Majumdar et al., 1989). Sequence-specific antisense inhibitory effects have been observed for a series of antisense phosphorothioates when tested for their ability to inhibit HIV-induced syncytia formation and protein p24 synthesis in cells infected by HIV (Agrawal et al., 1988). In this case, sequence-specificity was observed when the oligomers were used at low (0.6 μM) concentrations but not at high (5–25 μM) concentrations. More recent experiments have suggested the possibility that oligonucleotide phosphorothioates may have utility in treating HIV-infected cells and that they may have therapeutic utility (Agrawal & Tang, 1992; Lisziewicz et al., 1992, 1993).

α-Anomeric Oligonucleotides. α-Anomeric oligonucleotides (**11**) represent a rather unique class of oligonucleotide analogue in which α-anomeric nucleosides are linked by phosphodiester internucleotide bonds. These oligomers are prepared from β-cyanoethyl phosphoramidite derivatives of the protected α-nucleosides using standard automated synthetic procedures. Oligomers containing both deoxyribose and ribose sugars have been prepared (Debart et al., 1992; Morvan et al., 1988). Despite the presence of the phosphodiester linkage, α-anomeric oligonucleotides are extraordinarily resistant to nuclease hydrolysis (Cazenave et al., 1987; Durand et al., 1988; Morvan et al., 1988; Praseuth et al., 1987; Sun et al., 1992; Thuong et al., 1987). For example, α-anomeric oligodeoxyribonucleotides are hydrolyzed approximately 30 times more slowly than normal, β-anomeric oligodeoxyribonucleotides.

α-Oligonucleotides form duplexes with both single-stranded DNA and RNA (Morvan et al., 1991). In contrast to normal oligonucleotides, the orientation of the sugar–phosphate backbone of α-oligomer is parallel to that of the sugar phosphate backbone of the target strand in the duplex. α-Oligomer/RNA duplexes appear to be more stable than α-oligomer/DNA duplexes (Thuong et al., 1987), and the stabilities of both types of duplexes are similar to those formed by normal β-oligonucleotides (Morvan et al., 1993).

Duplexes formed between α-oligodeoxyribonucleotides and RNA are not substrates for RNase H. When tested in cell-free systems, it was found that α-oligomers could inhibit translation of β-globin mRNA when targeted to the 5′-end and initiation codon regions of the mRNA, but not when they were targeted to the coding region of the mRNA (Bertrand et al., 1989; Boiziau et al., 1991). This suggests that the ribosome is capable of displacing the α-oligomer from the RNA during the elongation step of translation. Although extensive biological testing has not yet been carried out, antisense α-oligomers targeted against vesicular stomatitis virus mRNA or β-globin mRNA were not effective inhibitors in cell culture experiments (Boiziau & Toulme, 1991; Leonetti et al., 1988). However, α-oligomers targeted against the reverse transcriptase primer binding site of murine leukemia virus were found to inhibit viral replication (Lavignon et al., 1992).

Oligonucleoside Methylphosphonates. Oligonucleoside methylphosphonates (**12**) contain methylphosphonate internucleotide bonds. Unlike the phosphodiester or phosphorothioate linkage, the methylphosphonate linkage is nonionic. Thus these oligomers are neutral electronically under physiological conditions. A number of other nonionic phosphorous-containing oligomers have also been prepared (Miller, 1992). These include oligonucleotide alkylphosphotriesters (Miller et al., 1977) and oligonucleotide phosphoramidates (Froehler et al., 1988; Jager et al., 1988).

X-ray crystallographic structural studies on methylphosphonate dimers has shown that the methylphosphonate linkage has a geometry which is very similar to that of the normal phosphodiester linkage (Chacko et al., 1983; Han et al., 1990). As shown in Fig. 12-9, the linkage can exist in an Rp or Sp configuration. Methylphosphonate oligomers are readily prepared in automated DNA synthesizers (Miller et al., 1991) using suitably protected nucleoside methylphosphonamidite synthons (Agrawal & Goodchild, 1987; Dorman et al., 1984; Jager & Engles, 1984). In most respects, the synthesis is analogous to that used to prepare oligodeoxyribonucleotides. However, because the methylphosphonate linkage can be cleaved by bases, the oligomers cannot be deprotected by prolonged treatment with ammonium hydroxide. Rather, deprotection is carried out by first treating the oligomer briefly with ammonium hydroxide followed by treatment with ethylenediamine (Hogrefe et al., 1993). This method results in complete removal of the protecting groups from the bases as well as cleavage of the oligomer from the support. As is the case with oligonucleotide phosphorothioates, it is not yet possible to synthesize the methylphosphonate oligomers in a stereoselective manner on the support. Therefore, the oligomer is obtained as a mixture of 2^n diastereoisomers, where n is the number of methylphosphonate linkages in the oligomer.

Unlike the oligonucleotide phosphorothioates and α-anomeric oligonucleotides, oligonucleoside methylphosphonates appear to be completely resistant to hydrolysis by endo-and exonucleases (Agarwal & Riftina, 1979; Agrawal et al., 1988; Miller et al., 1979). Hybrids formed between the oligomer and RNA are not hydrolyzed by RNase H (Walder, 1988). Recent studies have focused on the preparation of "chimeric" oligomers which contain four to six contiguous phosphodiester linkages surrounded by methylphosphonate linkages on either side. When bound to RNA, such chimeric molecules do provide substrates for RNase H cleavage of the RNA strand of the hybrid (Giles & Tidd, 1992; Larrouy et al., 1992). Of course, the phosphodiester bonds of these oligomers would be expected to be susceptible to cleavage by endonucleases, but they still may be stable in the intracellular environment where endonuclease activity is low.

Oligodeoxyribonucleoside methylphosphonates form stable duplexes with both single-stranded DNA and RNA target molecules (Froehler et al., 1988; Lin et al., 1989; Murakami et al., 1985; Quartin & Wetmur, 1989; Sarin et al., 1988) and have also been observed to form triplexes with double-stranded DNA oligonucleotide targets (Callahan et al., 1991). The T_ms of duplexes formed between methylphosphonate oligomers and DNA targets are generally similar to those of duplexes formed by oligodeoxyribonucleotides. The stabilities of methylphosphonate-containing duplexes are much less sensitive to ionic strength conditions than are the corresponding oligodeoxyribonucleotide duplexes. Thus, methylphosphonate oligomers can form duplexes with complementary DNA oligonucleotides in the absence of salt. This reduced sensitivity is most likely due to the reduced charge repulsion between backbone of the nonionic methylphosphonate oligomer and the negatively charged phosphodiester backbone of the target.

A number of theoretical (Ferguson & Kollman, 1991; Hausheer et al., 1992) and experimental studies have shown that the stability of duplex formation depends upon the configuration of the methylphosphonate linkage. For example, a dT_{10} oligomer was synthesized which contained a backbone having either (1) alternating Rp methylphosphonates and phosphodiester linkages or (2) alternating Sp methylphosphonates and phosphodiester linkages (Miller et al., 1980). The Rp isomer formed much more stable complexes with both poly(dA) and poly(rA) than did the Sp isomer. Similar results have been obtained with dT_8 methylphosphonates in which all but the central methylphosphonate linkage had either an Rp or an Sp configuration (Lesnikowski et al., 1990).

A number of studies have shown that oligodeoxyribonucleoside methylphosphonates are effective inhibitors of protein synthesis in cell-free systems. For example, oligomers complementary to rabbit globin messenger RNA (mRNA) and to vesicular stomatitis virus mRNA were found to be inhibitory at 100 μM concentration when these mRNAs were translated in a rabbit reticulocyte lysate (Agris et al., 1986; Blake et al., 1985). Antisense methylphosphonate oligomers have also been demonstrated to inhibit virus replication and protein synthesis in cell culture when targeted against the mRNAs of vesicular stomatitis virus (Agris et al., 1986), herpes simplex virus (Kulka et al., 1989; Smith et al., 1986), and human immunodeficiency virus (Agrawal et al., 1988; Goodchild et al., 1988a,b; Zamecnik et al., 1986)). Antisense oligonucleoside methylphosphonates complementary to Ha-*ras* mRNA were also found to inhibit p21 synthesis in Ha-*ras* transformed mouse 3T3 cells in culture (Chang et al., 1991).

Analogues with Nonphosphorous Internucleotide Linkages. An attractive feature of the nonionic oligonucleotide analogues is their total resistance to nuclease hydrolysis. However, this advantage is offset somewhat by the presence of additional chiral centers in the sugar phosphate backbone which give rise to diastereomeric mixtures of products. To overcome this problem, nucleic acid chemists have investigated a variety of oligonucleotide analogues in which the nucleoside residues are joined by nonionic achiral nonphosphorous linkages. Examples of such analogues are shown in Fig. 12-10. These include carbonate analogues (**13**) (Jones & Tittensor, 1969; Mertes & Coates, 1969), carbamate analogues (**14**) (Coull et al., 1987; Gait et al., 1974; Mungall & Kaiser, 1977; Stirchak & Summerton, 1987), and analogues containing formacetal linkages (**15**) (Matteucci, 1990; Matteucci et al., 1991).

An even more radical change is represented by peptide nucleic acid (PNA) analogues (**16**) whose general structure is shown in Fig. 12-11. In these analogues both the sugar and phosphate have been replaced by an uncharged, peptide-like backbone (Egholm et al., 1992; Nielsen et al., 1991, 1993a). Although at first this structure ap-

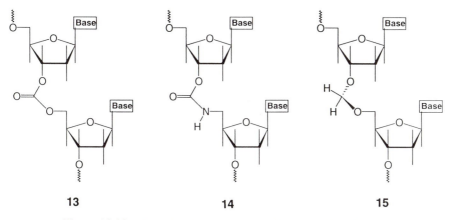

13   **14**   **15**

Figure 12-10. Non-phosphorus-containing oligonucleotide analogues.

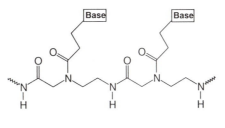

16

Figure 12-11. Backbone structure of a peptide nucleic acid (PNA) analogue.

pears to be completely unlike nucleic acids, PNAs are connected via six bonds per backbone unit, which is the same number as found in the sugar phosphate backbones of nucleic acids. As expected, PNAs are totally resistant to hydrolysis by nucleases. They are also resistant to hydrolysis by peptidases and proteases.

PNAs were originally designed to form triplexes with DNA, and molecular modeling experiments show that this analogue should be capable of forming triple-stranded structures with double-stranded DNA by binding in the major groove. Experiments show that PNAs containing six or eight thymine bases form stable complexes with the double-stranded target, $dA_{10} \cdot dT_{10}$. It was also found that these PNAs formed duplexes with dA_{10} having T_ms comparable to those observed with the duplex target. This suggested that the PNA binds to the duplex target by a strand displacement mechanism, rather than through triplex formation. Experiments involving chemical and enzymatic probing, footprinting, and photoaffinity cleavage are in agreement with this binding mode (Nielsen et al., 1991).

Recent experiments have shown that PNAs are capable of inhibiting cleavage of DNA by a restriction enzyme (Nielsen et al., 1993b). PNAs were also shown to be capable of binding to RNA. As a result of such binding, reverse transcription and in vitro translation of the RNA template were inhibited (Hanvey et al., 1992).

Oligonucleotide Conjugates. Considerable effort has been expended on improving the efficacy of antisense oligonucleotides and oligonucleotide analogues. These efforts have focused on conjugating the oligomers with functional groups that are designed either to increase the stability of the duplex formed between the oligomer and its target or to carry out chemical reactions that covalently link the oligomer to the target or cleave the target (Goodchild, 1990). A wide variety of functional groups have been attached to oligonucleotides, including intercalators, photoreactive crosslinking groups, alkylating groups, and metal chelating groups.

Intercalating Groups. Acridine is a well-known tricyclic heteroaromatic compound that binds to double-stranded nucleic acids by intercalation. Derivatives of acridine have been attached to the 5'- and 3'-ends of oligonucleotides (Asseline et al., 1984, 1985; Hélène et al., 1985, Lancelot et al., 1985; Thuong et al., 1987). Figure 12-12 shows 6-chloro-9-aminoacridine conjugated to the 5'-end of an oligomer via a five-carbon linker arm (**17**). Both normal oligodeoxyribonucleotides and α-anomeric oligodeoxyribonucleotides have been derivatized with acridine (Boiziau et al., 1992; Durand et al., 1993).

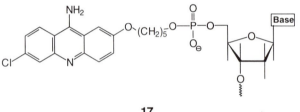

17

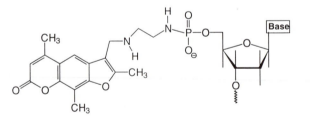

18

A   B

Figure 12-12. Acridine- and psoralen-derivatized oligonucleotides.

The linker arm allows the acridine ring to intercalate between base pairs formed between the oligonucleotide and its target as shown schematically in A and B of Fig. 12-12. As a consequence of intercalation, the T_m of the oligomer/target duplex increases relative to that of the nonderivatized oligomer. For example, d(Tp)$_7$TpEt, an octathymidylate whose 3'-terminal phosphate is derivatized with an ethyl group, forms a duplex with d(Ap)$_7$A whose T_m is 10°C. When the ethyl group of d(Tp)$_7$TpEt is replaced by 6-amino-6-chloroacridine linked through the phosphoryl group by a six-carbon linker arm, the T_m of the duplex increases to 20°C (Thuong et al., 1987). Similar increases are seen with other oligomer sequences. Although the increase in T_m imparted by the acridine group is rather dramatic for shorter oligomers, less of a change is observed as the length of the oligomer increases.

Acridine-derivatized oligodeoxyribonucleotides show increased resistance to hydrolysis by exonucleases. Of course the internucleotide linkages are still susceptible to cleavage by endonucleases. Oligodeoxyribonucleotides derivatized with acridine are capable of inhibiting mRNA translation in cell-free systems (Toulme et al., 1986). They have also shown inhibitory effects when targeted against sequences in trypanosomes in cell culture (Pascolo et al., 1993).

Psoralen. Derivatives of psoralen have been conjugated to oligodeoxyribonucleotides (Pieles & Englisch, 1989; Teare & Wollenzien, 1989; Woo & Hopkins, 1991) and oligonucleoside methylphosphonates (Bhan & Miller, 1990; Lee et al., 1988). For example, 4-aminoethyl(aminomethyl)-4,5',8-trimethylpsoralen, (ae)AMT, has been linked to the 5'-end of oligonucleoside methylphosphonates via a phosphoramidate bond as shown in structure (**18**) (Fig. 12-12). Psoralen has also been placed at internal positions in methylphosphonate oligomers using synthetic non-nucleotide linker groups (Reynolds et al., 1992). Like acridine, psoralen binds to double-stranded nucleic acids via intercalation. When linked to the end of oligomer, the psoralen moiety can stack on the last base pair formed by the oligomer and its target as shown in A of Fig. 12-12. In this binding mode the six-membered pyrone ring of the psoralen partially intercalates with the next base in the target. If this base is a pyrimidine, a light-activated 2 + 2 cycloaddition reaction can take place between the double bond of the pyrone ring and the 5,6 double bond of the pyrimidine base. The photoreaction which occurs when the complex is irradiated with ultraviolet (UV) light (365 nm) results in the formation of a cyclobutane bridge between the psoralen and the pyrimidine. This adduct formation covalently crosslinks the oligomer to its target. The crosslink is chemically stable under physiological conditions, but can be reversed by irradiation with light (254 nm). The psoralen ring of (ae)AMT can also adopt a binding mode in which the ring is completely intercalated, as shown in B of Fig. 12-12. Crosslinking to pyrimidine bases in the target strand can also occur in this binding mode.

Psoralen-derivatized oligodeoxyribonucleotides and oligonucleoside methylphosphonates have been shown to crosslink effectively in vitro to complementary mRNA sequences. Because the oligomer becomes covalently crosslinked to the mRNA, it cannot be displaced by ribosomes. For example, (ae)AMT-derivatized methylphosphonate oligomers complementary to sequences in the coding regions of rabbit α- and β-globin mRNA crosslink in a sequence-selective manner (Kean et al., 1988). As a consequence of crosslinking, in vitro translation of globin mRNA is selectively inhibited. Non-crosslinked oligomers did not have any effect on the translation reaction. As described below, psoralen-derivatized methylphosphonates have also been shown to have effects in cell culture and in animals.

Oligodeoxyribopyrimidines derivatized with psoralen are capable of forming triplexes with double-stranded DNA targets (Giovannangeli et al., 1992). In this case, the psoralen was attached to the 5'-end of the oligomer via a linker arm connected to the 8-position of the psoralen ring. This point of attachment allows both the furan and the pyrone rings to participate in cycloaddition reactions with pyrimidine bases in the DNA target. Thus irradiation resulted in the formation of crosslinks between the psoralen and the two strands of the DNA target. Triplex formation followed by crosslinking has been shown to be effective in preventing in vitro transcription of DNA (Duvalvalentin et al., 1992).

Alkylating Groups. Although psoralen-derivatized oligonucleotides can crosslink efficiently to either single-stranded RNA targets or to double-stranded DNA, they require activation by long-wavelength UV light to do so. Functional groups which can crosslink through chemical reactions have also been developed. Two examples of such groups are shown in Fig. 12-13. Oligonucleotides have been derivatized with aromatic 2-chloroethylamino groups as shown in structure **19** (Knorre & Vlassov, 1985; Vlassov, 1993). This alkylating agent is similar to the antitumor drug chlorambucil, and it reacts by formation of a highly reactive aziridinium intermediate. When bound to its complementary target, the alkylating group of the oligomer is positioned such that it can react with the N7 atom of a guanine in the target strand, thus forming a crosslink. Oligopyrimidines which carry this group are also able to crosslink to DNA via triplex

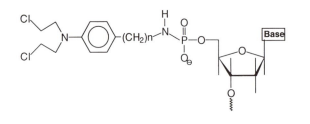

19

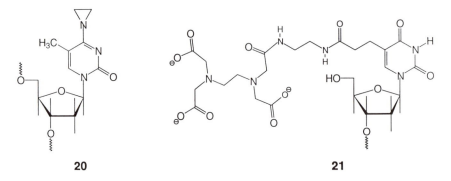

20   **21**

Figure 12-13. Oligonucleotides derivatized with functional groups capable of crosslinking or cleaving target RNAs and DNAs.

formation (Brossalina et al., 1991). Although these alkylating groups can lead to high levels of crosslinking, their high reactivity is also a potential problem because they are prone to crosslink at sites other than their intended target (Gorshkova et al., 1986). In addition, the alkylating group itself can react with solvent, which leads to deactivation.

A somewhat different type of alkylating group is represented by 5-methyl-$N4,N4$-ethanocytosine (**20**), shown in Fig. 12-13. Oligomers carrying this modified base are only capable of reacting with a complementary guanine base when the oligomer binds to its target (Webb & Matteucci, 1986). Protonation of the N4 nitrogen produces a highly reactive aziridinium ring which alkylates the N7 position of guanine. The half-life for reaction in single-stranded targets is approximately 30 days. However, much greater reactivity is observed when the modified C is incorporated into a triplex-forming oligonucleotide (Shaw et al., 1991b).

Cleaving Groups. Oligonucleotides have also been equipped with functional groups which are designed to cleave their complementary targets. For example, oligomers have been derivatized with the metal chelator ethylenediaminetetraacetate (EDTA). This group has been attached to either the 5′-end of the oligomer via a methylene side chain (Boutorin et al., 1984; Chu & Orgel, 1985; Lin et al., 1989) or, as shown in Fig. 12-13, to the 5-position of deoxyuridine (**21**) (Dreyer & Dervan, 1985). In the presence of Fe(II) and a reducing agent, the EDTA group produces hydroxyl radicals; these abstract hydrogens from the sugar residues of the target strand, thus leading to strand scission. EDTA-derivatized oligonucleotides can cleave both single-stranded and double-stranded targets. The latter reaction has been quite useful in developing reagents that can produce site-specific strand breaks in large genomic DNA (Dervan, 1992; Strobel et al., 1991). EDTA-driven affinity cleavage reactions have proven extraordinarily useful in determining binding constants between third-strand oligonucleotides and their double-stranded DNA targets (Singleton & Dervan, 1992).

Other radical-producing functional groups have been attached to oligonucleotides. These include 1,10-phenanthroline, which in the presence of copper can cleave both DNA and RNA strands (Chen & Sigman, 1986; Chen et al., 1993; Sigman et al., 1993), and porphyrins, which produce oxygen and hydroxyl radicals when irradiated with UV light (Le Doan et al., 1987). Although oligomers derivatized with radical-generating groups have been used quite successfully in in vitro experiments, it remains to be seen whether they can be used as antisense agents in cell culture experiments. Oligomers of this type that generate diffusible oxygen radicals are potentially toxic to cells. In addition, the oligomers themselves tend to undergo autodegradative reactions that could limit their useful lifetime in cell culture, where the rates of diffusion of the oligomers into the cells is rather long.

Biological Activity of Antisense/Antigene Oligonucleotides

Antisense and antigene oligonucleotides have been shown to be effective inhibitors of mRNA translation and DNA transcription, respectively, in cell-free systems. The ultimate goal is to use these oligomers as therapeutic agents. Toward this end, the uptake of oligonucleotides by cells in culture has been examined, as well as their ability to inhibit gene expression in cultured cells. A number of laboratories are now beginning to extend these studies to animal model systems.

Uptake of Oligonucleotides by Cells in Culture. Negatively charged oligodeoxyribonucleotides and oligonucleotide phosphorothioates are taken up by mammalian cells in culture (Crooke, 1993a; Geselowitz & Neckers, 1992; Hawley & Gibson, 1992; Iverson et al., 1992; Jaroszewski & Cohen, 1991; Loke et al., 1989; Marti et al., 1992; Stein et al., 1988a; Yakubov et al., 1989; Zhao et al., 1993). Uptake can be monitored by using radioactively labeled oligomers or oligomers derivatized with fluorescent tags such as fluorescein or rhodamine (see Chapter 8). In general, uptake of these oligomers is both time- and temperature-dependent. It appears that at least some cells have receptors that are responsible for the uptake of the oligomers. These receptors range in size from 34 kD to 90 kD and are located on the surface of the cell. The natural biological role of these receptors is unknown. Thus the oligomers do not simply diffuse into the cells, but rather are taken up by an active, energy-dependent process called *absorptive endocytosis* or *receptor-mediated endocytosis*. Cells such as HL60 cells can also take up oligomers by a non-receptor-mediated process called *fluid-phase pinocytosis* (Stein et al., 1993). In this case the cells engulf a small volume of the culture medium. Any solutes which happen to dissolved in the culture medium are also ingested.

Once inside the cells, radioactivity from oligodeoxynucleotides is found distributed in both the nucleus and cytoplasm. Approximately 20% of the radioactivity is associated with the nucleus and another 50% with the vesicles, mitochondria, and lysosomes found in the cytoplasm. A similar distribution is found for the oligonucleotide phosphorothioates. Examination of fluorescently tagged phosphorothioate oligomers shows that the majority of the fluorescence accumulates in vesicular structures resembling endosomes, which are located outside the nucleus. It is interesting to note that when fluorescently tagged phosphodiester or phosphorothioate 28-mers are microinjected into African green monkey kidney cells, the fluorescence rapidly migrates to the nucleus and appears to be associated with small nuclear ribonucleoproteins (Chin et al., 1990; Leonetti et al., 1991).

Oligodeoxyribonucleoside methylphosphonates (Marcus-Sekura et al., 1987; Miller et al., 1981; Vasanthakumar & Ahmed, 1989) and chimeric oligomers containing methylphosphonate and phosphodiester linkages (Spiller & Tidd, 1992) are also taken up by mammalian cells in culture. However, methylphosphonate oligomers longer than four nucleoside units are not taken up by bacterial cells (Jayaraman et al., 1981). It appears that the bacterial cell wall acts as a "molecular sieve" and excludes oligomers above a certain molecular weight from entry. Uptake of phosphodiester or phosphorothioate oligonucleotides is not inhibited by the presence of methylphosphonate oligomers of the same chain length. This suggests that the nonionic methylphosphonate oligomers are taken up by a mechanism distinct from that of the charged oligonucleotides and not involving receptor-mediated endocytosis.

Examination of the uptake of fluorescein or rhodamine-conjugated oligonucleoside methylphosphonates by fluorescence microscopy showed that the oligomers are taken up by an energy-dependent process (Shoji et al., 1991). The fluorescently tagged oligomers appear to accumulate primarily in vesicles within the cytoplasm, although low levels of fluorescence are also observed throughout the cytoplasm and nucleus (Shoji et al., 1991; Thaden & Miller, 1993). When the oligomers are microinjected into cells, they, like the phosphodiester and phosphorothioate oligomers, accumulate in the nucleus (Chin et al., 1990; Clarenc et al., 1993). Unlike the charged oligomers, however, the methylphosphonate oligomers appear to associate with chromatin within the nucleus.

When an oligonucleotide phosphorothioate 20 nucleotides in length was administered to mice interperitoneally or intravenously, approximately 30% of the dose was excreted within 24 hr (Agrawal et al., 1991). The oligomer appeared to accumulate preferentially in the liver and kidneys. Interperitoneal injection gave rise to extensive degradation and modification of the oligomer.

In contrast to this behavior, a tritium-labeled methylphosphonate 12 nucleotides in length and containing a single phosphodiester linkage at the 5'-end of the oligomer was rapidly excreted ($t_{1/2} = 17$ min) after intervenous injection into a mouse (Chen et al., 1990). However, the oligomer was also rapidly distributed to various organs and tissues including the kidney, liver, lung, muscle, and spleen with a half-life of 6 min. Radioactive material recovered from these organs or from the urine was found in two forms: intact oligomer and oligomer lacking the 5'-terminal nucleotide unit. The latter oligomer most likely resulted from exonuclease cleavage of the 5'-terminal phosphodiester linkage. Significant, however, was the lack of degradation of the methylphosphonate linkages in the oligomer.

Specific Examples of Antisense/Antigene Oligonucleotide Activity in Cell Culture and in Mice. A variety of antisense and antigene oligonucleotides and oligonucleotide analogues have been tested in cell culture for their inhibitory effects on cellular and viral genes. A number of monographs have been published that describe these types of experiments (Cohen, 1989; Crooke & Lebleu, 1993; Erickson & Izant, 1992; Murray, 1992). This section focuses on several experiments designed to examine the biological activities of psoralen-derivatized oligonucleotide analogues.

In the test tube, oligonucleoside methylphosphonates derivatized with psoralen selectively bind and crosslink to mRNA when irradiated with long-wavelength UV light. Such oligomers can also function selectively in cell culture. Oligomers complementary to the coding region of human Ha-*ras* mRNA were found to be capable of inhibiting *ras* p21 protein synthesis in *ras* transformed NIH 3T3 cells in culture (Chang et al., 1991). Two oligomers, d-TpCCTCCTGGCC, *ras* T, and d-TpTCCTCCAGGCC, *ras* A, were synthesized. The underline indicates the positions of the methylphosphonate linkages. *Ras* T is complementary to nucleotide sequences surrounding the 61st amino acid codon of normal human c-Ha-*ras* mRNA, whereas *ras* A is complementary to a similar region of c-Ha-*ras* mRNA found in a human lung carcinoma. The single nucleotide difference in the two oligomers corresponds to an A to T point mutation found in the second nucleotide of the 61st codon of the c-Ha-*ras* gene. The oligomers were further derivatized with psoralen as shown in structure **18** (Fig. 12-12).

Cell lines derived from NIH 3T3 cells were developed which carry multiple copies of c-Ha-*ras* DNA having a point mutation in the 61 codon (453 cells) or having a normal 61st codon and a point mutation in the 12th codon (RS504 cells). Thus the *ras* T oligomer is complementary to *ras* mRNA in the RS504 cells and *ras* A is complementary to *ras* mRNA in the 453 cells. The differences in amino acid sequences caused by the mutations results in the production of *ras* p21 proteins whose electrophoretic mobilities are different on polyacrylamide gels.

Cultures containing a mixture of 454 and RS504 cells were incubated with either psoralen-derivatized *ras* T or *ras* A for 48 hr and the cells were then irradiated at 365 nm using a laser. The irradiated cells were then labeled with [^{35}S]methionine, and the amounts of "normal" and "mutant" p21 proteins were analyzed after immunoprecipitation and by gel electrophoresis. Up to 90% inhibition was observed when the cells were irradiated in the presence of 15 μM oligomer, and no inhibition was observed in oligomer-treated, unirradiated cells. Inhibition was dependent upon the concentration and sequence of the oligomer. *Ras* T oligomer specifically inhibited p21 derived from the RS504 cells, whereas *ras* A specifically inhibited p21 derived from the 453 cells. This selectivity suggested that the oligomers were capable of distinguishing a single base mismatch when binding to the *ras* mRNA.

An antisense psoralen-derivatized oligonucleoside methylphosphonate has also been shown to be effective in inhibiting herpes simplex virus (HSV) replication in a mouse model system (Kulka et al., 1993). The oligonucleoside methylphosphonate, a 12-mer, was complementary to the acceptor splice junction of HSV-1 immediate-early mRNAs 4 and 5. The underivatized oligomer and the psoralen-derivatized oligomer were previously shown to be capable of inhibiting HSV-1 replication in HSV-infected cells in culture (Kulka et al., 1989).

Mice were infected in both ears with either HSV-1 or HSV-2. A total of 26 mice were then given a topical application of the psoralen-derivatized oligomer suspended in polyethylene glycol at an approximate concentration of 50 μM. Application was initiated at 0 hr postinfection, and two more applications were made at 4 hr and 8 hr postinfection. Immediately after each application the ears were irradiated with 365-nm UV light for 10 min. Single daily treatments and irradiations were continued for 5 more days. For each animal, one ear was treated with the oligomer and the other ear was treated with saline in polyethylene glycol. The virus titers in the treated and nontreated ears were then assayed 6 days postinfection. The psoralen-derivatized oligomer reduced the virus titer in the HSV-1 infected mice 91% relative to the saline-treated control. In contrast the oligomer only reduced virus titer by 27% in the HSV-2-infected mice. The sequence of the oligomer-binding site in the splice acceptor region of HSV-2 differs from that of HSV-1 by five nucleotides, and thus the oligomer would be expected to be less effective against HSV-2.

An antigene oligodeoxyribopyrimidine derivatized with psoralen was shown to be effective at inhibiting gene transcription both in vitro and in cells in culture as a result of triplex formation (Grigoriev et al., 1993). The oligomer, a 15-mer, contained thymidine and 5-methyldeoxycytidine and was derivatized at the 5′-end with psoralen. The oligomer was targeted to a homopurine sequence found in the promoter region of the α-subunit of the interleukin-2 receptor gene. The homopurine sequence overlaps a binding site for NF-κB, a transcription factor which activates the promoter. The promoter sequence was incorporated into a plasmid in such a way that it could drive the expression of a chloramphenicol acetyltransferase (CAT) gene. A similar plasmid was also created in which several of the purines in the homopurine sequence of the promoter were substituted by pyrimidines. This modified promoter could not serve as a binding site for the psoralen-derivatized 15-mer.

Binding between the 15-mer and the promoter in vitro was demonstrated by DNase I footprinting experiments. The modified promoter was not protected by the 15-mer in these experiments. Irradiation of the plasmid in the presence of the 15 mer, followed by restriction endonuclease digestion, produced fragments which contained crosslinked oligomer. These experiments served to establish the fidelity of binding and the ability of the oligomer to crosslink to its target.

In vitro transcription of the plasmid in the presence of NF-κB was inhibited when the plasmid was irradiated in the presence of the oligomer. Inhibition was not observed when the irradiation was not performed, and inhibition not was observed when the plasmid carrying the modified promoter was irradiated in the presence of the oligomer.

Plasmids carrying the wild-type promoter or the modified promoter were irradiated in the presence of increasing concentrations of the oligomer in vitro. The plasmids were then transfected into a human tumor T-cell line by electroporation and the CAT activity was assayed. Increasing concentrations of the oligomer resulted in increased inhibition of CAT activity in cells transfected with the wild-type plasmid but not in cells transfected with the mutated plasmid. The amount of inhibition was consistent with the extent of crosslinking observed in vitro. Similar inhibitory activity was observed when cells were transfected with the plasmid and the oligomer and then irradiated. Again CAT activity in cells transfected by the plasmid carrying the modified promoter was not inhibited after irradiation.

Potential Therapeutic Applications of Antisense/Antigene Oligonucleotides. The results obtained with antisense and antigene oligonucleotides and oligonucleotide analogues in cell culture demonstrate that these agents can be used to regulate gene expression specifically in biological systems. Experiments in animal model systems, although limited, are encouraging and suggest that such oligomers may have therapeutic potential.

However, much remains to be learned before these agents can be used in a clinical setting. Oligonucleotides are rather complex molecules, and methods to prepare them on a scale sufficient for extensive animal and clinical testing are only now being worked out (Zon & Geiser, 1991). The toxicity, pharmacology, and immunogenicity of oligonucleotides and oligonucleotide analogues is only now beginning to be examined. Methods must be found to increase their efficacy, in order for these molecules to be useful at clinically relevant doses. Further modifications of the oligomers, such as conjugation with crosslinking or cleaving groups, may help to solve this problem. In addition, methods for site-specific delivery of the oligomers need to be developed. Topical application of oligomers appears to be a feasible method and would be applicable in certain situations such as herpetic infections.

It is obvious that significant challenges remain to be overcome before these compounds can realize their full potential as therapeutic agents. Nevertheless, antisense/antigene oligonucleotides hold out the intriguing and very real possibility of becoming drugs that can be designed in a rational manner to have a very high degree of selectivity.

Summary

Oligonucleotides have been designed to selectively suppress genetic information in living cells. Antisense oligonucleotides form duplexes with the initiation codon and coding regions of mRNA, or the splice junction regions of precursor mRNA, and consequently inhibit mRNA translation or splicing. Antigene oligonucleotides bind in the major groove of double-stranded DNA through the formation of a triple-stranded complex. At present, such binding is limited to homopurine tracts. However, such tracts are often found in the promoter regions of genes, and antigene oligonucleotides can be designed to regulate gene expression directly by suppressing gene transcription.

Oligodeoxyribonucleotides contain phosphodiester linkages which are hydrolyzed by cellular exo- and endonucleases, and this limits their effectiveness in biological systems. Replacement of oligomer phosphodiester linkages with non-natural linkages such as phosphorothioate or methylphosphonate enhances their resistance to nuclease degradation and thus their antisense or antigene activity. Conjugation with functional groups which allow the oligomers to crosslink to or cleave their targets further enhances their effectiveness.

Antisense/antigene oligonucleotides and oligonucleotide analogues are taken up by cells in culture and are distributed to various tissues when injected into mice. Such oligomers exhibit biological activity in a variety of cell culture systems, and antisense oligomers have demonstrated activity in a number of animal model systems. These experiments suggest that antisense/antigene oligomers can be used to specifically regulate gene expression in biological systems. In addition, these compounds may serve as the basis for the development of highly selective therapeutic agents.

Catalytic RNA

Francis J. Schmidt

Cells can function as chemical engines because they possess *enzymes*, catalysts capable of greatly enhancing the rate of specific chemical reactions. Enzymes are highly specialized protein molecules.
Lehninger, 1975

RNA enzymes (ribozymes) fit the above definition, with the obvious exception that they are not proteins. Therefore, the same questions that motivate the study of protein enzymology, including those listed below, motivate RNA enzymology. What is the nature of the binding of enzyme and substrates? How does the structure of the catalyst contribute to its activity? What chemical forces contribute to the rate enhancement of an enzymatic reaction? What is the detailed chemical mechanism of the catalytic event? How widely are enzymes of this type distributed in nature?

Other questions derive from the fact that these enzymes are made of RNA. What are the contributions of Watson–Crick and noncanonical structures to the overall secondary structure of the molecule? How do metal ions bind specifically to the active site of the catalyst? How is it that large domains of an RNA enzyme can be added or substracted evolutionarily without loss of function?

This chapter does not presume to provide answers to these questions, or even to suggest that all questions have been posed. Rather, the intent is to provide a catalogue of the best-known RNA catalysts and to point out the potential relevance of RNA catalysis to the great "chicken and egg" problem of molecular biology: which came first, genes or enzymes, heredity or metabolism? It is not useful to pretend that the problem was solved by the discovery of ribozymes, but it is necessary to discuss the problem in light of that discovery and of the rapid pace of RNA research in general.

History and Background

Molecular biology contains a singularity, implied by the one-way nature of the central dogma: DNA makes RNA, which, in turn, makes protein. This relationship presumably arose through natural selection of genomes encoding fitter catalysts. Natural selection requires three steps: variation, selection, and propagation. Variation and selection are easy: Proteins differing in amino acid sequence will likely have different catalytic properties, some of which may be selected for the task at hand. The difficulty arises when we try to imagine how, in the absence of a cell, a "fitter" protein can promote the replication of a nucleic acid genome encoding that fitter protein. It was recognized early that one way in which the paradox could be resolved would be to demonstrate a direct link between nucleic acids and catalysis.

Demonstration of such a link would seem to require a "molecular fossil"—that is, some nucleic acid in contemporary biochemistry able to link the informational (nucleic acid) and metabolic (catalytic) domains. Given the power of protein catalysts, however, there was little reason a priori to expect that direct experimentation would ever reveal such a molecule. Any molecule linking the two domains would more likely be a "molecular dodo," long since driven into extinction by the more efficient catalysts of the present era.

Early speculations on the evolution of the genetic code included attempts to identify structural fits between codons and their cognate amino acids, but there seemed to be little way in which the regular structure of DNA triplets, especially within double-helical DNA, could show complementarity with the variety of structural features associated with amino acid side chains. As the structures of transfer RNA (tRNA) species became known, it was thought that these molecules, with their abundance of modified bases, could have had some structural relationship, now obscured, to their cognate amino acids. Speculation changed to the idea that translation could have originated in an RNA-dominated context, and a logical extension of this is that RNA species could have catalytic as well as coding roles.

Several systems lent credence to these speculations, if not proof. Ribosomal RNA modification determines resistance to some antibiotics, and a single endonucleolytic cleavage of rRNA by colicin E3 inactivates translation. The tRNA processing enzyme ribonuclease P was known to possess an intrinsic RNA component (Guthrie & Atchison, 1980; Kole & Altman, 1979). Finally, there was some evidence for an RNA component of metabolic enzymes, although this was controversial and ultimately disproved. While these facts provided some support for the involvement of RNA in enzyme activity, direct proof for RNA catalysis was lacking. It was not clear that essentiality implied any function other than that of providing a scaffold for protein catalysts. Even if these systems had once been RNA-catalyzed, there was always the possibility that RNA's catalytic function could have been supplanted by more efficient protein enzymes as soon as cells were available to spatially connect enzyme phenotype with nucleic acid genotype. Indeed, if metabolic reactions ever were RNA-catalyzed, the vast majority of them are not RNA-catalyzed at present.

The link between the genomic and catalytic domains was provided when RNA was shown to be chemically competent for catalysis, an observation causing a paradigm shift in models of prebiotic evolution (Kruger et al., 1982). The finding arose from studies of intron processing. The large ribosomal RNA transcript from *Tetrahymena thermophila* contains a single 414-nt insert which, in contrast to mRNA introns at that time, was shown in 1980 to be removed accurately and efficiently in vitro in a simple reaction. Fractionation of the nuclear processing extract soon revealed that no extract was necessary: Removal of the intron could be accomplished in buffer with only a guanosine nucleotide or nucleoside as a cofactor. In an experiment recalling the demonstration of the role of DNA in transformation by Avery et al. (1994), addition of purified protease to the processing reaction did not inhibit the reaction. Thus, intron removal required only RNA (quickly termed a "ribozyme") and a small molecule. It was an electrifying discovery, opening up new chemistry and solving, at least in principle, the dilemma of how coding, replicative and catalytic functions could coevolve. The phrase, "RNA world," became shorthand for the model of precellular evolution that this discovery implied (Gesteland & Atkins, 1993).

What was the RNA world like? First, it was relatively brief (Joyce & Orgel, 1993). There is some evidence for Calvin-cycle photosynthesis in the oldest known rocks (3.8×10^9 years ago), and credible bacterial microfossils are observed in rocks 3.1×10^9 years old. Thus, only 1–1.5 billion years were available for the evolution of cellular life after the accretion of the planet 4.7×10^9 years ago. Statistical analyses by Eigen and coworkers (Eigen, 1992) indicate that RNA sequences and the genetic code are "not older than, but almost as old as, our planet." So one can conclude that there was RNA present before there were cells but that cells arose almost immediately afterward, on a geological time scale. At the time of the putative RNA world, however, Earth was inhospitable to macromolecules. Its surface was reducing and aqueous, with high temperature and pressure (500–1000°C, 500 atm). Although it is possible to synthesize bases, sugars, and amino acids under such conditions, an aqueous environment probably precluded the condensation of monomers into polymers in free solution. Indeed, glycosidic bond formation under such conditions is problematic and it is not clear how nucleosides became available for condensation into RNA. RNA is also easily hydrolyzed in alkaline environments due to the presence of a 2'-OH group. These considerations make it unlikely that the RNA world existed in solution. Models involving surface chemistry have been proposed to deal with this paradox, but experimental proof is not presently available (Waechterhaueser, 1988). Whether involving RNA or not, our picture of "prebiotic life" is still unsatisfying.

The Systems of RNA Catalysis

All known, naturally occurring RNA catalytic activities involve RNA substrates. There are no satisfying explanations for our failure to identify RNA catalysts that work on other substrates or, for that matter, why ribozymes still exist. It may be, as Cech and Bass (1986) have suggested, that RNA is uniquely qualified to carry out reactions on RNA. In this view, other RNA catalysts would have been supplanted by the more efficient protein catalysts as life evolved. Alternatively, RNA catalysis and protein metabolism could have coevolved, with the connection being supplied by the primitive genetic apparatus, probably tRNA (Eigen, 1992).

Group I Introns. These were the first catalytic RNAs to be discovered; they occur both in eukaryotes and in prokaryotes. The basic reaction catalyzed by these introns is a transesterification whereby the phosphodiester linkage between the exon and intron sequences at the 5'-junction is replaced by a guanosine nucleotide to make a new phosphodiester bond. In a second transesterification step, the new 3'-end of the exon is joined to its downstream counterpart. Overall, the intron has a G added to its 5'-end and the intron is excised from the rRNA precursor. Although in vivo the intron is not truly catalytic (i.e., it does not turn over), the intron can be structured to recognize and transesterify exogenous substrates. The transesterification activity of a Group I intron can be used to carry out a "synthetic" reaction whereby nucleotides can be joined together with the elimination of 5'-terminal G residues (see below).

Figure 13-1 shows an accepted core structure and nomenclature for Group I introns (Burke et al., 1987; Michel & Westhof, 1990). Three types of structures are defined:

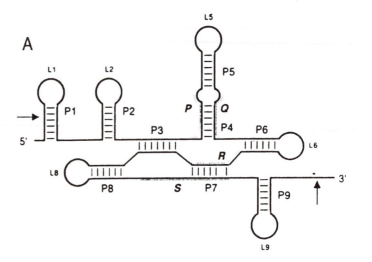

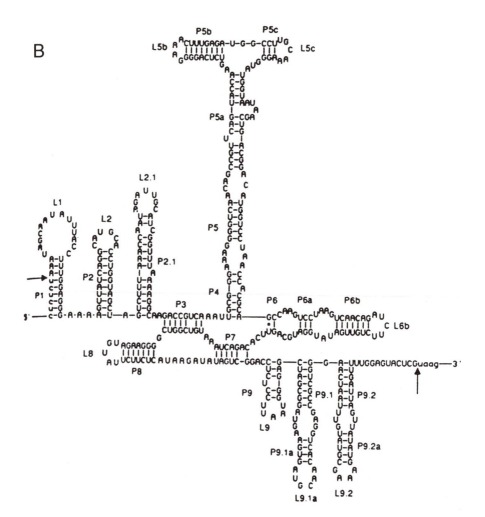

P regions are base-paired, L regions are unpaired loops at the ends of P regions, and J regions join the other structural elements together. Thus, for example, J1/2 would be bounded by the P1 and P2 helices (cf. Young et al., 1991). Although they are shown in Fig. 13-1 as unpaired, L and J regions undoubtedly participate in tertiary interactions (e.g., Michel & Westhof, 1990). In addition (somewhat confusingly) the areas of conserved primary sequence are designated *P*, *Q*, *R*, and *S*, as indicated in the figure. This latter nomenclature is used less often as the secondary structure of the enzyme becomes more defined.

The conventions for naming nonconserved stems are as follows. Extended stems are designated, for example, P5a, P5b, and so on, indicating that they extend the conserved shorter P5 stem. Where a nonconserved stem is found, it is named from the conserved stem immediately 5′ to the conserved sequence (e.g., P9.1 and P9.2) in the *Tetrahymena* intron. Other interactions involving tertiary hydrogen bonding have been described—for example, the P11 pairing for a subset of Group I introns (Jaeger et al., 1991).

Group I introns are found in nature as self-splicing introns. In some cases they encode a primitive reverse transcriptase, allowing them to be genetically mobile. The native intron does not fit the definition of a true catalyst because it does not turn over. Structurally modified forms of the intron can, however, catalyze a variety of reactions (e.g., hydrolysis, ligation, transesterification, and phosphotransfer) with multiple turnovers, as shown in Fig. 13-2. In all these reactions, the oligonucleotide substrates are recognized by base-pairing with a guide sequence as part of the P1 stem.

Group II Introns and the Spliceosome. These are primarily found in fungal mitochondria and carry out a transesterification reaction, not to an external nucleotide but to the 2′-OH of an A residue in the intron. This reaction is of interest because it is an exact analogue of the eukaryotic splicing reaction carried out by the ribonucleoprotein complex called the *spliceosome* (Wise, 1993). Group II introns are not as well characterized as are Group I introns; they require nonphysiological concentrations of ionic strength and temperature to function in vitro. Both Group I and Group II introns exhibit "a-type" cleavage, whereby the 5′-phosphate at the scissile bond is transferred to the splicing branch point.

Current models propose that the RNA components of the spliceosome adopt a base-paired structure similar to the structure of Group II introns (Guthrie, 1991; Lesser & Guthrie, 1993; Sontheimer & Steitz, 1993; Steitz, 1992; Wise, 1993). Watson–Crick interactions position the intron G and the acceptor 2′-OH group of the lariat A residue, while the internal "guide sequence" of a loop in U5 RNA (spliceosome) or Domain ID (Group II) binds to both the exon–intron junctions (Fig. 13-3). U6 RNA, the most highly conserved of the spliceosome RNAs, is likely a key catalytic player, and recent results in another splicing system have shown a transient, covalent linkage of the splicing substrates to this molecule (Yu et al., 1993).

Hairpins, Hammerheads, and HDV. Viroids and virusoids are RNA-based pathogens. Viroids are capable of causing diseases by themselves, while virusoids are generally

---

Figure 13-1. Secondary structure in Group I introns. (**A**) The conserved core common to Group Ia. (**B**) Secondary structure of the *Tetrahymena* intron. The arrow in the P1 structure shows the site of transesterification. Reprinted from Burke et al., 1987, with permission.

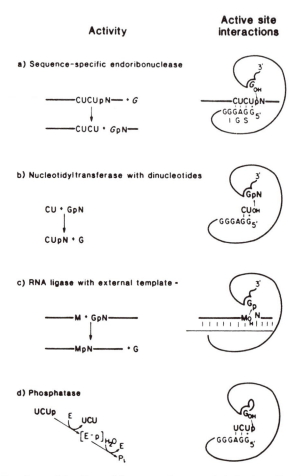

Figure 13-2. Reactions of the Group I introns and truncated derivatives. Reprinted from Cech, 1990, with permission.

satellites of larger RNA viruses. During RNA replication, either of these RNAs is cleaved from a large multimeric precursor to single genome molecules by an RNA re-action (Symons, 1992). In contrast to the reactions of Group I and Group II introns, cleavage is "b-type,"—that is, yielding 3'-phosphate and 5'-hydroxyl termini. This process has been well studied; it involves nucleophilic attack of the 2'-OH group on the phosphorus atom, yielding a pentacoordinate intermediate which resolves to the cleavage products. The surprising aspect of these catalytic RNAs is that so little RNA information is required for hammerhead cleavage. The consensus structure is shown in Fig. 13-4. Only 13 nucleotides are required for activity, but these must be bounded by base-paired regions. Two base-paired regions position the scissile bond; the third region, the "hammerhead," must contain enough base pairs to form a stable structure (Tuschl & Eckstein, 1993). Apparently, formation of the structure strains the scissile bond and favors cleavage at this site (Long & Uhlenbeck, 1993).

Recently, a crystallographic structure has been obtained for the hammerhead ribozyme (Pley et al., 1994). It reveals a number of important structural features of the catalytic RNA. First, the molecule is a Y-shaped entity, with domains I and II of the hammerhead forming the branches of the Y and the third domain forming the stem (see Fig. 13-4). Non-Watson–Crick hydrogen bonds constrain the conserved catalytic core; these are indicated by the lines in Fig. 13-4A. The conserved CUGA sequence between domains I and II adopts the identical conformation as the T-loop of tRNA[Phe]. Only one metal-binding site was observed in the structure near the scissile bond: Mn^{2+} and Cd^{2+} interacted with a phosphate of the conserved A at the base of stem II and with N7 of the adjacent G. These nucleotides are part of the boxed GAR sequence in Fig. 13-4A. It is unclear whether or how this mode of binding relates to the participation of Mg^{2+} in the normal reaction, since Mg^{2+} does not ordinarily coordinate to heterocyclic atoms (see Chapter 9). In three dimensions, however, the bound metal is close

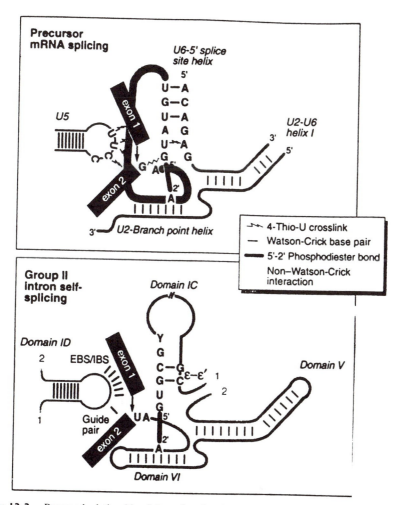

Figure 13-3. Proposed relationship of the active site of Group II introns with the eukaryotic spliceosome core. Reprinted from Wise, 1993, with permission. Copyright 1993 by the AAAS.

(A)

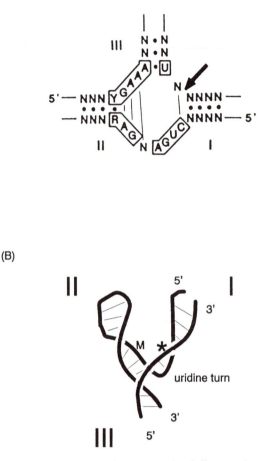

(B)

Figure 13-4. (A) The consensus structure of the hammerhead ribozyme (reprinted from Symons, 1992, with permission). Only a relatively few nucleotides in the hammerhead structure are required for catalysis. The crystal structure determined by Pley et al. (1994) revealed that these required bases form hydrogen bonds (indicated by lines between the residues). (B) Schematic depiction of the tertiary structure of the hammerhead ribozyme as determined by Pley et al. (1994). The "substrate" strand is depicted as lying in front of the hammerhead; the relative positions of the scissile bond (*) and the metal-binding site (M) are indicated. The "uridine turn" analogous to that found in the T loop of tR-NA[Phe] is indicated in the figure. See also Cech & Uhlenbeck (1994) for discussion of the structure.

to the site of cleavage; the relative positions are indicated in Fig. 13-4B. It is likely that the details of ion–ribozyme interaction are different during Mg-assisted catalysis, since the crystal structure requires some slight rearrangement before the position of the metal ion could contribute to the known "in-line" attack on the scissile phosphate to generate a pyramidal intermediate. Nevertheless, the structure of Pley et al. constitutes an important milestone in our understanding of the structural biology of RNA.

Hairpins occur in another group of self-cleaving RNAs. These RNAs require two separate helical domains and two single-stranded regions for activity. Again the scissile bond is cleaved to leave a 2,3'-cyclic phosphate. In common with the structure of the hammerhead, the scissile bond is unpaired and is bounded by helical regions; however, a second domain is required for activity. In vitro selection experiments (see Chapter 14) have shown that active ribozymes have 16 invariant nucleotides while another eight of the positions do not allow all four nucleotides. At four of the invariant positions the 2'-OH groups are essential as well, at least at low Mg^{2+} concentrations. The consensus structure derived from the in vitro selection/mutagenesis analysis is shown in Fig. 13-5 (Benzal-Herranz et al., 1993; Chowrira et al., 1993).

Finally, the hepatitis delta virus, a satellite of hepatitis B virus, contains an internal cleavage site which can function in vitro. This structure is unusual in that it contains a pseudoknot motif [a pseudoknot is a structure in which Watson–Crick pairs form between regions bounded by other helices; for example, the loop sequences of two stem-loops could base-pair (Perotta & Been, 1991; Pleij & Bosch, 1989; see also the RNase P RNA structure discussed below)].

Even Smaller Self-Cleaving RNA Species. How much information is required for ribozyme activity? Not much. The first small sequence to be identified was that of Dange et al. (1990), who examined whether the 15 nucleotides released by circularization of the *Tetrahymena* intron could exhibit ribozyme activity. An RNA containing these nu-

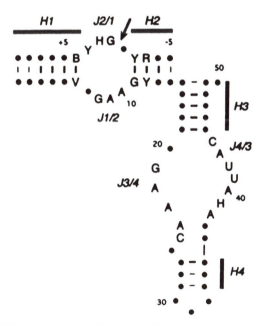

Figure 13-5. The hairpin ribozyme requires a particular structure and some specific nucleotides for activity. Key to the IUB ambiguity codes: Dots represent any nucleotide, dashes represent required pairings, V is "not U" (A,C, or G allowed), Y is a pyrimidine (U or C), R is a purine (A or G), B is "not A" (U,C or G allowed), and H is "not G" (A,C, or U allowed). Reprinted from Benzal-Herranz et al., 1993, with permission.

cleotides was cleaved in vitro in a unimolecular, Mn^{2+}-dependent reaction. Cleavage was b-type, yielding a $2',3'$-cyclic phosphate and a free $5'$-OH group. Even less information than a 15-mer directs specific cleavage: Although the reaction has not been characterized kinetically, formation of a complex between the pentanucleotide GAAACp and the trinucleotide UUU led to release of G from the former substrate (Kazakov & Altman, 1992). Mn^{2+} is key in these reactions. In contrast to Mg^{2+}, it has the capacity to coordinate with the ring nitrogens of the bases (perhaps N7 of the purines in this case), which may position the scissile bond for cleavage. The exact structure of the active RNA is not clear, nor are all helical regions of AAA and UUU hydrolyzed in the presence of Mn^{2+}.

An effort to define the requirements for metal ion-induced cleavage of RNA has been made by using combinatorial techniques (Chapter 14). Pan and Uhlenbeck (1992) studied the sequence-specificity of the Pb^{2+}-induced cleavage of yeast $tRNA^{Phe}$ by randomizing 9 or 10 positions in the D-loop where cleavage occurs. Several small variant loops identified in the whole tRNA were made in isolation. These loops were cleaved in the presence of Pb^{2+}. In a second step, the $2',3'$-cyclic phosphate intermediate was resolved to a $3'$-phosphate product. A variety of sequences share the ability to be cleaved by Pb^{2+}. This result implies that many of the potential sequences in an RNA are competent for a particular chemical reaction, a concept with implications for models explaining the origin of life (Kauffman, 1993).

Ribonuclease P RNA—A True Catalyst. Unlike the self-splicing and self-cleavage systems discussed above, the RNA subunit of *Escherichia coli* RNase P RNA, sometimes called M1 RNA, exhibits multiple turnovers in vitro and in vivo. The normal substrates for RNase P are the tRNA precursors with extended $5'$ terminal sequences. RNase P cleaves these precursors endonucleolytically to produce mature $5'$-termini. In vivo, the holoenzyme exists as a complex of protein and RNA bound to the bacterial membrane (Srivastava et al., 1991). In vitro, the *E. coli* enzyme contains (1) a single RNA molecule that is 377 nucleotides long and (2) a protein subunit that is 119 amino acids in length. The RNA subunits of *E. coli* and *Bacillus subtilis*, as well as those of closely related bacteria, have been shown to be catalytically active in vitro (Guerrier-Takada et al., 1983). Other bacterial RNase P RNAs, although possessing recognizable structural features similar to other RNase P RNA species, have not been demonstrated to be catalytically active, but this negative result is obviously not definitive.

Comparison of the structures of RNase P RNA species has led to a minimal or consensus core structure containing only conserved domains (Fig. 13-6). This molecule is enzymatically active in vitro, although not in vivo (Ramamoorthy & Schmidt, unpublished results; Waugh et al., 1989). In a gap-scan deletion experiment, Waugh & Pace (1993) made small deletions which collectively removed each of the nucleotides in *E. coli* RNase P. All the deleted derivatives retained some activity, indicating that there is no single essential active site residue.

Ribosomal RNA. All of the ribozymes discussed above utilize nucleic acid substrates, either in self-splicing reactions or in catalysis. The discovery of RNA catalysis by Cech and coworkers immediately suggested the possibility of RNA-catalyzed peptide synthesis; there were undoubtedly many late-night experiments that tried to demonstrate peptide synthesis by RNA. They didn't work, largely because the structure of ribosomal RNA (rRNA) is so complex, with many potentially misfolded species being formed

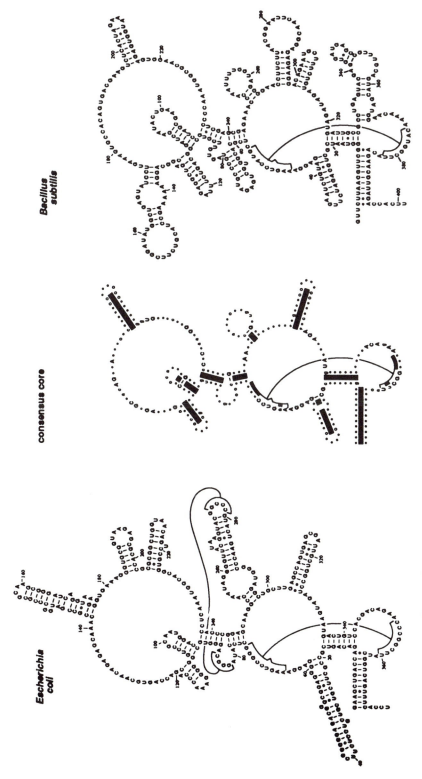

Figure 13-6. Secondary structure of RNase P RNA. Long-range (pseudoknot) pairings in the structure are shown by lines that cross from one domain to another. The consensus core structure (Min1 RNA of Waugh et al., 1989) designates invariant nucleotides in capitals and those that are at least 90% conserved in lower case. Note the bulge U at residue 69 in the pseudoknot. Reprinted from Brown & Pace, 1992, with permission.

Escherichia coli

consensus core

Bacillus subtilis

under the conditions of synthesis. The demonstration that almost all of the protein sub-units could be extracted from the ribosome, leaving most of the protein synthetic activity intact, required the use of rRNA from *Thermus aquaticus*, a bacterium living in hot springs and chosen because its rRNA was presumably more robust (Noller et al., 1992). This synthetic activity remaining after extensive proteolysis and extraction was sensitive to antibiotics known to have rRNA targets. Simultaneously, Cech and cowork-ers showed that the *Tetrahymena* Group I intron could catalyze the hydrolysis of aminoacyl linkages at the end of tRNA (Piccirilli et al., 1992), lending some credence to the idea that the primitive protein-synthetic activity had in fact resided in an RNA molecule.

Catalytic RNA—Methods and Lessons

Identification of RNA Catalysts. Proving the existence of an RNA subunit in an enzyme can be difficult. Physical means can be used first. For example, *E. coli* RNase P activity sediments with a greater buoyant density than do (protein) marker enzymes in an ultracentrifuge; however, some RNase P species, even though they contain RNA, have a density characteristic of pure protein. Conversely, the mere association of RNA with an enzymatic activity, even after several purification steps, can be fortuitous.

RNA catalysts were first identified by purification and enzyme sensitivity. Disso-ciation of the RNA component of a reaction mixture from all the detectable protein left activity in the RNA fraction. This was demonstrated in the first two cases of RNA catalysis: Group I introns (Kruger et al., 1982) and RNase P (Guerrier-Takada et al., 1983).

In both these systems, protease treatment failed to interfere with the RNA-catalyzed activity. The converse experiment, using RNase to destroy the catalytic activity, is somewhat more problematic. Obviously, RNase destroys ribozymes. However, when the same RNA molecule is both substrate and catalyst for the reaction, the results of the experiment cannot be interpreted. Furthermore, even when the substrate and cata-lyst are different molecules, as for RNase P, the essentiality of an RNA does not im-ply its catalytic role; RNAs could be a scaffold, allowing protein subunit(s) to assem-ble into an active cluster.

A more convincing demonstration of RNA catalysis is available if one can synthe-size catalytically active RNA in vitro. This can often be done for RNA species con-taining up to a few hundred nucleotides. RNA polymerase systems derived from phages T7 and SP6 transcribe cloned DNA sequences efficiently in vitro. For larger RNAs, such as ribosomal RNA, the problem becomes much more problematic because these RNA polymerases can terminate at specific sequences or simply dissociate from a large template before reaching the end. Even more difficult to overcome is a variant of the folding problem: A large RNA will perforce contain a number of regions comple-mentary to any small subsequence. Thus, a sequence that forms base pairs with a "noncognate" sequence may be trapped in an inactive conformation. Selection of the right conformation may be a function of protein-mediated folding and rearrangements during assembly of the active RNA–protein complex in vivo or in vitro. This is a vex-ing problem in demonstrating conclusively the catalytic role of ribosomal RNA. In vitro assembly of active *E. coli* 50S ribosomal particles requires a heating step at an intermediate stage of the pathway. Analogous treatment of a full-length rRNA transcript (>2900 nucleotides in *E. coli*) would likely result in having the RNA locked into one or more inactive conformers. Thus, the dilemma is that if a particular protein must bind to the RNA in order for the RNA to be active, how can one conclude that the RNA component is the catalyst? Alternatively, nuclease sensitivity of an enzyme can be used to show the requirement for a (potentially catalytic) RNA subunit; however, an essential RNA component can be insensitive to nuclease because it is masked by protein components of the enzyme. Subsequent genetic or molecular biological analysis could show the requirement for an RNA subunit. A case in point is the RNase P from the archaebacterium *Sulfolobus acidocaldarius* (LaGrandeur et al., 1993). The lesson, of course, is that negative results must be interpreted with caution.

Determining RNA Structure: The Case of RNase P. Given the involvement of an RNA in an enzymatic reaction, it is of great interest to understand its structure. The paradigm motif of nucleic acid structure is Watson–Crick base-pairing, so the first step in understanding the mechanism of a catalytic RNA is to define the pairings available in the RNA sequence. This can't be done by simple inspection in most cases. RNA pairings can be quite short, with as few as three base pairs determining a stable structure. Even a naive calculation would indicate that any given trinucleotide sequence could occur about half a dozen times in the *Tetrahymena* intron on a random basis ($4 \times 4 \times 4 = 64$ trinucleotides; $414/64 = 6.5$). Thus, if we begin with a trinucleotide and look for complementary sequences, we could have a variety of potential pairings. Furthermore, noncanonical Watson–Crick pairings (e.g., G·U) are common in RNAs. Which of the many possible pairings reflects the actual pairing in the catalytic RNA? In the absence of other information there is no possible way of knowing. There are four principal ways of deciding on secondary structure elements in an RNA: *free energy minimization, chemical and enzymatic probing, phylogenetic comparison,* and *mutational analysis.*

Free energy minimization (Zuker, 1989) assumes that the native structure of an RNA corresponds to the lowest free energy state available to the RNA. Thus, if one can determine the free energy changes involved in forming base pairs of each type, one can sum them to determine the preferred secondary structure of the RNA. This would appear at first glance to be fairly easy, since there are only four types of base pairs, but the values for an individual base pair are context- and solvent-dependent. Therefore, the energies of base-pair formation must be determined for a matrix of nearest-neighbor pairs. Any solution depending on a summation of free energies will only be as accurate as the individual free energy values, and the error of the total is the sum of the errors at each step. It is also important to note that these values are always determined in the absence of divalent cation, simply because bound Mg^{2+} can increase the thermal stability of a helix so markedly that it typically doesn't melt below the boiling point of water. Thus a preferred magnesium-binding site might exhibit a somewhat different structure than that determined by simple base-pairing considerations. Additionally, some tetranucleotide sequences, especially UUCG and GAGA, impart extra stability to helices in which they are present. These "tetraloops" contain extensive non-Watson–Crick interactions and hydrogen bonds among the loop nucleotides (Chastain & Tinoco, 1991). Finally, prediction of the most stable structure is difficult computationally for large ribozymes because the time required for the computation increases as the cube of chain length.

In practice, the programs build a family of possible structures whose predicted free energies are within about 10% of each other. The performance of these programs, while improving, is still problematic; not even all transfer RNA cloverleafs can be predicted with certainty.

The second method for determining secondary structure involves *chemical and enzymatic probing*. It is based on the assumption that when nucleotides are involved in a base-pairing interaction, they will be relatively less accessible to digestion with ribonuclease or reaction with small molecules. Conversely, there are also reagents that specifically recognize double-stranded regions of RNA (see Chapters 1 and 5).

Phylogenetic covariation analysis (James et al., 1989) requires only sequence information to determine a structure. As such it can be applied in the absence of detailed functional information about an RNA species. Furthermore, it can be applied to quite complex RNAs such as ribosomal RNA.

The only requirement is for several homologous sequences to be known. "Homologous" in this context implies functional as well as sequence similarity. Given a set of such homologous RNAs, they are then aligned so that the maximum similarity is obtained. This determination requires some judgment on the part of the investigator. Any two RNA sequences can be aligned "completely" if enough gaps are introduced into the lineup (see Fig. 13-7), so some care must be taken to evaluate the statistical significance of the alignment. Usually, a region of maximal evolutionary conservation is taken as the starting point for the alignment.

Once a family of sequences is aligned, regions of potential Watson–Crick complementarity are searched to see if a residue varies between or among members of the family. Then the sequences are searched to see if another change occurs between the same family members so that base-pairing is maintained. For example, a change of G to A between two RNAs must be balanced by a complementary change of C to U between the two molecules in another position. Two independent changes in the same helical region of the molecule are necessary to consider a change as proven.

An exercise in this analysis is given in Fig. 13-8. Using the sequence set given, it is possible to identify two long-range base pairing interactions in *E. coli* RNase P RNA (M1 RNA). For example, the sequence GGGC (nt 82–85 in *E. coli* RNase P RNA) is changed to GAGC in RNAs from *Thermotoga maritima* and *Streptomyces bikiniensis*.

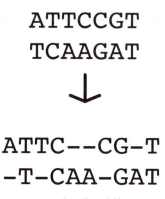

Figure 13-7. Two unrelated sequences can be aligned if enough gaps are allowed. This is a general property of alignment programs and is usually accounted for by setting a gap penalty in the scoring matrix so that too many gaps abort a structural comparison.

	helix 82/276[†]		helix 66/353	
eubacterial consensus[¥]	GGGC	GCCC	AAGuCCGGG[§]	CCCGGCUU
Thermotoga maritima	G**A**GC	GC**U**C	AAGuCCGG**A**	**U**CCGGCUU
Deinococcus radiodurans	GGGC	GCCC	AAGuCCGGG	CCCGGCUU
Streptomyces bikiniensis	G**A**GC	GC**U**C	A**C**GuCCGGG	CCCGGC**G**U
Rhodospirillum rubrum	G**AA**C	G**UU**U	AAGuCCGGG	CCCGGCUU
Agrobacterium tumefaciens	**AAA**U	G**UUU**	AAGuCCGGG	CCCGGCUU
Alcaligenes eutrophus	GGGC	GCCC	AAGuCCGG**A**	**U**CCGGCUU
Thiobacillus ferrooxidans	GGGC	GCCC	AAGuCCGGG	CCCGGCUU
γδ-protobacteria	GGGC	GCCC	AAGuCCGGG	CCCGGCUU
Bacillus brevis	no homolog[‖]		AAGuCC**AG**G	CC**U**GGCUU
Other *Bacillus* spp.			AAGuCC**AUG**	C**AU**GGCUU

Figure 13-8. Phylogenetic analyses showing pseudoknot formation in RNase P RNAs. The pairings are shown in Fig. 13-6. Reprinted from Haas et al., 1991, with permission. Copyright 1991 by the AAAS.

The observation that the complementary sequence GCCC becomes GCUC in the corresponding region (nt 276–279) of these RNAs supports the assignment of these nucleotides to a base-paired structure. The predicted structures may be checked by examining the secondary structure of *E. coli* RNA given in Fig. 13-6.

Covariations of the non-Watson–Crick type can also be detected. This has been spectacularly successful, first in the analysis of rRNA by Woese and coworkers (Woese & Pace, 1993) and later in the analysis of Group I introns by Michel & Westhof (1990).

Phylogenetic analysis is limited only by the number of covariations that can be identified. This points out a major limitation of the analysis: If a sequence does not vary, it cannot be structured. Fortunately, many base pairs in RNAs do covary; the identity of nucleotides in a sequence is less important than the fact of base-pair formation. What kind of sequence would fail to vary? If a particular nucleotide were essential for catalytic function, it most likely would be conserved among all the RNAs in the dataset. Thus, some of the most interesting parts of a catalytic RNA are inaccessible to study by this means. For example, the G nucleoside binding site of the Group I introns was too tightly conserved to be identified by covariation analysis (Michel et al., 1989). Furthermore, it is possible for base pairs to open during catalysis or assembly of the enzyme (see below), and there is no way that phylogenetic covariation analysis can detect these events.

The fourth method involves *mutational analysis* of RNA structure. This method, while logically consistent with phylogenetic covariation, uses a different experimental approach (Haas et al., 1991). Mutations are generated, by either directed or random mutagenesis, within an RNA structure. These mutations cause a loss of function. Regain of function can often be accomplished by a restoration of base-pairing in the RNA. Thus, if, for example, a G·C base pair is mutagenized so that it becomes an A·C mismatch, restoration of function could occur by a second mutagenic event to make an A·U base pair, since Watson–Crick geometry would be restored.

Several conditions are necessary to apply this protocol to the identification of previously unsuspected base pairs. First, it is necessary to have a system to identify functional RNA species. In the case of RNase P, an *E. coli* indicator strain with a deficient protein subunit is phenotypically corrected by introducing the gene for the wild-type

RNA subunit on a multicopy plasmid (Motamedi et al., 1982). Experimentally, the indicator strain is unable to grow at 42°C in the absence of the plasmid. The wild-type RNA, produced in high copy, allows growth of the plasmid at the higher temperature. At 30°C, the strain is able to grow whether or not the plasmid is present.

In order to produce mutants, we take advantage of the properties of hydroxylamine. Hydroxylamine chemically modifies C residues in DNA. The modified C, when used as a template for DNA polymerase during replication, specifies an A residue instead of the normal G. Subsequent replication of the duplex leads to the transition of a G·C base pair into an A·T base pair. When the mutated *rnp* gene is transcribed, a G or C of the wild-type RNA is replaced by A or U, respectively, in the mutant. This mutant is detected by its inability to support growth of the indicator *E. coli* strain at 42°C.

By itself, the initial mutation wouldn't tell much; there are too many ways to ruin a good enzyme for any one of them to be very interesting. All that the first mutation reveals is that the affected residue has some importance for the structure, stability, function, or biosynthesis of RNase P RNA. The structural information comes through a second round of mutagenesis. The plasmid containing the mutated RNA gene is treated with hydroxylamine and transformed into the indicator strain. Recombinant colonies which have received a functional copy of the RNase P RNA gene are identified by their ability to grow at 42°C. Because hydroxylamine is a "monodirectional mutagen," it cannot induce back-reversion of the original mutation which is still present in these second generation plasmids. The further mutational event identified is a second G → A or C → U mutation that restores base-pairing in the helical region disrupted by the first mutation. This RNA is now functional because it allows growth of the indicator bacterial strain. Functionality of the RNA has been restored by reconstructing base-pairing within the region disrupted by the first mutation. Overall, a G·C base pair has been converted to an A·U. Sequence analysis identifies the base-paired residues within the RNA molecule.

Although the mutagenic analysis is similar to phylogenetic covariation in the types of information obtained, there are subtle differences. The base pairs revealed by mutational analysis are all essential for function of the RNA, at least in the assay system. In this sense the information given by mutational analysis is more robust than that given by phylogenetic analysis. On the other hand, nonessential base pairs are inaccessible to mutagenic analysis. Phylogenetic analysis cannot identify structure in domains of an RNA that are unique to an organism because sequences must be compared.

A further advantage of mutagenesis is the ability to identify new structural events in a catalytic RNA (Morse and Schmidt, 1993). During the analysis of mutations affecting *E. coli* RNase P RNA structure we observed that several mutations were reverted phenotypically by one of two transition mutations, either $C_{32} \rightarrow U$ or $G_{48} \rightarrow A$. These two nucleotides occur in the same stem of RNase P, although because of the limited phylogenetic data in this region it was not entirely clear that they were base-paired. The functions of these two nucleotides were examined more closely. There were too many primary mutations reverted by this change for all of them to interact spatially; rather, we hypothesized that disruption of the C_{32}–G_{48} base pair in some way activated the RNA molecule. This proved to be the case. Two RNA sequences were constructed with only a single mutation, either $C_{32} \rightarrow U$ or $G_{48} \rightarrow A$. RNA containing a disruption of the C_{32}–G_{48} base pair supported growth of the indicator strain better than RNA containing a Watson–Crick base pair (either G·C or A·U) at this position. It is likely that the base-pair disruption allows more facile assembly of the active RNase

P ribonucleoprotein complex in vivo. Steady-state kinetic parameters, K_m and V_{max}, of the RNA-catalyzed reaction were indistinguishable from the wild-type values. It should be noted, however, that steady-state analyses do not describe all aspects of an enzymatic reaction.

Minimal Structures and Evolution of RNA Domains. Macromolecular evolution is characterized by two processes: (1) the substitution of single-sequence elements and (2) the insertion of domains within a previous sequence. The latter events are more rare because they are more likely to disrupt an essential structure of the macromolecule. Given this premise, the essential catalytic portion of a molecule should only include domains that are conserved phylogenetically. Thus, minimal structures can be derived for Group I introns and for RNase P RNAs that do not include nonconserved elements; as predicted, these are catalytically active (Fig. 13-6). For example, the domain-spanning nucleotides 62–80 of *Bacillus subtilis* RNase P RNA is not present in *E. coli* RNase P RNA, nor for that matter in any eubacterial RNase P RNA other than those from *Bacillus* species. As expected, this domain can be deleted from the minimal sequence without loss of catalytic activity. Similarly, *E. coli* RNase P RNA contains nucleotides from positions 205–225 that can be deleted while maintaining catalytic activity, even though mutations in this region destroy the ability of the RNA to support growth of the indicator host. This observation implies that in vivo and in vitro activities of the catalytic RNA are somewhat different. For example, nucleotides 205–225 might be required for association with the protein subunit of the holoenzyme.

Mutational analysis of RNase P RNA and phylogenetic sequence comparison both indicated that the loop spanning nucleotides 276–279 of *E. coli* RNase P RNA was base paired to nucleotides 82–85, forming a pseudoknot (Haas et al., 1991). Because disruption of this structure led to the loss of function in *E. coli*, the pseudoknot must be considered essential in vivo. Why, then, isn't the pseudoknot conserved phylogenetically? The apparent paradox can potentially be resolved if we consider the three-dimensional structure of this region of RNase P RNAs. Figure 13-9 shows models of these domains with the phosphodiester backbone of the RNA represented as a continuous line of pencils. The difference between the hairpin of *Bacillus* RNA and the pseudoknot characteristic of RNAs from other genera is simply a discontinuity along one strand. Viewed from the perspective of a protein or substrate molecule, the stem-loop and pseudoknot would present a similar surface for interaction.

We know too little of the details of large RNAs to decide if this proposed example of the conservation of tertiary structure in the absence of primary sequence conservation is a general feature of catalytic RNA structure, but the speculation is interesting. The ability to conserve tertiary structure even though primary structure is altered by substitution, deletion or insertion, may be a key factor in the development of RNA catalysts.

The Protein Subunit. Even though RNase P RNA is capable of carrying out the catalytic reaction in vitro, the reaction requires concentrations of monovalent and divalent cations well above the physiological range. In vivo and in vitro, catalysis at physiological concentrations of cations requires the protein subunit of RNase P.

Genes encoding RNase P protein subunits have been identified by several strategies. First, they can complement *E. coli rnp* mutations (Morse & Schmidt, 1992). For example, a chromosomal library of *Streptomyces* DNA was transformed into *E. coli*

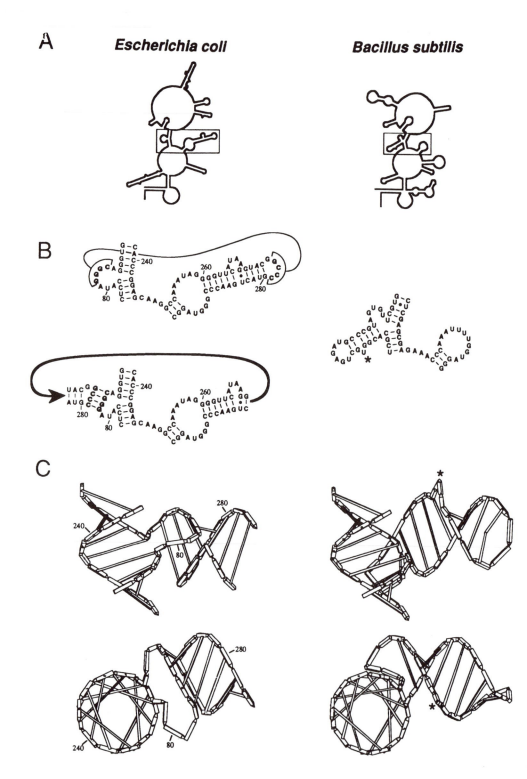

FS101, the same indicator strain used in mutational analysis of the RNA subunit described above. Cloning and sequencing of the complementing DNA revealed that the sequence of part of the predicted protein product was highly homologous to ribosomal protein L34; however, the database search did not identify a homologue of the RNase P protein subunit. In other bacteria, the sequence encoding L34 is closely linked to the *rnpA* gene encoding the RNase P subunit, prompting a more intensive search which revealed the presence of the RnpA translation product. Second, the sequences encoding L34-RnpA are closely linked to the origin of chromosomal replication in eubacteria. RnpA sequences are often identified fortuitously as a result of analyses of origin regions (Calcutt & Schmidt, 1992; Hansen et al., 1985).

Phylogenetic analysis of a number of bacterial RNase P proteins is shown in Fig. 13-10. It is striking how poorly these proteins are conserved, accounting for the failure of database searches to identify the *Streptomyces* sequence, as described above. The main conserved features of the proteins involve a central core region with six conserved arginine and lysine residues. Only 30% of the amino acids are identical between *E. coli* and *S. bikiniensis* RnpA proteins. In contrast the L34 proteins of these species, which are co-transcribed with their respective *rnpA* genes in both organisms, have greater than 65% identical amino acids. Even within the conserved region there are some striking differences—for example, arginine-to-leucine replacements. It may also be noted that the protein is fairly hydrophobic, consistent with its role in anchoring the RNA subunit in the membrane (Srivastava et al., 1991). The only characterized *rnpA* mutant has an arginine-to-histidine replacement in one of the invariant positions. The affected arginine of wild-type RnpA protein is circled in Fig. 13-10. The mutant protein has a lower affinity for M1 RNA (Baer et al., 1989).

The evolutionary maxim that what is important is conserved has the corollary that what varies is not important. The fact that the *Streptomyces* and *E. coli* proteins are functionally interchangeable despite their different sequences implies that most of the sequence is not very important. In molecular evolutionary terms, this means that a variety of proteins could augment the activity of ribozymes; for example, any general RNA binding motif in a protein could serve to stabilize or enhance the activity of a ribozyme. This is a general phenomenon: An evolutionarily unrelated protein enhanced the hammerhead ribozyme reaction in a nonphysiological (bacterial) cell (Tsuchihashi et al., 1993).

What is the molecular function of the protein subunit of RNase P? It seems to act as a general counterion, allowing the approach of the RNA substrate to the active site of the enzyme. Alternatively, the phenotypes of the U_{32} and G_{48} mutations suggest that it facilitates formation of the active conformation of the ribozyme. Under steady-state conditions, the protein subunit also facilitates product release (Reich et al., 1988), the rate-limiting step, but there is little further information available on this point.

Figure 13-9. A model accounting for the conservation of function near the catalytic core of *E. coli* and *B. subtilis* RNase P RNAs. (**A**) The regions of potentially conserved secondary structure are boxed in the model of the overall molecule. (**B**) An alternate representation of the base-paired regions in the *E. coli* pseudoknot compared with the stem-loop from *B. subtilis*. (**C**) In the views at the lower portion of the figure, the pseudoknot (*E. coli*) and stem-loop (*B. subtilis*) are proposed to adopt similar three-dimensional structures. Reprinted from Haas et al., 1991, with permission.

```
Streptomyces bikiniensis           M L P T E N R L R R R E D F A T A V R R G R   R A G R P L L V V H L
Micrococcus luteus                 M L P R D R R V R T P A E F   R H L G R T G T R A G R R T V V V S V
Bacillus subtilis              M S H L K K R N R L K K N E D F   Q K V F K H G T S V A N R Q F V L Y T
Proteus mirabilis        M V K L A F P R E L R L L T P K H F   N F V F Q Q R Q R A S S P E V T I L G
Escherichia coli         M V K L A F P R E L R L L T P S Q F   T F V F Q Q P Q R A G T P Q I T I L G

R S G A T D P H A P G E S A P P T R A G F V V S K A   V G G A V V R N Q V K R R L R H L V C D R L S A L
A T D P N Q T R S T S P S A P R P R A G F V V S K A   V G N A V T R N R V K R R L R A V V A E   Q M R L
L D Q P E N D E L                 R V G L S V S K K   I G N A V M R N R I K R L I R Q A F L E E K E R L
R Q N E L G H P                   R I G L T I A K K N V K R A H E R N R I K R L A R E Y F R L H Q H Q L
R L N S L G H P                   R I G L T V A K K N V R R A H E R N R I K R L T R E S F R L R Q H E L

P P G S L   V V V     R A L P G A G D A D H A Q L A R D L D A A L Q R L L G G G T R
P P L R D L P V L V Q V R A L P A A A E A D Y A L L R R E T V G A L G K A L K P H L P A A S G H A
K E K D Y   I I I       A R K P A S Q L T Y E E T K K S L Q H   L F R K S S L Y K K S S S K
P A M A P   V V L       V R K G V A E L D N H Q L T E V L G K   L W R R H C R L A Q K S
P A M D F   V V V       A K K G V A D L D N R A L S E A L D K   L W R R H C R L A R G S
```

Figure 13-10. Lineup of the amino acid sequences of RNase P protein subunits. Identical amino acids are contained in boxes and chemically similiar amino acids are indicated by dots. Reprinted from Morse & Schmidt, 1992, with permission.

Substrate Features Involved in Interaction with RNase P RNA. Both in vivo and in vitro, the best substrates for *E. coli* RNase P, either the holoenzyme or the catalytic RNA, have a 3'-terminal CCA sequence characteristic of mature tRNAs (McClain et al., 1987; Schmidt et al, 1976; Vold & Green, 1988). In vivo, tRNAs whose CCA terminus is not transcribed have this sequence synthesized before cleavage by RNase P. Preference for the CCA terminus of the RNA is more pronounced in the RNA-catalyzed reaction than in the reaction catalyzed by holoenzyme (Green & Vold, 1988; McClain et al., 1987); however, substrates lacking CCA termini are still cleaved at higher concentrations of Mg^{2+}. Mg^{2+} apparently forms extra contacts that allow the substrate lacking CCA to be recognized by the RNase P RNA (Green & Vold, 1988; Nichols & Schmidt, 1988).

The exact mechanistic step affected by the CCA sequence has not yet been determined. Although not completely characterized at present, the CCA trinucleotide may have one or more unique structural properties that contributes to the observed effects. It is known, for example, that the presence of the 3'-terminal CCA sequence stabilizes the double helical portion of a tRNA acceptor stem, raising the melting temperature by 5°C (Limmer et al., 1993), but itself is relatively unstructured. Interestingly, RNAs made by $Q\beta$ replicase in de novo synthetic reactions all contain CCA termini (Biebricher et al., 1993).

Besides the CCA sequence, the minimal substrate for RNase P RNA cleavage contains two of the four helices of the tRNA cloverleaf (McClain et al., 1987): a 7-bp acceptor stem and a T stem-loop. In the three-dimensional structure of tRNA, these two domains form a continuous helical stack on the "top half" of the molecule; interference with stacking impairs cleavage. Although the 7-bp acceptor stem is characteristic of most mature tRNAs, there are exceptions: The tRNA[His] precursor from *Bacillus* contains 8 bp. The "extra" nucleotide, a G, is base-paired to a C upstream from the CCA sequence (see Fig. 13-11). If this base pair is disrupted by mutation, cleavage occurs at the "normal" site leaving a 7-bp acceptor stem (Green & Vold, 1988). This effect is Mg^{2+}-dependent. At higher concentrations, Mg^{2+} apparently allows the formation of an appropriate structure of the stem and RNA-catalyzed cleavage occurs to leave an 8-bp stem (Carter et al., 1990). This example of the dependence of RNA cleavage on both substrate sequence and conformation (especially as affected by Mg^{2+}) is apparently general. For example, the substitution of deoxynucleotides immediately 5' to the scissile bond affects the chemical step of cleavage so that the reaction requires three Mg^{2+} ions rather than the two required for cleavage of a ribonucleotide linkage (Smith & Pace, 1993). Thus, hydrated Mg^{2+} "substitutes" for a ribose 2'-OH group.

Further complicating the models of substrate recognition are sequence effects. These are somewhat surprising, given that there is no recognizable consensus sequence at the substrate cleavage site. Nonetheless, studies with a single precursor molecule, that of tRNA[Tyr]$_{SuIII}$, indicate that important nucleotides are located at positions -2, -1 and $+1$ relative to the cleavage site, as well as at the base of the T stem (Svard & Kirsebom, 1993). It is unclear how this anomaly will be resolved; perhaps, as in some examples of promoter recognition, there are "forbidden" nucleotides for RNase P interaction.

Features of the Enzyme Involved in Substrate Recognition. RNase P RNA, like protein enzymes, has an active site. Photoactivatible tRNAs, when used as substrate analogues, modify a small number of the nucleotides in RNase P RNA (Fig. 13-12; Nolan

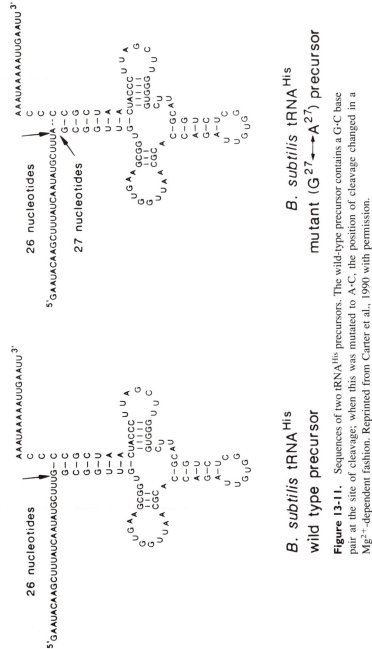

B. subtilis tRNA^{His} wild type precursor

B. subtilis tRNA^{His} mutant ($G^{27} \longleftrightarrow A^{27}$) precursor

Figure 13-11. Sequences of two tRNA^{His} precursors. The wild-type precursor contains a G·C base pair at the site of cleavage; when this was mutated to A·C, the position of cleavage changed in a Mg^{2+}-dependent fashion. Reprinted from Carter et al., 1990 with permission.

et al., 1993). In the first application of the technique, the 5′-end of mature tRNA was used as a substrate analogue. The greatest level of modification (primary labeling) was at A_{248}, and secondary labeling was at nucleotides 330–333. In the structure shown in Fig. 13-12, these residues are potentially brought close together by the pseudoknot between residues 66–74 and 353–360. In an extension of the technique, the polymerase

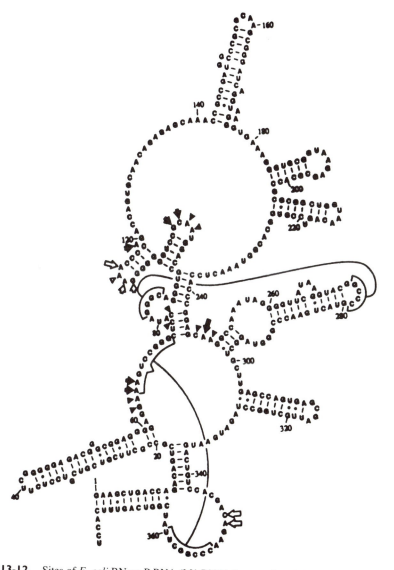

Figure 13-12. Sites of *E. coli* RNase P RNA (M1 RNA) that are photolabeled by tRNA precursor analogues. The arrows show regions of the ribozyme that interact with substrate; the extent of reaction is indicated by the length of the arrow(head). Closed and open arrows refer to crosslinking observed using substrate analogues with reactive moieties at different nucleotide positions. Reprinted from Nolan et al., 1993, with permission. Copyright 1993 by the AAAS.

chain reaction was used to generate circularly permuted templates from a tandemly repeated tRNA gene. When the transcripts were transcribed in vitro with T7 RNA polymerase, the substrate analogue had a 5′-end at either nucleotide 64 or 53 of the tRNA sequence, while the usual 3′-CCA sequence and scissile site were tethered together. Photoaffinity labeling of nucleotides 53 and 64 indicated that the central core and the loops from nucleotides 90–100 were labeled. These conserved loops may have an unusual structure, as *E. coli* RNA polymerase pauses during in vitro transcription of this region (Ramamoorthy & Schmidt, 1991). Additionally, nucleotide 92 has been proposed as an important "discriminator site" in the structure (Forster & Altman, 1990).

Structure and Catalysis: Group I Introns. Any model of Group I intron structure and catalysis must account for the fact that the intron participates in at least three ways: It binds oligonucleotides at the scissile bond, effects a transesterification reaction using guanosine residues, and releases product. The following discussion highlights essential lessons learned from the analyses of Group I intron structure and function; most of the analyses of RNase P RNA described above have also been carried out on Group I molecules. These are not described again unless they illustrate some unique feature of these latter ribozymes.

In addition to positioning the scissile bond, the intron must also bind the G residue for the initial transesterification reaction. Binding is specific for G ($K_m = 20$ μM). A variety of guanosine analogues bind to the G site of the *Tetrahymena* intron (Bass & Cech, 1984). From the structures of the analogues tested, it can be concluded that the transesterification reaction requires both ribose hydroxyl groups (all ribose substitutions are inactive) and a hydrogen bond acceptor at position 2 of the purine ring ($N2$-methylguanosine is acceptable but $N2,N2$-dimethylguanosine is not). If there is a substituent at C6 of the purine ring, it needs to be an H-bond acceptor (isoguanosine does not support splicing). The structure–function analysis is consistent with the finding by Yarus (1988) that arginine is a competitive inhibitor of guanosine binding. This observation implicated the same amino and ring nitrogen atoms as did the study of the guanosine analogues, since the guanidinium group of arginine and the guanosine ring are able to contribute H-bond donors with identical geometry (see Fig. 13-13).

The G-binding site within the intron was found by a series of related logical steps (Michel et al., 1989) (Fig. 13-13). First, the binding site must be conserved among Group I introns. [The ability to incorporate labeled G into purified cellular RNA has been used as a screening assay to identify organisms containing these species (Gott et al., 1986)]. Second, the Group I reaction involves not only G attack on the 5′-end of the intron but also attack of the 3′-terminal G of the intron on the cyclization site— that is, the phosphodiester bond between nucleotides 15 and 16. The 3′-terminal G of the intron and the attacking guanosine nucleotide must therefore be able to bind to the same site. A search for phylogenetically conserved residues within the catalytic core of the intron suggested a G·C base pair; when this pairing was changed to an A·U, the K_m of the intron for guanosine decreased by two orders of magnitude. This by itself would not necessarily be convincing. The key to extracting useful information from the mutant was to resurrect the enzyme activity by taking advantage of the stereochemistry of forming a hydrogen-bonded triple between the guanosine and a G·C base pair (Fig. 13-13). When the G·C was replaced by an A·U, N1 of the attacking guanosine would have to interact with the hydrogen-bond donor of the exocyclic amino group

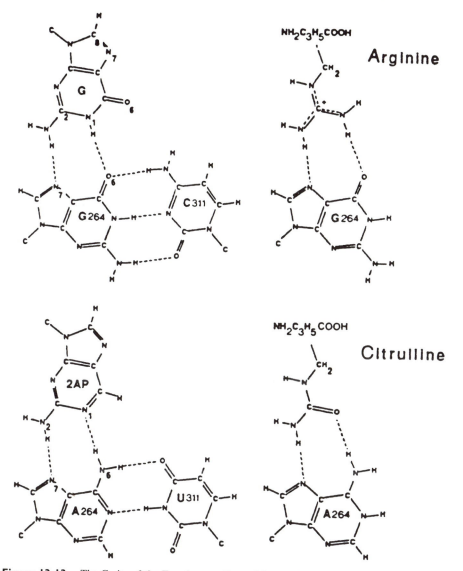

Figure 13-13. The G site of the *Tetrahymena* Group I intron. The contacts with a G·C base pair (wild-type) made by arginine and guanosine, as well as those with an A·U base pair (mutant) made by citrulline and 2-aminopurine, are shown. Reprinted from Michel et al., 1989, with permission.

of the base-paired A. On the other hand, N1 of 2-aminopurine, the ribonucleoside of which was known to support splicing, is an acceptor of H-bonds. In fact, 2-amino-purine ribonucleoside was capable of supporting ligation of the mutant intron under conditions where guanosine would not. Furthermore, arginine inhibition of splicing did not occur, since arginine forms the same H-bond system as does guanosine; rather, cit-rulline, in which an oxygen replaces a NH group of arginine, was found to inhibit the reaction dependent on 2-aminopurine. This combination of structure–function analyses on both the enzyme and its small molecule substrate firmly established the identity and nature of the G-binding site.

Irreversibility of Intron Excision. Self-splicing of the *Tetrahymena* intron requires no external energy source, only a G nucleoside. How, then, can the process be essentially irreversible? The irreversibility must be a function of the intron itself and is provided by successive transesterification reactions; that is, the phosphodiester bond at the intron–exon junction is exchanged for another phosphodiester linkage. Mechanistically, the first reaction is a nucleophilic attack by the 3'-OH group of the G nucleoside on the phosphodiester bond of the exon–intron junction. This results in a free 3'-OH on the exon which then attacks the downstream intron–exon junction, leading to formation of a complete exon and release of the intron. If the intron were now to remain intact, it could, in principle, reverse the reaction by using its free 3'-OH group to attack the joined exon segments in what would constitute simple reversal of the forward reaction. In many, but not all, Group I species a second reaction, namely intron cyclization, renders the intron catalytically incompetent and ensures irreversibility (Fig. 13-14). According to this process, the free 3'-OH group of the intron attacks an internal phosphodiester bond. This leads to a circular intron, lacking either 15 or 19 residues from the 5'-end of the intron. The former ("L-15") ribozyme can be linearized in a further transesterification step and recircularize to release four more nucleotides. This last form ("L-19"), which lacks 19 5'-terminal nucleotides, can function as a true hydrolytic enzyme, exhibiting catalysis in vitro (Cech & Bass, 1986; Zaug et al., 1984).

Substrate Binding by Base Pairing. Within the full intron, or with an exogenous substrate, the sites of reaction are determined by Watson–Crick or wobble base-pairing between an internal guide sequence (part of the P1 stem) and the scissile sequence. Scissile sequences are relatively susceptible to hydrolysis in the absence of nucleoside, indicating that the secondary and tertiary structures of the intron not only position the bond for hydrolysis but also facilitate the reaction. One key to this facilitation arises from unusual base-pairing. The last nucleotide of the exon is a U paired with a G; formation of an ordinary Watson–Crick base pair at this site inactivates the self-splicing process.

Group I enzymes all recognize and position their substrates by base-pairing. This mechanism has consequences for the accuracy and efficiency of catalysis by the L-19 intron species. In the steady state, the rate of reaction is usually determined by prod-

---

Figure 13-14. A variety of successive transesterificiations catalyzed by the *Tetrahymena* Group I intron lead to intron excision and truncation. The first two transesterification reactions lead to the replacement of the 5' exon-intron u-A bond with **G**-A. Then the free 3' OH group of the exon displaces the terminal **G** leading to exon ligation. The linear intron (L-IVS), free of the exon, can undergo further transesterification reactions leading to the circularized intron species C-IVS and C'-IVS, together with the small 15-mer and 19-mer hairpins. C-IVS can reopen if a suitable oligonucleotide binds to the guide sequence at the active site. The linear Group I intron used for most studies of catlysis in *trans* (see Fig. 13-2) is a transcript of the intron that lacks the 3' terminal 21 nucleotides. Intron sequences are in capital letters, exon sequences are in lower case, and the attacking G nucleotide is shown in bold. Reprinted from Cech & Bass, 1986, with permission.

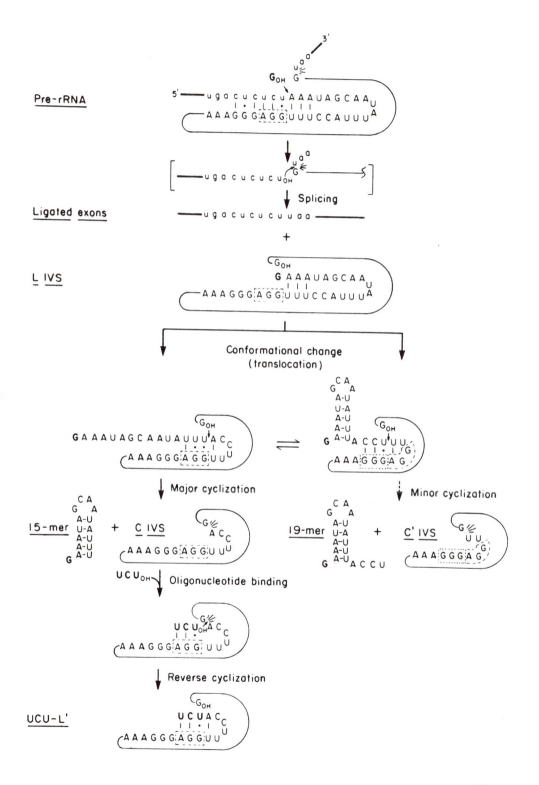

uct dissociation; chemical steps of catalysis are not observed. It follows that a mutation that reduces the affinity of product for enzyme will increase the rate of reaction. Thus, an enhancement in the catalytic rate constant k_{cat} (i.e., the rate when substrate is saturating) is achieved by mutations in the $J_{1/2}$ region which decrease the affinity of the product oligonucleotide for the enzyme. Because substrate and product both bind by base-pairing to the guide sequence, these mutations also decrease the affinity of the substrate for the ribozyme.

Under low substrate conditions, where formation of the enzyme–substrate complex is rate-limiting, the extremely tight binding of substrate oligonucleotide to the guide sequence (K_d in the nanomolar range) of the wild-type ribozyme means that the substrate never diffuses from the enzyme and all bound molecules are cleaved. Under these conditions, there is relatively little specificity of the wild-type ribozyme in that mismatched substrates are cleaved almost as fast as matched substrates.

Specificity under relatively low substrate conditions is increased by the same $J_{1/2}$ mutation that decreases overall affinity for the RNA substrate and product. A general decrease in RNA affinity means that the mismatched substrate has the opportunity to release from the enzyme before chemical cleavage while the matched substrate still is bound tightly. Thus, a second step, k_{off}, is introduced into the rate expression for the enzyme (Fig. 13-15).

While specificity of the ribozyme is increased by the mutation in $J_{1/2}$, fidelity, defined as the ability to cleave fully matched substrates at the correct position, is reduced (Herschlag, 1991; Young et al, 1991). Besides Watson–Crick base-pairing of the substrate to the P1 region, tertiary interactions occur between 2'-OH groups on the two nucleotides at positions immediately upstream from the transesterified nucleotide and the ribozyme (position A_{302}, in the $J_{7/8}$ region) (see Fig. 13-16). Because the $J_{1/2}$ mutation is unable to form tertiary interactions well, bound substrate is positioned inaccurately for the transesterification step.

Synthesis by Transesterification. Transesterification reactions are used occasionally in organic synthesis, as in the synthesis of polyvinyl alcohol from polyvinyl acetate. In this reaction, polyvinyl acetate, made by condensation of vinyl acetate, is treated with acidic methanol. The methyl acetate formed by the transesterification reaction is continuously distilled off, leaving the nonvolatile polyvinyl alcohol in quantitative yield. Because one ester bond is exchanged for another, the reaction must be entropically driven—by distillation in the above example. A further feature of transesterification reactions is that the number average polymerization of a polymer system

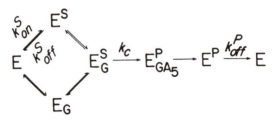

Figure 13-15. Kinetic reaction scheme for the site-specific endonuclease by transesterification reaction of the *Tetrahymena* intron. Reprinted from Young et al., 1991, with permission.

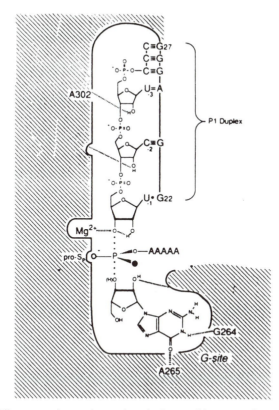

Figure 13-16. Ribozyme–substrate interactions in the transition state of a transesterification catalyzed by the *Tetrahymena* intron. Reprinted from Cech et al., 1992, with permission.

does not change, since one bond between monomeric elements is exchanged for another.

Transesterification for synthesis of polynucleotides is carried out by the L-19 Group I intron and derivatives. The transesterification substrate is prepared with an extra G at the 5'-end. Base-pairing of this substrate to the guide sequence allows the transesterification reaction to proceed with the release of guanosine. The entropy to drive the reaction comes from release of the nucleoside.

In a further modification of this reaction, it was possible for Doudna et al. (1991) to remove the covalent linkage between the substrate and catalytic portions of the molecule. As would be expected, this reaction was much less efficient than ligation to an intramolecular substrate; however, the "ribozyme-in-pieces" could be improved by mutagenesis and selection (Green & Szostak, 1992). The reaction still required base-pairing of the substrate oligonucleotide to a template. This was followed by transesterification to yield a joined product.

Can this bit of molecular sorcery be applied to the models for prebiotic evolution? Clearly there had to be a replication system for the first genomes. This could be envisioned as involving the successive transesterification of shorter oligonucleotides by a Group I-type molecule. The difficulty with this model seems to be the replication of the replicator itself. Group I enzymes require a complex secondary and tertiary structure for activity. Simultaneously removing this structure (to allow it to be a template) and stabilizing the structure (to allow it to be a replicator) would seem to be a fairly daunting prospect. On the other hand, if the replicator and replicated molecules were distinct species, replicators would have to evolve twice, so that more than one molecule of the replicator would have to be available. There would seem to be two ways out of this dilemma. The first replicator molecule could be a repeated sequence, which would allow it to possess substrate and catalytic domains simultaneously. Although there is speculation that the tRNAs contain an underlying GNC motif suggesting evolution from an initial trinucleotide repeat (Eigen, 1992), no catalytic RNAs have been analyzed for this property. Additionally, it is possible that RNA catalysis was more common than we think. Among the Group I ribozymes, relatively few nucleotides are absolutely conserved and these seem to be involved in directly positioning either the attacking guanosine or the substrate oligonucleotide for attack. Most of the other individual nucleotides are less conserved. This means that, consistent with studies of mutated and truncated RNase P RNAs, most of the ribozyme is important for its secondary and tertiary structure, rather than catalysis. Because random RNAs will spontaneously associate into internally base-paired structures, there may have been many individual RNAs in a primordial RNA mix that fit the role of primitive catalyst. The existence of a relatively limited set of secondary structures would have to have obtained in spite of the vanishingly low probability of there being more than one copy of a single sequence more than 100 bases in length.

Because the synthetic reaction is favored by rapid product release, which is enhanced by inaccurate base-pairing (see above), synthesis of RNA by Group I-type enzymes may be inherently error-prone. Rapid evolution requires error-prone replication to proceed (Eigen, 1992), so this property of ribozymes is consistent with their proposed role in prebiotic evolution.

The Chemistry of RNA Catalysis

All known ribozymes require metal ions to cleave or exchange phosphodiester bonds. Because proteins are much more accurate and rapid catalysts, it has been suggested that some unique chemical features of this system has allowed ribozymes to occupy this ecological niche away from direct competition from proteins.

The Importance of Being Phosphate. In a seminal article, Westheimer (1987) discussed unique properties of phosphate esters and anhydrides that make them unique players in biochemistry. When a phosphate is cleaved from an ester or an anhydride linkage, the free energy change is negative, on the order of 1–4 kJ, respectively. This corresponds to an equilibrium constant of 10^3–10^8 in the direction of hydrolysis. This large free energy of hydrolysis is due, among other things, to the fact that phosphates are negatively charged at neutral pH. Hydrolysis allows the negative charges in a nucleic acid or phosphodiester to be separated and then further delocalized by resonance. Paradoxically, phosphate esters are kinetically stable, because the same localized negative charge repels nucleophiles such as OH^- that would otherwise attack the phosphorus atom. The detailed intermediate forms of phosphate that occur during ribozyme catalysis are unexplored.

The Importance of Being Magnesium. Of all the metal ions available to serve as co-factors, Mg^{2+} is preferred by the naturally occurring ribozymes. This is despite the widespread availability of other metals in the environment and the widespread use of other metals (e.g. Zn) by protein enzymes. The RNase P reaction can use Ca^{2+} instead of Mg^{2+} as a cofactor for cleavage, although the rate of cleavage is lower by a factor of about 10^4 (Smith & Pace, 1993). Likewise, the Group I intron reaction can employ Ca^{2+}, and variants of the ribozyme have been isolated that can use Ca^{2+} more efficiently than the wild-type molecule (Lehman & Joyce, 1993). If Mg^{2+} is not unique, it certainly is favored. Why?

One hypothesis is historical: At the time of the RNA world, Mg^{2+} was more available in the environment. Ribozymes still use Mg ions because any mutation that made the reactions work in the presence of another metal would have yielded a less fit ribozyme than would the Mg-dependent molecule. This argument cannot be dismissed without a more thorough understanding of the early Earth and the conditions which led to the rise of the first catalysts. It should be noted that all cells actively exclude Ca^{2+} ions; this may reflect its relatively lower concentration at the time metabolism arose, prior to the first cellular forms.

Another idea invokes the unique properties of magnesium. First, Mg^{2+} lowers the apparent pK_a of water from 15.4 to 11.4. If this is significant for the chemistry of catalysis, then ribozyme reactions should be accelerated at higher pH. This seems to be the case, although the earlier literature is confusing. The confusion arises from the rate-limiting step (k_{cat}) of the steady-state reaction being product release; this has a very weak pH-dependence. On the other hand, single-turnover experiments (Smith & Pace, 1993) indicated a unimolecular dependence of the rate constant for the chemical step of cleavage on $[OH^-]$. These same experiments showed a bimolecular dependence of the rate on $[Mg^{2+}]$. Any mechanism for ribozyme-catalyzed hydrolysis should therefore predict the observed kinetic dependence of the chemical step on two metal ions, one hydroxide, and substrate. Phosphate reactions relevant to ribozyme action have been incompletely modeled, but Herschlag & Jencks (1990) determined that hydrolysis of phosphorylated pyridines by $Mg(OH)^+$ showed pH-dependence similar to that later found for the RNase P reaction. They ascribed two roles to the different Mg ions: one to complex with substrate and the other to act as a chemical catalyst, giving a linear rate increase with concentration.

Mg^{2+} has another effect. The elimination of essential features for substrate binding can be overcome by increased $[Mg^{2+}]$ (Green & Vold, 1988; Nichols & Schmidt, 1988; Smith & Pace, 1993). The mechanism by which this occurs is undefined but is unlikely to be merely the dissipation of electrostatic repulsion because monovalent cations and other divalent cations are not effective. Either the substrate requires a conformational change that is Mg-dependent or the hydrated ion forms some sort of bridged complex between substrate and ribozyme, compensating for lack of other contacts.

A General Chemical Model for Catalysis. Because Group I and RNase P RNA ribozymes kinetically require two Mg ions for catalysis, any transition-state model must have this feature. Also, a model should account for the inversion of configuration at phosphorus, which is a feature of ribozyme-mediated transformations (Cech, 1990). Analogies with the known mechanisms of RNA hydrolysis by RNase A and hydroxide ion provide a reasonable picture of "b-type" cleavage (i.e., that yielding a 2',3'-cyclic phosphate product or intermediate).

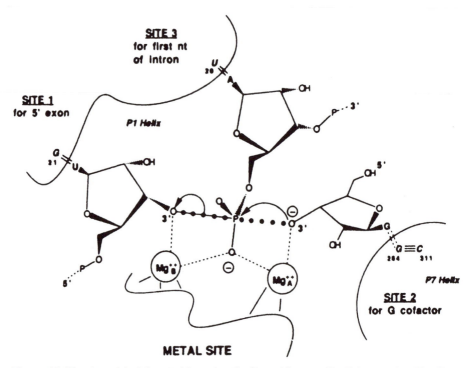

Figure 13-17. A model of the transition state of a Group I intron self-splicing reaction. The G cofactor at the right is positioned to attack the scissile bond and deprotonated by the Mg^{2+} at the right (A), acting as a Lewis acid. The scissile bond is destabilized by the coordination to the Mg^{2+} at the left (B). A somewhat more general variant of this reaction scheme has been proposed by Pyle (1993). Reprinted from Steitz & Steitz, 1993, with permission.

The mechanism of "a-type" cleavage, to yield 3'-hydroxyl and 5'-phosphate groups, is more obscure. The crystallographic structure of a protein exhibiting a-type cleavage activity, namely the editing (3' → 5') hydrolytic activity of *E. coli* DNA polymerase, provides an analogy for this process (Pyle, 1993; Steitz & Steitz, 1993). In this proposed transition state, two metal ions coordinate to one lone-pair oxygen of the scissile phosphate, thus sharing the negative charge. One of these metals also contacts the 3'-oxygen of the scissile bond; the other coordination sites are supplied by water, which in turn is hydrogen-bonded to amino acid side-chain residues. This is analogous to the "binding Mg" described by Herschlag & Jencks (1990). The other metal ion positions the nucleophilic OH^- for attack at phosphorus, analogous to the "catalytic $MgOH^+$" of Herschlag & Jencks (1990). This scheme (Fig. 13-17) readily accounts for inversion at phosphorus and can accommodate varied nucleophiles such as 2'-OH groups of ribose (Group II introns) and the transesterification intermediate (Group I introns), in addition to OH^- (RNase P). With slight modification, it also can account for b-type cleavage characteristic of hammerheads, hairpins, and delta ribozymes (see above). The mechanism must be tested by kinetic means such as isotope effects before it can be accepted, and thereby contribute to an understanding of the nature of RNA catalysis at a chemical level.

Summary

Catalytic RNA enzymes, also known as *ribozymes*, are hypothesized to form a link between genetic and metabolic information systems. Naturally occurring ribozymes primarily catalyze phosphodiester bond formation and cleavage, although there are indications that the ribosomal peptidyl transferase reaction may also be RNA-catalyzed. Ribozyme activity depends on the secondary and tertiary structure of the enzyme, metal ion cofactors, and, in vivo, protein cofactors. Methods of studying ribozyme reactions are discussed as applied to the Group I introns and ribonuclease P RNA. The commonalities of ribozyme reactions have led to proposals that all phosphodiester bond cleavage and ligation reactions may be explained by the formation of a pyramidal phosphate intermediate. Metal ions, water, or nucleoside cofactors participate in the resolution of the intermediate.

The Polymerase Chain Reaction and In Vitro Random Selection

Reuel B. Van Atta and Sam J. Rose

The introduction of the polymerase chain reaction (PCR) and its extension into numerous subprocedures has unquestionably been the most useful addition to the molecular biologists' technical capabilities in the last 10 years. Use of PCR has allowed scientists to tackle experiments literally impossible to perform before the invention of in vitro DNA amplification technology. PCR has rapidly become such a fundamental technique that a journal has been introduced exclusively devoted to documenting the expanding methods and applications; the journal *PCR Methods and Applications* began publication in 1991.

In the section of this chapter devoted to PCR, the initial focus is on the technique itself—that is, the biochemical and mechanical requirements for carrying out PCR. A selection of applications follows that is intended to give the reader a flavor of this broadly useful procedure. An exhaustive review of the literature on PCR will not be found here. Indeed, a review of PCR reviews would be a challenging undertaking due to the explosion of research papers using and referencing PCR. Finally, at the end of this section is an appendix which focuses on the practical aspects of getting the reaction to run smoothly and reproducibly in the laboratory.

The basis for Darwinian evolution might aptly be described as in vivo random selection. Random point mutations arising within the DNA genome caused by inaccuracies occurring during replication or by environmental factors translate into alterations within the functional elements thus encoded. For the most part, the altered phenotypes incurred impose effects which are deleterious to the survival of the organism and are subsequently eliminated by environmental selective pressures. Occasionally, mutations promote phenotypes which are neutral or confer enhanced survivability upon the organism and thus are incorporated into the genotype and passed on to successive progeny. Natural evolution is a slow process in which phenotypic variation occurs over the time course of many generations. Because new genotypes derive from mutations in preexisting genomic DNA, new phenotypes are heavily influenced by wild type.

In the laboratory, the study of the mechanism by which mutations in genotype influence phenotype is normally undertaken in a manner that reflects natural processes. To evaluate the importance of regions within a particular sequence, mutations are introduced at specific nucleotide positions in the DNA and the performance of the resulting phenotype is evaluated. While this method of "site-directed mutagenesis" often provides valuable information pertaining to sequence–function relationships, the process is labor-intensive and gross changes in phenotype are rarely observed.

A new technique termed in vitro random selection was originally developed as an extension of site-directed mutagenesis. The primary advantage over past mutagenic techniques lies in the ability to screen, simultaneously, a myriad of randomized sequences, instead of a few point mutations; in vitro random selection does not rely on any preconceived idea of which sequences are potentially functional. In general, in vitro strategies follow a pathway composed of random mutation, selection, and amplification steps, although the exact procedure varies depending on the prospective application.

The Polymerase Chain Reaction

Before beginning a discussion of the theory and technical aspects of PCR, it is important to point out what is *not* known about in vitro nucleic acid amplification. The physical biochemistry of DNA polymerase and its interactions with various DNA templates (or DNA "targets") as a function of buffer, temperature, and primer composition is poorly understood. Increasing this uncertainty is the relative lack of rigorous enzymology applied to the thermostable polymerases used in PCR. TAQ DNA polymerase is by far the most extensively studied polymerase enzyme for PCR in use today (Lawyer et al., 1993). Therefore, a rational algorithm for choosing a set of reaction conditions for application to a novel nucleic acid target sequence that will guarantee successful PCR amplification does not yet exist. Frustratingly, it is not always possible to guarantee successful DNA amplification by adopting a set of reaction conditions from another laboratory which work perfectly well on the identical DNA target. Discrepancies in the results between labs can arise due to the narrow tolerances in maintaining the proper reaction conditions that are inherent to certain PCR experiments (Rychlik et al., 1990). Experience in our laboratory has taught us that when anyone begins PCR, even when published experimental conditions are consulted for a specific DNA target, one should view the exercise as an *experiment* wherein the PCR reaction conditions should be varied *in the very first trial*. At the very least, the experimenter should include testing a range of magnesium salt concentrations (see Appendix).

Figure 14-1 illustrates the basic protocol of PCR. A double-stranded DNA target is shown (stage 1) together with two oligodeoxyribonucleotides (oligomers or "primers") A and B (Birkenmeyer & Mushahwar, 1991). Theoretically, the DNA target may need to be present only as a *single* molecule in the reaction in order to initiate amplification. The primers A and B are designed to anneal specifically to the two complementary DNA target strands and are typically present in excess at a concentration between 0.1 and 1.0 μM each. Heating the solution containing the DNA target to approximately 92°C–95°C denatures the DNA, rendering it single-stranded. As the reaction is cooled, perhaps to 55°C, the high primer concentration drives a rapid annealing between the two primers and their respective complementary DNA sequences, reaching half-saturation of the DNA target strands in seconds (stage 2). Certain DNA polymerases can recognize and bind to this primer/template complex and utilize deoxyribonucleoside triphosphates (dNTPs) to initiate synthesis of a complementary DNA strand extending from the 3'-terminus of the primers A and B (stage 3). The temperature of the

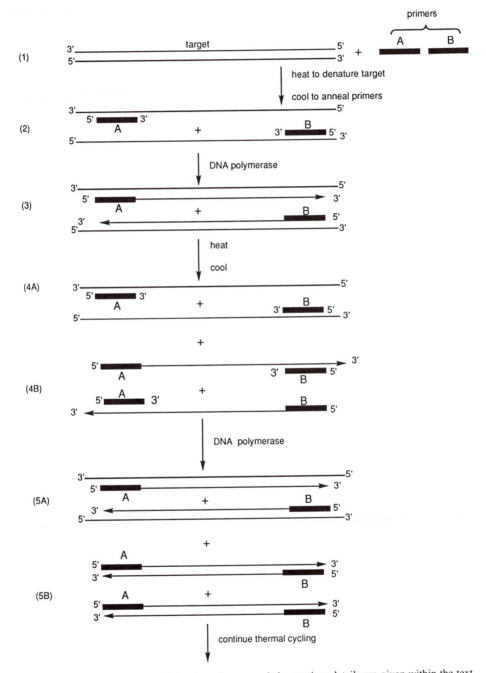

Figure 14-1. Basic protocol for the polymerase chain reaction; details are given within the text.

reaction is usually raised to between 72°C and 75°C, which is the optimum temperature for thermophilic DNA polymerase activity. Sufficient time is allowed for extending the new DNA strands past the primer annealing sites before the process is repeated. A second round of temperature elevation denatures the product DNA strands. In stage 4A, the primers A and B again find the original DNA templates and bind to them, just as in stage 2. Additionally, the two primer-extended DNA products from stage 3 anneal with primers A and B (stage 4B). Another round of polymerization extends the 3'-ends of the complementary strands along these primed templates, producing the DNA products shown in stages 5A and 5B.

To summarize, primer extension at stage 3 created *two* double-stranded DNA regions between the two priming oligomers from the *single* initial duplex originally present at stage 1. At stage 5, *four* double-stranded products had been synthesized from the two double-stranded DNA products present at stage 3. This three-temperature regimen comprised of steps involving DNA denaturation, primer annealing, and primer extension will, in theory, double the number of DNA template strands after each round of thermal cycling, resulting in an exponential expansion in the number of DNA templates. The dominant product of this "chain reaction" will be double-stranded DNA with a length extending from the 5'-terminus of primer A to the 5'-terminus of primer B. This molecule is referred to as the "amplicon"; if the DNA duplex present at stage 1 was part of a mixture, amplification of this particular DNA duplex can be viewed as a purification of a very specific (and perhaps rare) DNA fragment of interest from a complex population of nucleic acid sequences originally present in the PCR reaction mixture.

By performing a sufficient number of thermal cycles, the target DNA will multiply until the amplicon concentration reaches approximately 10–100 nM. This plateau is reached due to the combination of a number of factors which include (1) a limiting amount of DNA polymerase (typically about 10^{11}–10^{12} molecules), (2) the diminution in polymerase activity caused by repeated exposure to high temperatures, and (3) the competition of primer annealing to template versus template annealing to complementary template. The concentration of primers and dNTP's is not limiting in standard PCR.

While the three temperature regimen for PCR is the standard protocol, cycling between a DNA denaturation temperature and a lower temperature suitable for *both* annealing and primer extension is appropriate for many DNA targets. Numerous thermostable DNA polymerases are now commercially available for catalyzing PCR that are capable of withstanding repeated cycling at elevated temperatures. A variety of thermal cycling instruments are commercially available which are user-programmed and efficiently transfer heat to and from the plastic tubes or trays containing the PCR reaction mixtures; some of these have very short temperature cycling periods.

Polymerase Chain Reaction Applications. The first published application of PCR appeared in December 1985 (Saiki et al., 1985). This paper described the amplification of human β-globin genomic sequences for the diagnosis of sickle cell anemia. In the 9 years subsequent to this publication, the use of PCR to address a wide variety of research and commercial applications has exploded. There are several reasons why PCR has made such a profound impact on the way laboratories manipulate nucleic acid sequences.

First, PCR provides the researcher with an ability to seek out and amplify minute quantities of specific nucleic acid sequences present in a complex mixture of nontarget polynucleotides. Traditional molecular recombinant techniques involving ligation of DNA into cloning vehicles and subsequent screening of phage or bacterial libraries can, in many instances, be replaced by a particular PCR technique. PCR requires only partial knowledge of the sequence of the target polynucleotide in order to permit the design of primers. The technique is fast; an amplification experiment can be completed within several hours. Screening of the PCR-amplified DNA using nucleic acid probes or by restriction enzyme digestion can be completed in a day, thereby achieving what would have otherwise taken a week or more with traditional cloning techniques (Arnheim & Erlich, 1992). Because of the high input capacity of modern thermal cyclers, many individual PCR experiments may also be performed simultaneously.

Second, PCR can be very tolerant of the purity of the DNA template it is used to amplify. DNA in crude cell lysates from a variety of sources have proven to be excellent templates (White, 1993) without additional purification, further decreasing the work required to search for polynucleotide targets.

Third, PCR has been applied successfully to reverse transcriptase (RT)-derived complementary DNA (cDNA), fundamentally extending the range of PCR substrates to RNA targets; a whole range of gene expression studies become feasible when RT is combined with PCR. Recently, RT activity was shown to be an inherent property of a thermostable DNA polymerase, thus allowing cDNA synthesis and PCR amplification to be carried out with a single-enzyme, single-test-tube format (Myers & Gelfand, 1991).

Finally, procedures for the *quantitation* of polynucleotide targets are beginning to emerge. Estimation of the copy number of a specific gene as well as the level of RNA (especially mRNA) is of direct interest in both basic and clinical research. The challenge of imposing analytical criteria on the PCR amplification process requires strict attention to internal or external standards, an understanding of cycle-to-cycle efficiency of amplicon doubling, and a knowledge of the linear range of exponential amplification. The latter point is essential for an accurate assessment of the original target number based on extrapolation from PCR amplification data (Ferre, 1992).

Human Gene Linkage Analysis. To determine the linkage of genes in higher eukaryotes, two procedures have been available to researchers; these are the analysis of the progeny of selected matings or calculation of linkage relationships by pedigree analysis. Obviously, only the latter method can be applied to humans. DNA extracted from members of a family tree, combined with restriction fragment length polymorphism (RFLP) studies, has provided the foundation for construction of a human linkage map (Donis-Keller et al., 1987). When we employ this technique, genetic differences of 1 centimorgan (about 1 million base pairs in humans) or greater are the limit that can be measured with statistical reliability.

PCR allows measurement of genetic recombination over much shorter distances, using individual sperm (single haploid genome molecules) as DNA targets. A large number of meiotic products (sperm) can be examined from a single individual, allowing recombination frequency determination between genetic markers that are physically very close. This method is called "sperm typing" (Fig. 14-2).

Experimentally, PCR typing of sperm can be performed as follows (Li et al., 1988). Individual sperm are delivered to tubes or microtiter wells; in the latter case a fluorescence-activated cell sorter can deliver sperm into a 96-well microtiter plate. Sperm are lysed and amplified in a single solution, where two pairs of PCR primers

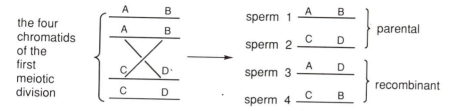

Figure 14-2. PCR analysis of individual sperm. PCR primers flanking the loci A, B, C, and D will reveal which DNA sequences are linked together in individual sperm. Four combinations are possible if recombination takes place (sperm 1–4). Two sperm are parental (A-B, C-D) and two are recombinant (A-D, C-B).

amplify two DNA loci simultaneously. Co-amplification of the two loci demonstrates that in this specific sperm, both gene regions are linked. In Fig. 14-2, genetic loci A or B can be combined with either C or D, indicating whether recombination has taken place. If identification of a locus is possible through amplicon length differences, direct gel electrophoretic analysis of amplicons will reveal which loci are amplified. Dot blot methods with allele-specific oligonucleotide probes can also detect single-base substitutions if this is necessary for identification of the locus. Once a sufficient number of individual sperm have been typed, the recombination frequency between the amplified loci can be determined. Determination of map order in three point crosses is also possible. The possibility of experimental error needs to be monitored carefully. For example, a sample tube may contain more than one sperm, or a region being studied could be poorly amplified. Methods for dealing with these complications are being examined (Cui et al., 1989).

Sperm typing allows quick identification of recombinant hot spots in male meiosis and will distinguish differences between individuals for recombination frequency in the same interval of time. Sperm typing can, in principle, be applied to mapping DNA polymorphisms in any species where haploid gametes can be obtained. Research in this area has concentrated on male meiosis, but linkage mapping in oocytes is also possible (Cui et al., 1992).

Until recently, sperm typing could reliably amplify only three loci simultaneously. Once a sperm had been analyzed in this way, it could not be studied again. A new procedure, *primer extension preamplification* (PEP), allows at least 30 copies (on average) of all DNA sequences present in an individual cell to be produced before PCR examination of genetic loci. This will allow many PCR reactions from a single cells' genetic complement. Briefly, PEP involves the application of random sequence primers 15 nt in length, followed by 50 primer extension temperature cycles upon the single sperm DNA target molecule. Most sequences present in the target DNA will anneal with some subset of the primer sequences, resulting in their copying and amplification (Zhang et al., 1992). At this point the reaction mixture can be subdivided. Multipoint genetic mapping should now be possible by applying this PEP technique prior to the PCR.

Ancient DNA. The PCR is capable of amplifying minute quantities of intact target DNA in the presence of nontarget DNA. A target often chosen for the amplification of aged DNA samples is an informative region of *mitochondrial* DNA because of its relative abundance in tissues when compared to genomic DNA sequences. Whatever target is chosen, two potential problems need to be addressed (Innis et al., 1990). First, the sensitivity of PCR could allow amplification of contaminating contemporary DNA, thereby compromising the authenticity of the amplicons. Second, inevitable modifications of the aged template DNA together with components present in the extracts of old tissue could inhibit activity or affect the accuracy of the DNA polymerase. Controls such as PCR amplification carried out on mock extractions of negative targets, as well as the examination of multiple tissue samples, are absolutely necessary for generating convincing sequence data from ancient DNA samples. If amplicons significantly longer than 200 base pairs are amplified, it is almost certainly due to contaminating contemporary DNA, since ancient DNA is too damaged to support extended-length PCR products. Two examples of PCR applied to aged DNA follow.

Extinct Marsupial Wolf. The extinct marsupial wolf of Australia, *Thylacinus cynocephalus*, has a controversial genetic lineage. Systematists have placed this animal, alternately, with the Australian or South American carnivorous marsupials on the basis of morphology. To help resolve this uncertainty, museum specimens of the extinct wolf collected in the late nineteenth and early twentieth centuries were used to supply samples of hide from which DNA suitable for PCR amplification was obtained. A region of the 12S ribosomal RNA mitochondrial gene was amplified and sequenced for comparison to representatives of six genera of extant marsupials.

A detailed analysis of the deduced DNA sequence phylogenetic tree showed that the marsupial wolf DNA sequences fell well within the radiation of Australian marsupials and statistically outside that of the South American carnivorous marsupial sequences. Combined with data for determining relatedness based upon immunological distances for the protein albumin, these molecular data strongly support an Australian origin. This result contrasts with the morphological evidence; based on a comparison of tooth characteristics and pelvic shape, the favored hypothesis is that this wolf is more closely related to an extinct group of South American marsupials. Thus the marsupial wolf appears to be an example of convergent evolution between South American and Australian carnivorous marsupials, a view which is strongly supported by sequence analysis of PCR-generated DNA (Thomas et al., 1989).

A 7000-Year-Old Human Brain. The study of "molecular anthropology" based upon the recovery of DNA sequence information from both ancient human remains and present-day populations is being strongly affected by PCR. Anthropologically informative sequences can be extracted from human tissues and amplified by PCR even in the event that the DNA is badly degraded. Amplification of mitochondrial DNA sequences between 50 and 200 bases in length yields sufficient sequence information to allow delineation of maternal lineages.

Human tissue is sometimes preserved for long periods of time in anaerobic peaty sediments. Brain tissue DNA was extracted from one such preserved human burial recently uncovered in Florida (Pääbo et al., 1988). Carbon radiodating of neural tissue placed the age of the remains at 6860 ± 110 years old. Following collagenase, protease, and phenol–chloroform treatment of the tissue, the researchers succeeded in recovering enough DNA for PCR analysis. One technical trick used was the inclusion of unusually high concentrations of bovine serum albumin and DNA polymerase during PCR amplification of the extracted DNA, which together were required to overcome an enzyme inhibitory factor present in the sample.

Incorporated in this study were appropriate negative DNA controls as well as multiple extractions of samples derived from different tissues. Furthermore, it was shown that PCR amplification resulting in DNA longer than 200 base pairs was not observed using these samples, in contrast to contemporary brain DNA where 2000-base-pair fragments are amplified routinely. As stated earlier, this feature of limited amplicon length provides circumstantial evidence for the ancient origins of the target DNA.

A phylogenetic analysis of sequence polymorphisms from the brain of this Middle Archaic Indian yielded an unexpected result. This individual does not belong to the mitochondrial lineage of Asian populations and also does not appear to be related to present-day American Indians in the southwestern United States. In fact, the mitochondrial DNA sequences revealed by PCR had not yet been described in the New World. The authors concluded that the colonization of the New World involved a minimum of three maternal lineages; this individual represented a newly found third lineage.

Genetic Disease Diagnosis. The first PCR amplification of genomic DNA involved a 110-base-pair region of a globin-coding sequence for the diagnosis of sickle-cell anemia (Saiki et al., 1985). These initial studies have been extended to other hemoglobinopathies and many other genetic lesions (Erlich & Arnheim, 1992). PCR offers a number of advantages compared to standard protocols for genetic disease diagnosis. PCR is rapid and can be expanded to allow processing of large numbers of samples. In addition, one requires only minuscule amounts of target DNA; for some targets, multiple pairs of PCR primers can amplify multiple diagnostic loci in a single reaction. The sensitivity of PCR is now being exploited to allow genetic disease diagnosis prior to embryo implantation and pregnancy. Single blastomeres for PCR-diagnostic analysis can be removed from human embryos at the six- or eight-cell stage without damaging embryo development (Chong et al., 1993; Handyside et al., 1990).

Screening for cystic fibrosis is a classic example of a case in which the application of PCR is of obvious benefit. Cystic fibrosis (CF) is an autosomal recessive disorder with an incidence of approximately 1 in 2000 live births and a carrier frequency of 1 in 20. The disease locus has been mapped to a gene on chromosome 7 coding for a protein of 1480 amino acids. Approximately 70% of CF mutations are a deletion of three base pairs in exon 10, resulting in the loss of a phenylalanine at a key position.

PCR primers flanking this deletion have been used to amplify and rapidly identify this specific mutation (Ballabio et al., 1990). One PCR variation employs allele-specific primers, one of which corresponds to the wild-type allele, while the other is specific for the deletion allele. Two simultaneous PCR amplifications are performed, each using a common primer and one of the two allele-specific primers. Gel electrophoresis is used to determine which of the alleles is present based on the size of the replicon—that is, 76 or 79 base pairs for the deletion mutant or wild-type, respectively. An extension of this technique involves (1) labeling the amplicons using allele-specific fluorescence-tagged primers and (2) separating the PCR products on an automatic DNA sequencing apparatus. Based on the presence of the appropriate fluorescent dye in the amplicon, the carrier status of the individual can be identified directly as normal homozygous, affected homozygous, or heterozygous.

The basis for another PCR variation for CF diagnosis involves the introduction of restriction enzyme recognition sites using cleverly engineered primers. One of the PCR primers encodes a unique *Mnl*I restriction site within an 86-base-pair amplicon if the wild-type allele is present in the sample during PCR amplification. Challenge with *Mnl*I digests the amplicon, resulting in two DNA fragments of 56 and 30 base pairs. Amplification using this primer on individuals homozygous for the deletion mutation will result in an 83-base-pair amplicon lacking the *Mnl*I site; only intact, full-length DNA is observed. Heterozygotes for the deletion exhibit all three fragments. The advantage of this technique is that it requires only two primers and no special primer labels. The results are evident upon inspection of agarose gels of the *Mnl*I-treated amplicons stained with ethidium bromide (Friedman et al., 1990).

Both of these tests and variations thereof are applicable to prenatal testing, newborn screening, and disease-carrier detection. Commercial centers involved in large-scale testing will need to be alert to the possibility of misdiagnosis due to amplicon contamination (see Appendix), since false-negative *and* false-positive results are possible unless rigorous attention to controls is assured.

Infectious Disease Diagnosis. An area where DNA amplification technology is beginning to make an important clinical impact is in human infectious disease diagnostics. For certain target organisms, present-day technology is inadequate to meet medical needs. Current methods for diagnosing viral or bacterial infections often requires culturing the pathogen for detection and identification, a process that can be slow and technically difficult because of the fastidious growth conditions required; for highly contagious pathogens, in vitro growth creates a potentially dangerous environment for the health care worker. Direct detection sensitivity is often inadequate even when isotopic or chemiluminescent enhancement techniques are employed. With bacterial pathogens that possess any of the above characteristics, such as the mycobacteria (tuberculosis and related diseases), *Borrelia burgdorferi* (Lyme disease) and *Legionella pneumophila* (pneumonia), the clinical utility of PCR has been demonstrated by providing faster, safer, and more specific diagnostic assays. These attributes extend to the diagnosis of fungal and parasitic infections as well (de Bruijn, 1988). As the appropriate gene targets are identified, PCR amplification of antibiotic resistance genes could greatly facilitate the diagnosis of particularly virulent antibiotic-resistant strains of bacterial pathogens, which are seen with increasing frequency within the general population.

An assay which tests the limits of PCR sensitivity, specificity, and quantitative capability is the detection of human immunodeficiency virus (HIV) RNA from the blood of infected individuals. Detection and quantitation of the viral RNA is important as a tool in monitoring the progression of disease and the efficacy of antiviral therapy. The viral titer in blood can be very low during much of the course of HIV infection. Depending on the method of virus isolation, PCR-based analysis can be compromised by significant amounts of human genomic DNA present in the sample. The challenge to PCR lies in detecting and selectively amplifying small numbers of viral RNA molecules (100–10,000) present among a larger number of nontarget nucleic acids, in a reproducible and quantitative manner such that the physician has reliable information on which to base therapeutic decisions.

Several research groups are currently exploring methods for simple and rapid quantitation of HIV viremia. For example, Holodniy and coworkers (Holodniy et al., 1991) have used an assay in which the HIV viral RNA is converted to DNA using reverse transcriptase (RT), then amplified by PCR (RT/PCR). A region of the viral *gag* gene is the target for amplification. Known quantities of a *gag* target RNA were mixed with HIV-seronegative blood serum and used to produce a standard curve of virus titer versus color production. The standards were assayed in parallel with samples of serum from patients infected with HIV. After extraction of the blood-borne viral RNA with guanidinium isothiocyanate and incubation with RT, the resulting cDNA was amplified in a reaction in which one of the PCR primers was biotinylated. The amplified DNA was heat-denatured, and the single biotinylated strand was captured on avidin-coated beads. This bead-bound, single-stranded DNA was then hybridized to an oligodeoxynucleotide probe labeled with the enzyme horseradish peroxidase (HRP), and the resulting complex of bead–amplicon–HRP probe was detected by HRP catalysis of a color-producing substrate in a microtiter well. The amount of RNA in the sample, and thus the concentration of virus, was proportional to the absorbance measured for the colored enzymatic product, which could be quantitated on an automated microtiter plate reader.

This RT/PCR procedure for detection of the RNA genome of HIV was standardized with known copy numbers of a *gag* gene transcript. Values obtained for this assay revealed a linear relationship between the measured absorbance and the RNA target at a concentration range of 1000 to 50,000 molecules per sample, although the useful limit of detection was as low as 10 copies per sample. The RT/PCR assay was compared to a standard tissue culture growth assay for HIV and was found to be predictive of the HIV copy number in vivo. Accurate results were obtained in only 2 days using the RT/PCR protocol compared to a waiting period of several days or weeks for the tissue culture-based assay. The authors concluded that direct detection of viral RNA in blood serum could improve the ability to monitor the course of HIV infection and provide an additional marker of disease progression and anti-HIV drug efficacy (see Fig. 14-3). Recently, nearly complete automation was achieved with a PCR-based quantitative procedure for detecting HIV proviral DNA within infected cells (Holodniy et al., 1992). Total time for completion of the assay is 8 hours with an accurate detection of HIV in samples containing between 25 and 10,000 copies of HIV DNA.

Studies such as these provide a glimpse of the future prospects for commercialization and eventual widespread adoption of PCR technology, namely through improvements in speed, selectivity, and sensitivity in the diagnosis of infectious microorganisms. PCR is likely to provide similar advantages in the field of cancer diagnosis as well (Lyons, 1992).

In Vitro Random Selection

A scheme outlining in vitro random selection as it pertains to nucleic acids research is outlined in Fig. 14-4. This technique involves the preparation of DNA libraries, usually via chemical synthesis and amplification, that contain an internal region composed of either a predetermined target sequence, which has been partially randomized with incorrect bases, or a sequence of completely random nucleotide distribution. Flanking sequences typically encode PCR amplification primer sites and endonuclease recognition sites for cloning. If required, RNA may be prepared by transcription with T7 RNA polymerase from a suitably randomized DNA template, while methods are available for isolation of single-stranded DNA. The randomized region is often a subset of some

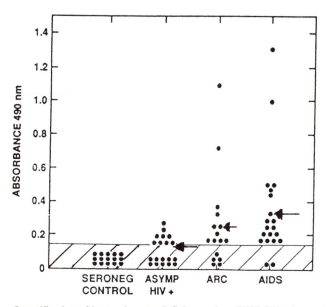

Figure 14-3. Quantification of human immunodeficiency virus (HIV) RNA in serum based on absorbance in 15 seronegative control subjects and 55 patients in different stages of HIV infection [asymptomatic, AIDS-related complex (ARC), and AIDS]. The arrow indicates the mean of the absorbance values for each category of patient; the horizontal line indicates the diagnostic cutoff value. Absorbance readings above the cutoff value are considered positive for viral detection. Reprinted from Holodniy et al., 1991, with permission.

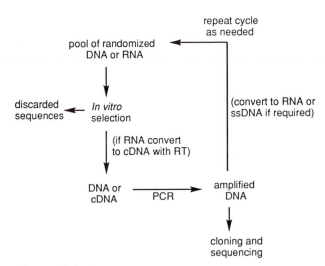

Figure 14-4. Outline of general in vitro random selection protocol.

larger established nucleotide sequence of interest; however, prior knowledge of functional sequence requirements is not always essential. The oligonucleotide pool is then subjected to a type of selective pressure. This may take the form of binding to a specific ligand, involvement as a substrate or catalyst in a chemical or biochemical reaction, or some other set of parameters which permit differentiation of functional and nonfunctional sequences. Typically, only a small fraction of the initial oligonucleotide pool is recovered, especially during the early rounds of selection; therefore, the enriched DNA or cDNA sequences are multiplied by PCR amplification. When dealing with RNA, amplification necessitates conversion to cDNA by reverse transcriptase following the selection step. The selection cycle is repeated as deemed necessary depending on the particular objective. Theoretically, a single sequence which optimizes the specific phenotypic property of interest may be obtained by reiterating the selection cycle and increasing the level of stringency during the selection step. Alternatively, if the goal is to sample a multitude of sequences with varying levels of functional competency, a single round of selection may suffice. The nucleotide composition of individual oligomers derived from active and inactive subpopulations may be secured following any chosen selection round by ligating the PCR-amplified DNA into a plasmid and sequencing the individual clones. More detailed examples of specific applications are provided in the following sections.

One of the first uses found for in vitro selection was the creation of new promoter recognition sequences for an *Escherichia coli* RNA polymerase (Horwitz & Loeb, 1986). Previously, base modifications of the promoter sequence that altered transcriptional regulation were identified by site-directed mutagenesis performed on wild type; activity analysis was limited to a few simultaneous base substitutions. By replacing key blocks within the promoter region with DNA of random sequence, new promoter sequences were identified that initiated transcription more strongly than the original. Promoters were linked to the tetracycline resistance gene, and selection was based on assaying for bacterial strains resistant to the antibiotic.

Selection methods based on antibiotic resistance preclude delineation of the degree to which the various steps involved in transcription signaling influence selection. By focusing on protein–DNA binding as a sole criterion for selection, sequence recognition properties have been defined by screening for high-affinity sites within random DNA sequences. In early examples the putative consensus binding sequence obtained from wild type was confirmed or redefined by in vitro random selection (Oliphant et al., 1989); subsequently, DNA recognition sequences have been identified for DNA-binding proteins in cases where the binding domain had not been assigned previously (Funk et al., 1992). The key step in the selection process is fractionation of the bound and unbound DNA. This may be accomplished by subjecting the randomized DNA pool to a protein affinity column or by mixing DNA with protein and isolating the DNA–protein adduct by gel electrophoresis, immunoprecipitation, or capture on nitrocellulose. Where the initial quantity of DNA is sufficient to ensure adequate recovery following several rounds of selection, ligation into plasmid for individual sequence determination may be performed directly; however, in most circumstances, selection is combined with PCR amplification.

A major advance in the field was provided by the invention of the SELEX procedure (Systematic Evolution of Ligands by Experimental Enrichment), which established a broader range of applications for in vitro random selection (Tuerk & Gold, 1990). In its initial form, a set of experiments was designed to study the interaction of an RNA-binding protein with its hairpin RNA target. The key feature of this methodology involved (1) transcription of RNA from a randomized DNA template prior to a protein-binding selection step and (2) the conversion of RNA to cDNA with reverse transcriptase subsequent to that step. This innovation allowed the selection of nucleic acids other than double-stranded DNA. Recognition sequences have been established for a number of RNA-binding proteins (Schneider et al., 1993; Tsai et al., 1991), including an interesting example in which an RNA epitope has been defined for an antibody implicated in autoimmune disease (Tsai et al., 1992). In each case, SELEX was performed within the context of a specific RNA–protein interaction; however, it is clear that SELEX will produce RNA molecules that recognize and bind any number of potential ligands.

Employing a SELEX-type procedure, Szostak and coworkers have examined the interaction of polynucleotides with small molecules. Beginning with a large pool of single-stranded oligomers of random sequence, DNA and RNA molecules, termed *aptamers*, were selected based on specific binding to a variety of organic dyes (Ellington & Szostak, 1990, 1992). Invariant nucleotides comprising a consensus sequence were identified within the secondary structure of several aptamers generated by folding minimization. Stereoselective RNA aptamers have been prepared recently that distinguish between the *R*- and *S*-isomers of tryptophan (Famulok & Szostak, 1992).

In vitro random selection has been explored as a method of identifying potential oligonucleotide-based therapeutic agents. For example, SELEX was used to obtain high-affinity RNA ligands that bind to proteins essential for HIV-1 replication (Tuerk et al., 1993) including Tat, Rev, and reverse transcriptase (RT), with the expectation that RNA binding would inhibit normal protein function. In the case of RT a pseudoknot motif binds with high affinity, inhibits enzyme activity, and is specific for HIV-1 RT (Tuerk et al., 1992), properties which are deemed essential to the development of an anti-HIV drug. It is perhaps expected that functional groups within the active site of a polymerase responsible for binding and aligning an RNA template would also attract competitive RNA inhibitors. However, it is surprising that a DNA aptamer has been found that binds to and inhibits thrombin (Bock et al., 1992), a protein not normally associated with nucleic acids biochemistry. Ecker et al. (1993) have chosen an alternate approach to the "traditional" in vitro selection methods for drug discovery. Combinatorial nucleotide libraries were synthesized in which one position was initially defined within an otherwise random sequence. With each successive round of selection and resynthesis, an additional position within the oligomer was defined until the active "pharmacophore" was obtained. This method obviates PCR amplification or other procedures involving polymerases, and it allowed the introduction of non-natural nucleotides for the first time.

The selection criteria for the study of autocatalytic RNAs has been based on their performance in catalytic reactions. Pan and Uhlenbeck (1992) have studied the well-known site-specific, Pb^{2+}-mediated tRNA cleavage reaction. Beginning with circularized versions of tRNA randomized at several positions, RNA molecules which underwent cleavage at alternate sites were isolated. Selection was based on separation of the inactive, intact circularized RNA from the linearized, active species. A number of new tRNA structures were active even though they lacked several highly conserved nucleotides previously thought to be essential for activity. However, of greater interest is the fact that the majority of active RNA molecules did not fold into the standard tRNA conformations, indicating that this may be a method for identifying new ribozyme structures. The importance of nucleotide positions located within the catalytic and substrate components of a hairpin ribozyme system was likewise established by in vitro random selection (Berzal-Herranz et al., 1992). The cleavage–ligation selection protocol did not require physical separation of the active and inactive species, because each subset could be amplified independently with separate sets of PCR primers.

Catalytic competency was restored to a ribozyme of the Group I intron family that had been rendered dysfunctional by deletion of the stem-loop structure important for catalytic activity (Green & Szostak, 1992). While the truncated RNA exhibited negligible activity in a selection assay for catalyzing exon ligation, mutations were identified which compensated functionally for the lost sequences. The protocol of "in vitro evolution" utilized in this experiment differed from the protocols described above in that the entire wild-type RNA was mutagenized at a low rate during each selection cycle. Separation of active and inactive ribozymes was accomplished by affinity purification based on binding unique sequences introduced into the active RNAs during ligation. In a parallel experiment, a Group I intron was selected for improved efficiency in utilizing a DNA substrate (Beaudry & Joyce, 1992). In vitro evolution was effected by introducing mutations into the core of the ribozyme at the level of the synthetic template and during amplification of cDNA through mutagenic PCR.

Specific Applications of In Vitro Random Selection to Problems in Nucleic Acids Biochemistry

In this section several of the examples alluded to in the overview are discussed in greater detail. These include examples involving the interaction of DNA and RNA with their respective binding proteins. The goal is to provide an appreciation of the potential applications and the nature of the experimental protocols of in vitro random selection described in the recent literature. Often apparently similar random selection techniques have been applied to solving a diverse variety of problems. The range of applications appears to be limited only by the ingenuity and imagination of the researcher.

A DNA-Binding Protein. Past methods for elucidating the binding-site size and sequence requirements of DNA-binding proteins relied implicitly on identification of the wild-type binding site. If binding-site data were available from a variety of organisms, the sequence requirements could be determined by phylogenetic comparisons. Sequences within the binding site essential to function are generally invariant or at least partially conserved between species.

Alternatively, wild-type sequence may be subjected to site-directed mutagenesis, where altering the identity of an individual nucleotide at a specific position is correlated with its effect on protein function. Both methods, however, have potential drawbacks. In some cases the number of examples found in nature is limited, making comparisons incomplete. Mutagenesis is often time-consuming and the assessment is biased toward the initial wild-type sequence. In either case, potential binding sequences unrelated to wild-type sequences would be missed.

By employing in vitro random selection—that is, presenting the protein with DNA substituted with a random distribution of nucleotides—all potential protein-binding sequences may be screened simultaneously. Indeed, selection may proceed without prior knowledge of wild-type binding sequence; therefore, the experimental outcome is not unduly influenced in this direction. Olifant and colleagues (Oliphant & Struhl, 1987; Oliphant et al., 1989) were one of the first groups of researchers to apply random techniques to the problem of formulating a consensus sequence for a DNA-binding protein. A summary of their work follows.

In many yeast strains, the DNA-binding protein GCN4 activates amino acid synthesis by binding to the promoter region of amino acid biosynthetic genes and initiating transcription under conditions of amino acid starvation. The DNA-binding domain resides within the C-terminus of the protein; this domain is also associated with a leucine zipper motif implicated in dimerization of GCN4 when bound to DNA, and a 25-residue region that is thought to supply the protein–DNA contacts essential for binding. In previous studies the putative, optimal 9-nt binding sequence (ATGA-(C/G)TCAT) of the *his*3 gene was identified by mutagenesis and found to differ from the wild-type sequence by a single base pair. Because of the dimeric nature of the protein and the apparent symmetry of the DNA-binding site, it was originally proposed that the dimeric protein recognized two identical, adjacent half sites. Inversion of the central C/G base pair, however, resulted in a diminution in binding, providing evidence for an inherent asymmetry within the GCN4-binding site. Random selection was used to investigate both the relative importance of the various nucleotide positions as well as the suspected asymmetric nature of the association of GCN4 with the *his*3 promoter. Because this method relies on identifying potential binding sites within the context of a completely random set of sequences, the results are independent of, and not influenced by, the earlier mutagenesis studies on wild-type sequences.

Generation of double-stranded DNA by mutually primed enzymatic synthesis required preparation by chemical synthesis of an oligomer 40 nucleotides in length. This contained a 23-nt section of random sequence comprised of all four bases, flanked by *Bam*HI and *Pst*I recognition sites (Fig. 14-5). The 40-mers were annealed by hybridizing through the self-complementary *Pst*I site to form a self-priming dimer, and the 3′-termini were extended using Klenow polymerase with concomitant incorporation of ^{32}P radiolabel. The DNA was subjected to an initial selection step as described below in order to reduce the total number of restriction sites, thereby reducing the quantity of restriction enzymes required. The DNA dimer was finally digested with *Bam*HI and *Pst*I to produce two identical duplexes ready for additional rounds of selection and cloning.

To perform in vitro selection, the GCN4 protein was coupled to cyanogen bromide-activated Sepharose and the duplex DNA, prepared as described above, was applied to a GCN4-Sepharose column. In the initial screening round, the column was subjected to three successive washes at low (100 mM NaCl) and one wash at medium (400 mM NaCl) salt concentration to remove poorly binding sequences. The remaining protein-bound DNA was eluted at high salt (1 M NaCl). (Elution conditions had been optimized previously with plasmid restriction fragments containing the authentic *his*3 promoter.) Based on the measured radioactivity associated with the DNA, the amount of DNA recovered comprised ~2% of that originally retained on the column under conditions of low stringency (100 mM NaCl). Following enzymatic digestion, the DNA was subjected to chromatography three additional times as described above, yielding ~0.01% of the total input DNA. The amount of DNA obtained by this method was sufficient to permit the subcloning of DNA directly into a plasmid for sequencing analysis, without the need for prior amplification. A blue-to-white reverse screen was em-

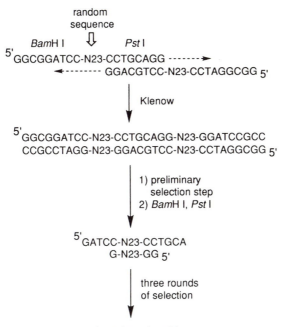

Figure 14-5. Generation of double-stranded DNA containing a 23-nt random sequence by mutually primed extension with Klenow polymerase, followed by selection by binding to GCN4-derivatized Sepharose and cloning of individual sequences.

ployed to reduce the ligation background. The modified cloning vector chosen contained an insert in the polylinker that resulted in a frameshift, thus rendering the *lac Z* gene dysfunctional (white colonies). Insertion of the DNA fragment obtained by selection restored the correct reading frame for β-galactosidase translation, and colonies with successful ligations could be identified on XGAL plates (blue colonies). Forty-three individual positive clones were sequenced by the chain termination method from their respective single-stranded DNAs obtained from superinfection with bacteriophage.

Comparison of the individual DNAs for sequence homologies indicated that 40 of the clones contained the sequence TGA(C/G)TCA, while the sequences in the remaining three clones deviated by a single nucleotide. Arbitrary alignment of the 7-nt core with C at the center position revealed that homology within the DNA sequences extended to 3 nt on either side of the primary region, although the degree of sequence conservation in these flanking regions was not as strict as that found in the central core. The new, expanded version of the binding site was given as $c_g g_t a_g$TGA(C/G)TCA$t_c c_a g_c$, where the lowercase letters represent the preferred sequences in the flanking regions that were symmetrically related (subscript letters represent the minor component). Also, the three deviations from the consensus in the 7-nt core occurred to one side of center, providing additional evidence for an overall asymmetry within the binding site.

The results obtained by in vitro random selection extended our understanding of the GCN4-*his*3 promoter interaction gleaned from mutagenesis. The overall binding site size was shown to extend to 13 nt, while the essential, most conserved core was redefined as 7 nt, which is smaller than what had been reported in previous studies. In addition, the suspected asymmetry of the promoter was verified. Surprisingly, variations in sequence within the *his3* promoter discovered through in vitro selection were less frequent than those observed in wild-type yeast *his3* promoters, indicating a higher degree of rigor in the experimental selection procedure than is required for effective translation initiation in nature. By lowering the stringency of the selection procedure, other functional DNA sequences with diminished binding capacity have been identified (Mavrothalassitis et al., 1990).

SELEX and an RNA-Binding Protein. In the previous example an in vitro selection technique was presented in which a DNA-binding protein consensus sequence was elucidated by employing in vitro techniques based on repetitive selection from an initial random sequence. As described, this method is limited in scope and may be applied only to analyzing protein-binding sites consisting of short stretches of duplex DNA. Recently, Tuerk & Gold (1990) developed a more versatile in vitro selection method entitled *Systematic Selection of Ligands by External Enrichment (SELEX)*. In the initial study, the interaction of mRNA with its binding protein was investigated; however, this method is general and has been applied to the study of other binding interactions involving polynucleotides.

Autoregulation of bacteriophage T4 DNA polymerase (gp43) occurs through interaction of the protein with the mRNA that encodes it. The protein-binding site includes a hairpin structure (see Fig. 14-6) within the context of a larger 36-nt segment upstream of the encoding region, which overlaps the Shine and Delgarno sequences. Binding of gp43 presumably inhibits translation by interfering with the interaction of mRNA at the ribosome. The distantly related bacteriophage RB69 is regulated in an analogous fashion; while some differences in sequence exist within the 3'- and 5'-end regions and helix of the 36-nt protein-binding site, the single-stranded loop within the hairpin is identical in both strains. Because the affinity of T4 polymerase for the 36-nt binding site is only marginally ($\sim 10^3$- to 10^4-fold) greater than nonspecific binding, the authors hoped to examine the extent to which evolutionary pressure influences the degree of sequence invariance observed within the hairpin loop common to both bacteriophage systems. A scheme describing the SELEX method is shown in Fig. 14-6.

A 110-nt single-stranded DNA template was prepared by ligation of three synthetic oligomers in which the middle oligomer encoded the 36-nt protein-binding sequence; the 8-nt hairpin loop was substituted with a random distribution of nucleotides, such that 4^8 (65,536) individual species were theoretically present. One of the overlapping oligomers (I) used in the ligation also contained the T7 promoter to be used in the subsequent transcription step and sufficient overlap of the 3'-end of the template for use as one of the PCR primers. Another oligomer (V) also served as the 3'-primer for reverse transcription, PCR amplification, and eventually DNA sequencing steps. Transcription of a 92-nt RNA with T7 RNA polymerase was initiated by annealing the randomized template to oligomer I, which contained the T7 promoter sequence and which partially overlapped with the region encoding the transcript. For comparison, an RNA which carried the wild-type sequence was also transcribed from a suitable template. The RNA transcripts containing the randomized sequences were then mixed with gp43 at protein concentrations of 0.03, 0.03, and 0.3 μM and RNA in 10-, 1000-, and 100-fold excess (experiments A, B, and C), respectively, initiating a competition between

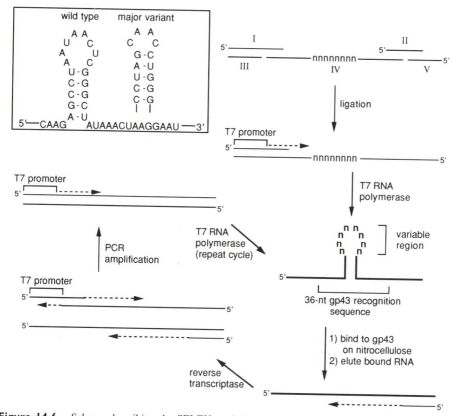

Figure 14-6. Scheme describing the SELEX method. Wild-type and major variant hairpin structures contained in the 36-nt gp43-binding site are outlined within the box. Broken arrows indicate direction of polymerization in the following step. Darkened lines represent RNA.

the various individual RNA sequences for the protein-binding site. The RNA–protein complex was captured on a nitrocellulose filter, thereby separating the protein-bound RNA from unbound RNA. The RNA was eluted from the filter, and a fraction of the recovered material was converted into cDNA with reverse transcriptase using as primer oligomer V from the original ligation. The bulk of the RNA was set aside to follow the sequence convergence through successive rounds of selection. With the addition of the 5′-primer (I) containing the T7 promoter, the cDNA was amplified by PCR through 30 cycles with *Taq* polymerase, providing full-length duplex DNA, which in turn served as a template for RNA transcription thus completing the cycle.

Initially, the affinity of gp43 for RNA composed entirely of wild-type sequence within the hairpin loop was 60-fold greater than for the corresponding RNA substituted with random sequence at these positions; however, after four rounds of selection, binding was identical (K_d ~5×10^{-9}). The progression of one experiment (B) was followed by dideoxy sequencing of the RNA recovered from filter binding and analyzing by gel electrophoresis. For the initial transcript and through the first two rounds of selection, the region of the gel encoding the RNA loop presented a nonspecific ladder-type pattern. However, after the third round a specific pattern began to emerge that was complete by the fourth round; additional rounds of selection produced no apparent changes. Sequence analysis of the fourth-round RNAs independently derived from experiments A, B, and C yielded the identical sequence pattern, A(A,G)-(U,C)AAC(U,C)(U,C), where four nucleotide positions were invariant and four positions were composed primarily of two bases. Theoretically, the 8-nt consensus sequence could comprise 24 different sequences; however, normalizing the band intensities through radioanalytical imaging revealed a bias of between 1.1 and 1.6 for the wild-type over alternative sequences in the three separate experiments. Identification of the exact nucleotide composition required a survey of the sequences of the individual RNAs. To this end, the cDNAs from the fourth round of selection of experiment B were PCR-amplified with primers containing *Hin*dIII and *Bam*HI endonuclease restriction sites. The DNA was then ligated into an appropriate plasmid and 20 individual clones were sequenced.

Although 16 different individual sequences were possible based on the gel electrophoretic data, only two were prevalent. One was the wild-type sequence (AAUAACUC, nine clones) and the other, termed the *major variant* (AGCAACCU, eight clones), was mutated at four positions. Three other clones were single-base mutational variants of these two species (AAUGACUC, AAUAACUU, and AGC-GACCU). As expected, the binding affinity of the protein for the wild-type variant and for the major variant were similar ($K_d \sim 4.8 \times 10^{-9}$) based on nitrocellulose filter binding assays with radiolabeled transcripts; in a direct competition between wild-type and major variant RNAs for binding to gp43 protein, the dissociation for the wild-type variant was determined to be twice that of the major variant. The dissociation constants for the three minor variants were larger than the two predominant species, as expected based on their low abundance in the sequence pool. An interesting potential structural difference between the two sequences can be found in the apparent degree of base-pairing within the hairpin (see Fig. 14-6). While the 8-nt sequence in the wild-type variant was expected to be single-stranded, the major variant was thought to contain two additional base pairs, based on standard RNA secondary structure predictions, thereby lengthening the stem and shortening the loop of the hairpin. If these proposed structures are correct, then binding to protein represents a relatively rare event in that two nonidentical oligonucleotide structures recognize the same protein-binding site through presumably nonequivalent contacts. Alternatively, the wild-type sequence may form additional A·C and A·U base pairs, or conversely the G·C and A·U base pairs of the major variant may denature upon binding to protein, making the two sequences structurally similar. An understanding of the exact nature of the interaction must await structural characterization.

Given the rather marginal binding preference (2.5 kcal) for wild-type over random sequence, it was somewhat surprising to find that of the potential 65,536 individual RNA species originally screened, the preferred sequences were limited to only two distinct species. However, the authors have shown that sequence conservation is a direct consequence of evolutionary optimization of the binding interaction between the RNA hairpin and the protein. In this respect, SELEX appears to mimic natural evolutionary processes; however, results are achieved within the short time frame of the SELEX experiment. In addition to the wild-type sequence, SELEX has identified an alternate sequence not seen in nature. This may be structurally different from, but appears to function as well as, the wild-type sequence, thus providing new information related to understanding the interaction between the RNA and the protein responsible for binding.

Therapeutics Discovery. Random selection techniques have been applied successfully to resolving the binding sequence requirements for proteins in which association with DNA or RNA is an essential element of protein function. Perhaps a more interesting question relates to the likelihood of discovering oligonucleotides which interact with proteins that are not normally associated with nucleic acids chemistry. Furthermore, it would be advantageous, from a therapeutic standpoint, if the binding domain resided at or near the active site of the protein, thus affording the potential to inhibit protein function. Recently a series of experiments was designed to address just this question (Bock et al., 1992). Through a variation of the aforementioned SELEX technology, a unique oligomer was discovered that binds to and inhibits thrombin. Thrombin was selected as a therapeutic target based on its importance as a locus for interdiction in the development of anti-clotting agents and its accessibility by large molecules. By initially choosing a blood protein to evaluate the utility of the procedure, potential problems resulting from the impermeability of cell membranes to polynucleotide-based therapeutic agents were avoided.

A pool of DNA oligonucleotides ($\sim 10^{13}$ sequences) was synthesized that contained a central randomized region 60 nt in length, flanked by endonuclease recognition sites for cloning and primer sites for PCR. Production of single-stranded DNA was accomplished by amplification with a 5′-biotinylated primer and binding the DNA to an avidin–agarose column, followed by base denaturation and elution of the complementary, unbound single-stranded oligomer. Thrombin, a glycoprotein, was immobilized via the sugar moiety to a concanavalin A (lectin)–agarose column and the DNA pool was applied. Pretreatment of the DNA pool by passing through underivatized con A–agarose eliminated sequences which bound to the column media itself. After washing away unbound DNA, the thrombin–DNA adducts were eluted from the column by addition of α-methyl mannoside, which displaced the sugar portion of the protein. Phenol extraction to eliminate the protein, along with amplification of the DNA by PCR, completed the cycle. By subjecting the initial DNA pool to five selection cycles, the fraction of applied DNA retained by the column rose from 0.01% to 40%. Binding of the oligomers was specific to thrombin; the DNA did not bind to other proteins tested. Upon completion of the final round of selection the DNA pool was cloned and sequenced in order to identify, within the internal section originally containing the 60-nt random sequence, regions of homology between the individual DNA oligomers.

The results were quite extraordinary. The majority of the clones could be aligned precisely along two adjacent, nearly identical hexamer sequences which were linked by a short tract of nucleotides. The consensus sequence was represented as GGNTGGN_{2-5}GGNTGG, where the identity frequency of N was T > G ~ A and N was 2–5, but most often 3. Exceptions included a few clones that were single nucleotide variants and others which possessed a single version of the hexamer. When tested for activity against thrombin, both the full-length DNA oligomer derived from one of the clones and a synthetic 15-mer, GGTTGGTGTGGTTGG, inhibited the thrombin-catalyzed conversion of fibrinogen to fibrin in vitro and in a clotting assay with human plasma. As might be expected, the 15-mer was more potent than the full-length DNA or GGTTGG added alone. Fifty percent inhibition in a clotting assay was achieved at a DNA concentration of 25 nM, a potency well within a useful therapeutic range.

Recently, two groups working independently (Macaya et al., 1993; Wang et al., 1993) have elucidated by 2-D NMR the three-dimensional conformation of the 15-mer described above (Fig. 14-7). The structure is unique in nucleic acid chemistry, but is reminiscent of other multi-guanosine-containing oligomers where hydrogen bonding patterns include G-quartets. The mechanism by which this "chair" structure binds to thrombin and mediates inhibition of thrombin activity is the subject of ongoing investigation.

Investigation of a Hairpin Ribozyme. *Ribozyme* is a generic term for autocatalytic RNAs that undergo self-splicing or self-cleavage in the presence of metal ions. There is a self-cleaving RNA hairpin motif within the satellite RNA of the tobacco ring spot mosaic virus. The noncontiguous nature of the essential nucleotides allows for delineation of the catalytic (ribozyme) and target (substrate) strands. In fact, these two components may be supplied in *trans*, permitting multiple ribozyme turnover when sufficient substrate is present. Current methods for identifying nucleotides essential for activity include mutagenesis and phylogenetic sequence comparisons; however, the first method is time-consuming while the second method is often precluded by the limited number of such ribozyme structures known in nature. An in vitro random selection method has been devised to study the degree to which sequence variation is tolerated within essential components of the hairpin ribozyme system (Berzal-Herranz et al., 1992). The method described relies on the reversible nature of the cleavage reaction when substrate is supplied in excess.

A duplex DNA template was constructed such that the transcribed RNA contained the target and catalytic components of the hairpin motif connected by a short intervening tract of cytidines. Terminal 3'- and 5'-sequences supported initiation sites for reverse transcription and PCR amplification, as well as restriction sites for cloning. The authors chose to examine the effect on the reaction rate of sequence variation introduced, alternately within a 4-nt region of the substrate encompassing the cleavage site and within the 7-nt, single-stranded region of the ribozyme previously suggested to undergo conformational changes concomitant with substrate binding (Fig. 14-8).

In the first experiment, nucleotides 38–44 were randomized in the DNA template and the resulting RNA transcript was subjected to conditions which promoted cleavage of the substrate located at the 3'-end of the RNA. The reaction mixture was then

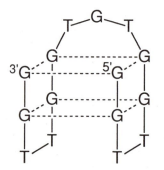

Figure 14-7. Proposed structure for 15-mer DNA that inhibits thrombin.

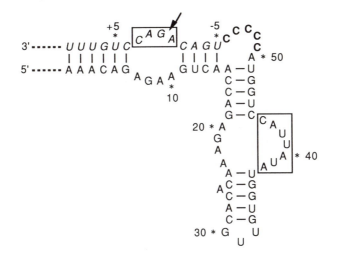

Figure 14-8. Two-dimensional representation of the hairpin RNA. The ribozyme sequence is designated by normal lettering, the substrate sequence is denoted by italics, and the pentacytidine linker is represented by boldface. The site of cleavage–ligation is designated with an arrow. Areas within the ribozyme and substrate studied by in vitro random selection are enclosed within a box.

incubated with a mixed DNA–RNA hybrid; a 5′-RNA oligomer composed of nucleotides G_{+1}–U_{+8} of the substrate was covalently attached to a 3′-DNA oligomer that encoded a new PCR primer-binding site different in sequence composition from the original transcript. Accordingly, only active RNA molecules that participated in the cleavage–ligation reaction were amplified by the new 3′- and original 5′- PCR primers. Inactive transcripts were amplified with the original primer sets. The cleavage–ligation selection process was performed only once. In this instance the goal of the experiment was not sequence optimization to acheive maximum reaction rates, but instead to sample a spectrum of sequence patterns possessing varying degrees of activity. Selection with excessive stringency would have eliminated sequences of lower overall activity. After conversion of RNA to DNA with reverse transcriptase, followed by PCR amplification with the appropriate set of primers, the DNA was cloned and sequenced. Of the 20 clones selected from the pool of inactive RNAs, a random sequence was found within the internal loop with all four bases represented at each position. Of the 30 clones selected from the pool of active RNAs, 11 were identical to the wild type, while the remaining clones contained single (14) or double (5) point mutations, primarily at position U_{39}. Four positions, A_{38}, U_{41}, U_{42}, and C_{44}, were invariant. The frequency at which each mutant was represented in the cloned population was proportional to its catalytic activity in a cleavage assay.

Analysis of the four positions within the RNA that encompassed the cleavage site was carried out in a manner similar to that described above, but with one important exception: Because the 3′-end of the substrate was lost subsequent to cleavage, selection pressure at positions $+1$, $+2$, and $+3$ was effected during ligation by introduction of a DNA–RNA hybrid randomized at those positions. Sequence examination of 15 inactive clones revealed a random distribution of all four bases at all four positions, except that in no case did G occupy position $+1$. Conversely, of the 33 active clones sequenced, the $+1$ position was occupied exclusively by G, while the three remaining positions contained a distribution of all four bases. The fact that G was always present at the position $3'$ to the scissile phosphate supported an independent study indicating that the exocyclic amino group of guanosine is intimately involved at the cleavage site (Chowrira et al., 1991).

These experiments demonstrate the effectiveness of a single selection cycle in providing valuable information on the consensus sequence of a small autocatalytic RNA in the absence of phylogenetic data and without the tedium involved in site-directed mutagenesis.

Polynucleotides That Ligate Small Molecules. Until recently it was assumed that molecular recognition of and chemical transformation of small organic molecules in biological systems was the exclusive domain of proteins. However, this view has changed radically with the recent discovery of autocatalytic RNAs. Progress toward understanding the mechanism of the Group I introns has provided a useful comparison between nucleic acid-based and amino acid-based enzymes. Self-splicing of the *Tetrahymena* rRNA is initiated by binding of free guanosine at the "G-binding site." Perhaps not unlike the forces involved in protein–ligand binding, G-specific recognition relies on H-bonding with the conserved G·C base pair, in addition to interactions with the neighboring functional groups found within the unique tertiary structure that defines the putative binding pocket. Arginine also binds to the G-binding site, to the exclusion of other amino acids, with a slight preference for the naturally occurring enantiomer. The three-base sequence located at the binding site corresponds to one of the arginine codon triplets in all cases studied, providing intriguing evidence that physicochemical interactions may have been a determinant in the origins of the genetic code. Therefore, it is of great interest to establish methods that permit investigation of the forces involved in coordinating small molecules. Szostak and coworkers have published several studies describing the discovery of DNA and RNA molecules that recognize and ligate specific small molecules (Ellington & Szostak, 1990, 1992; Famulok & Szostak, 1990).

A synthetic DNA ~155 nt in length was prepared that contained a 100-nt internal random sequence flanked by PCR primer sites, restriction sites for cloning, and a T7 promoter (Fig. 14-9). The complexity of the RNA pool resulting from transcription of the PCR-amplified DNA template was estimated at $\sim10^{13}$ different sequences, a small fraction of the total possible. In the first step, the RNA was applied to a beaded agarose column to which one of six dyes was immobilized (Cibacron Blue 3GA, Reactive Red 120, etc.). Unbound RNA was washed through the column with buffer, after which RNA that bound to the column was eluted with water. The "bound" fraction was converted into its DNA complement with reverse transcriptase, and the resulting cDNA was amplified by PCR. RNA was transcribed once again from the resulting duplex DNA, and the selection cycle was repeated. From sampling the RNA after each round of selection it was estimated that $<0.1\%$ of the RNA bound to the column initially; however, after five rounds of selection the retention level increased to $>50\%$. Some RNA pools exhibited considerable cross-reactivity, either by binding to other dyes of similar structure or by binding to the column matrix itself, while other RNA pools recognized exclusively the selected dye. Upon completion of the final selection cycle the PCR-amplified DNA was cloned and the sequences were analyzed. Comparisons of

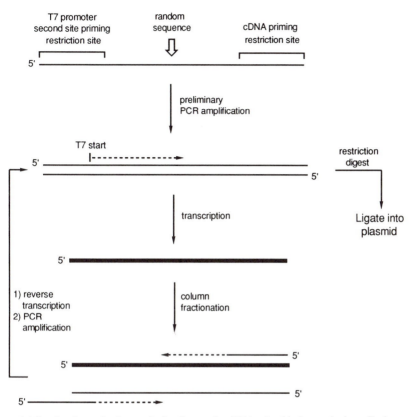

Figure 14-9. In vitro selection cycle for discovering RNAs that bind organic dyes. Broken arrows indicate direction of polymerization in the following step. Darkened lines represent RNA.

the individual clones revealed that for the most part each RNA oligomer or "aptamer" comprised a different sequence, although some similarities within short regions were noted when comparing aptamers originating from the same pool. It was estimated, based on the observed level of duplication, that each RNA pool contained between 10^2 and 10^3 RNA molecules of unique sequence, suggesting the existence of a sizable number of individual sequences within the population capable of folding into structures with an affinity for the dye ligand. Overall, this represented a $\sim 10^{10}$-fold enrichment over the initial RNA pool. To ascertain the identity of nucleotides important to binding, individual RNAs were mutagenized at each position such that the new transcripts retained 85% of the correct wild-type sequence, while an equal mix of the other three bases constituted the remaining fraction. Following the protocol of selection, cloning, and sequencing, it was discovered that 28 positions were completely conserved among 20 mutagenized clones derived from an aptamer which bound Reactive Blue (three or four conserved positions would be expected by chance); this invariant set of nucleotides thus constituted a binding consensus sequence. Furthermore, an unrelated clone derived from the same RNA pool exhibited similar conserved regions. Although the secondary structure of this particular family of aptamers was visualized through energy minimization techniques, it was clear that additional tertiary interactions were required for coordination of the dye molecule. The dissociation constant of the dye–aptamer complex was estimated at <100 μM, a value not unlike those typically encountered in protein chemistry.

In a related study, single-stranded DNA aptamers possessing comparable dye-binding characteristics were generated employing a similar selection methodology in concert with asymmetric PCR amplification. Initial attempts at isolating DNA pools with specific binding properties were unsuccessful; the pools exhibited substantial cross-reactivity. This problem was overcome by including a negative selection step in the protocol in which the DNA was passed through a precolumn containing a noncognate dye immobilized on an agarose support. Contaminating duplex DNA was removed by gel electrophoresis. An 18-nt stem-loop consensus sequence was identified in five of eight aptamers sequenced from the Reactive Green-binding pool; when produced synthetically the oligomer bound with high avidity (K_d 33 μM). RNA identical in sequence failed to bind to the column; the presence of the additional 2′-OH group interfered in some way.

These experiments demonstrate the potential for generating DNA or RNA molecules that bind ligands other than proteins. By employing similar methodologies, the potential exists for producing DNA or RNA molecules that recognize and bind any number of ligands.

Synthetic Methods Based on Systematic Elimination of Nonoptimal Sequences. Examples of in vivo selection have been presented which rely on PCR amplification of the small fraction of polynucleotides of interest isolated during the selection process. Involvement of a polymerase limits the chemical composition of the polynucleotides examined by this method to those naturally occurring forms of nucleic acids. However, from the standpoint of medicinal applications, oligonucleotide analogues modified at the phosphodiester linkage have proven to be more robust in biological systems. The chemical modification confers enhanced nuclease resistance, providing a potential advantage when screening for oligonucleotide-based therapeutic agents in vivo. Additionally, it may be of interest to prepare oligomers that are not limited in composition to standard nucleotides, but are composed of chemically altered bases or sugar moieties or perhaps substituted with other functional groups entirely.

A combinatorial technique termed *synthetic unrandomization of randomized fragments (SURF)* has been described recently that circumvents amplification procedures entirely (Ecker et al., 1993). Oligonucleotides are screened directly during an iterative synthetic protocol in which nonoptimal sequences are systematically eliminated. The approach is similar to current strategies employed for screening peptide libraries. In the following examples, 2′-*O*-methyl and phosphorothioate analogues are the subject of the investigation; however, in principle this technique may be applied to any oligonucleotide congener as long as the monomers employed are compatible with solid-phase synthesis.

In the first step of the general procedure, four sets of oligonucleotides are synthesized with random sequences throughout except for one unique position at or near the center of the oligomer. At this position, each of the four oligomer pools contains one of the four different bases. All four sets of pools are subjected to a form of selective pressure. The pool which produces the best results is designated as the "winner," and the identity of that unique position is retained. In subsequent rounds, oligomer pools are synthesized such that neighboring positions are fixed or "unrandomized" until the optimal sequence is determined.

To test the method, 2'-O-methyl oligonucleotide analogue libraries were evaluated for binding to an RNA hairpin found within H-ras mRNA (Fig. 14-10). Previously, nuclease digestion probing revealed a 16-nt unpaired loop sequence as a potential target for antisense hybridization. Four pools of 9-mers were synthesized, each containing exclusively one of the four bases at the center position; each of the four pools was tested for binding to the hairpin RNA. The oligomer pool containing C at the unique position was identified as tightest binding; therefore, a second set of four oligomer pools was synthesized, each with a C at the center position and one of the four bases occupying a second nearest-neighbor position. The process was repeated through nine rounds of selection until the identity of all positions was fixed. As expected, this oligomer sequence was complementary to a 9-nt sequence within the hairpin loop. Between the first and ninth rounds of selection the affinity of the oligomer pools for the hairpin increased by three orders of magnitude.

In order to test the feasibility of employing SURF to identify potential therapeutic agents, nuclease-resistant phosphorothioates were screened by this procedure for antiviral activity in vivo. From an initial synthetic library of potentially 65,536 different oligophosphorothioates 8 nt in length, an oligomer containing G at nearly every position was found to be active against herpes simplex virus type I in cell culture. Further screening of oligomers containing various-size blocks of contiguous guanosines suggested that tetraguanosine was the minimum sequence required for activity. The most potent oligomer tested had an IC_{50} of 0.8 μM and exhibited superior cell viability when compared with acyclovir, a nucleoside analogue antiviral agent currently in use. The authors suggested that, in vivo, the active form of the drug may in fact exist as a tetramer in which a guanosine from each strand, through alternative base-pairing interactions, participates to form a cyclic structure. This motif is repeated fourfold along the length of the tetramer, with each ring stacked upon the next. The similarities between the solution NMR structure of the oligomer found to inhibit thrombin and the

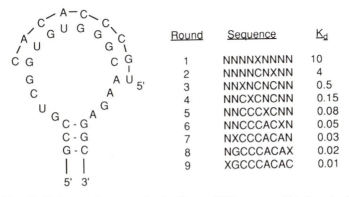

Round	Sequence	K_d
1	NNNNXNNNN	10
2	NNNNCNXNN	4
3	NNXNCNCNN	0.5
4	NNCXCNCNN	0.15
5	NNCCCXCNN	0.08
6	NNCCCACXN	0.05
7	NXCCCACAN	0.03
8	NGCCCACAX	0.02
9	XGCCCACAC	0.01

Figure 14-10. (**Left**) Proposed structure for the H-ras mRNA target and binding site for the winning 9-nt oligomer. (**Right**) Sequence progression for successive rounds of the SURF procedure. The value of K_d shown is for the optimal nucleotide found for position X, as determined in the following round.

tetramer structure proposed here for the active form of the complex should not go un-noticed.

Summary

In vitro random selection should continue to find utility as a means for elucidating the identity of the consensus sequence for DNA- and RNA-binding proteins. In addition, random evolutionary selection in many cases will supplant site-directed mutagenesis as the preferred method for evaluating the effect of mutations introduced over a wide range of sequence.

However, the major advances may come with the discovery of new and unique ligands that bind nucleic acids. Szostak and coworkers have shown that RNA and DNA form architectures which are capable of binding organic molecules. In much the same way that the immune system generates antibodies against a particular antigen, it may be possible to generate "RNA antibodies" that recognize and bind any number of structures. In a fashion analogous to that of catalytic antibodies, generation of RNA against molecules which mimic transition-state intermediates may lead to new classes of ribozymes that catalyze organic transformations other than phosphoryl transfer.

In the past, nucleotide-based theraputic agents have been limited to oligonucleotides which inhibit protein translation by binding to DNA or RNA targets through traditional base-pairing interactions. However, through in vitro natural selection, oligonucleotides have been discovered that bind to and inhibit protein function based on new three-dimensional structures and what will surely prove to be nontraditional protein–nucleotide interactions. Clearly, this provides fundamentally new avenues of investigation to search for pharmaceutical agents based on nucleic acids.

Appendix: Optimization of the Polymerase Chain Reaction

A number of experimental factors affect the efficiency of the PCR; however, by employing standard protocols a robust PCR will usually evolve prior to implementing systematic optimization of the experimental conditions. It is rumored that occasionally, despite all efforts to ensure success, a PCR primer pair just never "works" to produce an amplification. To lessen the chances of failure and assuming that the researcher has the necessary DNA synthesis resources, several pairs of primers should be prepared with any new DNA target; computer programs are available to help design PCR primers for specific DNA sequences. PCR trials with appropriate pairs of primers could then proceed in parallel to ensure a successful PCR result.

Several commercial thermostable DNA polymerases are available which are able to withstand the rigors of PCR. Choosing an enzyme that will best support a particular PCR application will involve weighing the importance of the various enzymatic properties inherent to a particular polymerase, including $5' \rightarrow 3'$ exonuclease activity, $3' \rightarrow 5'$ exonuclease activity, increased thermostability, reverse transcriptase activity, and price. Whichever enzyme is selected, the first trial of PCR should include a Mg^{2+} titration ranging in concentration from 0.5 to 5 mM in 0.5 mM steps. This will ensure that the available Mg^{2+} concentration is adequate to fulfill the Mg^{2+} activity requirement of the polymerase; both the dNTPs and nucleic acids present in the sample bind and compete for divalent metal ions.

In many laboratories, temperature cycling is generally first attempted using a protocol like the following: (1) 1 min at 94°C for denaturation, (2) 1 min at 55°C for primer annealing, and (3) 2 min at 72°C for primer extension. An experimentally derived algorithm for choosing an optimal annealing temperature for a given set of primers has been published (Rychlik et al., 1990). The range of optimal temperatures for primer extension activity of all commercial thermostable DNA polymerases currently available does not differ significantly from the temperature indicated above.

The number of thermal cycles required to detect an amplicon depends not only on the number of starting DNA target molecules and the DNA detection method, but also on the efficiency of amplification. A mathematical description of the yield obtained by PCR amplification is $Y_n = (1 + e)^n$, where Y_n is the multiplication factor of the amplicon above the initial target concentration after n thermal cycles at a designated amplification efficiency (e) (Mullis, 1991). Efficiency is a value between 0 and 1, which will vary from cycle to cycle and refers to the fraction of DNA targets in each cycle which are available for primer extension and actually become duplicated. Efficiencies in excess of 90% are easily attained in optimized PCR reactions in the early rounds of thermal cycling, but inevitably decrease in later rounds as the amplicon concentration plateaus.

When attempting PCR on a sample containing only a few target molecules, especially when the target resides within or is mixed with a complex DNA sequence, special precautions are necessary. When there are fewer than 1000 DNA target molecules, a large number of thermal cycles (>30) are required for sufficient amplification. If the primers happen to misprime at nontarget sequences, the amplification of these false sequences will compete in subsequent rounds of cycling with the desired target amplicon, sometimes resulting in a failed amplification of the desired sequence. The propensity for this occurrence increases with the complexity of the starting DNA mixture and when a high number of thermal cycles is required. A technique developed to help overcome this limitation is commonly referred to as "hot-start" PCR (Chou et al., 1992). This procedural modification entails withholding at least one reagent from the mixture until the reaction reaches a temperature above the annealing temperature. This practice increases the stringency of the annealing step because mismatched primer–DNA template complexes are destabilized at high temperatures relative to perfectly matched sequences. By reducing the number of misprimed nontarget sequences, gains can be expected in reproducibility, sensitivity, and specificity of the PCR experiment.

Several commercial sources supply reagents which are designed to facilitate "hot-start" setups. We have found that the following protocol allows a reliable "hot start" when Eppendorf-type tubes are used and the reaction volume is 50 microliters or larger. Initially all the PCR reagents are mixed *except* the DNA polymerase, which is aliquoted onto the upper portion of the Eppendorf tube, where it will remain as long as the tube surface stays dry. The tube is gently placed in the thermal cycler and heated to the denaturing temperature. Each tube is then quickly removed from the cycler, tapped with the fingers to mix the drop of polymerase with the rest of the reaction, and quickly returned to the thermal cycler. This procedure does not allow the reaction to cool; by depriving the reaction of polymerase below the annealing temperature, the goal of preventing mispriming is achieved.

The sensitivity of the PCR is such that a goal of detecting a single molecule of DNA target is not unreasonable. Therefore, if amplicon molecules from a previous PCR experiment contaminate a new reaction which employs the same or a nested primer pair, an amplicon product will arise irrespective of whether authentic target was present in the reaction mixture. A test which must be incorporated into the PCR protocol for each primer pair is a control in which no target is present. Failure to produce the expected amplicon indicates that the reaction is free of contaminants. Conversely, a successful amplification indicates that contamination of this reaction tube with target or amplicon has taken place. This is an indication that contaminating DNA may be the basis of amplifications in other reactions tubes as well.

A number of strategies have been adopted to limit amplicon contamination, including the following:

Control PCR reactions lacking target DNA should be assembled together, at the same time as all reactions that are using the same primers, and in an area removed from where amplified products are made and processed.

The PCR reagents should be prepared in areas separate from where the PCR experiments are carried out. Reagents should be aliquoted to small volumes suitable for about 50 reactions to avoid contamination of the original stocks. If contamination appears to be a problem, it is wise to simply discard the working stocks of reagents and start again with fresh unopened tubes.

New glassware and disposable plasticware should be used for all PCR work. Positive displacement pipettes should be employed when handling PCR reagents to decrease pipet contamination.

Experience shows that the source of most contamination problems derives from the experimenter performing the PCR reactions. The PCR practitioner collects amplicons on his/her body and clothing and sheds amplicon molecules onto work surfaces. The control(s) lacking target can become positive merely by being handled by the person performing the PCR experiments. In a laboratory routinely performing high-performance PCR (target concentrations of 1000 molecules or fewer) a near "clean-room" discipline must be maintained. In addition to a lab coat, coverings must be worn on the hair, nose, and chin, then discarded or cleaned between PCR preparations that use the same primer sets or targets. Concentrated solutions of DNA targets or amplicons must not be allowed to enter the work environment where PCR reactions are assembled.

Enzymatic and chemical methods are being developed that allow differentiation of amplicon and target DNA by specifically disabling the contaminating amplicon from acting as a substrate. Currently the preferred "amplicon sterilization" technique substitutes dUTP for dTTP as one of the polymerase substrates, which is then incorporated into the amplicon. In subsequent PCR reactions a pre-cycling incubation allows the enzyme uracil DNA glycosylase (UDG) to hydrolyze the uracil base from single- or double-stranded DNA (Longo et al., 1990), thus degrading any amplicons remaining from a previous experiment and precluding their amplification. Because most sources of target DNAs will not contain U residues, they are not a substrate for UDG activity and remain intact. Commercial kits for the express purpose of amplification sterilization are available, and their use should be considered in laboratories requiring high-performance PCR.

References

Chapter 1

Abrams, J. J., & Murrer, B. A. (1993) *Science 261*, 725–730.

Altman, S. (1984) *Cell 36*, 237–239.

Altman, S., Baer, M., Guerrier-Takada, C., & Vioque, A. (1986) *Trends Biochem. Sci. 11*, 515–518.

Amero, S. A., Raychaudhuri, G., Cass, C. L., van Venrooij, W. J., Habets, W. J., Krainer, A. R., & Beyer, A. L. (1992) *Proc. Natl. Acad. Sci. U.S.A. 89*, 8409–8413.

Arnott, S., Hukins, D. W. L., Dover, S. D., Fuller, W., & Hodgson, A. R. (1973) *J. Mol. Biol. 81*, 107–122.

Bailly, C., & Henichart, J.-P. (1991) *Bioconjugate Chem. 2*, 379–393.

Barbier, B., & Brack, A. (1988) *J. Am. Chem. Soc. 110*, 6880–6882.

Barton, J. K. (1986) *Science 233*, 727–734.

Basile, L. A., & Barton, J. K. (1989) in *Metal Ions in Biological Systems* Vol. 25 (Sigel, H., & Sigel, A., Eds.), pp. 31–103, Marcel Dekker, New York.

Benner, S. A., Ellington, A. D., & Tauer, A. (1989) *Proc. Natl. Acad. Sci. U.S.A. 86*, 7054–7058.

Berman, H. M., & Young, P. R. (1981) *Annu. Rev. Biophys. Bioeng. 10*, 87–114.

Bizanek, R., McGuinness, B. F., Nakanishi, K., & Tomasz, M. (1992) *Biochemistry 31*, 3084–3091.

Brimacombe, R. (1988) *Biochemistry 27*, 4207–4214.

Brown, R. S., Dewan, J. C., & Klug, A. (1985) *Biochemistry 24*, 4785–4801.

Burkhoff, A. M., & Tullius, T. D. (1987) *Cell 48*, 935–943.

Calladine, C. R. (1982) *J. Mol. Biol. 161*, 343–352.

Carter, B. J., de Vroom, E., Long, E. C., van der Marel, G. A., van Boom, J. H., & Hecht, S. M. (1990) *Proc. Natl. Acad. Sci. U.S.A.* 87, 9373–9377.

Caruthers, M. H., Barone, A. D., Beaucage, S. L., Dodds, D. R., Fisher, E. F., McBride, L. J. Matteucci, M., Stabinsky, Z., & Tang, J. Y. (1987) *Methods Enzymol. 154*, 287–313.

Cech, T. R. (1987) *Science 236*, 1532–1539.

Cech, T. R., & Bass, B. L. (1986) *Annu. Rev. Biochem. 55*, 599–629.

Cedar, H. (1988) *Cell 53*, 3–4.

Chen, X., Rokita, S. E., & Burrows, C. J. (1991) *J. Am. Chem. Soc. 113*, 5884–5886.

Chen, X., Woodson, S. A., Burrows, C. J., & Rokita, S. E. (1993) *Biochemistry 32*, 7610–7616.

Chow, C. S., & Barton, J. K. (1990) *J. Am. Chem. Soc. 112*, 2839–2841.

Chow, C. S., Behlen, L. S., Uhlenbeck, O. C., & Barton, J. K. (1992) *Biochemistry 31*, 972–982.

Clore, G. M., & Gronenborn, A. M. (1989) *CRC Crit. Rev. Biochem. 24*, 479–654.

Cohen, J. S. (1989) *Oligodeoxynucleotides—Antisense Inhibitors of Gene Expression*, CRC Press, Boca Raton, FL.

Coll, M., Ayamami, J., van der Marel, G. A., van Boom, J. H., Rich, A., & Wang, A. H.-J. (1989) *Biochemistry 28*, 310–320.

Crick, F. (1970) *Nature 227*, 561–563.

Dahlberg, A. E. (1989) *Cell 57*, 525–529.

Dam, E., Pleij, K., Draper, D. (1992) *Biochemistry 31*, 11665–11676.

Dervan, P. B. (1986) *Science 232*, 464–471.

Dervan, P. B. (1991) *Methods Enzymol. 208*, 497–515.

Dickerson, R. E., Drew, H. R., Conner, B. N., Wing, R. M., Fratini, A. V., & Kopka, M. L. (1982) *Science 216*, 475–485.

Drew, H. R., & Travers, A. A. (1984) *Cell 37*, 491–502.

Drew, H. R., McCall, M. J., & Calladine, C. R. (1990) in *DNA Topology and Its Biological Effects* (Cozzarelli, N. R., & Wang, J. C., Eds.). Cold Spring Harbor Press, Cold Spring Harbor, NY.

Ellenberger, T. E., Brandl, C. J., Struhl, K., & Harrison, S. C. (1992) *Cell 71*, 1223–1237.

EMBO Workshop (1989). *EMBO J. 8*, 1–4.

Fkyerat, A., Demeunynck, M., Constant, J.-F., Michon, P., & Lhomme, J. (1993) *J. Am. Chem. Soc. 115*, 9952–9959.

Fleisher, M. B., Mei, H.-Y., & Barton, J. K. (1988) *Nucleic Acids Mol. Biol. 2*, 65–84.

Frederick, C. A., Quigley, G. J., Teng, M.-K., Coll, M., van der Marel, G. A., van Boom, J. H., Rich, A., & Wang, A. H.-J. (1989) *Eur. J. Biochem. 181*, 295–307.

Gait, M. J., Ed. (1984) *Oligonucleotide Synthesis. A Practical Approach*, IRL Press, Washington, D.C.

Gale, E. F., Cundliffe, E., Reynolds, P. E., Richmond, M. H., & Waring, M. J. (1972) *The Molecular Basis of Antibiotic Action*, Wiley, London.

Gao, X., & Patel, D. J. (1989) *Q. Rev. Biophys. 22*, 93–138.

Geierstanger, B. H., Kagawa, T. F., Chen, S.-L., Quigley, G. J., & Ho, P. S. (1991) *J. Biol. Chem. 266*, 20185–20191.

Gesteland, R., & Atkins, J., Eds. (1993) *The RNA World*, Cold Spring Harbor Press, Cold Spring Harbor, NY.

Goldberg, I. (1991) *Acc. Chem. Res. 24*, 191–198.

Gosh, P. K., Reddy, V. B., Piatak, M., Lebowitz, P., & Weissman, S. M. (1980) *Methods Enzymol. 65*, 580–595.

Goyne, T. E., & Sigman, D. S. (1987) *J. Am. Chem. Soc. 109*, 2846–2848.

Guschlbauer, W., Chantot, J.-F., & Thiele, D. (1990) *J. Biomol. Struct. Dynamics 8*, 491–511

Hagerman, P. J. (1986) *Nature 321*, 449–450.

Hall, R. H. (1971) *The Modified Nucleosides in Nucleic Acids*, Columbia University Press, New York.

Hecht, S. M. (1986) *Acc. Chem. Res. 19*, 383–391.

Hertzberg, R. P. & Dervan, P. B. (1984) *Biochemistry 23*, 3934–3945.

Holliday, R. (1989) *Sci. Am. 260*, 60–73.

Holmes, C. E., Carter, B. J., & Hecht, S. M. (1993) *Biochemistry 32*, 4293–4307.

Hou, Y.-M., Francklyn, C., & Schimmel, P. (1989) *Trends Biochem. Sci. 14*, 233–237.

Hsieh, C.-H., & Griffith, J. D. (1988) *Cell 52*, 535–544.

Htun, H., & Dahlberg, J. E. (1988) *Science 241*, 1791–1796.

Hunter, W. N., D'Estainlot, B. L., & Kennard, O. (1989) *Biochemistry 28*, 2444–2451.

Ide, H., Akamatsu, K., Kimura, Y., Michiue, K., Makino, K., Asaeda, A., Takamori, Y., & Kubo, K. (1993) *Biochemistry 32*, 8276–8283.

Itakura, K., Possi, J. J., & Wallace, B. R. (1984) *Annu. Rev. Biochem. 53*, 323–356.

Jaworski, A., Hsieh, W.-T., Blaho, J. A., Larson, J. E., & Wells, R. D. (1987) *Science 238*, 773–777.

Kagawa, T. F., Geierstanger, B. H., Wang, A. H.-J., & Ho, P.S. (1991) *J. Biol. Chem. 266*, 20175–20184.

Kalnik, M. W., Chang, C.-N., Johnson, F., Grollman, A. P., & Patel, D. J. (1989) *Biochemistry 28*, 3373–3383.

Kennard, O., & Hunter, W. N. (1989) *Q. Rev. Biophys. 22*, 327–379.

Kennard, O., & Hunter, W. N. (1991) *Angew. Chem. Int. Ed. Eng. 30*, 1254–1277.

Khorana, H. G. (1979) *Science 203*, 614–625.

Koo, H.-S., & Crothers, D. M. (1988) *Proc. Natl. Acad. Sci. U.S.A. 85*, 1763–1767.

Kopka, M. L., Yoon, C., Goodsell, D., Pjura, P., & Dickerson, R. E. (1985) *Proc. Natl. Acad. Sci. U.S.A. 82*, 1376–1380.

Latham, J. A., & Cech, T. R. (1989) *Science 245*, 276–282.

Lawley, P. D., & Brookes, P. (1963) *Biochem. J. 89*, 127–128.

Lee, M. D., Ellestad, G. A., & Borders, D. B. (1991) *Acc. Chem. Res. 24*, 235–243.

Lerman, L. S. (1961) *J. Mol. Biol. 3*, 18–30.

Lilley, D. M. J. (1989) *Chem. Soc. Rev. 18*, 53–83.

Lilley, D. M. J., & Dahlberg, J. E., Eds. (1992) *Methods in Enzymology*, Vol. 211, Academic Press, New York.

Long, E. C., & Barton, J. K. (1990) *Acc. Chem. Res. 23*, 271–273.

Maniatis, T., Fritsch, E. F., & Sambrook, J. (1982) *Molecular Cloning. A Laboratory Manual*, Cold Spring Harbor Press, Cold Spring Harbor, NY.

Martin, R. B. (1985) *Acc. Chem. Res. 18*, 32–38.

Maxam, A. M., & Gilbert, W. (1977) *Proc. Natl. Acad. Sci. U.S.A. 74*, 560–564.

Maxam, A. M., & Gilbert, W. (1980) *Methods Enzymol. 65*, 499–560.

Milman, G., Langridge, R., & Chamberlin, M. J. (1967) *Proc. Natl. Acad. Sci. U.S.A. 57*, 1804–1810.

Morrow, J. R., Buttrey, L. A., Shelton, V. M., & Berback, K. A. (1992) *J. Am. Chem. Soc. 114*, 1903–1905.

Murphy, F. L., & Cech, T. R. (1993) *Biochemistry 32*, 5291–5300.

Nathans, D., & Smith, M. O. (1975) *Annu. Rev. Biochem. 44*, 273–293.

Nielsen, P. E. (1991) *Bioconjugate Chem. 2*, 1–12.

Ollis, D. L., & White, S. W. (1987) *Chem. Rev. 87*, 981–995.

Otwinowski, Z., Schevitz, R. W., Zhang, R.-G., Lawson, C. L. L., Joachimiak, A., Marmorstein, R. Q., Luisi, B. F., & Sigler, P. B. (1988) *Nature 335*, 321–329.

Pil, P. M., & Lippard, S. J. (1992) *Science 256*, 234–237.

Pindur, U., Haber, M., & Sattler, K. (1993) *J. Chem. Ed. 70*, 263–272.

Ptashne, M. (1987) *A Genetic Switch*, Blackwell Scientific Publications, Palo Alto, CA.

Pyle, A. M. (1993) *Science 261*, 709–714.

Pyle, A. M., & Barton, J. K. (1990) *Prog. Inorg. Chem. 38*, 413–475.

Rajagopol, P., & Feigon, J. (1989) *Nature 339*, 637–640.

Rich, A. (1977) *Acc. Chem. Res. 10*, 388–396.

Rich, A., Nordhiem, A., & Wang, A. H.-J. (1984) *Annu. Rev. Biochem. 53*, 791–846.

Rould, M. A., Perona, J. J., Soll, D., & Steitz, T. A. (1989) *Science 246*, 1135–1142.

Saenger, W. (1984) *Principles of Nucleic Acid Structure*, Springer-Verlag, New York.

Schimmel, P. (1987) *Annu. Rev. Biochem. 56*, 125–158.

Schimmel, P. (1989) *Cell 58*, 9–12.

Seeman, N. C., Rosenberg, J. M., & Rich, A. (1976) *Proc. Natl. Acad. Sci. U.S.A. 73*, 804–808.

Sharp, P. A. (1987) *Science 235*, 766–771.

Sherman, S. E., & Lippard, S. J. (1987) *Chem. Rev. 87*, 1153–1181.

Sigman, D. S. (1986) *Acc. Chem. Res. 19*, 180–186.

Smith, A. J. H. (1980) *Methods Enzymol. 65*, 560–580.

Sprinzl, M., Dank, N., Nock, S., & Schon, A. (1991) *Nucleic Acids Res. 19*, 2127–2171.

Steitz, T. A. (1990) *Q. Rev. Biophys. 23*, 205–280.

Stern, S., Powers, T., Changchien, L.-I., & Noller, H. F. (1989) *Science 244*, 783–790.

Stern, M. K., Bashkin, J. K., & Sall, E. D. (1990) *J. Am. Chem. Soc. 112*, 5357–5359.

Stroynowski, I., Kuroda, M., & Yanofsky, C. (1983) *Proc. Natl. Acad. Sci. U.S.A. 80*, 2206–2210.

Stubbe, J., & Kozarich, J. W. (1987) *Chem. Rev. 87*, 1107–1136.

Sugiura, Y., Shiraki, T., Konishi, M., & Oki, T. (1990) *Proc. Natl. Acad. Sci. U.S.A. 87*, 3831–3835.

Sundquist, W. I., & Klug, A. (1989) *Nature 342*, 825–829.

Tabernero, L., Verdaguer, N., Coll, M., Fita, I., van der Marel, G. A., van Boom, J. H., Rich, A., & Ayamami, J. (1993) *Biochemistry 32*, 8403–8410.

Tomasz, M., Lipman, R., McGuinness, B. F., & Nakanishi, K. (1988) *J. Am. Chem. Soc. 110*, 5892–5896.

Tullius, T. D., & Dombroski, B. A. (1986) *Proc. Natl. Acad. Sci. U.S.A. 83*, 5469–5473.

Tullius, T. D., Dombroski, B. A., Churchill, M. E. A., & Kam, L. (1987) *Methods Enzymol. 155*, 537–559.

van de Ven, F. J. M., & Hilbers, C. W. (1988) *Eur. J. Biochem. 178*, 1–38.

Wang, A. H.-J., Quigley, G. J., Kolpak, F. J., Crawford, J. L., van Boom, J. H., van der Marel, G., & Rich, A. (1979) *Nature 282*, 680–686.

Wang, A. H.-J., Fujii, S., van Boom, J. H., van der Marel, G. A., van Boeckel, S. A. A., & Rich, A. (1982) *Nature 299*, 601–604.

Waring, M. J. (1970) *J. Mol. Biol. 54*, 247–279.

Watson, J. D., Hopkins, N. H., Roberts, J. W., Steitz, J. A., & Weiner, A. M. (1987) *Molecular Biology of the Gene*, 4th ed. Benjamin/Cummings, Menlo Park, CA.

Weiss, R., Teich, N., Varmus, H. E., & Coffin, J. (1985) *RNA Tumor Viruses*, Cold Spring Harbor Press, Cold Spring Harbor, NY.

Wells, R. D., & Harvey, S. C. (1987) *Unusual DNA Structures*, Springer, New York.

Wells, R. D., Collier, D. A., Hanvey, J. C., Shimizu, M., & Wohlrab, F. (1988) *FASEB J. 2*, 2939–2949.

Westhof, E., Dumas, P., & Moras, D. (1985) *J. Mol. Biol. 184*, 119–145.

Wilson, W. D., & Jones, R. L. (1982) in *Intercalation Chemistry* (Whittingham, M. S., & Jacobson, A. J., Eds.), pp. 445–501, Academic Press, New York.

Wing, R., Drew, H., Takano, T., Broka, C., Tanaka, S., Itakura, K., & Dickerson, R. E. (1980) *Nature 287*, 755–758.

Wu, H.-M., & Crothers, D. M. (1984) *Nature 308*, 509–513.

Wüthrich, K. (1986) *NMR of Proteins and Nucleic Acids*, Wiley, New York.

Yanofsky, C. (1988) *J. Biol. Chem. 263*, 609–612.

Yuan, R. (1981) *Annu. Rev. Biochem. 50*, 285–315.

Yoshinari, K., Yamazaki, K., & Komiyama, M. (1991) *J. Am. Chem. Soc. 113*, 5899–5901.

Zacharias, W., Jaworski, A., Larson, J. E., & Wells, R. D. (1988) *Proc. Natl. Acad. Sci. U.S.A. 85*, 7069–7073.

Zimmer, C., & Wahnert, U. (1986) *Prog. Biophys. Mol. Biol. 47*, 31–112.

Zucker, M. (1989) *Science 244*, 48–52.

Chapter 2

Adamiak, R. W., Barciszewska, M. Z., Biala, E., Grzeskowiak, K., Kierzek, R., Kraszewski, A., Markiewicz, W. T., & Wiewiorowski, M. (1976) *Nucleic Acids Res. 3*, 3397–3408.

Adams, S. P., Kavka, K. S., Wykes, E. J., Holder, S. B., & Galluppi, G. R. (1983) *J. Am. Chem. Soc. 105*, 661–663.

Agrawal, S. (1992) *Trends Biotechnol. 10*, 152–158.

Agrawal, S., & Sarin, P. S. (1991) *Adv. Drugs Del. Rev. 6*, 251–270.

Akhtar, S., Basu, S., Wickstrom, E., & Juliano, R. L. (1991) *Nucleic Acids Res. 19*, 5551–5559.

Amarnath, V., & Broom, A. D. (1977) *Chem. Rev. 77*, 183–217.

Andrus, A., & Beaucage, S. L. (1988) *Tetrahedron Lett. 29*, 5479–5482.

Andrus, A., Efcavitch, J. W., McBride, L. J., & Giusti, B. (1988) *Tetrahedron Lett. 29*, 861–864.

Arentzen, R., & Reese, C. B. (1977) *J. Chem. Soc. Perkin Trans. 1*, 445–460.

Arnhelm, N., & Erlich, H. (1992) *Annu. Rev. Biochem. 61*, 131–156.

Barone, A. D., Tang, J. -Y., & Caruthers, M. H. (1984) *Nucleic Acids Res. 12*, 4051–4061.

Beaton, G., Brill, W. K.-D., Grandas, A., Ma, Y.-X., Nielsen, J., Yau, E., & Caruthers, M. H. (1991) *Tetrahedron 47*, 2377–2388.

Beaucage, S. L. (1984) *Tetrahedron Lett. 25*, 375–378.

Beaucage, S. L., & Caruthers, M. H. (1981) *Tetrahedron Lett. 22*, 1859–1862.

Beaucage, S. L., & Iyer, R. P. (1992) *Tetrahedron 48*, 2223–2311.

Beaucage, S. L., & Iyer, R. P. (1993) *Tetrahedron 49*, 6123–6194.

Beijer, B., Sulston, I., Sproat, B. S., Rider, P., Lamond, A. I., & Neuner, P. (1990) *Nucleic Acids Res. 18*, 5143–5151.

Bentrude, W. G., Sopchik, A. E., & Gajda, T. (1989) *J. Am. Chem. Soc. 111*, 3981–3987.

Berlin, Y. A., Chakhmakhcheva, O. G., Efimov, V. A., Kolosov, M. N., & Korobko, V. G. (1973) *Tetrahedron Lett.*, 1353–1354.

Berner, S., Mühlegger, K., & Seliger, H. (1989) *Nucleic Acids Res. 17*, 853–864.

Böhringer, M. P., Graff, D., & Caruthers, M. H. (1993) *Tetrahedron Lett. 34*, 2723–2726.

Brill, W. K.-D., Yau, E. K., & Caruthers, M. H. (1989) *Tetrahedron Lett. 30*, 6621–6624.

Brown, D. M., Magrath, D. I., Nielson, A. H., & Todd, A. R. (1956) *Nature 177*, 1124–1125.

Caruthers, M. H. (1981) in *Recombinant DNA, Proceedings of the Third Cleveland Symposium on Macromolecules* (Walton, A. G., Ed.), pp. 261–272, Elsevier, Amsterdam.

Caruthers, M. H. (1985) *Science 230*, 281–285.

Caruthers, M. H., Beaucage, S. L., Becker, C., Efcavitch, W., Fisher, E. F., Galluppi, G., Goldman, R., deHaseth, P., Martin, F., Matteucci, M., & Stabinsky, Y. (1982) in *Genetic Engineering: Principles and Methods*, Vol. 4 (Setlow, J. K., & Hollaender, A., Eds.), pp. 1–17, Plenum, New York.

Caruthers, M. H., Barone, A. D., Beaucage, S. L., Dodds, D. R., Fisher, E. F., McBride, L. J., Matteucci, M., Stabinsky, Z., & Tang, J.-Y. (1987a) *Methods Enzymol. 154*, 287–313.

Caruthers, M. H., Kierzek. R., & Tang, J. Y. (1987b) in *Biophosphates and Their Analogues—Synthesis, Structure Metabolism and Activity* (Bruzik, K. S., & Stec, W. J., Eds.), pp. 3–21, Elsevier, Amsterdam.

Caulfield, T. J., Prasad, C. V. C., Prouty, C. P., Saha, A. K., Sardaro, M. P., Schairer, W. C., Yawman, A., Upson, D. A., & Kruse, L. I. (1993) *Bioorg Med. Chem. Lett. 3*, 2771–2776.

Chandrasegaran, S., Murakami, A., & Kan, L.-S. (1984) *J. Org. Chem. 49*, 4951–4957.

Chattopadhyaya, J. B., & Reese, C. B. (1979) *Tetrahedron Lett. 20*, 5059–5062.

Christodoulou, C. (1993) in *Methods in Molecular Biology*, Vol. 20: *Protocols for Oligonucleotides and Analogs* (Agrawal, S., Ed.), pp. 19–31, Humana Press, Totowa.

Christodoulou, C., Agrawal, S., & Gait, M. J. (1986) *Tetrahedron Lett. 27*, 1521–1522.

Chur, A., Holst, B., Dahl, O., Valentin-Hansen, P., & Pedersen, E. B. (1993) *Nucleic Acids Res. 21*, 5179–5183.

Claesen, C., Tesser, G. I., Dreef, C. E., Marugg, J. E., van der Marel, G. A., & van Boom, J. H. (1984) *Tetrahedron Lett. 25*, 1307–1310.

Cohen, J. S. (1991) *Antiviral Res. 16*, 121–133.

Crea, R, Kraszewski, A., Hirose, T., & Itakura, K. (1978) *Proc. Natl. Acad. Sci. U.S.A. 75*, 5765–5769.

Cullis, P.M. (1984) *J. Chem. Soc. Chem. Commun.*, 1510–1512.

Cusack, N. J., Reese, C. B., & van Boom, J. H. (1973) *Tetrahedron Lett.*, 2209–2212.

Dahl, B. H., Nielsen, J., & Dahl, O. (1987) *Nucleic Acids Res. 15*, 1729–1743.

Damha, M. J., Giannaris, P. A., & Zabarylo, S. V. (1990) *Nucleic Acids Res. 18*, 3813–3821.

Damha, M. J., & Ogilvie, K. K. (1993) in *Methods in Molecular Biology*, Vol. 20: *Protocols for Oligonucleotides and Analogs* (Agrawal, S., Ed.), pp. 81–114, Humana Press, Totowa, NJ.

Daub, G. W., & van Tamelen, E. E. (1977) *J. Am. Chem. Soc. 99*, 3526–3528.

deHaseth, P. L., Goldman, R. A., Cech, C. L., & Caruthers, M. H. (1983) *Nucleic Acids Res. 11*, 773–787.

de Rooij, J. F. M., Wille-Hazeleger, G., van Deursen, P. H., Serdijn, J., & van Boom, J. H. (1979) *Rec. Trav. Chim. Pays-Bas 98*, 537–548.

de Vroom, E., Spierenburg, M. L., Dreef, C. E., van der Marel, G. A., & van Boom, J. H. (1987) *Rec. Trav. Chim. Pays-Bas 106*, 65–66.

Dorman, M. A., Noble, S. A., McBride, L. J., & Caruthers, M. H. (1984) *Tetrahedron 40*, 95–102.

Dörper, T., & Winnacker, E.-L. (1983) *Nucleic Acids Res. 11*, 2575–2584.

Drahos, D., Galluppi, G. R., Caruthers, M., & Szybalski, W. (1982) *Gene 18*, 343–354.

Eadie, J. S., & Davidson, D. S. (1987) *Nucleic Acids Res. 15*, 8333–8349.

Eckstein, F. (1985) *Annu. Rev. Biochem. 54*, 367–402.

Efimov, V. A., Reverdatto, S. V., & Chakhmakhcheva, O. G. (1982) *Nucleic Acids Res. 10*, 6675–6694.

Efimov, V. A., Chakhmakhcheva, O. G., & Ovchinnikov, Y. A. (1985) *Nucleic Acids Res. 13*, 3651–3666.

Efimov, V. A., Kalinkina, A. L., & Chakhmakhcheva, O. G. (1993) *Nucleic Acids Res. 21*, 5337–5344.

Englisch, U., & Gauss, D. H. (1991) *Angew. Chem. Int. Ed. Engl. 30*, 613–629.

Erickson, B. W., & Merrifield, R. B. (1976) in *The Proteins* (Neurath, H., Hill, R. L., & Boeder, C.-L., Eds.), pp. 255–527, Academic Press, New York.

Farrance, I. K., Eadie, J. S., & Ivarie, R. (1989) *Nucleic Acids Res. 17*, 1231–1245.

Fourrey, J.-L., & Varenne, J. (1984) *Tetrahedron Lett. 25*, 4511–4514.

Fourrey, J. -L., & Varenne, J. (1985) *Tetrahedron Lett. 26*, 1217–1220.

Fourrey, J.-L., Varenne, J., Fontaine, C., Guittet, E., & Yang, Z. W. (1987) *Tetrahedron Lett. 28*, 1769–1772.

Froehler, B. C. (1986) *Tetrahedron Lett. 27*, 5575–5578.

Froehler, B., & Matteucci, M. D. (1983) *Tetrahedron Lett. 24*, 3171–3174.

Froehler, B. C., Ng, P. G., & Matteucci, M. D. (1986) *Nucleic Acids Res. 14*, 5399–5407.

Fromageot, H. P. M., Reese, C. B., & Sulston, J. E. (1968) *Tetrahedron 24*, 3533–3540.

Gait, M. J. (1980) in *Polymer-Supported Reactions in Organic Synthesis* (Hodge, P., & Sherrington, D. C., Eds.), pp. 435–456, John Wiley & Sons, New York.

Gao, W.-Y., Han, F.-S., Storm, C., Egan, W., & Cheng, Y.-C. (1992) *Mol. Pharmacol. 41*, 223–229.

Garegg, P. J., Lindh, I., Regberg, T., Stawinski, J., Strömberg, R., & Henrichson, C. (1986) *Tetrahedron Lett. 27*, 4055–4058.

Garegg, P. J., Regberg, T., Stawinski, J., & Strömberg, R. (1985) *Chemica Scripta 25*, 280–282.

Garegg, P. J., Regberg, T., Stawinski, J., & Strömberg, R. (1987a) *Nucleosides Nucleotides 6*, 655–662.

Garegg, P. J., Stawinski, J., & Strömberg, R. (1987b) *J. Org. Chem. 52*, 284–287.

Gasparutto, D., Livache, T., Bazin, H., Duplaa, A.-M., Guy, A., Khorlin, A., Molko, D., Roget, A., & Téoule, R. (1992) *Nucleic Acids Res. 20*, 5159–5166.

Giles, R. V., & Tidd, D. M. (1992) *Anti-Cancer Drug Design 7*, 37–48.

Goodarzi, G., Gross, S. C., Tewari, A., & Watabe, K. (1990) *J. Gen. Virol. 71*, 3021–3025.

Goodchild, J. (1990) *Bioconjugate Chem. 2*, 165–187.

Goodchild, J. (1992) *Nucleic Acids Res. 20*, 4607–4612.

Gough, G. R., Brunden, M. J., & Gilham, P. T. (1981) *Tetrahedron Lett. 22*, 4177–4180.

Green, R., Szostak, J. W., Benner, S. A., Rich, A., & Usman, N. (1991) *Nucleic Acids Res. 19*, 4161–4166.

Griffin, B. E., & Reese, C. B. (1964) *Tetrahedron Lett.*, 2925–2931.

Gröger, G., Ramalho-Ortigao, G., Steil, H., & Seliger, H. (1988) *Nucleic Acids Res. 16*, 7763–7771.

Gryaznov, S. M., & Letsinger, R. L. (1991) *J. Am. Chem. Soc. 113*, 5876–5877.

Hall, R. H., Todd, A., & Webb, R. F. (1957) *J. Chem. Soc.*, 3291–3296.

Harrison, S. C., & Aggarwal, A. K. (1990) *Annu. Rev. Biochem. 59*, 933–969.

Hayakawa, Y., Uchiyama, M., & Noyori, R. (1986) *Tetrahedron Lett. 27*, 4191–4194.

Heidenreich, O., & Eckstein, F. (1992) *J. Biol. Chem. 267*, 1904–1909.

Hering, G., Stîcklein-Schneiderwind, R., Ugi, I., Pathak, T., Balgobin, N., & Chattopadhyaya, J. (1985) *Nucleosides Nucleotides 4*, 169–171.

Hogrefe, R. I., Reynolds, M. A., Vaghefi, M. M., Young, K. M., Riley, T. A., Klem, R. E., & Arnold, L. J., Jr. (1993) in *Methods in Molecular Biology*, Vol. 20: *Protocols for Oligonucleotides and Analogs* (Agrawal, S., Ed.), pp. 143–164, Humana Press, Totowa.

Hostomsky, Z., Smrt, J., Arnold, L., Tocík, Z., & Paces, V. (1987) *Nucleic Acids Res. 15*, 4849–4856.

Idziak, I., Just, G., Damha, M. J., & Giannaris, P. A. (1993) *Tetrahedron Lett. 34*, 5417–5420.

Imbach, J.-L., Rayner, B., & Morvan, F. (1989) *Nucleosides Nucleotides 8*, 627–648.

Inoue, H., Hayase, Y., Imura, A., Iwai, S., Miura, K., & Ohtsuka E. (1987) *Nucleic Acids Res. 15*, 6131–6148.

Iribarren, A. M., Sproat, B. S., Neuner, P., Sulston, I., Ryder, U., & Lamond, A. I. (1990) *Proc. Natl. Acad. Sci. U.S.A. 87*, 7747–7751.

Itakura, K., Katagiri, N., & Narang, S. A. (1974) *Can. J. Chem. 52*, 3689–3693.

Iyer, R. P., Phillips, L. R., Egan, W., Regan, J. B., & Beaucage, S. L. (1990) *J. Org. Chem. 55*, 4693–4698.

Jones, S. S., & Reese, C. B. (1979) *J. Chem. Soc. Perkin Trans. 1*, 2762–2764.

Josephson, S., Lagerholm, E., & Palm, G. (1984) *Acta Chem. Scand., B38*, 539–545.

Kabachnik, M. M., Potapov, V. K., Shabarova, Z. A., & Prokofiev, M. A. (1971) *Dokl. Akad. Nauk SSSR 201*, 858–861.

Kamer, P. C. J., Roelen, H. C. P. F., van den Elst, H., van der Marel, G. A., & van Boom, J. H. (1989) *Tetrahedron Lett. 30*, 6757–6760.

Kaplan, B. E., & Itakura, K. (1987) in *Synthesis and Applications of DNA and RNA* (Narang, S. A., Ed.), pp. 9–45, Academic Press, Orlando, FL.

Katagiri, N., Itakura, K., & Narang, S. A. (1974) *J. Chem. Soc. Chem. Commun.*, 325–326.

Kawai, S. H., Wang, D., Giannaris, P. A., Damha, M. J., & Just, G. (1993) *Nucleic Acids Res. 21*, 1473–1479.

Kennard, O., & Hunter, W. N. (1989) *Q. Rev. Biophys. 22*, 327–379.

Kierzek, R., Caruthers, M. H., Longfellow, C. E., Swinton, D., Turner, D. H., & Freier, S. M. (1986) *Biochemistry 25*, 7840–7846.

Kierzek, R., Rozek, M., & Markiewicz, W. T. (1987) *Nucleic Acids Res. Symp. Ser. #18*, 201–204.

Koga, M., Moore, M. F., & Beaucage, S. L. (1991) *J. Org. Chem. 56*, 3757–3759.

Koga, M., Geyer, S. J., Regan, J. B., & Beaucage, S. L. (1993) *Nucleic Acids Res. Symp. Ser. 29*, 3–4.

Koga, M., Wilk, A., Moore, M. F., Scremin, C. L., Zhou, L., & Beaucage, S. L. (1995) *J. Org. Chem. 60*, 1520–1530.

Kössel, H., & Seliger, H. (1975) *Forstschr. Chem. Org. Naturst. 32*, 297–508.

Kumar, G., & Poonian, M. S. (1984) *J. Org. Chem. 49*, 4905–4912.

Latimer, L. J. P., Hampel, K., & Lee, J. S. (1989) *Nucleic Acids Res. 17*, 1549–1561.

Laurence, J., Sikder, S. K., Kulkosky, J., Miller, P., & Ts'O, P. O. P. (1991) *J. Virol. 65*, 213–219.

Leatherbarrow, R. J., & Fersht, A. R. (1986) *Protein Eng. 1*, 7–16.

Lebreton, J., DeMesmaeker, A., Waldner, A., Fritsch, V., Wolf, R. M., & Freier, S. M. (1993) *Tetrahedron Lett. 34*, 6383–6386.

Lee, H.-J., & Moon, S. -H. (1984) *Chem. Lett.*, 1229–1232.

Lehmann, C., Xu, Y.-Z, Christodoulou, C., Tan, Z.-K., & Gait, M. J. (1989) *Nucleic Acids Res. 17*, 2379–2390.

Letsinger, R. L., & Kornet, M. J. (1963) *J. Am. Chem. Soc. 85*, 3045–3046.

Letsinger, R. L., & Lunsford, W. B. (1976) *J. Am. Chem. Soc. 98*, 3655–3661.

Letsinger, R. L., & Mahadevan, V. (1965) *J. Am. Chem. Soc. 87*, 3526–3527.

Letsinger, R. L., & Mahadevan, V. (1966) *J. Am. Chem. Soc. 88*, 5319–5324.

Letsinger, R. L., & Miller, P. S. (1969) *J. Am. Chem Soc. 91*, 3356–3359.

Letsinger, R. L., & Ogilvie, K. K. (1967) *J. Am. Chem Soc. 89*, 4801–4803.

Letsinger, R. L., & Ogilvie, K. K. (1969) *J. Am. Chem Soc. 91*, 3350–3355.

Letsinger, R. L., Caruthers, M. H., & Jerina, D. M. (1967a) *Biochemistry 6*, 1379–1388.

Letsinger, R. L., Caruthers, M. H., Miller, P. S., & Ogilvie, K. K. (1967b) *J. Am. Chem. Soc. 89*, 7146–7147.

Letsinger, R. L., Ogilvie, K. K., & Miller, P. S. (1969) *J. Am. Chem Soc. 91*, 3360–3365.

Letsinger, R. L., Finnan, J. L., Heavner, G. A., & Lunsford, W. B. (1975) *J. Am. Chem. Soc. 97*, 3278–3279.

Lohrmann, R., & Khorana, H. G. (1966) *J. Am. Chem. Soc. 88*, 829–833.

Loke, S. L., Stein, C. A., Zhang, X.-H., Mori, K., Nakanishi, M., Subasinghe, C., Cohen, J. S., & Neckers, L. M. (1989) *Proc. Natl. Acad. Sci. U.S.A. 86*, 3474–3478.

Löschner, T., & Engels, J. W. (1988) *Nucleosides Nucelotides 7*, 729–732.

Maher, L. J., III, & Dolnick, B. J. (1988) *Nucleic Acids Res. 16*, 3341–3358.

Markiewicz, W. T. (1985) in *Natural Product Chemistry 1984* (Zalewski, R. I., & Skolik, J. J., Eds.), pp. 275–286, Elsevier, Amsterdam.

Markiewicz, W. T., & Wiewiórowski, M. (1978) *Nucleic Acids Res. Spec. Publ. 4*, s185–s188.

Marshall, W. S., & Caruthers, M. H. (1993) *Science 259*, 1564–1570.

Marshall, W. S., Beaton, G., Stein, C. A., Matsukura, M., & Caruthers, M. H. (1992) *Proc. Natl. Acad. Sci. U.S.A. 89*, 6265–6269.

Marugg, J. E., McLaughlin, L. W., Piel, N., Tromp, M., van der Marel, G. A., & van Boom, J. H. (1983) *Tetrahedron Lett. 24*, 3989–3992.

Marugg, J. E., Tromp, M., Jhurani, P., Hoyng, C. F., van der Marel, G. A., & van Boom, J. H. (1984) *Tetrahedron 40*, 73–78.

Matsukura, M., Zon, G., Shinozuka, K., Robert-Guroff, M., Shimada, T., Stein, C. A., Mitsuya, H., Wong-Staal, F., Cohen, J. S., & Broder, S. (1989) *Proc. Natl. Acad. Sci. U.S.A. 86*, 4244–4248.

Matteucci, M. D., & Caruthers, M. H. (1980) *Tetrahedron Lett. 21*, 719–722.

Matteucci, M. D., & Caruthers, M. H. (1981) *J. Am. Chem. Soc. 103*, 3185–3191.

McBride, L. J., & Caruthers, M. H. (1983) *Tetrahedron Lett. 24*, 245–248.

McCollum, C., & Andrus, A. (1991) *Tetrahedron Lett. 32*, 4069–4072.

Merrifield, R. B. (1962) *Fed. Proc. 21*, 412.

Michelson, A. M., & Todd, A. R. (1955) *J. Chem. Soc.*, 2632–2638.

Miller, P. S., & Ts'O, P. O. P. (1987) *Anti-Cancer Drug Design 2*, 117–128.

Moore, M. F., & Beaucage, S. L. (1985) *J. Org. Chem. 50*, 2019–2025.

Morvan, F., Rayner, B., Imbach, J.-L., Chang, D.-K., & Lown, J. W. (1986) *Nucleic Acids Res. 14*, 5019–5035.

Morvan, F., Rayner, B., Leonetti, J.-P., & Imbach, J.-L. (1988) *Nucleic Acids Res. 16*, 833–847.

Morvan, F., Rayner, B., & Imbach, J. -L. (1990) *Tetrahedron Lett. 31*, 7149–7152.

Morvan, F., Porumb, H., Degols, G., Lefebvre, I., Pompon, A., Sproat, B. S., Rayner, B., Malvy, C., Lebleu, B., & Imbach, J.-L. (1993) *J. Med. Chem. 36*, 280–287.

Nielsen, J., & Caruthers, M. H. (1988) *J. Am. Chem. Soc. 110*, 6275–6276.

Ogilvie, K. K. (1973) *Can. J. Chem. 51*, 3799–3807.

Ogilvie, K. K., & Entwistle, D. W. (1981) *Carbohydr. Res. 89*, 203–210.

Ogilvie, K. K., Beaucage, S. L., Schifman, A. L., Theriault, N. Y., & Sadana, K. L. (1978) *Can. J. Chem. 56*, 2768–2780.

Ogilvie, K. K., Usman, N., Nicoghosian, K., & Cedergren, R. J. (1988) *Proc. Natl. Acad. Sci. U.S.A. 85*, 5764–5768.

Olsen, D. B., Benseler, F., Aurup, H., Pieken, W. A., & Eckstein, F. (1991) *Biochemistry 30*, 9735–9741.

Ott, J., & Eckstein, F. (1984) *Nucleic Acids Res. 12*, 9137–9142.

Parker, R., Muhlrad, D., Deshler, J. O., Taylor, N., & Rossi, J. J. (1992) in *Gene Regulation: Biology of Antisense RNA and DNA* (Erickson, R. P., & Izant, J. G., Eds.), pp. 55–70, Raven Press, New York.

Perreault, J.-P., Wu, T.-F, Cousineau, B., Ogilvie, K. K., & Cedergren, R. (1990) *Nature 344*, 565–567.

Perreault, J.-P., Labuda, D., Usman, N., Yang, J.-H., & Cedergren, R. (1991) *Biochemistry 30*, 4020–4025.

Pieken, W. A., Olsen, D. B., Benseler, F., Aurup, H., & Eckstein, F. (1991) *Science 253*, 314–317.

Pon, R. T., Usman, N., Damha, M. J., & Ogilvie, K. K. (1986) *Nucleic Acids Res. 14*, 6453–6470.

Porritt, G. M., & Reese, C. B. (1989) *Tetrahedron Lett. 30*, 4713–4716.

Quartin, R. S., Brakel, C. L., & Wetmur, J. G. (1989) *Nucleic Acids Res. 17*, 7253–7262.

Rao, M. V., & Reese, C. B. (1989) *Nucleic Acids Res. 17*, 8221–8239.

Rao, T. S., Reese, C. B., Serafinowska, H. T., Takaku, H., & Zappia, G. (1987) *Tetrahedron Lett. 28*, 4897–4900.

Rao, M. V., Reese, C. B., & Zhengyun, Z. (1992) *Tetrahedron Lett. 33*, 4839–4842.

Rayner, B., Matsukura, M., Morvan, F., Cohen, J. S., & Imbach, J.-L. (1990) *C. R. Acad. Sci. Paris Ser. III 310*, 61–64.

Reese, C. B. (1978) *Tetrahedron 34*, 3143–3179.

Reese, C. B., & Pei-Zhuo, Z. (1993) *J. Chem. Soc. Perkin Trans. 1*, 2291–2301.

Reese, C. B., & Skone, P. A. (1984) *J. Chem. Soc. Perkin Trans. 1*, 1263–1271.

Reese, C. B., & Thompson, E. A. (1988) *J. Chem. Soc. Perkin Trans. 1*, 2881–2885.

Reese, C. B., & Ubasawa, A. (1980) *Tetrahedron Lett. 21*, 2265–2268.

Reese, C. B., & Yau, L. (1978) *Tetrahedron Lett.*, 4443–4446.

Reese, C. B., & Zard, L. (1981) *Nucleic Acids Res. 9*, 4611–4626.

Reese, C. B., Saffhill, R., & Sulston, J. E. (1970) *Tetrahedron 26*, 1023–1030.

Reese, C. B., Titmas, R. C., & Yau, L. (1978) *Tetrahedron Lett.*, 2727–2730.

Reese, C. B., Serafinowska, H. T., & Zappia, G. (1986) *Tetrahedron Lett. 27*, 2291–2294.

Regan, J. B., Phillips, L. R., & Beaucage, S. L. (1992) *Org. Prep. Proc. Int. 24*, 488–492.

Saha, A. K., Sardaro, M., Waychunas, C., Delecki, D., Kutny, R., Cavanaugh, P., Yawman, A., Upson, D. A., & Kruse, L. I. (1993) *J. Org. Chem. 58*, 7827–7831.

Sakatsume, O., Ohtsuki, M., Takaku, H., & Reese, C. B. (1989) *Nucleic Acids Res. 17*, 3689–3697.

Schwartz, M. E., Breaker, R. R., Asteriadis, G. T., deBear, J. S., & Gough, G. R. (1992) *Bioorg. Med. Chem. Lett. 2*, 1019–1024.

Scremin, C. L., Zhou, L., Srinivasachar, K., & Beaucage, S. L. (1994) *J. Org. Chem. 59*, 1963–1966.

Sekine, M., Masuda, N., & Hata, T. (1986) *Bull. Chem. Soc. Jpn. 59*, 1781–1789.

Seliger, H., Kotschi, U., Scharpf, C., Martin, R., Eisenbeiss, F., Kinkel, J. N., & Unger, K. K. (1989) *J. Chromatogr. 476*, 49–57.

Séquin, U. (1974) *Helv. Chim. Acta 57*, 68–81.

Shibahara, S., Mukai, S., Morisawa, H., Nakashima, H., Kobayashi, S., & Yamamoto, N. (1989) *Nucleic Acids Res. 17*, 239–252.

Shimidzu, T., & Letsinger, R. L. (1968) *J. Org. Chem. 33*, 708–711.

Shoji, Y., Akhtar, S., Priasamy, A., Herman, B., & Juliano, R. L. (1991) *Nucleic Acids Res. 19*, 5543–5550.

Sinha, N. D., Biernat, J., & Köster, H. (1984a) *Nucleosides Nucleotides 3*, 157–171.

Sinha, N. D., Biernat, J., McManus, J., & Köster, H. (1984b) *Nucleic Acids Res. 12*, 4539–4557.

Sonveaux, E. (1986) *Bioorg. Chem. 14*, 274–325.

Sproat, B. S., & Lamond, A. I. (1991) in *Oligonucleotides and Analogues: A Practical Approach* (Eckstein, F., Ed.), pp. 48–86, IRL Press, Oxford.

Sproat, B. S., Lamond, A. I., Beijer, B., Neuner, P., & Ryder, U. (1989) *Nucleic Acids Res. 17*, 3373–3386.

Stawinski, J., Hozumi, T., & Narang, S. A. (1976) *Can. J. Chem. 54*, 670–672.

Stawinski, J., Strömberg, R., Thelin, M., & Westman, E. (1988a) *Nucleosides Nucleotides 7*, 601–604.

Stawinski, J., Strömberg, R., Thelin, M., & Westman, E. (1988b) *Nucleosides Nucleotides 7*, 779–782.

Stawinski, J., Strömberg, R., Thelin, M., & Westman, E. (1988c) *Nucleic Acids Res. 16*, 9285–9298.

Stawinski, J., Thelin, M., & Zain, R. (1989) *Tetrahedron Lett. 30*, 2157–2160.

Stec, W. J., & Zon, G. (1984) *Tetrahedron Lett. 25*, 5279–5282.

Stec, W. J., Uznanski, B., Wilk, A., Hirschbein, B. L., Fearon, K. L., & Bergot, B. J. (1993) *Tetrahedron Lett. 34*, 5317–5320.

Sund, C., Agback, P., & Chattopadhyaya, J. (1993) *Tetrahedron 49*, 649–668.

Takaku, H., Yoshida, M., Kato, M., & Hata, T. (1979) *Chem. Lett.*, 811–814.

Tanaka, T., & Letsinger, R. L. (1982) *Nucleic Acids Res. 10*, 3249–3260.

Tanimura, H., Sekine, M., & Hata, T. (1987) *Chem. Lett.*, 1057–1060.

Taylor, N. R., Kaplan, B. E., Swiderski, P., Li, H., & Rossi, J. J. (1992) *Nucleic Acids Res. 20*, 4559–4565.

Tidd, D. M., & Warenius, H. M. (1989) *Br. J. Cancer 60*, 343–350.

Uhlmann, E., & Peyman, A. (1990) *Chem. Rev. 90*, 543–584.

Usman, N., Pon, R. T., & Ogilvie, K. K. (1985) *Tetrahedron Lett. 26*, 4567–4570.

van Boom, J. H., Burgers, P. M. J., van der Marel, G., Verdegaal, C. H. M., & Wille, G. (1977) *Nucleic Acids Res. 4*, 1047–1063.

Viari, A., Ballini, J. P., Vigny, P., Blonski, C., Dousset, P., & Shire, D. (1987) *Tetrahedron Lett. 28*, 3349–3352.

Vu, H., & Hirschbein, B. L. (1991) *Tetrahedron Lett. 32*, 3005–3008.

Vu, H., McCollum, C., Lotys, C., & Andrus, A. (1990) *Nucleic Acids Res. Symp. Ser. 22*, 63–64.

Wiesler, W. T., Marshall, W. S., & Caruthers, M. H. (1993) in *Methods in Molecular Biology, Vol. 20: Protocols for Oligonucleotides and Analogs* (Agrawal, S., Ed.), pp. 191–206, Humana Press, Totowa, NJ.

Williams, D. M., Pieken, W. A., & Eckstein, F. (1992) *Proc. Natl. Acad. Sci. U.S.A. 89*, 918–921.

Wolter, A., Biernat, J., & Köster, H. (1986) *Nucleosides Nucleotides 5*, 65–77.

Wu, T., & Ogilvie, K. K. (1990) *J. Org. Chem. 55*, 4717–4724.

Yang, J., Usman, N., Chartrand, P., & Cedergren, R. (1992) *Biochemistry 31*, 5005–5009.

Zarytova, V. F., & Knorre, D. G. (1984) *Nucleic Acids Res. 12*, 2091–2110.

Zon, G., Gallo, K. A., Samson, C. J., Shao, K., Summers, M. F., & Byrd, R. A. (1985) *Nucleic Acids Res. 13*, 8181–8196.

Chapter 3

Adams, R.L.P., Knowles, J., & Leader, D.B. (1986) *The Biochemistry of the Nucleic Acids*, p. 87, Chapman and Hall, New York.

Agrawal, S. (1992) *Trends Biotechnol. 10*, 152–158.

Ambrose, B. J. B., & Pless, R. C. (1987) *Methods Enzymol. 152*, 522–537.

Ausubel, F. M., Brent, R., Kingston, R. E., Moore, D. D., Seidman, J. G., Smith, J. A., & Struhl, K. (1989) *Current Protocols in Molecular Biology*, Greene Publishers and Wiley-Interscience, New York.

Balbas, P., Soberon, X., Merino, E., Zurita, M., Lomeli, H., Valle, F., Flores, N., & Bolivar, F. (1986) *Gene 50*, 3–40.

Berger, S. L. (1987) *Methods Enzymol. 152*, 49–54.

Berger, S. L., & Kimmel, A. R. (1987) *Methods Enzymol. 152*, 307–315.

Bernardi, A., & Bernardi, G. (1971) *The Enzymes 4*, 329–336.

Bitter, G. A. (1987) *Methods Enzymol. 152*, 673–683.

Blackburn, P., & Moore, S. (1982) *The Enzymes 15*, 317–433.

Boyer, P. D., Ed. (1982) *The Enzymes*, Vols. 14 and 15, Academic Press, New York.

Brennan, C. A., & Gumport, R. I. (1985) *Nucleic Acids Res. 13*, 8665–8684.

Brennan, C. A., Manthey, A. E., & Gumport, R. I. (1983) *Methods Enzymol. 100*, 38–51.

Brennan, C. A., Van Cleve, M. D., & Gumport, R. I. (1986) *J. Biol. Chem. 261*, 7270–7278.

Bruce, A., & Uhlenbeck, O. C. (1982a) *Biochemistry 21*, 855–861.

Bruce, A. G., & Uhlenbeck, O. C. (1982b) *Biochemistry 21*, 3921–3926.

Brutlag, D., & Kornberg, A. (1972) *J. Biol. Chem 247*, 241–248.

Butcher, S. E., & Burke, J. M. (1994) *Biochemistry 33*, 992–999.

Butler, E. T., & Chamberlin, M. J. (1982) *J. Biol. Chem. 257*, 5772–5778.

Cairns, J., Stent, G. S., & Watson, J. D., Eds. (1966) *Phage and the Origins of Molecular Biology*, Cold Spring Harbor Press, Cold Spring Harbor, NY.

Cameron, V., & Uhlenbeck, O. C. (1977) *Biochemistry 16*, 5120–5126.

Chamberlin, M. J. (1974) *Annu. Rev. Biochem. 43*, 721–775.

Chamberlin, M. J. (1982) *The Enzymes 15*, 61–86.

Chamberlin, M. J., & Ryan, T. (1982) *The Enzymes 15*, 87–108.

Chapeville, F., Lipmann, F., von Ehrenstein, G., Weisblum, B., Ray, W. J., Jr., & Benzer, S. (1962) *Proc. Natl. Acad. Sci. U.S.A. 48*, 1086–1092.

Chung, H. H., Benson, D. R., Cornish, V. W., & Schultz, P. G. (1993) *Proc. Natl. Acad. Sci. U.S.A. 90*, 10145–10149.

Cotton, F. A., & Hazen, E. E. (1971) *The Enzymes 4*, 153–176.

Cozzarelli, N. R., Kelly, R. B., & Kornberg, A. (1969) *J. Mol. Biol. 45*, 513–531.

Crouch, R. J., & Dirksen, M.-L. (1982) in *Nucleases* (Linn, S. M. & Roberts, R. J., Eds.), pp. 211–241, Cold Spring Harbor Press, Cold Spring Harbor, NY.

Cullen, B. R. (1987) *Methods Enzymol. 152*, 684–703.

David, M., Borasio, G. D., & Kaufman, G. (1982) *Virology 123*, 480–483.

Dente, L., Cesareni, G., & Cortese, R. (1983) *Nucleic Acids Res. 11*, 1645–1655.

Deutscher, M. P., & Kornberg, A. (1969) *J. Biol. Chem. 244*, 3029–3037.

Donis-Keller, H. (1980) *Nucleic Acids Res. 8*, 3133–3142.

Dotto, G. P., & Zinder, N. D. (1983) *Virology 130*, 252–256.

Dotto, G. P., Enea, V., & Zinder, N. D. (1981) *Virology 114*, 463–473.

Dotto, G. P., Horiuchi, K., & Zinder, N. D. (1984) *J. Mol. Biol. 172*, 507–521.

Dyson, A., & Harris, J., Eds. (1994) *Ethics and Biotechnology*, Routledge, London.

Ellington, A. D., & Szostak, J. W. (1990) *Nature 346*, 818–822.

Endlich, B., & Linn, S. (1981) *The Enzymes 14*, 137–156.

Engler, M. J., & Richardson, C. C. (1982) *The Enzymes 15*, 3–29.

Erlich, H. A. (1989) *PCR Technology: Principles and Applications for DNA Amplification*, Stockton Press, New York.

Erlich, H. A., Gibbs, R., & Kazazian, H. H., Eds. (1989) *Polymerase Chain Reaction*, Cold Spring Harbor Press, Cold Spring Harbor, NY.

Frank, R., Meyerhans, A., Schwellnus, K., & Blocker, H. (1987) *Methods Enzymol. 154*, 221–229.

Friedberg, E. C. (1985) *DNA Repair*, W. H. Freeman, New York.

Fuchs, L. Y., Covarrubias, L., Escalante, L., Sanchez, S., & Bolivar, F. (1980) *Gene 10*, 39–46.

Galas, D. J., & Schmitz, A. (1978) *Nucleic Acids Res. 5*, 3157–3170.

Geider, K., Hohmeyer, C., Haas, R., & Meyer, T. F. (1985) *Gene 33*, 341–349.

George, J., Nardone, G., & Chirikjian, J. G. (1985) *J. Biol. Chem. 260*, 14387–14392.

Gingeras, T. R., Myers, P. A., Olson, J. A., Hanberg, F. A., & Roberts, R. J. (1978) *J. Mol. Biol. 118*, 113–122.

Greene, J. R., & Guarente, L. (1987) *Methods Enzymol. 152*, 512–521.

Handel, T. M., Williams, S. A., & Degrado, W. F. (1993) *Science 261*, 879–885.

Harada, K., & Orgel, L. E. (1993) *Nucleic Acids Res. 21*, 2287–2291.

Haseltine, W. A. (1988) in *HIV and Other Highly Pathogenic Viruses* (Smith, R. A., Ed.), p. 25, Academic Press, San Diego, CA.

Hecht, S. M. (1992) *Acc. Chem. Res. 25*, 545–552, and references therein.

Herbert, R. A. (1992) *Trends Biotechnol. 10*, 395–402.

Horwitz, J. P., Chua, J., Curby, R. J., Tomson, A. J., DaRooge, M. A., Fisher, B. E., Mauricio, J., & Klundt, I. (1964) *J. Med. Chem. 7*, 574–575.

Kannan, P., Cowan, G. M., Daniel, A. S., Gann, A. A. F., & Murray, N. E. (1989) *J. Mol. Biol. 209*, 335–344.

Karkas, J. D. (1973) *Proc. Natl. Acad. Sci. U.S.A. 70*, 3834–3838.

Kelly, R. B., Atkinson, M. R., Huberman, J. A., & Kornberg, A. (1969) *Nature 224*, 495–501.

Kessler, C., Ed. (1993) *Nonradioactive Labeling and Detection of Biomolecules*, Springer-Verlag, New York.

Khorana, H. G. (1979) *Science 203*, 614–625.

Kim, Y., Grable, J. C., Love, R., Greene, P. J., & Rosenberg, J. M. (1990) *Science 249*, 1307–1309.

King, K., Benkovic, S. J., & Modrich, P. (1989) *J. Biol. Chem. 264*, 11807–11815.

Klenow, H. & Henningsen, I. (1970) *Proc. Natl. Acad. Sci. U.S.A. 65*, 168–175.

Kleppe, K., van de Sande, J. H., & Khorana, H. G. (1970) *Proc. Natl. Acad. Sci. U.S.A. 67*, 68–73.

Klett, R. P., Cerami, A., & Reich, E. (1968) *Proc. Natl. Acad. Sci. U.S.A. 60*, 943–950.

Kornberg, A. (1969) *Science 163*, 1410–1418.

Kornberg, A. (1981a) *The Enzymes 14*, 3–13.

Kornberg, A. (1981b) *DNA Replication*, Freeman, San Francisco.

Kornberg, A., & Baker, T. A. (1992) *DNA Replication*, 2nd ed., W. H. Freeman, New York.

Krieg, P. A., & Melton, D. A. (1987) *Methods Enzymol. 155*, 397–415.

Krieg, P. A., & Melton, D. A. (1984a) *Nature 308*, 203–206.

Krieg, P. A., & Melton, D. A. (1984b) *Nucleic Acids Res. 12*, 7057–7070.

Kumar, A., Tchen, P., Roullet, F., & Cohen, J. (1988) *Anal. Biochem. 169*, 376–382.

Kunkel, T. A. (1988) *Cell 53*, 837–840.

Laskowski, M. (1971) *The Enzymes 4*, 313–328.

Lehman, I. R. (1981a) *The Enzymes 14*, 15–37.

Lehman, I. R. (1981b) *The Enzymes 14*, 51–65.

Lesser, D. R., Kurpiewski, M. R., & Jen-Jacobson, L. (1990) *Science 250*, 776–786.

Levinson, A., Silver, D., & Seed, B. (1984) *J. Mol. Appl. Genet. 2*, 507–517.

Lewin, B. (1990) *Genes IV*, Oxford University Press, New York.

Lewis, M. K., & Burgess, R. R. (1982) *The Enzymes 15*, 110–153.

Lillehaug, J. R. (1977) *Eur. J. Biochem. 73*, 499–506.

Linn, S. M., & Roberts, R. J., Eds. (1982) *Nucleases*, Cold Spring Harbor Press, Cold Spring Harbor, NY.

Liuzzi, M., & Paterson, M. C. (1992) *J. Biol. Chem. 267*, 22421–22427.

Lockard, R., & Kumar, A. (1981) *Nucleic Acids Res. 9*, 5125–5140.

Lyttle, M. H., Wright, P. B., Sinha, N. D., Bain, J. D., & Chamberlin, A .R. (1991) *J. Org. Chem. 56*, 4608–4615.

Makula, R. A., & Meagher, R. B. (1980) *Nucleic Acids Res. 8*, 3125–3131.

Maxam, A. M., & Gilbert, W. (1977) *Proc. Natl. Acad. Sci. U. S. A. 74*, 560–564.

McClelland, M., & Nelson, M. (1985) *Nucleic Acids Res. 13*, 201–207.

Mead, D. A., Szczesna-Skorupa, E., & Kemper, B. (1985) *Nucleic Acids Res. 13*, 1103–1118.

Melton, D. A, Krieg, P. A., Rebagliati, M. R., Maniatis, T., Zinn, K., & Green, M. R. (1984) *Nucleic Acids Res. 12*, 7035–7056.

Mendel, D., Ellman, J. A., & Schultz, P. G. (1991) *J. Am. Chem. Soc. 113*, 2758–2760.

Mendel, D., Ellman, J. A., Chang, Z., Veenstra, D. L., Kollman, P. A., & Schultz, P. G. (1992) *Science 256*, 1798–1802.

Messing, J., Gronenborn, B., Muller-Hill, B., & Hofschneider, P. H. (1977) *Proc. Natl. Acad. Sci. U.S.A. 74*, 3642–3646.

Milligan, J. F., & Uhlenbeck, O. C. (1989) *Methods Enzymol. 180*, 51–62.

Mitsuhashi, S., Hashimoto, H., Iyobe, S., & Ionue, M. (1977) in *DNA Insertion Elements, Plas-*

mids, and Episomes (Bukhari, A. I., Shapiro, U. A., & Adhya, S. L., Eds.), pp. 136–146, Cold Spring Harbor Press, Cold Spring Harbor, NY.

Modrich, P., & Zabel, D. (1976) *J. Biol. Chem. 251*, 5866–5874.

Moffatt, B. A., Dunn, J. J., & Studier, F. W. (1984) *J. Mol. Biol. 173*, 265–269.

Moore, M. J., & Sharp, P. A. (1992) *Science 256*, 992–997.

Moore, S. (1981) *The Enzymes 14*, 281–296.

Noren, C. J., Anthony-Cahill, S. J., Griffith, M. C., & Schultz, P. G. (1989) *Science 244*, 182–188.

Norrander, J., Kempe, T., & Messing, J. (1983) *Gene 26*, 101–106.

Olsvik, O., Popovic, T., Skjerve, E., Cudjoe, K. S., Hornes, E., Ugelstad, J., & Uhlen, M. (1994) *Clin. Microbiol. Rev. 7*, 43–54.

O'Neil, K. T., Hoess, R. H., & DeGrado, W. F. (1990) *Science 249*, 774–778.

Oxender, D. L., & Fox, C. F. Eds. (1987) *Protein Engineering*, Alan R. Liss, New York.

Pick, L., & Hurwitz, J. (1986) *J. Biol. Chem. 261*, 6684–6693.

Polisson, C., & Morgan, R.D. (1988) *Nucleic Acids Res. 16*, 5205.

Portugal, F. H., & Cohen, J. S. (1980) *A Century of DNA: A History of the Discovery of the Structure and Function of the Genetic Substance*, MIT Press, Cambridge, MA.

Preston, B. D., Poiesz, B. J., & Loeb, L. A. (1988) *Science 242*, 1168–1171.

Ratliff, R. L. (1981) *The Enzymes 14*, 105–118.

Reyes, A. A., & Wallace, R. B. (1987) *Methods Enzymol. 154*, 87–93.

Richardson, C. C. (1981) *The Enzymes 14*, 299–314.

Roberts, J. D., Bebenek, K., & Kunkel, T. A. (1988) *Science 242*, 1171–1173.

Roberts, L. (1990) *Science 249*, 1503.

Robson, B., & Garnier, J. (1988) *Introduction to Proteins and Protein Engineering*, Elsevier, New York.

Roesser, J.R., Chorghade, M.S., & Hecht, S.M. (1986) *Biochemistry 25*, 6361–6365.

Romaniuk, P. J., & Uhlenbeck, O. C. (1983) *Methods Enzymol. 100*, 52–59.

Saiki, R. K., Gelfand, D. H., Stoffel, S., Scharf, S. J., Higuchi, R., Horn, G. T., Mullis, K. B., & Erlich, H. A. (1988) *Science 239*, 487–491.

Sambrook, J., Fritsch, E. F., & Maniatis, T. (1989) *Molecular Cloning: A Laboratory Manual*, 2nd ed., Cold Spring Harbor Press, Cold Spring Harbor, NY.

Sampson, J. R., & Uhlenbeck, O. C. (1988) *Proc. Natl. Acad. Sci. U.S.A. 85*, 1033–1037.

Sanger, F., Nicklen, S., & Coulson, A. R. (1977) *Proc. Natl. Acad. Sci. U.S.A. 74*, 5463–5467.

Schimmel, P. (1987) *Annu. Rev. Biochem. 56*, 125–158.

Schimmel, P., Soll, D., & Abelson, J. A. (1979) *Transfer RNA: Structure, Properties, and Recognition*, Cold Spring Harbor Press, Cold Spring Harbor, NY.

Sharp, P. A., Sugden, B., & Sambrook, J. (1973) *Biochemistry 12*, 3055–3063.

Shatzman, A. R., & Rosenberg, M. (1987) *Methods Enzymol. 152*, 661–672.

Shishido, K., & Ando, T. (1982) in *Nucleases* (Linn, S. M., & Roberts, R. J., Eds.), pp. 155–184, Cold Spring Harbor Press, Cold Spring Harbor, NY.

Silber, R., Malathi, V. G., & Hurwitz, J. (1972) *Proc. Natl. Acad. Sci. U.S.A. 69*, 3009–3013.

Tabor, S., & Richardson, C. C. (1987a) *J. Biol. Chem. 262*, 15330–15333.

Tabor, S., & Richardson, C. C. (1987b) *Proc. Natl. Acad. Sci. U.S.A. 84*, 4767–4771.

Takahashi, K., & Moore, S. (1982) *The Enzymes 15*, 435–468.

Troutt, A. B., McHeyzer-Williams, M. G., Pulendran, B., & Nossal, G. H. V. (1992) *Proc. Natl. Acad. Sci. U.S.A. 89*, 9823–9825.

Van Cleve, M. D., & Gumport, R. I. (1992) *Biochemistry 31*, 334–339.

Varmus, H., & Swanstrom, R. (1985a) in *Molecular Biology of RNA Tumor Viruses*, Vol. 2 (Weiss, R., Teich, N., Varmus, H., & Coffin, J. Eds.), pp. 75–134, Cold Spring Harbor Press, Cold Spring Harbor, NY.

Varmus, H. W., & Swanstrom, R. (1985b) in *Molecular Biology of Tumor Viruses*, Vol. 2: *RNA Tumor Viruses* (Weiss, R., Teich, N., Varmus, H., & Coffin, J. Eds.), pp. 369–512, Cold Spring Harbor Press, Cold Spring Harbor, NY.

Vieira, J., & Messing, J. (1982) *Gene 19*, 259–268.

von Hippel, P. H., Bear, D. G., Morgan, W. D., & McSwiggen, J. A. (1984) *Annu. Rev. Biochem. 53*, 389–446.

Walker, J. M., Ed. (1984) *Methods in Molecular Biology*, Vol. 2: *Nucleic Acids*, Humana Press, Clifton, NJ.

Watson, J. D., Hopkins, N. H., Roberts J. W., Steitz, J. A., & Weiner, A. M. (1987) in *Molecular Biology of the Gene*, 4th ed., Benjamin/Cummings Publishing, Menlo Park, CA.

Wells, R. D., Klein, R. D., & Singleton, C. K. (1981) *The Enzymes 14*, 157–191.

Wilson, G. A., & Young, F. E. (1975) *J. Mol. Biol. 97*, 123–125.

Wong-Staal, F. (1988) in *HIV and Other Highly Pathogenic Viruses* (Smith, R. A., Ed.), p. 33, Academic Press, San Diego, CA.

Wu, R., & Grossman, L. Eds. (1987) *Methods in Enzymology*, Vol. 153, Academic Press, New York.

Yanisch-Perron, C., Vieira, J., & Messing, J. (1985) *Gene 33*, 103–119.

Zagursky, R. J., & Berman, M. L. (1984) *Gene 27*, 183–191.

Zoller, M. J., & Smith, M. (1987) *Methods Enzymol. 154*, 329–350.

Chapter 4

Bancroft, D., Williams, L. D., Rich, A., & Egli, M. (1994) *Biochemistry 33*, 1073–1086.

Behlen, L. S., Sampson, J. R., DiRenzo, A. B., & Uhlenbeck, O. C. (1990) *Biochemistry 29*, 2515–2523.

Benevides, J. M., Wang, A. H.-J., Rich, A., Kyogoku, Y., van der Marel, G. A., van Boom, J. H., & Thomas, G. J., Jr. (1986) *Biochemistry 25*, 41–50.

Bingman, C. A., Jain, S., Jebaratnam, D., & Sundaralingam, M. (1989) in *Abstracts, Sixth Conversation in Biomolecular Stereodynamics*, Albany, NY, p. 28.

Bingman, C. A., Zon, G., & Sundaralingam, M. (1992) *J. Mol. Biol. 227*, 738–756.

Blundell, T. L., & Johnson, L. N. (1976) *Protein Crystallography*, Academic Press, New York.

Brennan, R. G., Westhof, E., & Sundaralingam, M. (1986) *J. Biomol. Struct. Dynamics 3*, 649–665.

Brown, D. G., Sanderson, M. R., Garman, E., & Neidle, S. (1992) *J. Mol. Biol. 226*, 481–490.

Brown, T., Kneale, G., Hunter, W. N., & Kennard, O. (1986) *Nucleic Acids Res. 14*, 1801–1809.

Brown, T., Leonard, G. A., Booth, E. D., & Chambers, J. (1989) *J. Mol. Biol. 207*, 455–457.

Brunger, A. T. (1992a) X-PLOR, Version 3.1, The Howard Hughes Medical Institute & Department of Molecular Biophysics & Biochemistry, Yale University, New Haven, CT.

Brunger, A. T. (1992b) *Nature 355*, 472–474.

Brunger, A. T. (1993) *Acta Crystallogr. D. 49*, 24–36.

Brunger, A. T., Kuriyan, J., & Karplus, M. (1987) *Science 235*, 458–460.

Calladine, C. R., (1982) *J. Mol. Biol. 161*, 343–352.

Carter, C. W. (1992) in *Crystallization of Nucleic Acids and Proteins: A Practical Approach* (Ducruix, A. & Giege, R., Eds.), pp. 47–72, Oxford University Press, New York.

Chevrier, B., Dock, A. C., Hatmann, B., Leng, M., Moras, D., Thuong, M. T., & Westhof, E. (1986) *J. Mol. Biol. 188*, 707–719.

Chattopadhyaya, R., Ikuta, S., Grzeskowiak, K., & Dickerson, R. E. (1988) *Nature 334*, 175–179.

Coll, M., Frederick, C. A., Wang, A. H.-J., & Rich, A. (1987) *Proc. Natl. Acad. Sci. U.S.A. 84*, 8385–8389.

Coll, M., Aymami, J., van der Marel, G. A., van Boom. J. H., Rich, A., & Wang, A. H.-J. (1989a) *Biochemistry 28*, 310–320.

Coll, M., Saal, D., Frederick, C. A., Aymami, J., Rich, A., & Wang, A. H.-J. (1989b) *Nucleic Acids Res. 17*, 911–923.

Conner, B. N., Yoon, C., Dickerson, J. L., & Dickerson, R. E. (1984) *J. Mol. Biol. 174*, 663–695.

Crawford, J. L., Kolpak, F. J., Wang, A. H.-J., Quigley, G. J., van Boom, J. H., van der Marel, G., & Rich, A. (1980) *Proc. Natl. Acad. Sci. U.S.A. 77*, 4016–4020.

Cruse, W. B. T., Salisbury, S. A., Brown, T., Cosstick, R., Eckstein, F., & Kennard, O. (1986) *J. Mol. Biol. 192*, 891–905.

Dickerson, R. E., et al. (1989) *EMBO J. 8*, 1–4.

DiGabriele, A. D., Sanderson, M. R., & Steitz, T. A. (1989) *Proc. Natl. Acad. Sci. U.S.A. 86*, 1816–1820.

Dock-Bregeon, A. C., Chevrier, B., Podjarny, A., Moras, D., deBear, J. S., Gough, G. R., Gilham, P. T., & Johnson, J. E. (1988) *Nature 335*, 375–378.

Drenth, J., (1994) *Principles of Protein Crystallography*, Springer-Verlag, New York.

Drew, H. R., & Dickerson, R. E. (1981a) *J. Mol. Biol. 151*, 535–556.

Drew, H. R., & Dickerson, R. E. (1981b) *J. Mol. Biol. 152*, 723–736.

Drew, H., & Travers, A. (1984) *Cell 37*, 491–501.

Drew, H., Takano, T., Tanaka, S., Itakura, K., & Dickerson, R. E. (1980) *Nature 286*, 567–573.

Drew, H., Samson, S., & Dickerson, R. E. (1982) *Proc. Natl. Acad. Sci. U.S.A. 79*, 4040–4044.

Eisenberg, D., & Hill, C. P. (1989) *Trends Biochem. Sci.* 14, 260–264.

Eisenstein, M., Hope, H., Haran, T. E., Frolow, F., Shakked, Z., & Rabinovich, D. (1988) *Acta Crystallogr. B44*, 625–628.

Fitzgerald, P. M. D. (1988) *J. Appl. Crystallogr. 21*, 273–278.

Frederick, C. A., Coll, M., van der Marel, G. A., van Boom, J. H., & Wang, A. H.-J. (1988) *Biochemistry 27*, 8350–8361.

Frederick, C. A., Williams, L. D., Ughetto, G., van der Marel, G. A., van Boom, J. H., Rich, A., & Wang, A. H.-J. (1990) *Biochemistry 29*, 2538–2549.

Fujii, S., Wang, A. H.-J., van Boom, J. H., & Rich, A. (1982) *Nucleic Acids Res. Symp. Series 11*, 109–112.

Fujii, S., Wang, A. H.-J., Quigley, G. J., Westernick, H., van der Marel, G. A., van Boom, J. H., & Rich, A. (1985) *Biopolymers 24*, 243–250.

Gao, Y.-G., Liaw, Y.-C., Li, Y.-K., van der Marel, G. A., van Boom, J. H., & Wang, A. H.-J. (1991a) *Proc. Natl. Acad. Sci. U.S.A. 88*, 4845–4849.

Gao, Y.-G., van der Marel, G. A., van Boom, J. H., & Wang, A. H.-J. (1991b) *Biochemistry 30*, 9922–9931.

Gao, Y.-G., Sriram, M., & Wang, A. H.-J. (1993) *Nucleic Acids Res. 21*, 4093–4101.

Gehring, K. Leroy, J. L., & Gueron, M. (1993) *Nature 363*, 561–565.

Geierstanger, B. H., Kagawa, T. F., Chen, S.-L., Quigley, G. J., & Ho, P. S. (1991) *J. Biol. Chem. 266*, 20185–20191.

Gessner, R. G., Quigley, G. J., Wang, A. H.-J., van der Marel, G. A., van Boom, J. H., & Rich, A. (1985) *Biochemistry 24*, 237–250.

Gessner, R. G., Frederick, C. A., Quigley, G. J., Rich, A., & Wang, A. H.-J. (1989) *J. Biol. Chem. 264*, 7921–7935.

Goodsell, D. S., Kopka, M. L., Cascio, D., & Dickerson, R. E. (1993) *Proc. Natl. Acad. Sci. U.S.A. 90*, 2930–2934.

Grzeskowiak, K., Yanagi, K., Prive, G. G., & Dickerson, R. E. (1991) *J. Biol. Chem. 266*, 8861–8883.

Haran, T. E., Shakked, Z., Wang, A. H.-J., Rich, A. (1987) *J. Biomol. Struct. Dynamics 5*, 199–217.

Hauptman, H. (1986) *Science 233*, 178–183.

Heinemann, U., & Alings, C. (1985) *J. Mol. Biol. 210*, 369–381.

Heinemann, U., & Hahn, M. (1992) *J. Biol. Chem. 267*, 7332–7341.

Heinemann, U., Lauble, H., Frank, R., & Bloeker, H. (1987) *Nucleic Acids Res. 15*, 9531–9550.

Hendrickson, W. A., Smith, J., & Sheriff, S. (1985a) *Methods Enzymol. 115*, 41–55.

Hendrickson, W. A., Smith, J., & Sheriff, S. (1985b) *Methods Enzymol. 115*, 252–270.

Ho, P. S., Frederick, C. A., Quigley, G. J., van der Marel, G. A., van Boom, J. H., & Rich, A. (1985) *EMBO J. 4*, 3617–3623.

Ho, P. S., Frederick, C. A., Saal, D., Wang, A. H.-J., & Rich, A. (1987) *J. Biomol. Struct. Dynamics 4*, 521–534.

Holbrook, S. R., Sussman, J. L., Warrant, R. W., Church, G. M., & Kim, S.-H. (1977) *Nucleic Acids Res. 8*, 2811–2820.

Holbrook, S. R., Dickerson, R. E., & Kim, S.-H. (1985) *Acta. Crystallogr. B41*, 255–262.

Hunter, W. N., Brown, T., & Kennard, O. (1986b) *J. Biomol. Struct. Dynamics 4*, 173–191.

Hunter, W. N., Kneale, G., Brown, T., Rabinovich, D., & Kennard, O. (1986c) *J. Mol. Biol. 190*, 605–618.

Hunter, W. N., Brown, T., Kneale, G., Anand, N. N., Rabinovich, D., & Kennard, O. (1987) *J. Biol. Chem. 21*, 9962–9970.

Hunter, W. N., D'Estaintot, B. L., & Kennard, O. (1989) *Biochemistry 28*, 2444–2451.

Jack, A., Ladner, J. E., Rhodes, D., Brown, R. S., & Klug, A. (1977) *J. Mol. Biol. 111*, 315–328.

Jain, S., Zon, G., & Sundaralingam, M. (1989) *Biochemistry 28*, 2444–2451.

Joshua-Tor, L., Rabinovich, D., Hope, H., Frolow, F., Apella, E., & Sussman, J. L. (1988) *Nature 334*, 82–84.

Kagawa, T. F., Geierstanger, B. H., Wang, A. H.-J., & Ho, P. S. (1991) *J. Biol. Chem. 266*, 20175–20184.

Kang, C.-H., Zhang, X., Ratliff, R., Moyzis, R., & Rich, A. (1992) *Nature 356*, 126–131.

Karle, J. (1986) *Science 232*, 837–843.

Karplus, M., & McCammon, J. A. (1983) *Annu. Rev. Biochem. 52*, 263–300.

Kennard, O., Cruse, W. B. T., Nachman, J., Prange, T., Shakked, Z., & Rabinovich, D. (1986) *J. Biomol. Struct. Dynamics 3*, 623–647.

Kim, J. L., Nikolov, D. B., & Burley, S. K. (1993a) *Nature 365*, 520–527.

Kim, Y., Geiger, J. H., Hahn, S., & Sigler, P. B. (1993b) *Nature 365*, 512–520.

Kirkpatrick, S., Gelatt, C. D., Jr., & Vecchi, M. P. (1983) *Science 220*, 671–680.

Klysik, J., Stirdivant, S. M., Larson, J. E., Hart, P. A., & Wells, R. D. (1981) *Nature 290*, 672–677.

Kneale, G., Brown, T., Kennard, O., & Rabinovich, D. (1985) *J. Mol. Biol. 186*, 805–814.

Kopka, M. L., Fratini, A. V., Drew, H. R., & Dickerson, R. E. (1983) *J. Mol. Biol. 163*, 129–146.

Ladd, M. F. C., & Palmer, R. A. (1993) *Structure Determination by X-Ray Crystallography*, third edition, Plenum Press, New York.

Larsen, T., Goodsell, D., Cascio, D., Grzeskowiak, K., & Dickerson, R. E. (1989) *J. Biomol. Struct. Dynamics 7*, 477–491.

Lauble H., Frank, R., Bloecker, H., & Heinemann, U. (1988) *Nucleic Acids Res. 16*, 7799–7816.

Marzilli, L. G. (1977) *Prog. Inorg. Chem. 23*, 255–378.

McCall, M., Brown, T., & Kennard, O. (1985) *J. Mol. Biol. 183*, 385–396.

McCall, M., Brown, T., Hunter, W. N., & Kennard, O. (1986) *Nature 322*, 661–664.

McPherson, A. (1982) *Preparation & Analysis of Protein Crystals*, Wiley, New York.

Metropolis, N., Rosenbluth, M., Rosenbluth, A., Teller, A., & Teller, E. (1953) *J. Chem. Phys. 21*, 1087–1092.

Miller, M., Harrison, R. W., Wlodawer, A., Apella, E., & Sussman, J. L. (1988) *Nature 334*, 85–86.

Nelson, H. C. M., Finch, J. T., Luisi, B. F., & Klug, A. (1987) *Nature 330*, 221–226.

Nordheim, A., Pardue, M. L., Lafer, E. M., Moller, A., Stollar, D., & Rich, A. (1981) *Nature 294*, 417–422.

Piccirilli, J. A., Vyle, J. S., Caruthers, M. H., & Cech, T. R. (1993) *Nature 361*, 85–88.

Pjura, P. E., Grzeskowiak, K., & Dickerson, R. E. (1987) *J. Mol. Biol. 197*, 257–271.

Prive, G. G., Heinemann, U., Chandrasegaran, S., Kan, L.-S., Kopka, M. L., & Dickerson, R.E. (1988) in *Structure and Expression*, Vol. 2 (Sarma, R. H., & Sarma M. H., Eds.), pp. 27–47, Adenine Press, New York.

Prive, G. G., Heinemann, U., Chandrasegaran, S., Kan, L.-S., Kopka, M. L., & Dickerson, R.E. (1987) *Science 238*, 498–504.

Prive, G. G., Yanagi, K., & Dickerson R. E. (1991) *J. Mol. Biol. 217*, 177–199.

Propst, C. L., & Perun, T. J. (1992) *Nucleic Acid Targeted Drug Design*, pp. 1–13, Marcel Dekker, New York.

Quigley, G. J., Teeter, M. M., & Rich, A. (1978) *Proc. Natl. Acad. Sci. U.S.A. 75*, 64–68.

Quintana, J. R., Grzeskowiak, K., Yanagi, K., & Dickerson, R. E. (1992) *J. Mol. Biol. 225*, 379–395.

Rabinovich, D., & Shakked, Z. (1984) *Acta Crystallogr. A40*, 195–200.

Rabinovich, D., Haran, T., Eisenstein, M., & Shakked, Z. (1988) *J. Mol. Biol. 200*, 151–161.

Ramakrishnan, B., & Sundaralingam, M. (1993) *J. Mol. Biol. 231*, 431–444.

Rhodes, D., & Klug, A. (1986) *Cell 46*, 123–132.

Rich, A., Nordheim, A., & Wang, A. H.-J. (1984) *Annu. Rev. Biochem. 53*, 791–846.

Robinson, H., & Wang, A. H.-J. (1993) *Proc. Natl. Acad. Sci. U.S.A. 90*, 5224–5228.

Robinson, H., van der Marel, G. A., van Boom, J. H., & Wang, A. H.-J. (1992) *Biochemistry 31*, 10510–10517.

Robinson, H., van Boom, J. H., & Wang, A. H.-J. (1994) *J. Am. Chem. Soc. 116*, 1565–1566.

Saenger, W. (1984) *Principles of Nucleic Acid Structure*, Springer-Verlag, New York.

Saenger, W., Hunter, W. N., & Kennard, O. (1986) *Nature 324*, 385–388.

Schevitz, R., Otwinowski, Z., Joachimiak, A., Lawson, C. L., & Sigler, P. B. (1985) *Nature 317*, 782–786.

Shakked, Z., Rabinovich, D., Kennard, O., Cruse, W. B. T., Salisbury, S. A., & Vishwamitra, M. A. (1983) *J. Mol. Biol. 166*, 183–201.

Shakked, Z., Guzikevichguerstein, G., Frolow, F., & Rabinovich, D. (1989) *Nature 342*, 456–460.

Shakked, Z., Guzikevichguerstein, G., Frolow, F., Rabinovich, D., Joachimiak, A., & Sigler, P. B. (1994) *Nature 368*, 469–473.

Singleton, C. K., Klysik, K., Stirdivant, S. M., & Wells, R. D. (1982) *Nature 299*, 312–316.

Smith, J. L. (1991) *Current Opin. Struct. Biol. 1*, 1002–1011.

Smith, J. L., Zaluzec, E. J., Wery, J. P., Niu, L., Switzer, R. L., Zalkin, H., & Satow, Y. (1994) *Science 264*, 1427–1433.

Sriram, M., van der Marel, G. A., Roelen, H. L. P. F., van Boom, J. H., & Wang, A. H.-J. (1992a) *EMBO J. 11*, 225–232.

Sriram, M., van der Marel, G. A., Roelen, H. L. P. F., van Boom, J. H., & Wang, A. H.-J. (1992b) *Biochemistry 31*, 11823–11834.

Steitz, T. A. (1990) *Q. Rev. Biophys. 23*, 205–280.

Stout, G. H., & Jensen, L.H. (1989) *X-Ray Structure Determination*, 2nd ed., John Wiley & Sons, New York.

Suck, D., & Oefner, C. (1986) *Nature 321*, 620–625.

Sundaralingam, M., & Jain, S. (1989) in *Abstracts, Sixth Conversation in Biomolecular Stereo-dynamics*, Albany, NY, p. 28.

Swaminathan, V., & Sundaralingam, M. (1979) *CRC Crit. Rev. Biochem. 6*, 245–336.

Teng, M., Usman, N., Frederick, C. A., & Wang, A. H.-J. (1988) *Nucleic Acids Res. 16*, 2671–2690.

Timsit, Y., Westhof, E., Fuchs, R. P. P., & Moras, D. (1989) *Nature 341*, 459–462.

Wang, A. H.-J. (1992) *Current Opin. Struct. Biol. 2*, 361–368.

Wang, A. H.-J., & Gao, Y.-G. (1990) *Methods 1*, 91–99.

Wang, A. H.-J., Quigley, G. J., Kolpak, F. J., Crawford, J. L., van Boom, J. H., van der Marel, G. A., & Rich, A. (1979) *Nature 282*, 680–686.

Wang, A. H.-J., Quigley, G. J., Kolpak, F. J., van der Marel, G. A., van Boom, J. H., & Rich, A. (1981) *Science 211*, 171–176.

Wang, A. H.-J., Fujii, S., van Boom, J. H., & Rich, A. (1982a) *Proc. Natl. Acad. Sci. U.S.A. 79*, 3968–3972.

Wang, A. H.-J., Fujii, S., van Boom, J. H., van der Marel, G. A., van Boeckel, S. A. A., & Rich, A. (1982b) *Nature 299*, 601–604.

Wang, A. H.-J., Hakoshima, T., van der Marel, G., van Boom, J. H., & Rich, A., (1984a) *Cell 37*, 321–331.

Wang, A. H.-J., Ughetto, G., Quigley, G. J., Hakoshima, T., van der Marel, G. A., van Boom, J. H., & Rich, A. (1984b) *Science 225*, 1115–1121.

Wang, A. H.-J., Gessner, R. V., van der Marel, G., van Boom, J. H., & Rich, A., (1985) *Proc. Natl. Acad. Sci. U.S.A. 82*, 3611–3615.

Wang, A. H.-J., Ughetto, G., Quigley, G. J., & Rich, A. (1987) *Biochemistry 26*, 1152–1163.

Wang, A. H.-J., Gao, Y.-G., Liaw, Y.-C., & Li, Y.-K. (1991) *Biochemistry 30*, 3812–3815.

Wang, A. H.-J., & Robinson, H. (1992) in *Nucleic Acid Targeted Drug Design* (Propst, C. L., & Perun, T. J., Eds.), pp. 17–64, Marcel Dekker, New York.

Westhof, E. (1987) *J. Biomol. Struct. Dynamics 5*, 581–600.

Westhof, E., Dumas, P., & Moras, D. (1985) *J. Mol. Biol. 184*, 119–145.

Williamson, J. R. (1993) *Current Opin. Struct. Biol. 3*, 357–362.

Wing, R. M., Drew, H. R., Takano, T., Broka, C., Tanaka, S., Itakura, K., & Dickerson, R. E. (1980) *Nature 287*, 755–758.

Wing, R. M., Pjura, P., Drew, H. R., & Dickerson, R. E. (1984) *EMBO J. 3*, 1201–1206.

Wittig, B., Wolf, S., Dorbic, T., Vahrson, W., & Rich, A. (1992) *EMBO J. 11*, 4653–4663.

Wyckoff, H. W., Hirs, C. H. W., & Timasheff, S. N., Eds. (1985) *Methods in Enzymology*, Vols. 114 and 115, Academic Press, New York.

Yanagi, K., Prive, G. G., & Dickerson, R. E. (1991) *J. Mol. Biol. 217*, 201–214.

Yoon, C., Prive, G. G., Goodsell, D. S., & Dickerson, R. E. (1988) *Proc. Natl. Acad. Sci. U.S.A. 85*, 6332–6336.

Yuan, H., Quintana, J., & Dickerson, R. E. (1992) *Biochemistry 31*, 8009–8021.

Zhang, S., Lockshin, C., Herbert, A., Winter, E., & Rich, A. (1992) *EMBO J. 11*, 3787–3796.

Chapter 5

Borowiec, J. A., Zhang, L., Sasse-Dwight, S., & Gralla, J. D. (1987) *J. Mol. Biol. 196*, 101–111.

Brenowitz, M., Senear, D. F., Shea, M. A., & Ackers, G. K. (1986) *Methods Enzymol. 130*, 132–181.

Burkhoff, A. M., & Tullius, T. D. (1987) *Cell 48*, 935–943.

Celander, D. W., & Cech, T. R. (1991) *Science 251*, 401–407.

Cheong, C., Varani, G., & Tinoco, I. (1990) *Nature 346*, 680–714.

Chow, C. S., & Barton, J. K. (1992) *Methods Enzymol. 212*, 219–242.

Churchill, M. E. A., Tullius, T. D., Kallenbach, N. R., & Seeman, N. C. (1988) *Proc. Natl. Acad. Sci. U.S.A. 85*, 4653–4656.

Cooper, J. P., & Hagerman, P. J. (1987) *J. Mol. Biol. 198*, 711–719.

Crothers, D. M., Haran, T. E., & Nadeau, J. G. (1990) *J. Biol. Chem. 265*, 7093–7096.

Dervan, P. B. (1986) *Science 232*, 464–471.

Dickerson, R. E. (1992) *Methods Enzymol. 211*, 67–110.

Diekmann, S., Mazzarelli, J. M., McLaughlin, L. W., von Kitzing, E., & Travers, A. A. (1992) *J. Mol. Biol. 225*, 729–738.

DiGabriele, A. D., & Steitz, T. A. (1993) *J. Mol. Biol. 231*, 1024–1039.

Dixon, W. J., Hayes, J. J., Levin, J. R., Weidner, M. F., Dombroski, B. A., & Tullius, T. D. (1991) *Methods Enzymol. 208*, 380–413.

Duckett, D. R., Murchie, A. I. H., Diekmann, S., von Kitzing, E., Kemper, B., & Lilley, D. M. J. (1988) *Cell 55*, 79–89.

Feigon, J., Sklenar, V., Wang, E., Gilbert, D. E., Macaya, R. F., & Schultze, P. (1992) *Methods Enzymol. 211*, 235–253.

Hatfull, G. E., Noble, S. M., & Grindley, N. D. F. (1987) *Cell 49*, 103–110.

Hayes, J., Kam, L., & Tullius, T. D. (1990a) *Methods Enzymol. 186*, 545–549.

Hayes, J. J., Tullius, T. D., & Wolffe, A. (1990b) *Proc. Natl. Acad. Sci. U.S.A. 87*, 7405–7409.

Kallenbach, N. R., Ma, R. -I, & Seeman, N. C. (1983) *Nature 305*, 829–831.

Kang, C., Zhang, X., Ratliff, R., Moyzis, R., & Rich, A. (1992) *Nature 356*, 126–131.

King, P. A., Jamison, E., Strahs, D., Anderson, V. E., & Brenowitz, M. (1993) *Nucleic Acids Res. 21*, 2473–2478.

Latham, J. A., & Cech, T. R. (1989) *Science 245*, 276–282.

Lilley, D. M. J., & Dahlberg, J. E., Eds. (1992) *Methods in Enzymology*, Vol. 212, Academic Press, San Diego.

Lutter, L. C. (1978) *J. Mol. Biol. 124*, 391–420.

Marini, J. C., Levene, S. D., Crothers, D. M., & Englund, P. T. (1982) *Proc. Natl. Acad. Sci. U.S.A. 79*, 7664–7668.

Maxam, A. M., & Gilbert, W. (1980) *Methods Enzymol. 65*, 499–560.

Mazzarelli, J. M., Rajur, S. B., Iadarola, P. L., & McLaughlin, L. W. (1992) *Biochemistry 31*, 5925–5936.

Metzler, W. J., Wang, C., Kitchen, D., Levy, R. M., & Pardi, A. (1990) *J. Mol. Biol.* 711–736.

Mueller, P. R., & Wold, B. (1989) *Science 246*, 780–786.

Nielsen, P. E., Jeppesen, C., & Buchardt, O. (1988) *FEBS Lett. 235*, 122–124.

Palecek, E. (1992a) *Methods Enzymol. 212*, 139–155.

Palecek, E. (1992b) *Methods Enzymol. 212*, 305–318.

Price, M. A., & Tullius, T. D. (1992) *Methods Enzymol. 212*, 194–219.

Robertus, J. D., Ladner, J. E., Finch, J. T., Rhodes, D., Brown, R. S., Clark, B. F., & Klug, A. (1974) *Nature 250*, 546–551.

Saiki, R. K., Gelfand, D. H., Stoffel, S., Scharf, S. J., Higuchi, R., Horn, G. T., Mullis, K. B., & Erlich, H. A. (1988) *Science 239*, 487–491.

Sasse-Dwight, S., & Gralla, J. D. (1988) *J. Mol. Biol. 202*, 107–119.

Schultze, P., Macaya, R. F., & Feigon, J. (1994) *J. Mol. Biol. 235*, 1532–1547.

Siebenlist, U., & Gilbert, W. (1980) *Proc. Natl. Acad. Sci. U.S.A. 77*, 122–126.

Sklenar, V., & Feigon, J. (1990) *Nature 345*, 836–838.

Smith, F. W., & Feigon, J. (1993) *Biochemistry 32*, 8682–8692.

Stubbe, J., & Kozarich, J. W. (1987) *Chem. Rev. 87*, 1107–1136.

Suddath, F. L., Quigley, G. J., McPherson, A., Sneden, D., Kim, J. J., Kim, S. H., & Rich, A. (1974) *Nature 248*, 20–24.

Sundquist, W. I., & Klug, A. (1989) *Nature 342*, 825–829.

Tullius, T. D. (1987) *Trends Biochem. Sci. 12*, 297–300.

Tullius, T. D. (1991) *Curr. Opin. Struct. Biol. 1*, 428–434.

Tullius, T. D., & Dombroski, B. A. (1985) *Science 230*, 679–681.

Walling, C. (1975) *Acc. Chem. Res. 8*, 125–131.

Wang, A. H. -J., Quigley, G. J., Kolpak, F. J., Crawford, J. L., van Boom, J. H., van der Marel, G., & Rich, A. (1979) *Nature 282*, 680–686.

Watson, J. D., & Crick, F. H. C. (1953) *Nature 171*, 737–738.

Wemmer, D. E. (1991) *Curr. Opin. Struct. Biol. 1*, 452–458.

Williamson, J. R., Raghuraman, M. K., & Cech, T. R. (1989) *Cell 59*, 871–880.

Wing, R., Drew, H., Takano, T., Broka, C., Tanaka, S., Itakura, K., & Dickerson, R. E. (1980) *Nature 287*, 755–758.

Yoon, C., Kuwabara, M. D., Law, R., Wall, R., & Sigman, D. S. (1988) *J. Biol. Chem. 263*, 8458–8463.

Chapter 6

Addess, K. J., & Feigon, J. (1994a) *Biochemistry 33*, 12386–12396.

Addess, K. J., & Feigon, J. (1994b) *Biochemistry 33*, 12397–12404.

Addess, K. J., Gilbert, D. E., Olsen, R. K., & Feigon, J. (1992) *Biochemistry 31*, 339–350.

Addess, K. J., Sinsheimer, J. S., & Feigon, J. (1993) *Biochemistry 32*, 2498–2508.

Batey, R. T., Inada, M., Kujawinski, E., Puglisi, J. D., & Williamson, J. R. (1992) *Nucleic Acids Res. 20*, 4515–4523.

Boelens, R., Scheek, R. M., Dijkstra, K., & Kaptein, R. (1985) *J. Magn. Reson. 63*, 378–386.

Boelens, R., Vuister, G. W., Koning, T. M. G., & Kaptein, R. (1989) *J. Am. Chem. Soc. 111*, 8525–8526.

Bothner-By, A. A., Stephens, R. L., Lee, J., Warren, C. D., & Jeanloz, R. W. (1984) *J. Am. Chem. Soc. 106* 811–813.

Chazin, W. J., Wüthrich, K., Rance, M., Hyberts, S., Denny, W. A., & Leupin, W. (1986) *J. Mol. Biol. 109*, 439–453.

Dieckmann, T., & Feigon, J. (1994) *Curr. Opin. Struct. Biol. 4*, 745–749.

Egli, M., Williams, L. D., Frederick, C. A., & Rich, A. (1991) *Biochemistry 30*, 1364–1372.

Feigon, J. (1993) *Curr. Biol. 3*, 611–613.

Feigon, J., Wright, J. M., Leupin, W., Denny, W. A., & Kearns, D. R. (1982) *J. Am. Chem. Soc. 104*, 5540–5541.

Feigon, J., Leupin, W., Denny, W. A., & Kearns, D. R. (1983) *Biochemistry 22*, 5943–5951.

Feigon, J., Leupin, W., Denny, W. A., & Kearns, D. R. (1984a) *J. Med. Chem. 27*, 450–465.

Feigon, J., Wang, A. H.-J., van der Marel, G. A., van Boom, J. H., & Rich, A. (1984b) *Nucleic Acids Res. 12*, 1243–1263.

Feigon, J., Sklenár, V., Wang, E., Gilbert, D. E., Macaya, R. F., & Schultze, P. (1992) *Methods Enzymol. 211*, 235–253.

Gao, X., & Patel, D. J. (1988) *Biochemistry 27*, 1744–1751.

Gao, Y.-G., Liaw, Y.-C., Robinson, H., & Wang, A. H.-J. (1990) *Biochemistry 29*, 10307–10316.

Gilbert, D. E., & Feigon, J. (1991) *Biochemistry 30*, 2483–2494.

Gilbert, D. E., van der Marel, G. A., van Boom, J. H., & Feigon, J. (1989) *Proc. Natl. Acad. Sci. U.S.A. 86*, 3006–3010.

Gorenstein, D.G. (1992) *Methods Enzymol. 211*, 254–286.

Haasnoot, C. A. G., Westerink, H. P., van der Marel, G. A., & van Boom, J. H. (1983) *J. Biomol. Struct. Dynamics 1*, 131–149.

Hare, D. R., Wemmer, D. E., Chou, S. H., Drobny, G., & Reid, B. R. (1983) *J. Mol. Biol. 171*, 319–336.

Jeener, J., Meier, B. H., Bachmann, P., & Ernst, R. R. (1979) *J. Chem. Phys. 71*, 4546–4553.

Katagiri, K., Yoshida, T., & Sato, K. (1975) in *Antibiotics III. Mechanism of Action of Antimicrobial and Antitumor Agents* (Corcoran, J., & Hahn, F. E., Eds.), pp. 234–251, Springer-Verlag, Berlin.

Kearns, D. R., Patel, D.J., & Shulman, R. G. (1971) *Nature 229*, 338–339.

Kellogg, G.W., Szewczak, A. A., & Moore, P. B. (1992) *J. Am. Chem. Soc. 114*, 2727–2728.

Klevit, R. E., Wemmer, D. E., & Reid, B. R. (1986) *Biochemistry 25*, 3296–3303.

Low, C. M. L., Drew, H. R., & Waring, M. J. (1984a) *Nucleic Acids Res. 12*, 4865–4879.

Low C. M. L., Olsen, R. K., & Waring, M. J. (1984b) *FEBS Lett. 176*, 414–419.

Low, C. M. L., Fox, K. R., Olsen, R. K., & Waring, M. J. (1986) *Nucleic Acids Res. 14*, 2015–2033.

Mooren, M. M. W., Hilbers, C. W., van der Marel, G. A., van Boom, J. H., & Wijmanga, S. S. (1991) *J. Magn. Reson. 94*, 101–111.

Nikonowicz, E.P., Sirr, A., Legault, P., Jucker, F.M., Baer, L.M. , & Pardi, A. (1992) *Nucleic Acids Res. 20*, 4507–4513.

Nilges, M., Habazettl, J., Brünger, A. T., & Holak, T. A. (1991) *J. Mol. Biol. 219*, 499–510.

Nunn, C. M., Van Meervelt, L., Zhang, S., Moore, M. H., & Kennard, O. (1991) *J. Mol. Biol. 222*, 167–177.

Olsen, R. K., Ramasamy, K., Bhat, K. L., Low, C. M. L., & Waring, M. J. (1986) *J. Am. Chem. Soc. 108*, 6032–6036.

Pardi, A., Walker, R., Rapoport, H., Wider, G., & Wüthrich, K. (1983) *J. Am. Chem. Soc. 105*, 1652–1653.

Pelton, J. G., & Wemmer, D. E. (1989) *Proc. Natl. Acad. Sci. U.S.A. 86*, 5723–5727.

Pelton, J. G., & Wemmer, D. E. (1990) *J. Am. Chem. Soc. 112*, 1393–1399.

Piotto, M. E., & Gorenstein, D. G. (1991) *J. Am. Chem. Soc. 113*, 1438–1440.

Powers, R., Olsen, R. K., & Gorenstein, D. G. (1989) *J. Biomol. Struct. Dynamics 7*, 515–553.

Rajagopal, P., & Feigon, J. (1989) *Nature (London) 339*, 637–640.

Rajagopal, P., Gilbert, D. E., van der Marel, G. A., van Boom, J. H., & Feigon, J. (1988) *J. Magn. Reson. 78*, 526–537.

Reid, B. R. (1987) *Q. Rev. Biophys. 20*, 1–34.

Scheek, R. M., Russo, N., Boelens, R., & Kaptein, R. (1983) *J. Am. Chem. Soc. 105*, 2914–2916.

Scheek, R. M., Boelens, R., Russo, N., van Boom, J. H., & Kaptein, R. (1984) *Biochemistry 23*, 1371–1376.

Schweizer, M. P., Chan, S. I., Helmkamp, G. K., & Ts'o, P. O. P. (1964) *J. Am. Chem. Soc. 86*, 696–701.

Sklenár, V., & Feigon, J. (1990a) *Nature 345*, 836–838.

Sklenár, V., & Feigon, J. (1990b) *J. Am. Chem. Soc. 112*, 5644–5645.

Ughetto, G., Wang, A. H.-J., Quigley, G. J., van der Marel, G. A., van Boom, J. H., & Rich, A. (1985) *Nucleic Acids Res. 13*, 2305–2323.

van Dyke, M. M., & Dervan, P. B. (1984) *Science 225*, 1122–1127.

van Wijk, J. Huckriede, B. D., Ippel, J. H., & Altona, C. (1992) *Methods Enzymol. 211*, 286–306.

Viswamitra, M. A., Kennard, O., Cruse, W. B. T., Egert, E., Sheldrick, G. M., Jones, P. G., Waring, M. J., Wakelin, L. P. G., & Olsen, R. K. (1981) *Nature 289*, 817–819.

Wang, A. H.-J., Ughetto, G., Quigley, G. J., Hakoshima, T., van der Marel, G. A., van Boom, J. H., & Rich, A. (1984) *Science 225*, 1115–1121.

Wang, A. H.-J., Ughetto, G., Quigley, G. J., & Rich, A. (1986) *J. Biomol. Struct. Dynamics 4*, 319–342.

Waring, M. J. (1979) in *Antibiotics* (Hahn, F. E., Ed.) Vol. 5, Part 2, pp. 173–194, Springer-Verlag, Berlin.

Widmer, H., & Wüthrich, K. (1987) *J. Magn. Reson. 7*, 316–336.

Williams, L. D., Egli, M., Gao, Q., & Rich, A. (1992) in *Structure and Function*, Vol. 1: *Nucleic Acids* (Sarma, R. H., & Sarma, M. H., Eds.), pp. 107–125, Adenine Press, Schenectady, NY.

Wüthrich, K. (1986) *NMR of Proteins and Nucleic Acids*, Wiley, New York.

Zhang, X., & Patel, D. J. (1990) *Biochemistry 29*, 9451–9466.

Chapter 7

Aboul-ela, F., Koh, D., & Tinoco, I. (1985) *Nucleic Acids Res. 13*, 4811–4824.

Aboul-ela, F., Bowater, R., & Lilley, D. M. J. (1992a) *J. Biol. Chem. 267*, 1776–1785.

Aboul-ela, F., Murchie, A. I. H., & Lilley, D. M. J. (1992b) *Nature 360*, 280–282.

Aboul-ela, F., Murchie, A. I. H., Homans, S. W., & Lilley, D. M. J. (1993) *J. Mol. Biol. 229*, 173–188.

Aboul-ela, F., Murchie, A. I. H, Norman, D. G., & Lilley, D. M. J. (1994) *J. Mol. Biol. 243*, 458–471.

Adam, S., Liquier, J., Taboury, J. A., & Taillandier, E. (1986) *Biochemistry 25*, 3220–3225.

Arnott, S., & Selsing, E. (1974) *J. Mol. Biol. 88*, 509–521.

Bednar, J., Dubochet, J., Murchie, A. I. H., and Lilley, D. M. J. (1995) Manuscript in preparation.

Behe, M., & Felsenfeld, G. (1981) *Proc. Natl. Acad. Sci. U.S.A. 78*, 1619–1623.

Bell, L. R., & Byers, B. (1979) *Proc Natl Acad Sci U.S.A. 76*, 3445–3449.

Benham, C. J. (1981) *J. Mol. Biol. 150*, 43–68.

Bernues, J., Beltran, R., Casanovas, J. M., & Azorin, F. (1989) *EMBO J. 8*, 2087–2094.

Bhattacharyya, A., & Lilley, D. M. J. (1989a) *Nucleic Acids Res. 17*, 6821–6840.

Bhattacharyya, A., & Lilley, D. M. J. (1989b) *J. Mol. Biol. 209*, 583–597.

Bhattacharyya, A., Murchie, A. I. H., & Lilley, D. M. J. (1990) *Nature 343*, 484–487.

Bhattacharyya, A., Murchie, A. I. H., von Kitzing, E., Diekmann, S., Kemper, B., & Lilley, D.M. J. (1991) *J. Mol. Biol. 221*, 1191–1207.

Bianchi, M. E., Beltrame, M., & Paonessa, G. (1989) *Science 243*, 1056–1059.

Bowater, R., Aboul-ela, F., & Lilley, D. M. J. (1991) *Biochemistry 30*, 11495–11506.

Bowater, R., Aboul-ela, F., & Lilley, D. M. J. (1994) *Nucleic Acids Res. 22*, 2042–2050.

Breslauer, K. J., Frank, R., Blöcker, H., & Marky, L. A. (1986) *Proc. Natl. Acad. Sci. U.S.A. 83*, 3746–3750.

Broker, T. R., & Lehman, I. R. (1971) *J. Mol. Biol. 60*, 131–149.

Brown, T., Hunter, W. N., Kneale, G., & Kennard, O. (1986) *Proc. Natl. Acad. Sci. U.S.A. 83*, 2402–2406.

Brown, T., Leonard, G. A., Booth, E. D., & Kneale, G. (1990) *J. Mol. Biol. 212*, 437–440.

Calladine, C. R., & Drew, H. R. (1992) *Understanding DNA*, Academic Press, London.

Cheong, C., & Moore, P. B. (1992) *Biochemistry 31*, 8406–8414.

Churchill, M. E., Tullius, T. D., Kallenbach, N. R., & Seeman, N. C. (1988) *Proc. Natl. Acad. Sci. U.S.A. 85*, 4653–4656.

Clegg, R. M., Murchie, A. I. H., Zechel, A., Carlberg, C., Diekmann, S., & Lilley, D. M. J. (1992) *Biochemistry 31*, 4846–4856.

Clegg, R. M., Murchie, A. I. H., Zechel, A., & Lilley, D. M. J. (1993) *Proc. Natl. Acad. Sci. U.S.A. 90*, 2994–2998.

Clegg, R. M., Murchie, A. I. H., & Lilley, D. M. J. (1994) *Biophys. J. 66*, 99–109.

Coll, M., Frederick, C. A., Wang, A. H.-J., & Rich, A. (1987) *Proc. Natl. Acad. Sci. U.S.A. 84*, 8385–8389.

Connolly, B., & West, S. C. (1990) *Proc. Natl. Acad. Sci. U.S.A. 87*, 8476–8480.

Connolly, B., Parsons, C. A., Benson, F. E., Dunderdale, H. J., Sharples, G. J., Lloyd, R. G., & West, S. C. (1991) *Proc. Natl. Acad. Sci. U.S.A. 88*, 6063–6067.

Connor, B. N., Takano, T., Tanaka, T., Itakura, K., & Dickerson, R. E. (1982) *Nature 295*, 294–299.

Cooper, J. P., & Hagerman, P. J. (1987) *J. Mol. Biol. 198*, 711–719.

Cooper, J. P., & Hagerman, P. J. (1989) *Proc. Natl. Acad. Sci. U.S.A. 86*, 7336–7340.

Cotton, R. G. H., Rodrigues, N. R., & Campbell, R. D. (1988) *Proc. Natl. Acad. Sci. U.S.A. 85*, 4397–4401.

Courey, A. J., & Wang, J. C. (1983) *Cell 33*, 817–829.

Dayn, A., Malkhosyan, S., & Mirkin, S. M. (1992) *Nucleic Acids Res. 20*, 5991–5997.

de los Santos, C., Rosen, M., & Patel, D. (1989) *Biochemistry 28*, 7282–7289.

de Massey, B., Studier, F. W., Dorgai, L., Appelbaum, F., & Weisberg, R. A. (1984) *Cold Spring Harbor Symp. Quant. Biol. 49*, 715–726.

Depew, R. E., & Wang, J. C. (1975) *Proc. Natl. Acad. Sci. U.S.A. 72*, 4275–4279.

Dervan, P. B. (1992) *Nature 359*, 87–88.

Diekmann, S. (1986) *FEBS Lett. 195*, 53–56.

Diekmann, S. (1987a) *EMBO J. 6*, 4213–4217.

Diekmann, S. (1987b) *Nucleic Acids Res. 15*, 247–265.

Diekmann, S., & Lilley, D. M. J. (1987) *Nucleic Acids Res. 14*, 5765–5774.

Diekmann, S., & Wang, J. C. (1985) *J. Mol. Biol. 186*, 1–11.

Diekmann, S., von Kitzing, E., McLaughlin, L., Ott, J., & Eckstein, F. (1987) *Proc. Natl. Acad. Sci. U.S.A. 84*, 8257–8261.

Diekmann, S., Mazzarelli, J. M., Mclaughlin, L. W., Vonkitzing, E., & Travers, A. A. (1992) *J. Mol. Biol. 225*, 729–738.

Drak, J., & Crothers, D. M. (1991) *Proc. Natl. Acad. Sci. U.S.A. 88*, 3074–3078.

Drew, H. R., & Travers, A. A. (1984) *Cell 37*, 491–502.

Duckett, D. R., & Lilley, D. M. J. (1990) *EMBO J. 9*, 1659–1664.

Duckett, D. R., & Lilley, D. M. J. (1991) *J. Mol. Biol. 221*, 147–161.

Duckett, D. R., Murchie, A. I. H., Diekmann, S., von Kitzing, E., Kemper, B., & Lilley, D. M. J. (1988) *Cell 55*, 79–89.

Duckett, D. R., Murchie, A. I. H., & Lilley, D. M. J. (1990) *EMBO J. 9*, 583–590.

Duckett, D. R., Murchie, A. I. H., Bhattacharyya, A., Clegg, R. M., Diekmann, S., von Kitzing, E., & Lilley, D. M. J. (1992) *Eur. J. Biochem. 207*, 285–295.

Duckett, D. R., Giraud Panis, M.-J. E., & Lilley, D. M. J. (1995) *J. Mol. Biol. 246*, 95–107.

Elborough, K. M., & West, S. C. (1990) *EMBO J. 9*, 2931–2936.

Ellison, M. J., Kelleher III, R. J., Wang, A. H.-J., Habener, J. F., & Rich, A. (1985) *Proc. Natl. Acad. Sci. U.S.A. 82*, 8320–8324.

Evans, T., Schon, E., Gora-Maslak, G., Patterson, J., & Efstratiadis, A. (1984) *Nucleic Acids Res. 12*, 8043–8058.

Fang, G. W., & Cech, T. R. (1993) *Cell 74*, 875–885.

Felsenfeld, G., Davies, D. R., & Rich, A. (1957) *J. Am. Chem. Soc. 79*, 2023–2024.

Fuller, F. B. (1971) *Proc. Natl. Acad. Sci. U.S.A. 68*, 815–819.

Furlong, J. C., Sullivan, K. M., Murchie, A. I. H., Gough, G. W., & Lilley, D. M. J. (1989) *Biochemistry 28*, 2009–2017.

Galazka, G., Palecek, E., Wells, R. D., & Klysik, J. (1986) *J. Biol. Chem. 261*, 7093–7098.

Gao, X., & Patel, D. (1987) *J. Biol. Chem. 262*, 16973–16984.

Gao, X., & Patel, D. J. (1988) *J. Am. Chem. Soc. 110*, 5178–5182.

Gellert, M., Lipsett, M. N., & Davies, D. R. (1962) *Proc. Natl. Acad. Sci. U.S.A. 48*, 2013–2018.

Gellert, M., Mizuuchi, K., O'Dea, M. H., Ohmori, H., & Tomizawa, J. (1979) *Cold Spring Harbor Symp. Quant. Biol. 43*, 35–40.

Gellert, M., O'Dea, M. H., & Mizuuchi, K. (1983) *Proc. Natl. Acad. Sci. U.S.A. 80*, 5545–5549.

Giraldo, R., & Rhodes, D. (1994) *EMBO J. 13*, 2411–2420.

Gohlke, C., Murchie, A. I. H., Lilley, D. M. J., & Clegg, R. M. (1994) *Proc. Natl. Acad. Sci. U.S.A. 91*, 11660–11664.

Gotoh, O., & Tagashira, Y. (1981a) *Biopolymers 20*, 1043–1058.

Gotoh, O., & Tagashira, Y. (1981b) *Biopolymers 20*, 1033–1042.

Gough, G. W., Sullivan, K. M., & Lilley, D. M. J. (1986) *EMBO J. 5*, 191–196.

Gough, G., W. & Lilley, D. M. J. (1985) *Nature 313*, 154–156.

Greaves, D. R., Patient, R. K., & Lilley, D. M. J. (1985) *J. Mol. Biol. 185*, 461–478.

Guéron, M., Kochoyan, M., & Leroy, J.-L. (1987) *Nature 328*, 89–92.

Guéron, M., Charratier, E., Kochoyan, M., & Leroy, J.-L. (1988) *Biochemistry 27*, 8894–8898.

Guo, Q., Seeman, N. C., & Kallenbach, N. R. (1989) *Biochemistry 28*, 2355–2359.

Guo, Q., Lu, M., Seeman, N. C., & Kallenbach, N. R. (1990) *Biochemistry 29*, 570–578.

Hagerman, P. J. (1985) *Biochemistry 24*, 7033–7037.

Haniford, D. B., & Pulleyblank, D. E. (1983) *Nature 302*, 632–634.

Haniford, D. B., & Pulleyblank, D. E. (1985) *Nucleic Acids Res. 13*, 4343–4363.

Hanvey, J. C., Shimizu, M., & Wells, R. D. (1988) *Proc. Natl. Acad. Sci. U.S.A. 85*, 6292–6296.

Hare, D., Shapiro, L., & Patel, D. J. (1986) *Biochemistry 25*, 7456–7464.

Hélène, C. (1991) *Anticancer Drug Dev. 6*, 569–584.

Henderson, E., Hardin, C. C., Wolk, S. K., Tinoco, I., Jr., & Blackburn, E. (1987) *Cell 51*, 899–908.

Hentschel, C. C. (1982) *Nature 295*, 714–716.

Herr, W. (1985) *Proc. Natl. Acad. Sci. U.S.A. 82*, 8009–8013.

Holliday, R. (1964) *Genet. Res., 5*, 282–304.

Hoogsteen, K. (1963) *Acta Crystallogr. 16*, 907–916.

Horowitz, D. S., & Wang, J. C. (1984) *J. Mol. Biol. 173*, 75–91.

Hsieh, C.-H., & Griffith, J. D. (1989) *Proc. Natl. Acad. Sci. U.S.A. 86*, 4833–4837.

Hsieh, P., Camerini-Otero, C. O., & Camerini-Otero, R. D. (1990) *Genes Dev. 4*, 1951–1963.

Hsu, P. L., & Landy, A. (1984) *Nature 311*, 721–726.

Htun, H., & Dahlberg, J. E. (1988) *Science 241*, 1791–1796.

Htun, H., & Dahlberg, J. E. (1989) *Science 243*, 1571–1576.

Htun, H., Lund, E., & Dahlberg, J. E. (1984) *Proc. Natl. Acad. Sci. U.S.A. 81*, 7288–7292.

Hunter, W. N., Brown, T., Anand, N. N., & Kennard, O. (1986a) *Nature 320*, 552–555.

Hunter, W. N., Kneale, G., Brown, T., Rabinovich, D., & Kennard, O. (1986b) *J. Mol. Biol. 190*, 605–618.

Husain, I., Griffith, J. D., & Sancar, A. (1988) *Proc. Natl. Acad. Sci. U.S.A. 85*, 2558–2562.

Iwasaki, H., Takahagi, M., Nakata, A., & Shinagawa, H. (1992) *Gene Dev. 6*, 2214–2220.

Iwasaki, H., Takahagi, M., Shiba, T., Nakata, A., & Shinagawa, H. (1991) *EMBO J. 10*, 4381–4389.

Jain, S. K., Inman, R. B., & Cox, M. M. (1992) *J. Biol. Chem. 267*, 4215–4222.

Jaworski, A., Hsieh, W.-T., Blaho, J. A., Larson, J. E., & Wells, R. D. (1988) *Science 238*, 773–777.

Jensch, F., & Kemper, B. (1986) *EMBO J. 5*, 181–189.

Jeyaseelan, R., & Shanmugam, G. (1988) *Biochem. Biophys. Res. Commun. 156*, 1054–1060.

Jin, R. Z., Gaffney, B. L., Wang, C., Jones, R. A., & Breslauer, K. J. (1992) *Proc. Natl. Acad. Sci. U.S.A. 89*, 8832–8836.

Johnston, B. H. (1988) *Science 241*, 1800–1804.

Johnston, B. H., & Rich, A. (1985) *Cell 42*, 713–724.

Joshua-Tor, L., Rabinovich, D., Hope, H., Frolow, F., Apella, E., & Sussman, J. L. (1988) *Nature 334*, 82–84.

Joshua-Tor, L., Frolow, F., Appella, E., Hope, H., Rabinovich, D., & Sussman, J. L. (1992) *J. Mol. Biol. 225*, 397–431.

Kallenbach, N. R., Ma, R.-I., & Seeman, N. C. (1983) *Nature 305*, 829–831.

Kalnik, M. W., Norman, D. G., Li, B. F., Swann, P. F., & Patel, D. J. (1990) *J. Biol. Chem. 265*, 636–647.

Kang, C., Zhang, X., Ratliff, R., Moyzis, R., & Rich, A. (1992a) *Nature 356*, 126–131.

Kang, S., Wohlrab, F., & Wells, R. D. (1992b) *J. Biol. Chem. 267*, 1259–1264.

Karlovsky, P., Pecinka, P., Vojtiskova, M., Makaturova, E., & Palecek, E. (1990) *FEBS Lett. 274*, 39–42.

Kemper, B., & Janz, E. (1976) *J. Virol. 18*, 992–999.

Kim, J., Cheong, C., & Moore, P. B. (1991) *Nature 351*, 331–332.

Kimball, A., Guo, Q., Lu, M., Cunningham, R. P., Kallenbach, N. R., Seeman, N. C., & Tullius, T. D. (1990) *J. Biol. Chem. 265*, 6544–6547.

Kleff, S., & Kemper, B. (1988) *EMBO J. 7*, 1527–1535.

Klug, A., Jack, A., Viswamitra, M. A., Kennard, O., Shakked, Z., & Steitz, T. A. (1979) *J. Mol. Biol. 131*, 669–680.

Kneale, G., Brown, T., Kennard, O., & Rabinovich, D. (1985) *J. Mol. Biol. 186*, 805–814.

Kohwi, Y., & Kohwi-Shigematsu, T. (1988) *Proc. Natl. Acad. Sci. U.S.A. 85*, 3781–3785.

Kohwi, Y., & Kohwi-Shigematsu, T. (1993) *J. Mol. Biol. 231*, 1090–1101.

Kohwi-Shigematsu, T., & Kohwi, Y. (1985) *Cell 43*, 199–206.

Koo, H.-S., Wu, H.-M., & Crothers, D. M. (1986) *Nature 320*, 501–506.

Kowalski, D., Natale, D. A., & Eddy, M. J. (1988) *Proc. Natl. Acad. Sci. U.S.A. 85*, 9464–9468.

Lafer, E. M., Möller, A., Nordheim, A., Stollar, B. D., & Rich, A. (1981) *Proc. Natl. Acad. Sci. U.S.A. 78*, 3546–3550.

Laughlan, G., Murchie, A. I. H., Norman, D. G., Moore, M. H., Moody, P. C. E., Lilley, D. M. J., & Luisi, B. (1994) *Science 265*, 520–524.

LeBlanc, D. A., & Morden, K. M. (1991) *Biochemistry 30*, 4042–4047.

Lee, F. S., & Bauer, W. R. (1985) *Nucleic Acids Res. 13*, 1665–1682.

Leontis, N. B., Kwok, W., & Newman, J. S. (1991) *Nucleic Acids Res. 19*, 759–766.

Leontis, N. B., Hills, M. T., Piotto, M., Malhotra, A., Nussbaum, J., & Gorenstein, D. G. (1993) *J. Biomol. Struct. Dynamics 11*, 215.

Levene, S. D., & Crothers, D. M. (1986) *J. Mol. Biol. 189*, 61–72.

Lilley, D. M. J. (1980) *Proc. Natl. Acad. Sci. U.S.A. 77*, 6468–6472.

Lilley, D. M. J. (1985) *Nucleic Acids Res. 13*, 1443–1465.

Lilley, D. M. J. (1992) *Nature 357*, 282–283.

Lilley, D. M. J., & Hallam, L. R. (1984) *J. Mol. Biol. 180*, 179–200.

Lipanov, A., Kopka, M. L., Kaczor-Grzeskowiak, M., Quintana, J., & Dickerson, R. E. (1993) *Biochemistry 32*, 1373–1389.

Liu, Z., Frantz, J. D., Gilbert, W., & Tye, B.-K. (1993) *Proc. Natl. Acad. Sci. U.S.A. 90*, 3157–3161.

Lloyd, R. G., & Sharples, G. J. (1993) *EMBO J. 12*, 17–22.

Longfellow, C. E., Kierzek, R., & Turner, D. H. (1990) *Biochemistry 29*, 278–285.

Lu, M., Guo, Q., Seeman, N. C., & Kallenbach, N. R. (1989) *J. Biol. Chem. 264*, 20851–20854.

Lu, M., Guo, Q., Pasternack, R. F., Wink, D. J., Seeman, N. C., & Kallenbach, N. R. (1990a) *Biochemistry 29*, 1614–1624.

Lu, M., Guo, Q., Seeman, N. C., & Kallenbach, N. R. (1990b) *Biochemistry 29*, 3407–3412.

Lu, M., Guo, Q., & Kallenbach, N. R. (1991) *Biochemistry 30*, 5815–5820.

Lu, M., Guo, Q., & Kallenbach, N. R. (1993) *Biochemistry 32*, 598–601.

Lumpkin, O. J., & Zimm, B. H. (1982) *Biopolymers 21*, 2315–2316.

Lyamichev, V. I., Mirkin, S. M., & Frank-Kamenetskii, M. D. (1985) *J. Biomol. Struct. Dynamics 3*, 327–338.

Lyamichev, V., Mirkin, S. M., & Frank-Kamenetskii, M. D. (1986) *J. Biomol. Struct. Dynamics 3*, 667–669.

Lyamichev, V., Mirkin, S. M., Frank-Kamenetskii, M. D., & Cantor, C. R. (1988) *Nucleic Acids Res. 16*, 2165–2178.

Lyamichev, V. I., Frank-Kamenetskii, M. D., & Soyfer, V. N. (1990) *Nature 344*, 568–570.

Mace, H. A. F., Pelham, H. R. B., & Travers, A. A. (1983) *Nature 304*, 555–557.

Marini, J. C., Levene, S. D., Crothers, D. M., & Englund, P. T. (1982) *Proc. Natl. Acad. Sci. U.S.A. 79*, 7664–7668.

McClellan, J. A., & Lilley, D. M. J. (1987) *J. Mol. Biol. 197*, 707–721.

McClellan, J. A., & Lilley, D. M. J. (1991) *J. Mol. Biol. 219*, 145–149.

McClellan, J. A., Palecek, E., & Lilley, D. M. J. (1986) *Nucleic Acids Res. 14*, 9291–9309.

McClellan, J. A., Boublikova, P., Palecek, E., & Lilley, D. M. J. (1990) *Proc. Natl. Acad. Sci. U.S.A. 87*, 8373–8377.

McKeon, C., Schmidt, A., & de Crombrugghe, B. (1984) *J. Biol. Chem. 259*, 6636–6640.

McLean, M. J., Lee, J. W., & Wells, R. D. (1988) *J. Biol. Chem. 263*, 7378–7385.

Milgram-Burkoff, A., & Tullius, T. D. (1987) *Cell 48*, 935–943.

Miller, M., Harrison, R. W., Wlodower, A., Apella, E., & Sussman, J. L. (1988) *Nature 334*, 85–86.

Mirkin, S. M., Lyamichev, V. I., Drushlyak, K. N., Dobrynin, V. N., Filippov, S. A., & Frank-Kamenetskii, M. D. (1987) *Nature 330*, 495–497.

Mizuuchi, K., Mizuuchi, M., & Gellert, M. (1982) *J. Mol. Biol. 156*, 229–243.

Møllegaard, N. E., Murchie, A. I. H., Lilley, D. M. J., & Nielsen, P. E. (1994) *EMBO J. 13*, 1508–1513.

Möller, A., Gabriels, J. E., Lafer, E. M., Nordheim, A., Rich, A., & Stollar, B. D. (1982) *J. Biol. Chem. 257*, 12081–12085.

Morden, K. M., Chu, Y. G., Martin, F. H., & Tinoco, I., Jr. (1983) *Biochemistry 22*, 5557–5563.

Morden, K. M., Gunn, B. M., & Maskos, K. (1990) *Biochemistry 29*, 8835–8845.

Morgan, A. R., & Wells, R. D. (1968) *J. Mol. Biol. 37*, 63–80.

Moser, H., & Dervan, P. B. (1987) *Science 238*, 645–650.

Mueller, J. E., Newton, C. J., Jensch, F., Kemper, B., Cunningham, R. P., Kallenbach, N. R., & Seeman, N. C. (1990) *J. Biol. Chem. 265*, 13918–13924.

Murchie, A. I. H., & Lilley, D. M. J. (1987) *Nucleic Acids Res. 15*, 9641–9654.

Murchie, A. I. H., & Lilley, D. M. J. (1989) *J. Mol. Biol. 205*, 593–602.

Murchie, A. I. H., & Lilley, D. M. J. (1993) *J. Mol. Biol. 233*, 77–85.

Murchie, A. I. H., & Lilley, D. M. J. (1994) *EMBO J. 13*, 993–1001.

Murchie, A. I. H., Clegg, R. M., von Kitzing, E., Duckett, D. R., Diekmann, S., & Lilley, D. M. J. (1989) *Nature 341*, 763–766.

Murchie, A. I. H., Carter, W. A., Portugal, J., & Lilley, D. M. J. (1990) *Nucleic Acids Res. 18*, 2599–2606.

Murchie, A. I. H., Portugal, J., & Lilley, D. M. J. (1991) *EMBO J. 10*, 713–718.

Murchie, A. I. H., Bowater, R., Aboul-ela, F., & Lilley, D. M. J. (1992) *Biochim. Biophys. Acta 1131*, 1–15.

Naylor, L. H., Lilley, D. M. J., & van de Sande, H. (1986) *EMBO J. 5*, 2407–2413.

Nejedly, K., Klysik, J., & Palecek, E. (1989) *FEBS Lett. 243*, 313–317.

Nelson, H. C. M., Finch, J. T., Luisi, B. F., & Klug, A. (1987) *Nature 330*, 221–226.

Nikonowicz, E., Roongta, V., Jones, C. R., & Gorenstein, D. G. (1989) *Biochemistry 28*, 8714–8725.

Nikonowicz, E. P., Meadows, R. P., & Gorenstein, D. G. (1990) *Biochemistry 29*, 4193–4204.

Nordheim, A., & Rich, A. (1983) *Proc. Natl. Acad. Sci. U.S.A. 80*, 1821–1825.

Nordheim, A., Lafer, E. M., Peck, L. J., Wang, J. C., & Rich, A. (1982) *Cell 31*, 309–318.

O'Connor, T. R., Kang, D. S., & Wells, R. D. (1986) *J. Biol. Chem. 261*,13302–13308.

Orr-Weaver, T. L., Szostak, J. W., & Rothstein, R. J. (1981) *Proc. Natl. Acad. Sci. U.S.A. 78*, 6354–6358.

Panayotatos, N., & Wells, R. D. (1981) *Nature 289*, 466–470.

Panyutin, I., Lyamichev, V., & Mirkin, S. M. (1985) *J. Biomol. Struct. Dynamics 2*, 1221–1234.

Parsons, C. A., Kemper, B., & West, S. C. (1990) *J. Biol. Chem. 265*, 9285–9289.

Parsons, C. A., Tsaneva, I., Lloyd, R. G., & West, S. C. (1992) *Proc. Natl. Acad. Sci. U.S.A. 89*, 5452–5456.

Patel, D. J., Kozlowski, S. A., Marky, L. A., Rice, J. A., Broka, L., Itakura, K., & Breslauer, K. J. (1982) *Biochemistry 21*, 445–451.

Peck, L. J., & Wang, J. C. (1981) *Nature 292*, 375–378.

Peck, L. J., & Wang, J. C. (1983) *Proc. Natl. Acad. Sci. U.S.A. 80*, 6206–6210.

Peck, L. J., Nordheim, A., Rich, A., & Wang, J. C. (1982) *Proc. Natl. Acad. Sci. U.S.A. 79*, 4560–4564.

Pohl, F. M., & Jovin, T. M. (1972) *J. Mol. Biol. 67*, 375–396.

Pöhler, J. R. G., Duckett, D. R., & Lilley, D. M. J. (1994) *J. Mol. Biol. 238*, 62–74.

Potter, H., & Dressler, D. (1976) *Proc. Natl. Acad. Sci. U.S.A. 73*, 3000–3004.

Potter, H., & Dressler, D. (1978) *Proc. Natl. Acad. Sci. U.S.A. 75*, 3698–3702.

Privé, G. G., Heinemann, U., Chandrasegaran, S., Kan, L.-S., Kopka, M. L., & Dickerson, R. E. (1987) *Science 238*, 498–504.

Pulleyblank, D. E., Shure, M., Tang, D., Vinograd, J., & Vosberg, H.-P. (1975) *Proc. Natl. Acad. Sci. U.S.A. 72*, 4280–4284.

Pulleyblank, D. E., Haniford, D. B., & Morgan, A. R. (1985) *Cell 42*, 271–280.

Quigley, G. J., Ughetto, G., van der Marel, G., van Boom, J. H., Wang, A. H.-J., & Rich, A. (1986) *Science 232*, 1255–1258.

Rahmouni, A. R., & Wells, R. D. (1989) *Science 246*, 358–363.

Rahmouni, A. R., & Wells, R. D. (1992) *J. Mol. Biol. 223*, 131–144.

Rajagopal, P., & Feigon, J. (1989a) *Biochemistry 28*, 7859–7870.

Rajagopal, P., & Feigon, J. (1989b) *Nature 339*, 637–640.

Rao, B. J., Dutreix, M., & Radding, C. M. (1991) *Proc. Natl. Acad. Sci. U.S.A. 88*, 2984–2988.

Rhodes, D., & Klug, A. (1980) *Nature 286*, 573–578.

Rice, J. A., & Crothers, D. M. (1989) *Biochemistry 28*, 4512–4516.

Rice, J. A., Crothers, D. E., Pinto, A. L., & Lippard, S. J. (1988) *Proc. Natl. Acad. Sci. U.S.A. 85*, 4158–4161.

Rich, A., Nordheim, A., & Wang, A. H. J. (1984) *Annu. Rev. Biochem. 53*, 791–846.

Riordan, F. A., Bhattacharyya, A., McAteer, S., & Lilley, D. M. J. (1992) *J. Mol. Biol. 226*, 305–310.

Rosen, M. A., & Patel, D. J. (1993) *Biochemistry 32*, 6576–6587.

Rosen, M. A., Live, D., & Patel, D. J. (1992a) *Biochemistry 31*, 4004–4014.
Rosen, M. A., Shapiro, L., & Patel, D. J. (1992b) *Biochemistry 31*, 4015–4026.
Schaeffer, F., Yeramian, E., & Lilley, D. M. J. (1989) *Biopolymers 28*, 1449–1473.
Sen, D., & Gilbert, W. (1988) *Nature 334*, 364–366.
Sen, D., & Gilbert, W. (1990) *Nature 344*, 410–414.
Shakked, Z., Rabinovich, D., Kennard, O., Cruse, W. B. T., Salisbury, S. A., & Viswamitra, M. A. (1983) *J. Mol. Biol. 166*, 183–201.
Sharples, G. J., & Lloyd, R. G. (1991) *J. Bacteriol. 173*, 7711–7715.
Shiba, T., Iwasaki, H., Nakata, A., & Shinagawa, H. (1993) *Mol. Gen. Genet. 237*, 395–399.
Sigal, N., & Alberts, B. (1972) *J. Mol. Biol. 71*, 789–793.
Singleton, C. K., Klysik, J., Stirdivant, S. M., & Wells, R. D. (1982) *Nature 299*, 312–316.
Singleton, C. K., Kilpatrick, M. W., & Wells, R. D. (1984) *J. Biol. Chem. 259*, 1963–1967.
Sklenar, V., & Feigon, J. (1990) *Nature 345*, 836–838.
Smith, F. W., & Feigon, J. (1992) *Nature 356*, 164–168.
Sobell, H. M. (1972) *Proc. Natl. Acad. Sci. U.S.A. 69*, 2483–2487.
Suggs, J. W., & Wagner, R. W. (1986) *Nucleic Acids Res. 14*, 3703–3716.
Sullivan, K. M., & Lilley, D. M. J. (1986) *Cell 47*, 817–827.
Sullivan, K. M., & Lilley, D. M. J. (1987) *J. Mol. Biol. 193*, 397–404.
Sullivan, K. M., & Lilley, D. M. J. (1988) *Nucleic Acids Res. 16*, 1079–1093.
Sullivan, K. M., Murchie, A. I. H., & Lilley, D. M. J. (1988) *J. Biol. Chem. 263*, 13074–13082.
Sundquist, W. I., & Heaphy, S. (1993) *Proc. Natl. Acad. Sci. U.S.A. 90*, 3393–3397.
Sundquist, W. I., & Klug, A. (1989) *Nature 342*, 825–829.
Symington, L., & Kolodner, R. (1985) *Proc. Natl. Acad. Sci. U.S.A. 82*, 7247–7251.
Timsit, Y., Westhof, E., Fuchs, R. P. P., & Moras, D. (1989) *Nature 341*, 459–462.
Trifonov, E. N., & Sussman, J. L. (1980) *Proc. Natl. Acad. Sci. U.S.A. 77*, 3816–3820.
Tsaneva, I. R., Muller, B., & West, S. C. (1992) *Cell 69*, 1171–1180.
Tsaneva, I. R., Muller, B., & West, S. C. (1993) *Proc. Natl. Acad. Sci. U.S.A. 90*, 1315–1319.
Ulanovsky, L., Bodner, M., Trifonov, E. N., & Choder, M. (1986) *Proc. Natl. Acad. Sci. U.S.A. 83*, 862–866.
van de Sande, J. H., Ramsing, N. B., Germann, M. W., Elhorst, W., Kalisch, B. W., von Kitzing, E., Pon, R. T., Clegg, R. M., & Jovin, T. M. (1988) *Science 241*, 551–557.
van den Hoogen, Y. T., van Beuzekom, A. A., van den Elst, H., van der Marel, G. A., van Boom, J. H., & Altona, C. (1988) *Nucleic Acids Res. 16*, 2971–2986.
Vinograd, J., Lebowitz, J., Radloff, R., Watson, R., & Laipis, P. (1965) *Proc. Natl. Acad. Sci. U.S.A. 53*, 1104–1111.
Viswamitra, M. A., Kennard, O., Jones, P. G., Sheldrick, G. M., Salisbury, S., Favello, L., & Shakked, Z. (1978) *Nature 273*, 687–688.
Vojtiskova, M., & Palecek, E. (1987) *J. Biomol. Struct. Dynamics 5*, 283–296.
Vojtiskova, M., Mirkin, S., Lyamichev, V., Voloshin, O., Frank-Kamenetskii, M., & Palecek, E. (1988) *FEBS Lett. 234*, 295–299.
Vologodskii, A. V., & Frank-Kamenetskii, M. D. (1983) *FEBS Lett. 160*, 173–176.
Voloshin, O. N., Mirkin, S. M., Lyamichev, V. I., Belotserkovskii, B. P., & Frank-Kamenetskii, M. D. (1988) *Nature 333*, 475–476.
von Kitzing, E., Lilley, D. M. J., & Diekmann, S. (1990) *Nucleic Acids Res. 18*, 2671–2683.
Waldman, A. S., & Liskay, R. M. (1988) *Nucleic Acids Res. 16*, 10249–10266.
Wang, A. H.-J., Quigley, G. J., Kolpak, F. J., Crawford, J. L., van Boom, J. H., van der Marel, G., & Rich, A. (1979) *Nature 282*, 680–686.
Wang, J. C., Peck, L. J., & Becherer, K. (1983) *Cold Spring Harbor Symp. Quant. Biol. 47*, 85–91.
Wang, Y., & Patel, D. J. (1992) *Biochemistry 31*, 8112–8119.
Wang, Y., de los Santos, C., Gao, X., Greene, K., Live, D., & Patel, D. J. (1991) *J. Mol. Biol. 222*, 819–832.
Wang, Y. H., & Griffith, J. (1991) *Biochemistry 30*, 1358–1363.

Wang, Y. H., Barker, P., & Griffith, J. (1992) *J. Biol. Chem. 267*, 4911–4915.
Webster, G. D., Sanderson, M. R., Skelly, J. V., Neidle, S., Swann, P. F., Li, B. F., & Tickle, I. J. (1990) *Proc. Natl. Acad. Sci. U.S.A. 87*, 6693–6697.
Welch, J. B., Duckett, D. R., & Lilley, D. M. J. (1993) *Nucleic Acids Res. 21*, 4548–4555.
Wemmer, D. E., Wand, A. J., Seeman, N. C., & Kallenbach, N. R. (1985) *Biochemistry 24*, 5745–5749.
Werntges, H., Steger, G., Riesner, D., & Fritz, H.-J. (1986) *Nucleic Acids Res. 14*, 3773–3790.
West, S. C., Parsons, C. A., & Picksley, S. M. (1987) *J. Biol. Chem. 262*, 12752–12758.
Williamson, J. R., Raghuraman, M. K., & Cech, T. R. (1989) *Cell 59*, 871–880.
Woodson, S. A., & Crothers, D. M. (1987) *Biochemistry 26*, 904–912.
Woodson, S. A., & Crothers, D. M. (1988) *Biochemistry 27*, 3130–3141.
Woodson, S. A., & Crothers, D. M. (1989) *Biopolymers 28*, 1149–1177.
Wu, H.-M., & Crothers, D. M. (1984) *Nature 308*, 509–513.
Yoon, C., Privé, G. G., Goodsell, D. S., & Dickerson, R. E. (1988) *Proc. Natl. Acad. Sci. U.S.A. 85*, 6332–6336.
Zacharias, W., Jaworski, A., Larson, J. E., & Wells, R. D. (1988) *Proc. Natl. Acad. Sci. U.S.A. 85*, 7069–7073.
Zahler, A. M., Williamson, J. R., Cech, T. R., & Prescott, D. M. (1991) *Nature 350*, 718–720.
Zakian, V. A. (1989) *Annu. Rev. Genet. 23*, 579–604.

Chapter 8

Agrawal, S., Christodoulou, C., & Gait, M. J. (1986) *Nucleic Acids Res. 14*, 6227–6245.
Allen, D. J., Darke, P. L., & Benkovic, S. J. (1989) *Biochemistry 28*, 4601–4607.
Alves, A. M., Holland, D., & Edge, M. D. (1989) *Tetrahedron Lett. 30*, 3089.
Ansorge, W., Sproat, B. S., Stegemann, J., & Schwager, C. (1986) *J. Biochem. Biophys. Methods 13*, 315–323.
Asseline, U., & Thuong, N. T. (1989) *Tetrahedron Lett. 30*, 2521–2524.
Asseline, U., & Thuong, N. T. (1990) *Tetrahedron Lett. 31*, 81–84.
Asseline, U., Delarue, M., Lancelot, G., Toulmé, F., Thuong, N. T., Montenay-Garestier, T., & Hélène, C. (1984) *Proc. Natl. Acad. Sci. U.S.A. 81*, 3297–3301.
Asseline, U., Bonfils, E., Kurfürst, R., Chassignol, M., Roig, V., & Thuong, N. T. (1992) *Tetrahedron 48*, 1233–1254.
Atherton, F. R., & Todd, A. R. (1947) *J. Chem. Soc.*, 674–678.
Atherton, F. R., Openshaw, H. T., & Todd, A. R. (1945) *J. Chem. Soc.*, 660–663.
Aurup, H., Tuschl, T., Benseler, F., Ludwig, J., & Eckstein, F. (1994) *Nucleic Acids Res. 22*, 20–24.
Bergstrom, D. E., & Ogawa, M. K. (1978) *J. Am. Chem. Soc. 100*, 8106–8112.
Bergstrom, D. E., & Ruth, J. L. (1976) *J. Am. Chem. Soc. 98*, 1587–1589.
Broker, T. R., Angerer, L. M., Yen, P. H., Hershey, D., & Davidson, N. (1978) *Nucleic Acids Res. 5*, 363–384.
Catalano, C. E., Allen, D. J., & Benkovic, S. J. (1990) *Biochemistry 29*, 3612–3621.
Chu, B. F. C., & Orgel, L. E. (1985) *Proc. Natl. Acad. Sci. U.S.A. 82*, 963–967.
Cocuzza, A. J. (1989) *Tetrahedron Lett. 30*, 6287–6290.
Connolly, B. A. (1987) *Nucleic Acids Res. 15*, 3131–3139.
Connolly, B. A., & Rider, P. (1985) *Nucleic Acids Res. 13*, 4485–4502.
Cosstick, R., & Eckstein, F. (1985) *Biochemistry 24*, 3630–3638.
Cosstick, R., McLaughlin, L. W., & Eckstein, F. (1984) *Nucleic Acids Res. 12*, 1791–1810.
Coull, J. M., Weith, H. L., & Bischoff, R. (1986) *Tetrahedron Lett. 27*, 3991–3994.
DeVos, M.-J., Cravador, A., Lenders, J.-P., Hovard, S., & Bollen, A. (1990) *Nucleosides Nucleotides 9*, 259–273.
Draper, D. E. (1984) *Nucleic Acids Res. 12*, 989–1002.

Dreyer, G. B., & Dervan, P. B. (1985) *Proc. Natl. Acad. Sci. U.S.A. 82*, 968–972.

Erlanson, D. A., Chen, L., & Verdine, G. L. (1993) *J. Am. Chem. Soc. 115*, 12583–12584.

Ermácora, M. R., Delfino, J. M., Cuenoud, B., Schepartz, A., & Fox, R. O. (1992) *Proc. Natl. Acad. Sci. U.S.A. 89*, 6383–6387.

Eshaghpour, H., Söll, D., & Crothers, D. M. (1979) *Nucleic Acids Res. 7*, 1485–1495.

Ferentz, A., & Verdine, G. L. (1992) *Nucleosides Nucleotides 11*, 1749–1763

Fidanza, J. A., & McLaughlin, L. W. (1989) *J. Am. Chem. Soc. 111*, 9117–9119.

Fidanza, J. A., Ozaki, H., & McLaughlin, L. W. (1992) *J. Am. Chem. Soc. 114*, 5509–5517.

Förster, T. (1965) in *Modern Quantum Chemistry*, Istanbul Lectures, Part III, (Sinanoglu, O. Ed.), Academic Press, New York and London, pp. 93–138.

Francois, J.-C., Saison-Behmoaras, T., Chassignol, M., Thuong, N. T., & Hélène, C. (1989) *J. Biol. Chem. 264*, 5891–5898.

Froehler, B. C. (1986) *Tetrahedron Lett. 27*, 5575–5578.

Gibson, K. J., & Benkovic, S. J. (1987) *Nucleic Acids Res. 15*, 6455–6467.

Gilliam, I. C., & Tener, G. M. (1986) *Anal. Biochem. 157*, 199–207.

Goodchild, J. (1990) *Bioconjugate Chem. 1*, 165–187.

Grachev, M. A., Kolocheva, T. I., Lukhtanov, E. A., & Mustaev, A. A. (1987) *Eur. J. Biochem. 163*, 113–121.

Guinosso, C. J., Hoke, G. D., Freier, S., Martin, J. F., Ecker, D. J., Mirabelli, C. K., Crooke, S. T., & Cook, P. D. (1991) *Nucleosides Nucleotides 10*, 259–263.

Gupta, K. C., Sharma, P., Kumar, P., & Sathyanarayana, S. (1991) *Nucleic Acids Res. 19*, 3019–3025.

Haralambidis, J., Chai, M., Tregear, G. W. (1987) *Nucleic Acids Res. 15*, 4857–4876.

Haralambidis, J., Angus, K., Pownall, S., Duncan, L., Chai, M., & Tregear, G. W. (1990) *Nucleic Acids Res. 18*, 501–505.

Harris, C. M., Zhou, L., Strand, E. A., & Harris, T. M. (1991) *J. Am. Chem. Soc. 113*, 4328–4329.

Hertzberg, R. P., & Dervan, P. B. (1982) *J. Am. Chem. Soc. 104*, 313–315.

Horne, D. A., & Dervan, P. B. (1990) *J. Am. Chem. Soc. 112*, 2435–2437.

Jablonski, E., Moomaw, E., Tullis, R. H., & Ruth, J. L. (1986) *Nucleic Acids Res. 14*, 6115–6128.

Jäger, A., Levy, M. J., & Hecht, S. M. (1988) *Biochemistry 27*, 7237–7246.

Kazimierczuk, Z., Cottam, H. B., Revankar, G. R., & Robins, R. K. (1984) *J. Am. Chem. Soc. 106*, 6379–6382.

Kempe, T., Sundquist, W. I., Chow, F., & Hu, S.-L. (1985) *Nucleic Acids Res. 13*, 45–57.

Kierzek, R., & Markiewicz, W. T. (1987) *Nucleosides Nucleotides 6*, 403–405.

Kouchakdjian, M., Marinelli, E., Gao, X., Johnson, F., Grollman, A., & Patel, D. (1989) *Biochemistry 28*, 5647–5657.

Kremsky, J. N., Wooters, J. L., Dougherty, J. P., Meyers, R. E., Collins, M., & Brown, E. L. (1987) *Nucleic Acids Res. 15*, 2891–2909.

Lakshman, M. K., Sayer, J. M., & Jerina, D. M. (1991) *J. Am. Chem. Soc. 113*, 6589–6594.

Lakshman, M. K., Sayer, J. M., & Jerina, D. M. (1992) *J. Org. Chem. 57*, 3438–3443.

Langer, P. R., Waldrop, A. A., Ward, D. C. (1981) *Proc. Natl. Acad. Sci. U.S.A. 78*, 6633–6637.

Leary, J. J., Brigati, D. J., & Ward, D. C. (1983) *Proc. Natl. Acad. Sci. U.S.A. 80*, 4045–4049.

Lee, H., Hinz, M., Stezowski, J. J., & Harvey, R. G. (1990) *Tetrahedron Lett. 31*, 6773–6777.

Letsinger, R. L., & Heavner, G. A. (1975) *Tetrahedron Lett. 16*, 147–150.

Letsinger, R. L., & Schott, M. E. (1981) *J. Am. Chem. Soc. 103*, 7394–7396.

MacMillan, A. M., & Verdine, G. L. (1990) *J. Org. Chem. 55*, 5931–5933.

MacMillan, A. M., & Verdine, G. L. (1991) *Tetrahedron 47*, 2603–2616.

Maggio, A. F., Lucas, M., Barascut, J.-L., & Imbach, J.-L. (1984) *Tetrahedron Lett. 25*, 3195–3198.

Maher, L. J., III, Wold, B., & Dervan, P. B. (1989) *Science 245*, 725–730.

Manoharan, M. (1993) in *Antisense Research and Applications* (Crooke, S. T., & Lebleu, B., Eds.), pp. 303–349, CRC Press, Boca Raton, FL.

Manoharan, M., Guinosso, C. J., & Cook, P. D. (1991) *Tetrahedron Lett. 32*, 7171–7174.

Manoharan, M., Johnson, L., Tivel, K. L., Springer, R. H., & Cook, P. D. (1993) *Bioorg. Med. Chem. Lett. 3*, 2765–2770.

McLaughlin, L. W., & Piel, N. (1984) in *Oligonucleotide Synthesis: A Practical Approach* (Gait, M. J., Ed.), pp. 199–218, IRL Press, Oxford.

McLendon, G. (1988) *Acc. Chem. Res. 21*, 160–167.

Meyer, R. B., Jr., Tabone, J. C., Hurst, G. D., Smith, T. M., & Gamper, H. (1989) *J. Am. Chem. Soc. 111*, 8517–8519.

Mitina, R. L., Mustaev, A. A., Zaychikov, E. F., Khomov, V. V., & Lavrik, O. I. (1990) *FEBS Lett. 272*, 181–183.

Mori, K., Subasinghe, C., & Cohen, J. S. (1989) *FEBS Lett. 249*, 213–218.

Moser, H., & Dervan, P. B. (1987) *Science 238*, 645–650.

Murchie, A. I., Clegg, R. M., von Kitzing, E., Duckett, D. R., Diekmann, S., & Lilley, D. M. J. (1989) *Nature 341*, 763–766.

Murphy, C. J., Arkin, M. R., Jenkins, Y., Ghatlia, N. D., Bossmann, S. H., Turro, N. J., & Barton, J. K. (1993) *Science 262*, 1025–1029.

Nelson, P. S., Sherman-Gold, R., & Leon, R. (1989) *Nucleic Acids Res. 17*, 7179–7183.

Nikiforov, T. T., & Connolly, B. A. (1992) *Nucleic Acids Res. 20*, 1209–1214.

O'Donnell, M. J., Hebert, N., & McLaughlin, L. W. (1994) *Bioorg. Med. Chem. Lett. 4*, 1001–1004.

Ozaki, H., & McLaughlin, L. W. (1992) *Nucleic Acids Res. 20*, 5205–5214.

Pendergrast, P. S., Chen, Y., Ebright, Y. W., & Ebright, R. H. (1992) *Proc. Natl. Acad. Sci. U.S.A. 89*, 10287–10291.

Pieles, U., Sproat, B. S., & Lamm, G. M. (1990) *Nucleic Acids Res. 18*, 4355–4360.

Praseuth, D., Perrouault, L., Le Doan, T., Chassignol, M. Thoung, N., & Helene, C. (1988) *Proc. Natl. Acad. Sci. U.S.A. 85*, 1349–1353.

Prober, J. M., Trainor, G. L., Dam, R. J., Hobbs, F. W., Robertson, C. W., Zagursky, R. J., Cocuzza, A. J., Jensen, M. A., & Baumeister, K. (1987) *Science 238*, 336–341.

Richardson, R. W., & Gumport, R. I. (1983) *Nucleic Acids Res. 11*, 6167–6184.

Roduit, J.-P., Shaw, J., Chollet, A., Chollet, A. (1987) *Nucleosides Nucleotides 6*, 349–352.

Roget, A., Bazin, H., & Teoule, R. (1989) *Nucleic Acids Res. 17*, 7643–7651. 46

Ruth, J. L. (1984) *DNA 3*, 123–125.

Sanger, F., Nicklen, S., & Coulson, A. R. (1977) *Proc. Natl. Acad. Sci. U.S.A. 74*, 5463–5467.

Schaap, A. P., Sandison, M. D., & Handley, R. S. (1987) *Tetrahedron Lett. 28*, 1159–1162.

Schultz, P. G., Taylor, J. S., & Dervan, P. B. (1982) *J. Am. Chem. Soc. 104*, 6861–6863.

Singh, D., Kumar, V., & Ganesh, K. N. (1990) *Nucleic Acids Res. 18*, 3339–3345.

Sinha, N. D., & Cook, R. M. (1988) *Nucleic Acids Res. 16*, 2659–2669.

Smith, L. M., Fung, S., Hunkapiller, M. W., Hunkapiller, T. J., & Hood, L. E. (1985) *Nucleic Acids Res. 13*, 2399–2412.

Smith, L. M., Sanders, J. Z., Kaiser, R. J., Hughes, P., Dodd, C., Connell, C. R., Heiner, C., Kent, S. B. H., & Hood, L. E. (1986) *Nature 321*, 674–679.

Soundararajan, N., & Platz, M. S. (1990) *J. Org. Chem. 55*, 2034–2044.

Spaltenstein, A., Robinson, B. H., & Hopkins, P. B. (1988) *J. Am. Chem. Soc. 110*, 1299–1301.

Sproat, B. S., Beijer, B., & Rider, P. (1987a) *Nucleic Acids Res. 15*, 6181–6196.

Sproat, B. S., Beijer, B., Rider, P., & Neuner, P. (1987b) *Nucleic Acids Res. 15*, 4837–4848.

Sproat, B. S., Lamond, A. I., Beijer, B., Neuner, P., & Ryder, U. (1989) *Nucleic Acids Res. 17*, 3373–3386.

Stein, C. A., Mori, K., Loke, S. L., Subasinghe, C., Shinozuka, K., Cohen, J. S., & Neckers, L. M. (1988) *Gene 72*, 333–341.

Strobel, S. A., Moser, H. E., & Dervan, P. B. (1988) *J. Am. Chem. Soc. 110*, 7927–7929.

Stryer, L., & Haugland, R. P. (1967) *Proc. Natl. Acad. Sci. U.S.A. 58*, 719–726.

Sung, W. L. (1981) *Nucleic Acids Res. 9*, 6139–6151.

Sung, W. L. (1982) *J. Org. Chem. 47*, 3623–3628.

Tang, J. Y., & Agrawal, S. (1990) *Nucleic Acids Res. 18*, 6461.

Taylor, J. S., Schultz, P. G., & Dervan, P. B. (1984) *Tetrahedron 40*, 457–465.

Telser, J., Cruickshank, K. A., Morrison, L. E., & Netzel, T. L. (1989a) *J. Am. Chem. Soc. 111*, 6966–6976.

Telser, J., Cruickshank, K. A., Morrison, L. E., Netzel, T. L., & Chan, C.-K. (1989b) *J. Am. Chem. Soc. 111*, 7226–7232.

Telser, J., Cruickshank, K. A., Schanze, K. S., & Netzel, T. L. (1989c) *J. Am. Chem. Soc. 111*, 7221–7226.

Thuong, N. T., & Asseline, U. (1991) in *Oligonucleotides and Analogues* (Eckstein, F., Ed.), pp. 283–306, Oxford University/IRL Press, New York.

Urdea, M. S., Warner, B. D., Running, J. A., Stempien, M., Clyne, J., & Horn, T. (1988) *Nucleic Acids Res. 16*, 4937–4956.

Vosberg, H. P., & Eckstein, F. (1977) *Biochemistry 16*, 3633–3640.

Wachter, L., Jablonski, J.-A., & Ramachandran, K. L. (1986) *Nucleic Acids Res. 14*, 7985–7994.

Webb, T. R., & Matteucci, M. D. (1986) *Nucleic Acids Res. 14*, 7661–7674.

Yamana, K., Ohashi, Y., Nunota, K., Kitamura, M., Nakano, H., Sangen, O., & Shimidzu, T. (1991) *Tetrahedron Lett. 32*, 6347–6350.

Zajc, B., Lakshman, M. K., Sayer, J. M., & Jerina, D. M. (1992) *Tetrahedron Lett. 33*, 3409–3412.

Zuckermann, R., Corey, D., Schultz, P. G. (1987) *Nucleic Acids Res. 15*, 5305–5321.

Chapter 9

Alexander, R. S., Kanyo, Z. F., Chirlian, L. E., & Christianson, D. W. (1990) *J. Am. Chem. Soc. 112*, 933–937.

Altman, S., Kirsemboom, L., & Talbot, S. (1993) *FASEB J. 7*, 7–14.

Anderson, J. A, Kuntz, G. P. P., Evans, H. H., & Swift, T. J. (1971) *Biochemistry 10*, 4368–4374.

Andronikashvili, E. L., & Mosulishvili, L. M. (1980) in *Metal Ions in Biological Systems* (Sigel, H., Ed.), Vol. 10, pp. 167–206, Marcel Dekker, New York.

Angelici, R. J. (1973) in *Inorganic Biochemistry* (Eichhorn, G. L., Ed.), pp. 63–101, Elsevier, Amsterdam.

Antao, V. P., Lai, S. Y., & Tinoco, I., Jr. (1991) *Nucleic Acids Res. 19*, 5901–5905.

Ascoli, F., Branca, M., Mancini, C., & Pispisa, B. (1973) *Biopolymers 12*, 2431–2434.

Atwood, L., & Haake, P. (1976) *Bioorg. Chem. 5*, 373–382.

Baker, B. F. (1993) *J. Am. Chem. Soc. 115*, 3378–3379.

Bancroft, D. P., Lepre, C. A., & Lippard, S. J. (1990) *J. Am. Chem. Soc. 112*, 6860–6871.

Barton, J. K., & Lippard, S.J. (1980) in *Nucleic Acid–Metal Ion Interactions* (Spiro, T. G., Ed.), pp. 31–113, Wiley, New York.

Basile, L. A., & Barton, J. K. (1989) in *Metal Ions in Biological Systems*, Vol. 25, (Sigel, H., & Sigel, A., Eds.), pp. 31–103, Marcel Dekker, New York.

Basile, L. A., Raphael, A. L., & Barton, J. K. (1987) *J. Am. Chem. Soc. 109*, 7550–7551.

Basolo, F., Gray, H.B., & Pearson, R. G. (1960) *J. Am. Chem. Soc. 82*, 4200–4210.

Bean, B. L., Koren, R., & Mildvan, A. S. (1977) *Biochemistry 16*, 3322–3333.

Been, M. D., & Cech, T. R. (1988) *Science 239*, 1412–1416.

Beese, L. S., & Steitz, T. A. (1991) *EMBO J. 10*, 25–33.

Behlen, L. S., Sampson, J. R., DiRenzo, A. B., & Uhlenbeck, O. C. (1990) *Biochemistry 29*, 2515–2523.

Bender, M. L., Bergeron, R. J., & Komiyama, M. (1984) *The Bioorganic Chemistry of Enzymatic Catalysis*, pp. 191–215, Wiley, New York.

Berggren, E., Nordenskiöld, L., & Braunlin, W. H. (1992) *Biopolymers 32*, 1339–1350.

Berman, H. M., & Shieh, H.-S. (1981) in *Topic in Nucleic Acid Structure* (Neidle, S., Ed.), pp. 17–32, Wiley, New York.

Bernués, J., Beltran, R., Casasnovas, J. M., & Azorin, F. (1990) *Nucleic Acids Res. 18*, 4067–4073.

Blum, B., Sturm, N. R., Simpson, A. M., & Simpson, L. (1991) *Cell 65*, 543–550.

Bose, R. N., Cornelius, R. D., & Viola, R. E. (1985) *Inorg. Chem. 24*, 3989–3996.

Bratty, J., Chartrand, P., Ferbeyre, G.R., & Cedergren, R. (1993) *Biochim. Biophys. Acta 1216*, 345–359.

Breslow, R., & Huang, D.-L. (1991) *Proc. Natl. Acad. Sci. U.S.A. 88*, 4080–4083.

Breslow, R., Huang, D.-L., & Anslyn, E. (1989) *Proc. Natl. Acad. Sci. U.S.A. 86*, 1746–1750.

Brown, R. S., Dewan, J. C., & Klug, A. (1985) *Biochemistry 24*, 4785–4801.

Bujalowski, W., Graeser, E., Laughlin, L. W., & Pörschke, D. (1986) *Biochemistry 25*, 6365–6371.

Burgers, P. M. J., & Eckstein, F. (1979) *Biochemistry 18*, 592–596.

Butzow, J. J., & Eichhorn, G. L. (1965a) *Biopolymers 3*, 79–94.

Butzow, J. J., & Eichhorn, G. L. (1965b) *Biopolymers 3*, 95–107.

Butzow, J. J., & Eichhorn, G. L. (1971) *Biochemistry 10*, 2019–2027.

Campbell, D. A., Kubo, K., Clark, C. G., & Boothroyd, J. C. (1987) *J. Mol. Biol. 196*, 113–124.

Cech, T. R. (1987) *Science 236*, 1532–1539.

Cech, T. R., & Bass, B. L. (1986) *Annu. Rev. Biochem. 55*, 599–629.

Cedergren, R., Lang, B. F., & Gravel, D. (1987) *FEBS Lett. 226*, 63–66.

Celander, D. W., & Cech, T. R. (1990) *Biochemistry 29*, 1355–1361.

Chan, S. I., & Nelson, J. H. (1969) *J. Am. Chem. Soc. 91*, 168–183.

Chatterji, D. (1988) *Biopolymers 27*, 1183–1186.

Chen, M. S. (1978) *Inorg. Perspect. Biol. Med. 1*, 217–222.

Cheng, A.-J., & Van Dyke, M. W. (1993) *Nucleic Acids Res. 21*, 5630–5635.

Chin, J. (1991) *Acc. Chem. Res. 24*, 145–152.

Chin, J., & Banaszczyk, M. (1989) *J. Am. Chem. Soc. 111*, 4103–4105.

Chin, J., & Zou, X. (1987) *Can. J. Chem. 65*, 1882–1884.

Chin, J. & Zou, X. (1988) *J. Am. Chem. Soc. 110*, 223–225.

Chowrira, B. M., Berzal-Herranz, A., & Burke, J. (1991) *Nature 354*, 320–322.

Chowrira, B. M., Berzal-Herranz, A., & Burke, J. (1993) *Biochemistry 186*, 1088–1095.

Christie, N. T., & Costa, M. (1983) *Biol. Trace Element Res. 5*, 55–71.

Chu, G. Y. H., Mancy, S., Duncan, S., & Tobias, R. S. (1978) *J. Am. Chem. Soc. 100*, 593–606.

Ciesiolka, J., Wrzesinski, J., Grniki, P., Podkowinski, J., & Krzyzosiak, W. J. (1989) *Eur. J. Biochem. 186*, 71–77.

Ciesiolka, J., Lorenz, S., & Erdmann, V. A. (1992) *Eur. J. Biochem. 204*, 583–589.

Cleare, M. J., & Hydes, P. C. (1980) in *Metal Ions in Biological Systems*, Vol. 11 (Sigel, H., Ed.), pp. 1–62, Marcel Dekker, New York.

Clement, R. M., Sturm, J., & Daune, M. P. (1973) *Biopolymers 12*, 405–421.

Cohn, M., Danchin, A., & Grunberg-Manago, M. (1969) *J. Mol. Biol. 39*, 199–217.

Collier, D. A., & Wells, R. D. (1990) *J. Biol. Chem. 265*, 10652–10658.

Cooperman, B. S. (1976) in *Metal Ions in Biological Systems*, Vol. 5, (Sigel, H., Ed.), pp. 79–125, Marcel Dekker, New York.

Cotton, F. A., & Wilkinson, G. (1988) *Advanced Inorganic Chemistry*, fifth edition, pp. 1285–1301, Wiley, New York.

Cowan, J. A. (1993) *J. Inorg. Biochem. 49*, 171–175.

Dahm, S. A. C., & Uhlenbeck, O.C. (1991) *Biochemistry 30*, 9464–9469.

Dahm, S. A. C., Derrick, W. B., & Uhlenbeck, O. C. (1993) *Biochemistry 32*, 13040–13045.

Dale, R. M. K., Martin, E., Livingston, D. C., & Ward, D. C. (1975) *Biochemistry 14*, 2447–2457.

Danchin, A. (1972) *Biochimie 54*, 333–337.

Danchin, A. (1975) *Biochimie 57*, 875–880.

Danchin, A., & Guéron, M. (1970) *J. Chem. Phys. 53*, 3599–3609.

Dange, V., Van Atta, R. B., & Hecht, S. M. (1990) *Science 248*, 585–588.

Darr, S. C., Brown, J. W., & Pace, N. R. (1992) *TIBS 17*, 178–182.

Daune, M. (1974) in *Metal Ions in Biological Systems* Vol. 3 (Sigel, H., Ed.), pp. 1–43, Marcel Dekker, New York.

David, S. S., & Que, L., Jr., (1990) *J. Am. Chem. Soc. 112*, 6455–6463.

Davies, J. F., II, Hostomska, Z., Hostomsky, Z., Jordan, S. R., & Matthews, D. A. (1991) *Science 252*, 88–94.

Deng, H. Y., & Termini, J. (1992) *Biochemistry 31*, 10518–10528.

Denman, R. B. (1993) *BioTechniques 15*, 1090–1094.

Dinter-Gottlieb, G. (1986) *Proc. Natl. Acad. Sci. U.S.A. 83*, 6250–6254.

Divakar, S., Vasudevachari, M. B., Antony, A., & Easwaran, K. R. K. (1987) *Biochemistry 26*, 3781–3785.

Dock-Bregeon, A. C., & Moras, D. (1987) *Cold Spring Harbor Symp. Quant. Biol. LII*, 113–121.

Doudna, J. A., Couture, S., & Szostak, J. W. (1991) *Science 251*, 1605–1608.

Doudna, J. A., Usman, N., & Szostak, J. W. (1993a) *Biochemistry 32*, 2111–2115.

Doudna, J. A., Grosshans, C., Gooding, A., & Kundrot, C. E. (1993b) *Proc. Natl. Acad. Sci. U.S.A. 90*, 7829–7833.

Downey, K. M., & So, A. G. (1989) in *Metal Ions in Biological Systems*, Vol. 25 (Sigel, H., & Sigel, A., Eds.), pp. 1–30, Marcel Dekker, New York.

Ehresmann, C., Baudin, F., Mougel, M., Romby, P., Ebel, J.-P., & Ehresmann, B. (1987) *Nucleic Acids Res. 15*, 9109–9128.

Eichhorn, G. L. (1962) *Nature 194*, 474–475.

Eichhorn, G. L. (1973) in *Inorganic Biochemistry*, Vol. 2 (Eichhorn, G. L., Ed.), pp. 1210–1243, Elsevier, Amsterdam.

Eichhorn, G. L. (1980) in *Metal Ions in Biological Systems*, Vol. 10 (Sigel, H., Ed.), pp. 1–21, Marcel Dekker, New York.

Eichhorn, G. L. (1981) in *Metal Ions in Genetic Information Transfer* (Eichhorn, G. L., & Marzilli, L. G., Eds.), pp. 1–46, Elsevier/North-Holland, New York.

Eichhorn, G. L., & Shin, Y. A. (1968) *J. Am. Chem. Soc. 90*, 7323–7328.

Eichhorn, G. L., Butzow, J. J., Clark, P., & Tarien, E. (1967) *Biopolymers 5*, 283–296.

Eichhorn, G. L., Tarien, E., & Butzow, J. J. (1971) *Biochemistry 10*, 2014–2019.

Eichhorn, G. L., Tarien, E., Elgavish, G. A., Butzow, J. J., & Shin, Y. A. (1981) in *Biomolecular Stereodynamics*, Vol. 2 (Sarma, R. A., Ed.), pp. 185–195, Adenine Press, New York.

Eisinger, J., Fawaz-Estrup, F., & Shulman, R. G. (1965) *J. Chem. Phys. 42*, 43–53.

Farkas, W. R. (1968) *Biochim. Biophys. Acta 155*, 401–409.

Favre, A., & Thomas, G. (1981) *Annu. Rev. Biophys. Bioeng. 10*, 175–195.

Fedor, M. J., & Uhlenbeck, O. C. (1990) *Proc. Natl. Acad. Sci. U.S.A. 87*, 1668–1672.

Felsenfeld, G., & Miles, H. T. (1967) *Annu. Rev. Biochem. 36*, 407–448.

Ferris, J. P., & Ertem, G. (1993) *J. Am. Chem. Soc. 115*, 12270–12275.

Flanagan, J. M., & Jacobson, K. B. (1987) *J. Chromatogr. 387*, 139–154.

Flanagan, J. M., & Jacobson, K. B. (1988) *Biochemistry 27*, 5778–5785.

Flessel, C. P., Furst, A., & Radding, S. B. (1980) in *Metal Ions in Biological Systems*, Vol. 10 (Sigel, H., Ed.), pp. 23–54, Marcel Dekker, New York.

Floro, N. A., & Wetterhahn, K. E. (1984) *Biochem. Biophys. Res. Commun. 124*, 106–113.

Forster, A. C., & Symons, R. H. (1987) *Cell 50*, 9–16.

Forster, A. C., Davis, C., Sheldon, C. C., Jeffries, A. C., & Symons, R. H. (1988) *Nature 334*, 265–267.

Fox, K. R. (1990) *Nucleic Acids Res. 18*, 5387–5391.

Franzen, J. S., Zhang, M., & Peebles, C. L. (1993) *Nucleic Acids Res. 21*, 627–634.

Freier, S. M., Kierzek, R., Jaeger, J. A., Sugimoto, N., Caruthers, M. H., Neilson, T., & Turner, D. H. (1986) *Proc. Natl. Acad. Sci. U.S.A. 83*, 9373–9377.

Frey, P. A., & Sammons, R. D. (1985) *Science 228*, 541–545.

Frey, C. M., & Stuehr, J. (1974) in *Metal Ions in Biological Systems*, Vol. 1 (Sigel, H., Ed.), pp. 51–116, Marcel Dekker, New York.

Fu, D.-J., & McLaughlin, L. W. (1992) *Biochemistry 31*, 10941–10949.

Fu, D.-J., Rajur, S. B., & McLaughlin, L. W. (1993) *Biochemistry 32*, 10629–10637.

Gao, Y.-G., Sriram, M., & Wang, A. H.-J. (1993) *Nucleic Acids Res. 21*, 4093–4101.

Gardiner, K. J., Marsh, T. L., & Pace, N. R. (1985) *J. Biol. Chem. 260*, 5415–5419.

Geierstanger, B. H., Kagawa, T. F., Chen, S.-L. Quigley, G. J., & Ho, P. S. (1991) *J. Biol. Chem. 266*, 20185–20191.

Gellert, R. W., & Bau, R. (1979) in *Metal Ions in Biological Systems*, Vol. 8 (Sigel, H., Ed.), pp. 2–55, Marcel Dekker, New York.

Gladchenko, G. O., Valeev, V. A., & Sorokin, V. A. (1982) *Studia Biophys. 87*, 237–238.

Gladchenko, G. O., Sorokin, V. A., Valeev, V. A., & Blagoi, Y. P. (1986) *Studia Biophys. 111*, 119–122.

Goodgame, D. M. L., Jeeves, I., Reynolds, C. D., & Skapski, A. C. (1975) *Nucleic Acids Res. 2*, 1375–1379.

Gornicki, P., Baudin, F., Romby, P., Wiewiorowski, M., Kryzosiak, W., Ebel, J.-P., Ehresmann, C., & Ehresmann, B. (1989) *J. Biomol. Struct. Dynamics 6*, 971–984.

Granot, J., & Kearns, D. (1982) *Biopolymers 21*, 219–232.

Granot, J., Feigon, J., & Kearns, D. R. (1982) *Biopolymers 21*, 181–201.

Green, R., & Szostak, J. W. (1992) *Science 258*, 1910–1914.

Gross, D. S., & Simpkins, H. (1981) *J. Biol. Chem. 256*, 9593–9598.

Grosshans, C. A., & Cech, T. R. (1989) *Biochemistry 28*, 6888–6894.

Grosshans, C. A., & Cech, T. R. (1991) *Nucleic Acids Res. 19*, 3875–3887.

Grunberg-Manago, M., Hoa, G. H. B., Douzou, P., & Wishina, A. (1981) in *Metal Ions in Genetic Information Transfer* (Eichhorn, G. L., & Marzilli, L. G., Eds.), pp. 193–232, Elsevier/North-Holland, New York.

Guerrier-Takada, C., Haydock, K., Allen, L., & Altman, S. (1986) *Biochemistry 25*, 1509–1515.

Haight, G. P., Jr. (1987) *Coord. Chem. Rev. 79*, 293–319.

Han, H., & Dervan, P. B. (1993) *Proc. Natl. Acad. Sci. U.S.A. 90*, 3806–3810.

Hardin, C. C., Watson, T., Corregan, M., & Bailey, C. (1992) *Biochemistry 31*, 833–841.

Hayashi, N., Takeda, N., Shiiba, T., Yashiro, M., Watanabe, K., & Komiyama, M. (1993) *Inorg. Chem. 32*, 3899–3900.

Haydock, K., & Allen, L. C. (1985) in *Molecular Basis of Cancer, Part A: Macromolecular Structure, Carcinogens & Oncogens (Progress in Clinical and Biological Research*, Vol. 172) (Rein, R., Ed.), pp. 87–98, Alan R. Liss, New York.

Heidenreich, O., Pieken, W., & Eckstein, F. (1993) *FASEB J. 7*, 90–96.

Hendry, P., & Sargeson, A. M. (1989) *J. Am. Chem. Soc. 111*, 2521–2527.

Herschlag, D., & Jencks, W. (1987) *J. Am. Chem. Soc. 109*, 4665–4674.

Herschlag, D., Piccirilli, J. A., & Cech, T. (1991) *Biochemistry 30*, 4844–4854.

Heus, H. A., & Pardi, A. (1991) *Science 253*, 191–194.

Hirao, I., Nishimura, Y., Naraoka, T., Watanabe, K., Arata, Y., & Miura, K.-I. (1989) *Nucleic Acids Res. 17*, 3891–3896.

Hirao, I., Nishimura, Y., Tagava, Y.-I., Watanabe, K., & Miura, K.-I. (1992) *Nucleic Acids Res. 20*, 3891–3896.

Hodgson, D. J. (1977) *Prog. Inorg. Chem. 23*, 211–254.

Holbrook, S. R., Sussman, J. L., Warrant, R. W., Church, G. M., & Kim, S. H. (1977) *Nucleic Acids Res. 4*, 2811–2820.

Holmes, C. E., & Hecht, S. M. (1993) *J. Biol. Chem. 268*, 25909–25913.

Holmes, C. E., Carter, B. J., & Hecht, S. M. (1993) *Biochemistry 32*, 4293–4307.

Hough, E., Hansen, L. K., Birknes, B., Jynge, K., Hansen, S., Hordirk, A., Little, C., Dodson, E. J., & Derewenda, Z. (1989) *Nature 338*, 357–360.

Hüttenhofer, A., Hudson, S., Noller, H. F., & Mascharak, P. K. (1992) *J. Biol. Chem. 267*, 24471–24475.

Ikenaga, H., & Inoue, Y. (1974) *Biochemistry 13*, 577–582.

Issaq, H. J. (1980) in *Metal Ions in Biological Systems*, Vol. 10 (Sigel, H., Ed.), pp. 55–93, Marcel Dekker, New York.

Izatt, R. M., Christiansen, J. J., & Rytting, J. H. (1971) *Chem. Rev. 71*, 439–481.

Jack, A., Ladner, J. E., Rhodes, D., Brown, R. S., & Klug, A. (1977) *J. Mol. Biol. 111*, 315–328.

Jack, T. R. (1979) in *Metal Ions in Biological Systems*, Vol. 8 (Sigel, H., Ed.), pp. 159–182, Marcel Dekker, New York.

Jencks, W. P. (1969) *Catalysis in Chemistry and Enzymology*, pp. 111–115, McGraw-Hill, New York.

Joyce, G. F. (1987) *Cold Spring Harbor Symp. Quant. Biol. LII*, 41–51.

Joyce, G. F. (1989) *Nature 338*, 217–224.

Kanaya, E., & Yanagawa, H. (1986) *Biochemistry 25*, 7423–7430.

Kang, S., & Wells, R. D. (1992) *J. Biol. Chem. 267*, 20887–20891.

Kao, T. H., & Crothers, D. M. (1980) *Proc. Natl. Acad. Sci. U.S.A. 77*, 3360–3364.

Karlik, S. J., Eichhorn, G. L., Lewis, P.N., & Crapper, D. R. (1980) *Biochemistry 19*, 5991–5998.

Karlin, K. (1993) *Science 261*, 701–708.

Karpel, R. L., Miller, N. S., Lesk, A. M., & Fresco, J. R. (1975) *J. Mol. Biol. 97*, 519–532.

Karpel, R. L., Miller, N. S., Lesk, A. M., & Fresco, J. R. (1980) *Biochemistry 19*, 504–512.

Kasyanenko, N. A., Sosa, G. S.-H., Uversky, V. N., & Frisman, E. V. (1987) *Mol. Biol. (Russ.) 21*, 140–146.

Kazakov, S., & Altman, S. (1991) *Proc. Natl. Acad. Sci. U.S.A. 88*, 9193–9197.

Kazakov, S., & Altman, S. (1992) *Proc. Natl. Acad. Sci. U.S.A. 89*, 7939–7943.

Kazakov, S., Astashkina, T., Mamaev, S., & Vlassov, V. (1988) *Nature 335*, 186–188.

Kazakov, S. A., & Hecht, S. M. (1994) in *Encyclopedia of Inorganic Chemistry* (King, R. B., Ed.), pp. 2697–2720, Wiley, New York.

Kesicki, E. A., DeRoch, M. A., Freeman, L. H., Walton, C. L., Harrey, D. F. & Trogler, W. (1993) *Inorg. Chem. 32*, 5851–5867.

Khan, B. T., Raju, M., & Zakeeruddin, S. M. (1988) *J. Chem. Soc. Dalton Trans.*, 67–71.

Kierzek, R. (1992) *Nucleic Acids Res. 20*, 5073–5084.

Kim, E. E., & Wyckoff, H. W. (1991) *J. Mol. Biol. 218*, 449–464.

Kim, S. H., & Martin, R. B. (1984) *Inorg. Chim. Acta 91*, 19–24.

Kochetkov, N. K., & Budovsky, E. I. Eds. (1972) *Organic Chemistry of Nucleic Acids*, Part B, pp. 496–504, Plenum Press, New York.

Kohwi, Y. (1989) *Nucleic Acids Res. 17*, 4493–4502.

Kohwi, Y., Wang, H., & Kohwi-Shigematsu, T. (1993) *Nucleic Acids Res. 21*, 5651–5655.

Kolasa, K. A., Morrow, J. R., & Sharma, A. P. (1993) *Inorg. Chem. 32*, 3983–3984.

Köpf-Mayer, P., & Köpf, H. (1987) *Chem. Rev. 87*, 1137–1152.

Krotz, A. H., Hudson, B. P., & Barton, J. K. (1993) *J. Am. Chem. Soc. 115*, 12577–12578.

Krzyzosiak, W. J., Marciniec, T., Wiewiorowski, M., Romby, P., Ebel, J. P., & Giegé, R. (1988) *Biochemistry 27*, 5771–5777.

Kuhsel, M. G., Strickland, R., & Palmer, J. D. (1990) *Science 250*, 1570–1573.

Laundon, C. H., & Griffith, J. D. (1987) *Biochemistry 26*, 3759–3762.

Labuda, D., Nicoghosian, K., & Cedergren, R. (1985) *J. Biol. Chem. 260*, 1103–1107.

Le, S.-Y., Chen, J.-H., & Maizel, J. V., Jr. (1993) *Nucleic Acids Res. 21*, 2173–2178.

Lehman, N., & Joyce, G. F. (1993) *Nature 361*, 182–185.

Leng, M. (1990) *Biophys. Chem. 35*, 155–163.

Leonard, G. A., Booth, E. D., Hunter, W. N., & Brown, T. (1992) *Nucleic Acids Res. 20*, 4753–4759.

Lepre, C. A., & Lippard, S. J. (1990) in *Nucleic Acids and Molecular Biology*, Vol. 4 (Eckstein, F., & Lilley, D. M. J., Eds.), pp. 9–37, Springer-Verlag, Berlin.

Liang, C., & Allen, L. C. (1987) *J. Am. Chem. Soc. 109*, 6449–6453.

Lickl, E., Alth, G., & Tuma, K. (1988) *Agr. Biol. Chem. 52*, 857–858.

Limmer, S., Hofmann, H.-P., Ott, G., & Sprinzl, M. (1993) *Prod. Natl. Acad. Sci. U.S.A. 90*, 6199–6202.

Lindahl, T., Adams, A., & Fresco, J. R. (1966) *Proc. Natl. Acad. Sci. U.S.A. 55*, 941–948.

Ling, E. C. H., Allen, G. W., & Hambley, T. W. (1993) *J. Chem. Soc. Dalton Trans.*, 3705–3710.

Lippard, S. J. (1991) in *Platinum & Other Metal Coordination Compounds in Cancer Chemotherapy* (Howell, S. B., Ed.), pp. 1–12, Plenum Press, New York.

Lippert, B., Arpalabti, J., Krizanovic, O., Micklitz, W., Schwarz, F., & Trötscher, G. (1988) in *Platinum & Other Metal Coordination Complexes in Cancer Chemotherapy* (Nicolini, M., Ed.), pp. 564–581, Nijhoff, Boston.

Lohrmann, R., & Orgel, L. E. (1978) *Tetrahedron 34*, 853–855.

Lohrmann, R., Bridson, P. K., & Orgel, L. E. (1980) *Science 208*, 1464–1465.

Long, D. M., & Uhlenbeck, O. C. (1993) *FASEB J. 7*, 25–30.

Lönnberg, H., & Apralahti, J. (1981) *Inorg. Chim. Acta 55*, 39–42.

Lönnberg, H., & Vihanto, P. (1981) *Inorg. Chim. Acta 56*, 157–161.

Loprete, D. M., & Hartman, K. A. (1993) *Biochemistry 32*, 4077–4082.

Luck, G., & Zimmer, C. (1972) *Eur. J. Biochem. 29*, 528–536.

Lyamichev, V. I., Voloshin, O. N., Frank-Kamenetskii, M. D., & Soyfer, V. N. (1991) *Nucleic Acids Res. 19*, 1633–1638.

Malkov, V. A., Voloshin, O. N., Soyfer, V. N., & Frank-Kamenetskii, M. D. (1993) *Nucleic Acids Res. 21*, 585–591.

Martin, R. B. (1964) Introduction to Biophysical Chemistry, p. 344, McGraw-Hill, New York.

Martin, R. B. (1983) in *Platinum, Gold & Other Metal Cherapeutic Agents* (Lippard, S. J., Ed.), ACS Symposium Series, Vol. 209, pp. 231–244, American Chemical Society, Washington, D.C.

Martin, R. B. (1985) *Acc. Chem. Res. 18*, 32–38.

Martin,R.B. (1986) in *Metal Ions in Biological Systems*, Vol. 20 (Sigel, H., Ed.), pp. 21–65, Marcel Dekker, New York.

Martin, R. B. (1990) in *Metal Ions in Biological Systems*, Vol. 26 (Sigel, H., Ed.), pp. 1–13, Marcel Dekker, New York.

Martin, R. B., & Mariam, Y. H. (1979) in *Metal Ions in Biological Systems*, Vol. 8 (Sigel, H., Ed.), pp. 57–124, Marcel Dekker, New York.

Marzilli, L. G. (1977) *Prog. Inorg. Chem. 23*, 255–378.

Marzilli, L. G. (1981) in *Metal Ions in Genetic Information Transfer* (Eichhorn, G. L., & Marzilli, L. G., Eds.), pp. 47–123, Elsevier, New York.

Marzilli, L. G., & Kistenmacher, T. J. (1977) *Acc. Chem. Res. 10*, 146–152.

Marzilli, L. G., Kistenmacher, T. J., & Eichhorn, G. L. (1980) in *Metal Ions in Biology*, Vol. 1 (Spiro, T. G., Ed.), pp. 180–250, Wiley, New York.

Massoud, S. S., & Sigel, H. (1988) *Inorg. Chem. 27*, 1447–1483.

Massoud, S. S., & Sigel, H. (1989) *Eur. J. Biochem. 179*, 451–458.

Maurel, M. C., & Décout, J. L. (1992) *J. Mol. Evol. 35*, 190–195.

Mei, H.-Y., Kaaret, T. W., & Bruce, T. C. (1989) *Proc. Natl. Acad. Sci. U.S.A. 86*, 9727–9731.

Milburn, R. M., Gautam-Basak, M., Tribolet, R., & Sigel, H. (1985) *J. Am. Chem. Soc. 107*, 3315–3321.

Mildvan, A. S., & Grisham, C. M. (1974) *Structure Bonding 20*, 1–21.

Mildvan, A. S., & Loeb, L. A. (1981) in *Metal Ions in Genetic Information Transfer* (Eichhorn G. L., & Marzilli, L. G., Eds.), pp. 103–123, Elsevier/North-Holland, New York.

Morales, M. J., Wise, C. A., Hollingsworth, M. J., & Martin, N. C. (1989) *Nucleic Acids Res. 17*, 6865–6881.

Morrow, J. R., & Trogler, W. C. (1988) *Inorg. Chem. 27*, 3387–3394.

Morrow, J. R., Buttrey, L. A., Shelton, V. M., & Berback, K. A. (1992) *J. Am. Chem. Soc. 114*, 1903–1905.

Murray, M. J., & Flessel, C. P. (1976) *Biochim. Biophys. Acta 425*, 256–261.

Nagesh, N., Bhargava, P., & Chatterji, D. (1992) *Biopolymers 32*, 1421–1424.

Narasimhan, V., & Bryan, A. M. (1984) *Inorg. Chim. Acta 91*, L39–L41.

Nejadly, K., Klysik, J., & Palecek, E. (1989) *FEBS Lett. 243*, 313–317.

Nieboer, E., & Sanford, W. E. (1985) *Rev. Biochem. Toxicol. 7*, 205–245.

Nishimura, S. (1979) in *Transfer RNA: Structure, Properties, & Recognition* (Schimmel, P. R., Söll, D., & Abelson, J. N., Eds.), pp. 547–549, Cold Spring Harbor Laboratory, Cold Spring Harbor, NY.

Ochiai, E.-I. (1988) *J. Chem. Ed. 65*, 943–946.

O'Halloran, T. V. (1989) in *Metal Ions in Biological Systems*, Vol. 25 (Sigel, H., Ed.), pp. 105–146, Marcel Dekker, New York.

O'Halloran, T. V. (1993) *Science 261*, 715–725.

Orgel, L. E. (1986) *J. Theor. Biol. 123*, 127–149.

Orgel, L. E. & Lohrmann, R. (1974) *Acc. Chem. Res. 7*, 368–377.

Orita, M., Nishikawa, F., Shimayama, T., Taira, K., Endo, Y., & Nishikawa, S. (1993) *Nucleic Acids Res. 21*, 5670–5678.

Pace, N. R. (1991) *Cell 65*, 531–533.

Pan, T., & Uhlenbeck, O. C. (1992) *Nature 358*, 560–563.

Pan, T., Gutell, R. R., & Uhlenbeck, O. C. (1991) *Science 254*, 1361–1364.

Pearson, R. G. (1966) *Science 151*, 172–177.

Pecoraro, V. L., Hermens, J. D., & Cleland, W. W. (1984) *Biochemistry 23*, 5262–5271.

Perreault, J.-P., Labuda, D., Usman, N., Yang, J.-H., & Cedergren, R. (1991) *Biochemistry 30*, 4020–4025.

Pett, V. B., Sorof, J. M., Fenderson, M. B., & Zeff, L. A. (1985) *Bioorg. Chem. 13*, 24–33.

Phillips, R. (1966) *Chem. Rev. 66*, 501–528.

Piccirilli, J. A., Vyle, J. S., Caruthers, M., & Cech, T. R. (1993) *Nature 361*, 85–88.

Pieken, W. A., Olsen, D. B., Benseler, F., Aurup, H., & Eckstein, F. (1991) *Science 253*, 314–317.

Pope, L. E., & Sigman, D. S. (1984) *Proc. Natl. Acad. Sci. U.S.A. 81*, 3–7.

Pörschke, D. (1979) *Nucleic Acids Res. 6*, 833–898.

Pörschke, D. (1986) in *Structure & Dynamics of RNA* (NATO Advanced Research Workshop, Series A: Life Science, Vol. 110) (van Knippenberg, P. H., & Hilbers, C. W., Eds.), pp. 77–85, Plenum Press, New York.

Prakash, A. S., Denny, W. A., Gourdi, T. A., Valu, K. K., Woodgate, P. D., & Wakelin, L. P. G. (1990) *Biochemistry 29*, 9799–9807.

Prasad, Y., Smith, J. B., Gottlieb, P. A., Bentz, J., & Dinter-Gottlieb, G. (1992) *Antisense Res. Dev. 2*, 267–277.

Pratviel, G., Duarte, V., Bernadou, J., & Meunier, B. (1993) *J. Am. Chem. Soc. 115*, 7939–7943.

Pyle, A. M. (1993) *Science 261*, 709–714.

Ramirez, F., & Marecek, J. F. (1980) *Pure Appl. Chem. 52*, 2213–2227.

Rao, K. S. J., Rao, B. S., Vishnuvardhan, D., & Prasad, K. V. S. (1993) *Biochim. Biophys. Acta 1172*, 17–20.

Rawji, G., Hediger, M., & Milburn, R. M. (1983) *Inorg. Chim. Acta 79*, 247–248.

Rawlings, J., Speckhard, D. C., & Cleland, W. W. (1993) *Biochemistry 32*, 11204–11210.

Reedijk, J. (1991) in *Platinum & Other Metal Coordination Compounds in Cancer Chemotherapy* (Howell, S. B., Ed.), pp. 13–23, Plenum Press, New York.

Reeves, R. H., Cantor, C., & Chambers, R. W. (1970) *Biochemistry 9*, 3993–4002.

Reich, C., Olsen, G. J., Pace, B., & Pace, N. R. (1988) *Science 239*, 178–181.

Reid, S. S., & Cowan, J. A. (1990) *Biochemistry 29*, 6025–6032.

Rialdi, G., Levy, J., & Biltonen, R. (1972) *Biochemistry 11*, 2472–2479.

Rigler, R., & Wintermeyer, W. (1983) *Annu. Rev. Biophys. Bioeng. 12*, 475–505.

Rordorf, B. F., & Kearns, D. R. (1976) *Biopolymers 15*, 1491–1504.

Rosenstein, S. P., & Been, M. D. (1990) *Biochemistry 29*, 8011–8016.

Rosental, L. S., & Nelson, D. J. (1986) *Inorg. Chim. Acta 125*, 89–95.

Ruffner, D. E., & Uhlenbeck, O. C. (1990) *Nucleic Acids Res. 18*, 6025–6029.

Ruffner, D. E., Dahm, S. C., & Uhlenbeck, O. C. (1989) *Gene 82*, 31–41.

Saenger, W. (1984) *Principles of Nucleic Acid Research*, pp. 201–219, Springer-Verlag, New York.

Samuni, A., Aranovitch, J., Godinger, D., Chevion, M., & Czapski, G. (1983) *Eur. J. Biochem. 137*, 119–124.

Sander, C. & Ts'o, P. O. P. (1971) *J. Mol. Biol. 55*, 1–21.

Sanger, H. L., Ramm, K., Domdey, H., Gross, H. J., Henco, K., & Riesner, D. (1979) *FEBS Lett. 99*, 117–122.

SantaLucia, J., Jr., Kierzek, R., & Turner, D. H. (1992) *Science 256*, 217–219.

Sarma, R. H., & Dhingra, M. M. (1981) in *Topics in Nucleic Acid Structure* (Neidle, S., Ed.), pp. 33–63, Wiley, New York.

Sawai, H., Shibusawa, T., & Kuroda, K. (1990) *Bull. Chem. Soc. Jpn. 63*, 1776–1780.

Sawata, S., Shimayama, T., Komiyama, M., Kumar, P. K. R., Nishikawa, S., & Taira, K. (1993) *Nucleic Acids Res. 21*, 5656–5660.

Scheller, K. H., & Sigel, H. (1983) *J. Am. Chem. Soc. 105*, 5891–5900.

Schimmel, R. R., & Redfield, A. G. (1980) *Annu. Rev. Biophys. Bioeng. 9*, 181–221.

Schnaith, L. M. T., Hanson, R. S., & Que, Jr., L. (1994) *Proc. Natl. Acad. Sci. U.S.A. 91*, 569–573.

Schoenknecht, T., & Diebler, H. (1993) *J. Inorg. Biochem. 50*, 283–298.

Sherman, S. E., & Lippard, S. J. (1987) *Chem. Rev. 87*, 1153–1181.

Shimazu, M., Shinozuka, K., & Sawai, H. (1993) *Angew. Chem. Int. Ed. Engl. 32*, 870–872.

Shin, Y. A. (1973) *Biopolymers 12*, 2459–2475.

Shin, Y. A., Heim, J. M., & Eichhorn, G. L. (1972) *Bioinorg. Chem. 1*, 149–163.

Sigel, H. (1969) *Angew. Chem. Int. Ed. Engl. 8*, 167–177.

Sigel, H. (1989) in *Metal–DNA Chemistry, ACS Symposium Series*, Vol. 402 (Tullius, T. D., Ed.), pp. 159–204, American Chemical Society, Washington, D.C.

Sigel, H. (1993) *Chem. Soc. Rev. 22*, 255–267.

Sigel, H., Hofstetter, F., Martin, R. B., Milburn, R. M., Scheller-Krattinger, V. & Scheller, K. H. (1984) *J. Am. Chem. Soc. 106*, 7935–7946.

Sigel, H., Tribolet, R., Malini-Balakrishanan, R., & Martin, R. B. (1987) *Inorg. Chem. 26*, 2149–2157.

Sirover, M. A., & Loeb, L. A. (1976) *Science 194*, 1434–1436.

Sissoëff, I., Grisvard, J., & Guillé, E. (1976) *Prog. Biophys. Mol. Biol. 31*, 165–199.

Sleeper, H. L., Lohrmann, R., & Orgel, L. E. (1979) *J. Mol. Evol. 13*, 203–214.

Sletten, E., & Lie, B. (1976) *Acta Crystallogr. Sect. B 32*, 3301–3302.

Slim, G., & Gait, M. J. (1991) *Nucleic Acids Res. 19*, 1183–1188.

Slim, G., & Gait, M. J. (1992) *Biochem. Biophys. Res. Commun. 32*, 605–609.

Smith, D., & Pace, N. R. (1993) *Biochemistry 32*, 5273–5281.

Smith, D., Burgin, A. B., Haas, E. S., & Pace, N. R. (1992a) *J. Biol. Chem. 267*, 2429–2436.

Smith, J. B., Gottlieb, P. A., & Dinter-Gottlieb, G. (1992b) *Biochemistry 31*, 9629–9635.

Smith, R. M., Martell, A. E., & Chen, Y. (1991) *Pure Appl. Chem. 63*, 1015–1080.

Sorokin, V. A., Blagoy, Y.P., Silina, L. K., Dalyan, E. B., Babayan, Y. S., & Aslanyan, V. M. (1982) *Mol. Biol. (Russ.) 16*, 1223–1232.

Stein, A., & Crothers, D. M. (1976) *Biochemistry 15*, 160–168.

Stern, M. K., Bashkin, J. K., & Sall, E. D. (1990) *J. Am. Chem. Soc. 112*, 5357–5359.

Steitz, T. A., & Steitz, J. A. (1993) *Proc. Natl. Acad. Sci. U.S.A. 90*, 6498–6502.

Streicher, B., von Ahsen, U., & Schroeder, R. (1993) *Nucleic Acids Res. 21*, 311–317.

Sugimoto, N., Kierzek, R., & Turner, D. H. (1988) *Biochemistry 27*, 6384–6392.

Sullivan, K. M., & Lilley, D. M. (1987) *J. Mol. Biol. 193*, 397–404.

Sumaoka, J., Yashiro, M., & Komiyama, M. (1992) *J. Chem. Soc., Chem. Commun.*, 1707–1708.

Sundaralingam, M., Rubin, J. R., & Cannon, J. F. (1984) *Int. J. Quantum Chem. 11*, 355–366.

Surratt, C. K., Carter, B. J., Payne, R. C., & Hecht, S. M. (1990) *J. Biol. Chem. 265*, 22513–22519.

Swaminathan, V., & Sundaralingam, M. (1979) *Crit. Rev. Biochem. 6*, 245–336.

Symons, R. H. (1992) *Annu. Rev. Biochem. 61*, 641–671.

Tafesse, F., Massoud, S. S., & Milburn, R. M. (1985) *Inorg. Chem. 24*, 2593–2594.

Tamm, C., Shapiro, H. S., & Chargaff, E. (1952) *J. Biol. Chem. 199*, 313–317.

Teeter, M. M., Quigley, G. J., & Rich, A. (1980) in *Metal Ions in Biology*, Vol. 1 (Spiro, T. G., Ed.), pp. 147–177, Wiley, New York.

Thomas, J. C., Schurr, J. M., Reid, B. R., Riberto, N. S., & Hare, D. R. (1984) *Biochemistry 23*, 5414–5420.

Thomas, T. J., & Bloomfield, V. A. (1988) *Biopolymers 22*, 1097–1106.

Thomas, T. J., & Messner, R. P. (1983) *Biochimie 70*, 221–226

Tinoco, I., Jr., Davis, P. W., Hardin, C. C., Puglisi, J. D., Walker, G. T., & Wyatt, J. (1987) *Cold Spring Harbor Symp. Quant. Biol. LII*, 135–146.

Toh, H., Imamura, A., & Kanda, K. (1987) *FEBS Lett. 219*, 279–282.

Ts'o, P. O. P. (1974) in *Basic Principles in Nucleic Acid Chemistry* (Ts'o, P. O. P., Ed.) pp. 453–584, Academic Press, New York.

Tu, A. T., & Heller, M. J. (1974) in *Metal Ions in Biological Systems*, Vol. 1 (Sigel, H., Ed.), pp. 1–49, Marcel Dekker, New York.

Tullius, T. D. (1989) in *Metal–Nucleotide Interactions*, ACS Symposium Series, Vol. 402, (Tullius, T. D., Ed.), pp. ix–x, 1–23, American Chemical Society, Washington, D.C.

Tuschl, T., Ng, M. M. P., Pieken, W., Benseler, F., & Eckstein, F. (1993) *Biochemistry 32*, 11658–11668.

Uchimaru, T., Uebaysi, M., Tanabe, K., & Taira, K. (1993) *FASEB J. 7*, 137–142.

Uhlenbeck, O. C. (1987) *Nature 328*, 596–600.

Usher, D. A., & McHale, A. H. (1976) *Proc. Natl. Acad. Sci. U.S.A. 73*, 1149–1153.

Usman, N., & Cedergren, R. (1992) *TIBS 17*, 334–339.

Utyanskaya, E. Z., Pavlovsky, A. G., Sosfenov, N. I., & Shilov, A. E. (1989) *Kinetics Catalysis (Russ.) 30*, 1343–1351.

Vallee, B. L., & Auld, D. S. (1993) *Proc. Natl. Acad. Sci. U.S.A. 90*, 2715–2718.

Van Atta, R. B., & Hecht, S. M. (1994) *Adv. Inorg. Biochem. 9*, 1–40.

Vary, C. P. H., & Vournakis, J. N. (1984a) *Nucleic Acids Res. 12*, 6783–6778.

Vary, C. P. H., & Vournakis, J. N. (1984b) *Proc. Natl. Acad. Sci. U.S.A. 81*, 6978–6982.

Vijay-Kumar, S., Sacore, T. D., & Sobell, H. M. (1984) *Nucleic Acids Res. 12*, 3649–3657.

Vincent, J. B., Crowder, M. W., & Averill, B. A. (1992) *TIBS 17*, 105–110.

Visscher, J. & Schwartz, A. W. (1992) *Nucleic Acids Res. 20*, 5749–5742.

Volbeda, A., Lahm, A., Sakiyama, F., & Suck, D. (1991) *EMBO J. 10*, 1607–1618.

Wacker, W. E. C., & Vallee, B. L. (1959) *J. Biol. Chem. 234*, 3257–3262.

Wacker, W. E. C., & Gordon, M. P., & Huff, J. W. (1963) *Biochemistry 2*, 716–719.

Walstrum, S. A., & Uhlenbeck, O. C. (1990) *Biochemistry 29*, 10573–10576.

Wang, J.-F., Downs, W. D., & Cech, T. R. (1993) *Science 256*, 504–508.

Wang, Y., & Van Ness, B. (1989) *Nucleic Acids Res. 17*, 6915–6926.

Werner, C., Krebs, B., Keith, G., & Dirheimer, G. (1976) *Biochim. Biophys. Acta 432*, 161–175.

Weser, J. U. (1968) *Structure Bonding 5*, 41–67.

Wherland, S., Deutsch, E., Eliason, J., & Sigler, P. B. (1973) *Biochem. Biophys. Res. Commun. 54*, 662–668.

Widom, J., & Baldwin, R. L. (1983) *Biopolymers 22*, 1595–1620.

Williams, D. M., Pieken, W., & Eckstein, F. (1992) *Proc. Natl. Acad. Sci. U.S.A. 89*, 918–921.

Williams, D. R. (1971) *The Metals of Life. The Solution Chemistry of Metal Ions in Biological Systems*, Van Nostrand Reinhold, London.

Williams, M. N., & Crothers, D. M. (1975) *Biochemistry 14*, 1944–1951.

Wintermeyer, W., & Zachau, H. G. (1973) *Biochim. Biophys. Acta 299*, 82–90.

Woodson, S. A., & Cech, T. R. (1991) *Biochemistry 30*, 2042–2050.

Wu, H.-N., & Lai, M. M. (1990) *Mol. Cell. Biol. 10*, 5575–5579.

Xu, M.-Q., Kathe, S. D., Goodrich-Blair, H., Nierzwicki-Bauer, S. A., & Shub, D. A. (1990) *Science 250*, 1566–1570.

Xu, Q., Deng, H., & Braunlin, W. H. (1993) *Biochemistry 32*, 13130–13137.

Yamada, A., Akasaka, K., & Hatano, H. (1976) *Biopolymers 15*, 1315–1331.

Yanagawa, H., Ogawa, Y., & Ueno, M. (1992) *J. Biol. Chem. 267*, 13320–13326.

Yang, J., Perreault, J.-P., Labuda, D., Usman, N. & Cedergren, R.J. (1990) *Biochemistry 29*, 11156–11160.

Yarus, M. (1993) *FASEB J. 7*, 31–39.

Yurgaitis, A. P., & Lazurkin, Y. S. (1981) *Biopolymers 20*, 967–975.

Zaug, A. J., & Cech, T. R. (1986) *Biochemistry 25*, 4478–4482.

Zaug, A. J., Kent, J. R., & Cech , T. R. (1985) *Biochemistry 24*, 6211–6218.

Zhou, X., Deng, Z., Firmin, J. L., Hopwood, D. A., & Kieser, T. (1988) *Nucleic Acids Res. 16*, 4341–4352.

Zou, Y., Van Houten, B., & Farrell, N. (1993) *Biochemistry 32*, 9632–9638.

Chapter 10

Alberts, B. M. (1984) *Cold Spring Harbor Symp. Quant. Biol. 49*, 1–12.

Barkley, M. D., Cheatham, S., Thurston, D. E., & Hurley, L. H. (1986) *Biochemistry 25*, 3021–3031.

Bhuyan, B. K., Newell, K. A., Crampton, S. L., & Von Hoff, D. D. (1982) *Cancer Res. 42*, 3532–3537.

Blum, R. H., & Carter, S. K. (1974) *Ann. Intern. Med. 80*, 249–259.

Brill, G. M., McAlpine, J. B., Whittern, D. N., & Buko, A. M. (1990) *J. Antibiot. 43*, 229–237.

Brown, D. G., Sanderson, M. R., Skelly, J. V., Jenkins, T. C., Brown, T., Garman, E., Stuart, D. I., & Neidle, S. (1990) *EMBO J. 9*, 1329–1334.

Coll, M., Aymami, J., van der Marel, G. A., van Boom, J. H., Rich, A., & Wang, A. H.-J. (1989) *Biochemistry 28*, 310–320.

De Clercq, E., & Dann, O. (1980) *J. Med. Chem. 23*, 787–795.

DeKoning, T. F., Postmus, R. J., Wallace, T. L., Kelly, R. G., & Lee, L. H. (1990) *Proc. Am. Assoc. Cancer Res. 31*, 348.

De Voss, J. J., Townsend, C. A., Ding, W.-D., Morton, G. O., Ellestad, G. A., Zein, N., Tabor, A. B., & Schreiber, S. L. (1990) *J. Am. Chem. Soc. 112*, 9669–9670.

Denny, W. A. (1989) *Anti-Cancer Drug Design 4*, 241–263.

Ding, W.-D., & Ellestad, G. A. (1991) *J. Am. Chem. Soc. 113*, 6617–6620.

Feigon, J., Denny, W. A., Leupin, W., & Kearns, D. R. (1984) *J. Med. Chem. 27*, 450–465.

Gao, X., & Patel, D. J. (1988) *Biochemistry 27*, 1744–1751.

Gilbert, D. E., van der Marel, G. A., van Boom, J. H., & Feigon, J. (1989) *Proc. Natl. Acad. Sci. U.S.A. 86*, 3006–3010.

Golik, J. C., Clardy, J., Dubay, G., Groenewold, G., Kawaguchi, H., Konishi, M., Krishnan, B., Ohkuma, H., Saitoh, K.-I., & Doyle, T. W. (1987a) *J. Am. Chem. Soc. 109*, 3461–3462.

Golik, J., Dubay, G., Groenewold, G., Kawaguchi, H., Konishi, M., Krishnan, B., Ohkuma, H., Saitoh, K.-I., & Doyle, T. W. (1987b) *J. Am. Chem. Soc. 109*, 3462–3464.

Graves, D. E., Pattaroni, C., Krishnan, B. S., Ostrander, J. M., Hurley, L. H., & Krugh, T. R. (1984) *J. Biol. Chem. 259*, 8202–8209.

Hanka, L. J., Dietz, A., Gerpheide, S. A., Kuentzel, S. L., & Martin, D. G. (1978) *J. Antibiot. 31*, 1211–1217.

Hansen, M., & Hurley, L. H. (1995) *J. Am. Chem. Soc. 117*, 2421–2429.

Hansen, M., Yung, S., & Hurley, L. H. (1995) *Chem. Biol. 2*, 229–240.

Hawley, R. C., Kiessling, L. L., & Schreiber, S. L. (1989) *Proc. Natl. Acad. Sci. U.S.A. 86*, 1105–1109.

Hertzberg, R. P., Hecht, S. M., Reynolds, V. L., Molineux, I. J., & Hurley, L. H. (1986) *Biochemistry 25*, 1249–1258.

Hurley, L. H. (1989) *J. Med. Chem. 32*, 2027–2033.

Hurley, L. H., Ed. (1992) *Advances in DNA Sequence Specific Agents*; JAI Press, Greenwich, CT.

Hurley, L. H., Reynolds, V. L., Swenson, D. H., Petzold, G. L., & Scahill, T. (1984) *Science* *226*, 843–844.

Hurley, L. H., Reck, T., Thurston, D. E., & Langley, D. R. (1988) *Chem. Res. Toxicol. 1*, 258–268.

Hurley, L. H., Warpehoski, M. A., Lee, C.-S., McGovren, J. P., Scahill, T. A., Kelly, R. C., Mitchell, M. A., Wicnienski, N. A., Gebhard, I., Johnson, P. D., & Bradford, V. S. (1990) *J. Am. Chem. Soc. 112*, 4633–4649.

Husain, I., Van Houten, B., Abdel-Monem, M., Thomas, D. C., & Sancar, A. (1985) *Proc. Natl. Acad. Sci. U.S.A. 82*, 6774–6778.

Islam, S. A., Neidle, S., Gandecha, B. M., Partridge, M., Patterson, L. H., & Brown, J. R. (1985) *J. Med. Chem. 28*, 857–864.

Jackson, M., Karwowski, J. P., Theriault, R. J., Hardy, D. J., Swanson, S. J., Barlow, G. J., Tillis, P. M., & McAlpine, J. B. (1990) *J. Antibiot. 43*, 223–228.

Jacobson, M. K., Twehous, D., & Hurley, L. H. (1986) *Biochemistry 25*, 5929–5932.

Kizu, R., Draves, P. H., & Hurley, L. H. (1993) *Biochemistry 32*, 8712–8722.

Kohn, K. W. (1975) in *Antibiotics III. Mechanism of Action of Antimicrobial and Antitumor Agents*, Vol. 3 (Corcoran, J. W., & Hahn, F. E., Eds.), pp. 3–11, Springer-Verlag, New York.

Kohn, K. W., Glaubiger, D., & Spears, C. L. (1974) *Biochim. Biophys. Acta 361*, 288–302.

Koo, H.-S., & Crothers, D. M. (1988) *Proc. Natl. Acad. Sci. U.S.A. 85*, 1763–1767.

Koo, H.-S., Wu, H.-M., & Crothers, D. M. (1986) *Nature 320*, 501–506.

Kopka, M. L., Yoon, C., Goodsell, D., Pjura, P., & Dickerson, R. E. (1985a) *J. Mol. Biol. 183*, 553–563.

Kopka, M. L., Yoon, C., Goodsell, D., Pjura, P., & Dickerson, R. E. (1985b) *Proc. Natl. Acad. Sci. U.S.A. 82*, 1376–1380.

Lambert, B., Jones, B. K., Roques, B. P., & Le Pecq, J.-B. (1989) *Proc. Natl. Acad. Sci. U.S.A. 86*, 6557–6561.

Lane, A. N., Jenkins, T. C., Brown, T., & Neidle, S. (1991) *Biochemistry 30*, 1372–1385.

Lee, C.-S., Sun, D., Kizu, R., & Hurley, L. H. (1991a) *Chem. Res. Toxicol. 4*, 203–213.

Lee, E. H., Kornberg, A., Hidaka, M., Kobayashi, T., & Horiuchi, T. (1989) *Proc. Natl. Acad. Sci. U.S.A. 86*, 9104–9108.

Lee, M. D., Dunne, T. S., Siegel, M. M., Chang, C. C., Morton, G. O., & Borders, D. B. (1987a) *J. Am. Chem. Soc. 109*, 3464–3466.

Lee, M. D., Dunne, T. S., Chang, C. C., Ellestad, G. A., Siegel, M. M., Morton, G. O., McGahren, W. J., & Borders, D. B. (1987b) *J. Am. Chem. Soc. 109*, 3466–3468.

Lee, M. D., Ellestad, G. A., & Borders, D. B. (1991b) *Acc. Chem. Res. 24*, 235–243.

Lehman, I. R. (1974) *Science 186*, 790–797.

Li, L. H., Swenson, D. H., Schpok, S. L., Kuentzel, S. L., Dayton, B. D., & Krueger, W. C. (1982) *Cancer Res. 42*, 999–1004.

Li, L. H., Kelly, R. C., Warpehoski, M. A., McGovren, J. P., Gebhard, I., & DeKoning, T. F. (1990) *Invest. New Drugs 9*, 137–148.

Li, L. H., DeKoning, T. F., Kelly, R. C., Krueger, W. C., McGovren, J. P., Padbury, G. E., Petzold, G. L., Wallace, T. L., Ouding, R. J., Prairie, M. D., & Gebhard, I. (1992) *Cancer Res. 52*, 4904–4913.

Lin, C. H., & Hurley, L. H. (1990) *Biochemistry 29*, 9503–9507.

Lin, C. H., Sun, D., & Hurley, L. H. (1991) *Chem. Res. Toxicol. 4*, 21–26.

Liu, L. F. (1989) *Annu. Rev. Biochem. 58*, 351–358.

Low, C. M., Drew, H. R., & Waring, M. J. (1984) *Nucleic Acids Res. 12*, 4865–4879.

Maine, I. P., Sun, D., Hurley, L. H., & Kodadek, T. (1992) *Biochemistry 31*, 3968–3975.

Matson, S. W. (1986) *J. Biol. Chem. 261*, 10169–10175.

McAllister, C. F., & Achberger, E. C. (1989) *J. Biol. Chem. 264*, 10451–10456.

Nelson, E. M., Tewey, K. M., & Liu, L. F. (1984) *Proc. Natl. Acad. Sci. U.S.A. 81*, 1361–1365.

Nicolaou, K. C., & Dai, W.-M. (1991) *Angew. Chem. 30*, 1387–1416.

Patel, D. J. (1974) *Biochemistry 13*, 2396–2402.

Patel, D. J., Kazlowski, S. A., Nordheim, A., & Rich, A. (1982) *Proc. Natl. Acad. Sci. U.S.A.* *79*, 1413–1417.

Pearl, L. H., Skelly, J. V., Hudson, B. D., & Neidle, S. (1987) *Nucleic Acids Res. 15*, 3469–3478.

Pelton, J. G., & Wemmer, D. E. (1989) *Proc. Natl. Acad. Sci. U.S.A. 86*, 5723–5727.

Petersheim, M., Mehdi, S., & Gerlt, J. A. (1984) *J. Am. Chem. Soc. 106*, 439–440.

Petrusek, R. L., Anderson, G. L., Garner, T. F., Quiton, F. L., Fannin, Q. L., Kaplan, D. J., Zimmer, S. G., & Hurley, L. H. (1981) *Biochemistry 20*, 1111–1119.

Pjura, P. E., Grzeskowiak, K., & Dickerson, R. E. (1987) *J. Mol. Biol. 197*, 257–271.

Pratt, W. B., Ruddon, R. W., Ensminger, W. D., & Maybaum, J. (1994) *The Anticancer Drugs*, Oxford University Press, New York.

Propst, C. L., & Perun, T. J., Eds. (1992) *Nucleic Acid Targeted Drug Design*, Marcel Dekker, New York.

Pullman, B., & Jortner, J., Eds. (1990) *Molecular Basis of Specificity in Nucleic Acid-Drug Interactions*, Kluwer Academic Publishers, Dordrecht, The Netherlands.

Quintana, J. R., Lipanov, A. A., & Dickerson, R. E. (1991) *Biochemistry 30*, 10294–10306.

Reynolds, V. L., Molineux, I. J., Kaplan, D., Swenson, D. H., & Hurley, L. H. (1985) *Biochemistry 24*, 6228–6237.

Reynolds, V. L., McGovren, J. P., & Hurley, L. H. (1986) *J. Antibiot. 39*, 319–334.

Sancar, A., & Rupp, W. D. (1983) *Cell 33*, 249–260.

Scahill, T. A., Jensen, R. M., Swenson, D. H., Hatzenbuhler, N. T., Petzold, G., Wierenga, W., & Brahme, N. D. (1990) *Biochemistry 29*, 2852–2860.

Sequin, U. (1986) *Fortschr. Chem. Org. Naturst. 50*, 57–122.

Shen, L. L., Mitscher, L. A., Sharma, P. N., O'Donnell, T. J., Chu, D. W. T., Cooper, C. S., Rosen, T., & Pernet, A. G. (1989) *Biochemistry 28*, 3886–3894.

Shore, D., Langowski, J., & Baldwin, R. (1981) *Proc. Natl. Acad. Sci. U.S.A. 78*, 4833–4837.

Stubbe, J., & Kozarich, J. W. (1987) *Chem. Rev. 87*, 1107–1136.

Sun, D., & Hurley, L. H. (1992a) *Biochemistry 31*, 2822–2829.

Sun, D., & Hurley, L. H. (1992b) *Anti-Cancer Drug Design 7*, 15–36.

Sun, D., & Hurley, L. H. (1992c) *J. Med. Chem. 35*, 1773–1782.

Sun, D., Hansen, M., Clement, J. J., & Hurley, L. H. (1993) *Biochemistry 32*, 8068–8074.

Sun, D., & Hurley, L. H. (1993) *J. Am. Chem. Soc. 115*, 5925–5933.

Sun, D., & Hurley, L. H. (1994a) *Gene 149*, 165–172.

Sun, D., & Hurley, L. H. (1994b) *Biochemistry 33*, 9578–9587.

Sun, D., & Hurley, L. H. (1995a) *J. Am. Chem. Soc. 117*, 2430–2440.

Sun, D., & Hurley, L. H. (1995b) *Chem. Biol. 2*, 457–469.

Tanaka, T., Hirama, M., Ueno, M., Imajo, S., Ishiguro, M., Mizugaki, M., Edo, K., & Komatsu, H. (1991) *Tetrahedron Lett. 32*, 3175–3178.

Teng, M.-K., Usman, N., Frederick, C. A., & Wang, A. H.-J. (1988) *Nucleic Acids Res. 16*, 2671–2690.

Thompson, A. S., & Hurley, L. H. (1995) *J. Mol. Biol. 252*, 86–101.

Thompson, A. S., Fan, J.-Y., Sun, D., Hansen, M., & Hurley, L. H. (1995) *Biochemistry 34*, 11005–11016.

Van Dyke, M. M., & Dervan, P. B. (1984) *Science 225*, 1122–1127.

Walker, S., Valentine, K. G., & Kahne, D. J. (1990) *J. Am. Chem. Soc. 112*, 6428–6429.

Walker, S., Yang, D., Kahne, D., & Gange, D. J. (1991) *J. Am. Chem. Soc. 113*, 4716–4717.

Wang, A. H.-J., Ughetto, G., Quigley, G. J., Hakoshima, T., van der Marel, G. A., van Boom, J. H., & Rich, A. (1984) *Science 225*, 1115–1121.

Warpehoski, M. A., & Bradford, V. S. (1988) *Tetrahedron Lett. 29*, 131–134.

Warpehoski, M. A., & Hurley, L. H. (1988) *Chem. Res. Toxicol. 1*, 315–333.

Yang, D., Kim, S. H., & Kahne, D. J. (1991) *J. Am. Chem. Soc. 113*, 4715–4716.

Yoshida, M., Banville, D. L., & Shafer, R. H. (1990) *Biochemistry 29*, 6585–6592.

Zakrzewska, K., & Pullman, B. J. (1986) *J. Biomol. Struct. Dynamics 4*, 127–136.

Zein, N., Sinha, A. M., McGahren, W. J., & Ellestad, G. A. (1988) *Science 240*, 1198–1201.

Zein, N., McGahren, W. J., Morton, G. O., Ashcroft, J., & Ellestad, G. A. (1989a) *J. Am. Chem. Soc. 111*, 6888–6890.

Zein, N., Poncin, M., Nilakantan, R., & Ellestad, G. A. (1989b) *Science 244*, 697–699.

Zein, N., Colson, K. L., Leet, J. E., Schroeder, D. R., Solomon, W., Doyle, T. W., & Casazza, A. M. (1993) *Proc. Natl. Acad. Sci. U.S.A. 90*, 2822–2826.

Chapter II

Aggarwal, A. K., Rodgers, D. W., Drottar, M., Ptashne, M., & Harrison, S. C. (1988) *Science 242*, 899–907.

Aiken, C. R., McLaughlin, L. W., & Gumport, R.I. (1991) *J. Biol. Chem. 266*, 19070–19078.

Allen, T. D., Wick, K. L., & Matthews, K. S. (1991) *J. Biol. Chem. 266*, 6113–6119.

Amouyal, M., & Buc, H. (1987) *J. Mol. Biol. 195*, 795–808.

Anderson, J. E., Ptashne, M., & Harrison, S. C. (1987) *Nature 326*, 846–852.

Anthony-Cahill, S. J., Benfield, P. A., Fairman, R., Wasserman, Z. R., Brenner, S. L., Stafford, III, W. F., Altenbach, C., Hubbell, W. L., & DeGrado, W. F. (1992) *Science 255*, 979–983.

Baer, M. E., & Sancar, G. B. (1993) *J. Biol. Chem. 268*, 16717–16724.

Bateman, E., & Paule, M. R. (1988) *Mol. Cell. Biol. 8*, 1940–1946.

Bayley, C. R., & Jones, A. S. (1959) *Trans. Faraday. Soc. 55*, 492.

Beamer, L. J., & Pabo, C. O. (1992) *J. Mol. Biol. 227*, 177–196.

Beer, M., Stern, S., Carmalt, D., & Hohlhenrich, K. H. (1966) *Biochemistry 5*, 2283–2288.

Blackwell, T. K., & Weintraub, H. (1990) *Science 250*, 1104–1110.

Blackwell, T. K., Kretzner, L., Blackwood, E. M., Eisenman, R. N., & Weintraub, H. (1990) *Science 250*, 1149–1151.

Blatter, E. E., Ebright, Y. W., & Ebright, R. H. (1992) *Nature 359*, 650–652.

Brennan, R. G., Roderick, S. L., Takeda, Y., & Matthews, B. W. (1990) *Proc. Natl. Acad. Sci. U.S.A. 87*, 8165–8169.

Brenowitz, M., Senear, D. F., Shea, M. A., & Ackers, G. K. (1986) *Proc. Natl. Acad. Sci. U.S.A. 83*, 8462–8466.

Brunelle, A., & Schleif, R. F. (1987) *Proc. Natl. Acad. Sci. U.S.A. 84*, 6673–6676.

Bushman, F. D., Anderson, J. E., Harrison, S. C., & Ptashne, M. (1985) *Nature 316*, 651–653.

Carey, J. (1989) *J. Biol. Chem. 264*, 1941–1945.

Chalepakis, G., Postma, J. P., & Beato, M. (1988) *Nucleic Acids Res. 16*, 10237–10247.

Chen, L., Oakley, M. G., Glover, J. N. M., Jain, J., Dervan, P. B., Hogan, P., Rao, A., & Verdine, G. L. (1995) *Current Biol. 5*, 882–889.

Chodosh, L. A., Carthew, R. W., & Sharp, P. A. (1986) *Mol. Cell. Biol. 6*, 4723–4733.

Churchill, M. E. A., & Travers, A. A. (1991) *Trends Biochem. Sci. 16*, 92–97.

Clark, K. L., Halay, E. D., Lai, E. S., & Burley, S. K. (1993) *Nature 364*, 412–420.

Clark, L., Nicholson, J., & Hay, R. T. (1989) *J. Mol. Biol. 206*, 615–626.

Dill, K. A. (1990) *Biochemistry 29*, 7133–7155.

Dunitz, J. D. (1994) *Science 264*, 670.

Dynan, W.S. (1989) *Cell 58*, 1–4.

Ebright, R. H. (1991) *Methods Enzymol. 208*, 620–640.

Ellenberger, T. E., Brandl, C. J., Struhl, K., & Harrison, S. C. (1992) *Cell 71*, 1223–1237.

Ellenberger, T., Fass, D., Arnaud, M., & Harrison, S. C. (1994) *Genes Dev. 8*, 970–980.

Ezaz-Nikpay, K., & Verdine, G. L. (1994) *Chem. Biol. 1*, 235–239.

Fairall, L., Schwabe, J. W. R., Chapman, L., Finch, J. T., & Rhodes, D. (1993) *Nature 366*, 483–487.

Feng, J.-A., Johnson, R. C., & Dickerson, R. E. (1994) *Science 263*, 348–355.

Ferré-D'Amaré, A. A., Prendergast, G., Ziff, E., & Burley, S. B. (1993) *Nature 363*, 38–45.

Furlong, J. C., & Lilley, D. M. J. (1986) *Nucleic Acids Res. 14*, 3995–4007.

Furlong, J. C., Sullivan, K. M., Murchie, A. I. H., Gough, G. W., & Lilley, D. M. J. (1989) *Biochemistry 28*, 2009–2017.

Galas, D. J., & Schmitz, A. (1978) *Nucleic Acids Res. 5,* 3157–3170.

Ghosh, G., Van Duyne, G., Ghosh, S., & Sigler, P. B. (1995) *Nature 373,* 303–310.

Gilbert, W., & Müller-Hill, B. (1966) *Proc. Natl. Acad. Sci. U.S.A. 56,* 1891–1898.

Gronostajski, R. M., Adhya, S., Nagata, K., Guggenheimer, R. A., & Hurwitz, J. (1985) *Mol. Cell. Biol. 5,* 964–971.

Ha, J.-H., Spolar, R. S., & Record, M. T., Jr. (1989) *J. Mol. Biol. 209,* 801–816.

Härd, T., Dahlman, K., Carlstedt-Duke, J., Gustafsson, J.-Å., & Rigler, R. (1990) *Biochemistry 29,* 5358–5364.

Harrison, S. C. (1991) *Nature 353,* 715–719.

Harshman, K. D., & Dervan, P. B. (1985) *Nucleic Acids Res. 13,* 4825–4835.

Hatfull, G. F., Noble, S. M., & Grindley, N. D. F. (1987) *Cell 49,* 103–110.

Hayashibara, K. C., & Verdine, G. L. (1991) *J. Am. Chem. Soc. 113,* 5104–5106.

Hayashibara, K. C. (1993) Ph. D. Thesis, Harvard University.

Hayashibara, K. C., & Verdine, G. L. (1992) *Biochemistry 31,* 11265–11273.

Hayes, J. J., & Tullius, T. D. (1989) *Biochemistry 28,* 9521–9527.

Hegde, R. S., Grossman, S. R., Laimins, L. A., & Sigler, P. B. (1992) *Nature 359,* 505–512.

Heitman, J., & Model, P. (1990) *EMBO J. 9,* 3369–3378.

Herr, W. (1985) *Proc. Natl. Acad. Sci. U.S.A. 82,* 8009–8013.

Heuer, C., & Hillen, W. (1988) *J. Mol. Biol. 202,* 407–415.

Hochschild, A., & Ptashne, M. (1986) *Cell 44,* 681–687.

Hogan, M. E., Roberson, M. W., & Austin, R. H. (1989) *Proc. Natl. Acad. Sci. U.S.A. 86,* 9273–9277.

Iverson, B. L., & Dervan, P. B. (1987) *Nucleic Acids Res. 15,* 7823–7830.

Jeppesen, C., & Nielsen, P. E. (1989) *Nucleic Acids Res. 17,* 4947–4956.

Joachimiak, A., Haran, T. E., & Sigler, P. B. (1994) *EMBO J. 13,* 367–372.

Johnson, A. D., Meyer, B. J., & Ptashne, M. (1979) *Proc. Natl. Acad. Sci. U.S.A. 76,* 5061–5065.

Jordan, S. R., & Pabo, C. O. (1988) *Science 242,* 893–899.

Jorgensen, W. L., & Pranata, J. (1990) *J. Am. Chem. Soc. 112,* 2008–2010.

Kalnik, M. W., Chang, C. N., Johnson, F., Grollman, A. P., & Patel, D. J. (1989) *Biochemistry 28,* 3373–3383.

Kao, J. Y., Goljier, I., Phan, T. A., & Bolton, P. H. (1993) *J. Biol. Chem. 268,* 17787–17793.

Kim, J. L., Nikolov, D. B., & Burley, S. K. (1993a) *Nature 365,* 520–527.

Kim, Y., Grable, J. C., Love, R., Green, P. J., & Rosenberg, J. M. (1990) *Science 249,* 1307–1309.

Kim, Y., Geiger, J. H., Hahn, S., & Sigler, P. B. (1993b) *Nature 365,* 512–520.

Kissinger, C. R., Liu, B., Martin-Blanco, E., Kornberg, T. B., & Pabo, C. O. (1990) *Cell 63,* 579–590.

Klemm, J. D., Rould, M. A., Aurora, R., Herr, W., & Pabo, C. O. (1994) *Cell 77,* 21–32.

Klimasauskas, S., Kumar, S., Roberts, R. J., & Cheng, X. (1994) *Cell 76,* 357–369.

Kneale, G. G. (1992) *Curr. Opin. Struct. Biol. 2,* 124–130.

Konig, P., & Richmond, T. J. (1993) *J. Mol. Biol. 233,* 139–154.

Kuwabara, M. D., & Sigman, D. S. (1987) *Biochemistry 26,* 7234–7238.

Larson, C. J., Zhang, X., & Verdine, G. L. (1995) submitted.

Lawson, C. L., & Carey, J. (1993) *Nature 366,* 178–182.

Lesser, D. R., Kurpiewski, M. R., & Jen-Jacobson, L. (1990) *Science 250,* 776–786.

Liu, F., & Green, M. R. (1990) *Cell 61,* 1217–1224.

Liu, J., Sodeoka, M., Lane, W. S., & Verdine, G. L. (1994) *Proc. Natl. Acad. Sci. U.S.A. 91,* 908–912.

Lowenthal, J. W., Ballard, D. W., Böhnlein, E., & Greene, W. C. (1989) *Proc. Natl. Acad. Sci. U.S.A. 86,* 2331–2335.

Luisi, B. F., Xu, W. X., Otwinowski, Z., Freedman, L. P., Yamamoto, K. R., & Sigler, P. B. (1991) *Nature 352,* 497–505.

Lyamichev, V. I., Mirkin, S. M., Danilevskaya, O. N., Voloshin, O. N., Balatskaya, S. V., Dobrynin, V. N., Filippov, S. A., & Frank-Kamenetskii, M. D. (1989) *Nature 339*, 634–637.

Ma, P. C. M., Rould, M. A., Weintraub, H., & Pabo, C. O. (1994) *Cell 77*, 451–459.

MacMillan, A. M., Lee, R. J., & Verdine, G. L. (1993) *J. Am. Chem. Soc. 115*, 4921–4922.

Mao, C., Carlson, N. G., & Little, J. W. (1994) *J. Mol. Biol. 235*, 532–544.

Marmorstein, R., Carey, M., Ptashne, M., & Harrison, S. C. (1992) *Nature 356*, 408–414.

Mascareñas, J. L., Hayashibara, K. C., & Verdine, G. L. (1993) *J. Am. Chem. Soc. 115*, 373–374.

Mazzarelli, J. M., Rajur, S. B., Iadarola, P. L., & McLaughlin, L. W. (1992) *Biochemistry 31*, 5925–5936.

McClarin, J. A., Frederick, C. A., Wang, B., Greene, P., Boyer, H. W., Grable, J., & Rosenberg, J. M. (1986) *Science 234*, 1526–1541.

McLaughlin, L. W., Benseler, F., Graeser, E., Piel, N., & Scholtissek, S. (1987) *Biochemistry 26*, 7238–7245.

Metzger, S., Halaas, J. L., Breslow, J. L., & Sladek, F. M. (1993) *J. Biol. Chem. 268*, 16831–16838.

Meyer, B. J., Maurer, R., & Ptashne, M. (1980) *J. Mol. Biol. 139*, 163–194.

Mitchell, P. J., & Tjian, R. (1989) *Science 245*, 371–378.

Müller, C. M., Rey, F. A., Sodeoka, M., Verdine, G. L., & Harrison, S. C. (1995) *Nature 373*, 311–317.

Nagain, K. (1992) *Curr. Opin. Struct. Biol. 2*, 131–137.

Nelms, K., & van Ness, B. (1990) *Mol. Cell. Biol. 10*, 3843–3846.

Newman, P. C., Williams, D. M., Cosstick, R., Seela, F., & Connolly, B. A. (1990) *Biochemistry 29*, 9902–9910.

Nielsen, P. E., Jeppesen, C., & Buchardt, O. (1988) *FEBS Lett. 235*, 122–124.

O'Halloran, T. V., Frantz, B., Shin, M. K., Ralston, D. M., & Wright, J. G. (1989) *Cell 56*, 119–129.

Omichinski, J. G., Clore, G. M., Schaad, O., Felsenfeld, G., Trainor, C., Appella, E., Stahl, S. J., & Gronenborn, A. M. (1993) *Science 261*, 438–446.

Otting, G., Qian, Y. Q., Billeter, M., Müller, M., Aff^lter, M., Gehring, W. J., & Wüthrich, K. (1990) *EMBO J. 9*, 3085–3092.

Otwinowski, Z., Schevitz, R. W., Zhang, R.-G., Lawson, C. L., Joachimiak, A., Marmorstein, R. Q., Luisi, B. F., & Sigler, P. B. (1988) *Nature 335*, 321–329.

Pabo, C. O., & Sauer, R. T. (1992) *Annu. Rev. Biochem. 61*, 1053–1095.

Pavletich, N. P., & Pabo, C. O. (1991) *Science 252*, 809–817.

Pavletich, N. P., & Pabo, C. O. (1993) *Science 261*, 1701–1707.

Raumann, B. E., Rould, M. A., Pabo, C. O., & Sauer, R. T. (1994) *Nature 367*, 754–757.

Record, M. T., Jr., Lohman, T. M., & deHaseth, P. L. (1976) *J. Mol. Biol. 107*, 145–158.

Richards, F. M. (1977) *Annu. Rev. Biophys. Bioeng. 6*, 151–176.

Rodgers, D. W., & Harrison, S. C. (1993) *Structure 1*, 227–240.

Russu, I. M. (1991) *Biotechnology 9*, 96–103.

Sasse-Dwight, S., & Gralla, J. D. (1988) *J. Mol. Biol. 202*, 107–119.

Sawadogo, M., & Roeder, R. G. (1985) *Cell 43*, 165–175.

Schultz, S. C., Shields, G. C., & Steitz, T. A. (1991) *Science 253*, 1001–1007.

Schwabe, J. W., Chapman, L., Finch, J. T., & Rhodes, D. (1993) *Cell 75*, 567–578.

Seela, F., & Driller, H. (1986) *Nucleic Acids Res. 14*, 2319–2332.

Seeman, N. C., Rosenberg, J. M., & Rich, A. (1976) *Proc. Natl. Acad. Sci. U.S.A. 73*, 804–808.

Shimon, L. J. W., & Harrison, S. C. (1993) *J. Mol. Biol. 232*, 826–838.

Siebenlist, U., & Gilbert, W. (1980) *Proc. Natl. Acad. Sci. U.S.A. 77*, 122–126.

Smith, S. A., Rajur, S. B., & McLaughlin, L. W. (1994) *Nature Struct. Biol. 1*, 18–22.

Somers, W. S., & Phillips, S. E. V. (1992) *Nature 359*, 387–393.

Spassky, A., & Sigman, D. S. (1985) *Biochemistry 27*, 8050–8056.

Spolar, R. S., & Record, M. T., Jr. (1994) *Science 263*, 777–784.

Staacke, D., Walter, B., Kisters-Woike, B., Wilcken-Bergmann, B. V., & Müller-Hill, B. (1990) *EMBO J. 9*, 1963–1967.

Steitz, T. A. (1990) *Q. Rev. Biophys. 23*, 205–280.

Suck, D. (1992) *Curr. Opin. Struct. Biol. 2*, 84–92.

Suck, D., & Oefner, C. (1986) *Nature 321*, 620–625.

Suck, D., Lahm, A., & Oefner, C. (1988) *Nature 332*, 464–468.

Thiesen, H.-J., & Bach, C. (1990) *Nucleic Acids Res. 18*, 3203–3209.

Tjian, R., & Maniatis, T. (1994) *Cell 77*, 5–8.

Travers, A. A. (1989) *Annu. Rev. Biochem. 58*, 427–452.

Travers, A. A. (1992) *Curr. Opin. Struct. Biol. 2*, 71–77.

Travers, A. A. (1993) *DNA–Protein Interactions*, Chapman and Hall, New York.

Tullius, T. D., & Dombrowski, B. A. (1985) *Science 230*, 679–681.

Van Dyke, M. W., & Dervan, P. B. (1983) *Nucleic Acids Res. 11*, 5555–5567.

Verdine, G. L. (1994) *Cell 76*, 197–200.

Vincze, A., Henderson, R. E. L., McDonald, J. J., & Leonard, N. J. (1973) *J. Am. Chem. Soc. 95*, 2677–2682.

Wick, K. L., & Matthews, K. S. (1991) *J. Biol. Chem. 266*, 6106–6112

Willis, M. C., Hicke, B. J., Uhlenbeck, O. C., Cech, T. R., & Koch, T. H. (1993) *Science 262*, 1255–1257.

Winkler, F., Banner, D., Oefner, C., Tsernoglou, D., Brown, R., Heathman, S., Bryan, R., Martin, P., Petratos, K., & Wilson, K. (1993) *EMBO J.* 12, 1781–1795.

Winkler, F. K. (1992) *Curr. Opin. Struct. Biol. 2*, 93–99.

Wolberger, C., Dong, Y., Ptashne, M., & Harrison, S. C. (1988) *Nature 335*, 789-795.

Wolberger, C., Vershon, A. K., Liu, B., Johnson, A. D., & Pabo, C. O. (1991) *Cell 67*, 517–536.

Wu, C. (1985) *Nature 317*, 84–87.

Zhang, H., Zhao, D., Revington, M., Lee, W., Jia., X., Arrowsmith, C., & Jardetzky, O. (1994) *J. Mol. Biol. 238*, 592–614.

Zinn, K., & Maniatis, T. (1986) *Cell 45*, 611–618.

Chapter 12

Agarwal, K. L., & Riftina, F. (1979) *Nucleic Acids Res. 6*, 3009–3024.

Agrawal, S., & Goodchild, J. (1987) *Tetrahedron Lett. 28*, 3539–3542.

Agrawal, S., & Tang, J. Y. (1992) *Antisense Res. Dev. 2*, 261–266.

Agrawal, S., Goodchild, J., Civeira, M. P., Thornton, A. H., Sarin, P. S., & Zamecnik, P. C. (1988) *Proc. Natl. Acad. Sci. U.S.A. 85*, 7079–7083.

Agrawal, S., Temsamani, J., & Tang, J. Y. (1991) *Proc. Natl. Acad. Sci. U.S.A. 88*, 7595–7599.

Agris, C., Blake, K., Miller, P., Reddy, M., & Ts'o, P. (1986) *Biochemistry 25*, 6268–6275.

Altman, S. (1990) *Angew. Chem. Int. Ed. Engl. 29*, 749–758.

Asseline, U., Delarue, M., Lancelot, G., Toulme, F., Thuong, N. T., Montenay-Garestier, T., & Hélène, C. (1984) *Proc. Natl. Acad. Sci. U.S.A. 81*, 3297–3301.

Asseline, U., Thuong, N. T., & Hélène, C. (1985) *J. Biol. Chem. 260*, 8936–8941.

Beal, P. A., & Dervan, P. B. (1991) *Science 251*, 1360–1363.

Becker, D., Meier, C., & Herlyn, M. (1989) *EMBO J. 8*, 3685–3690.

Bertrand, J. R., Imbach, J.-L., Paoletti, C., & Malvy, C. (1989) *Biochem. Biophys. Res. Commun. 164*, 311–318.

Bhan, P., & Miller, P. S. (1990) *Bioconjugate Chem. 1*, 82–88.

Bjergärde, K., & Dahl, O. (1991) *Nucleic Acids Res. 19*, 5843–5850.

Blake, K., Murakami, A., & Miller, P. (1985) *Biochemistry 24*, 6132–6138.

Boiziau, C., & Toulme, J.-J. (1991) *Biochimie 73*, 1403–1408.

Boiziau, C., Kurfurst, R., Cazenave, C., Roig, V., Thuong, N. T., & Toulme, J.-J. (1991) *Nucleic Acids Res. 19*, 1113–1119.

Boiziau, C., Thuong, N. T., & Toulme, J.-J. (1992) *Proc. Natl. Acad. Sci. U.S.A. 89*, 768–772.

Boutorin, A. S., Vlassov, V. V., Kazakov, S. A., Kutiavin, I. V., & Podyminogin, M. A. (1984) *FEBS Lett. 172*, 43–46.

Brill, W., Tang, J.-Y. , Ma, Y.-X., & Caruthers, M. H. (1989) *J. Am. Chem. Soc. 111*, 2321–2322.

Broitman, S. L., Im, D. D., & Fresco, J. R. (1987) *Proc. Natl. Acad. Sci. U.S.A. 84*, 5120–5124.

Brossalina, E., & Toulme, J. J. (1993) *J. Am. Chem. Soc. 115*, 796–797.

Brossalina, E. B., Demchenko, E. N., Vlassov, V. V., & Mamaev, S. V. (1991) *Antisense Res. Dev. 1*, 229–242.

Callahan, D. E., Trapane, T. L., Miller, P. S., Ts'o, P. O. P., & Kan, L. S. (1991) *Biochemistry 30*, 1650–1665.

Campbell, J. M., Bacon, T. A., & Wickstrom, E. (1990) *J. Biochem. Biophys. Methods 20*, 259–264.

Cazenave, C., Chevrier, M., Thuong, N. T., & Hélène, C. (1987) *Nucleic Acids Res. 15*, 10507–10521.

Cech, T. R. (1990) *Angew. Chem. Int. Ed. Engl. 29*, 759–768.

Cech, T. R., Herschlag, D., Piccirilli, J. A., & Pyle, A. M. (1992) *J. Biol. Chem. 267*, 17479–17482.

Chacko, K. K., Lindner, K., Saenger, W., & Miller, P. S. (1983) *Nucleic Acids Res. 11*, 2801–2814.

Chang, E. H., Miller, P. S., Cushman, C., Devadas, K., Pirollo, K. F., Ts'o, P. O. P., & Yu, Z. P. (1991) *Biochemistry 30*, 8283–8286.

Chastain, M., & Tinoco, I., Jr. (1992) *Nucleic Acids Res. 20*, 315–318.

Chastain, M., & Tinoco, I., Jr. (1993) in *Antisense Research and Applications*, (Crooke, S. T., & Lebleu, B., Eds.), pp. 55–66, CRC Press, Boca Raton, FL.

Chen, F.-M. (1991) *Biochemistry 30*, 4472–4479.

Chen, C.-H. B., & Sigman, D. S. (1986) *Proc. Natl. Acad. Sci. U.S.A. 83*, 7147–7151.

Chen, C.-H. B., Gorin, M. B., & Sigman, D. S. (1993) *Proc. Natl. Acad. Sci. U.S.A. 90*, 4206–4210.

Chen, T.-L., Miller, P. S., Ts'o, P. O. P., & Colvin, O. M. (1990) *Drug Metab. Dispos. 18*, 815–818.

Chin, D. J., Green, G. A., & Zon, G. (1990) *New Biol. 2*, 1091–1100.

Chu, B. C. F., & Orgel, L. E. (1985) *Proc. Natl. Acad. Sci. U.S.A. 82*, 963–967.

Chowrira, B. M., & Burke, J. M. (1992) *Nucleic Acids Res. 20*, 2835–2840.

Clarenc, J. P., Lebleu, B., & Leonetti, J. P. (1993) *J. Biol. Chem. 268*, 5600–5604.

Cohen, J. S., Ed. (1989) *Oligonucleotides, Antisense Inhibitors of Gene Expression*, Macmillan Press, London.

Cohen, J. S. (1991) *Pharmac. Ther. 52*, 211–225.

Cook, P. D. (1991) *Anti-Cancer Drug Design 6*, 585–607.

Cooney, M., Czernuszeqicz, G., Postel, E. H., Flint, S. J., & Hogan, M. E. (1988) *Science 241*, 456–459.

Coull, J. M., Carlson, D. V., & Weith, H. L. (1987) *Tetrahedron Lett. 28*, 745–748.

Crooke, S. T. (1992) *Bio/Technology 10*, 882–886.

Crooke, R. M. (1993a) in *Antisense Research and Applications* (Crooke, S. T., & Lebleu, B., Eds.), pp. 427–449, CRC Press, Boca Raton, FL.

Crooke, S. T. (1993b) *Antisense Res. and Dev. 3*, 1–2.

Crooke, S. T., & Lebleu, B. (1993) *Antisense Research and Applications*, CRC Press, Boca Raton, FL.

Davison, E. C., & Johnsson, K. (1993) *Nucleosides Nucleotides 12*, 237–243.

Debart, F., Rayner, B., Degols, G., & Imbach, J.-L. (1992) *Nucleic Acids Res. 20*, 1193–1200.

Dervan, P. B. (1989) in *Oligonucleotides, Antisense Inhibitors of Gene Expression* (Cohen, J. S., Ed.), pp. 197–210, Macmillan Press, London.

Dervan, P. (1992) *Nature 359*, 87–88.

Dolnick, B. J. (1990) *Biochem. Pharmacol. 40*, 671–675.

Dorman, M. A., Noble, S. A., McBride, L. J., & Caruthers, M. H. (1984) *Tetrahedron 49*, 95.

Dreyer, G. B., & Dervan, P. B. (1985) *Proc. Natl. Acad. Sci. U.S.A. 82*, 968–972.

Durand, M., Maurizot, J. C., Thuong, N. T., & Hélène, C. (1988) *Nucleic Acids Res. 16*, 5039–5053.

Durand, M., Peloille, S., Thuong, N., & Maurizot, J. (1992) *Biochemistry 31*, 9197–9204.

Durand, M., Maurizot, J. C., Asseline, U., Thuong, N. T., & Hélène, C. (1993) *Bioconjugate Chem. 4*, 206–211.

Duvalvalentin, G., Thuong, N. T., & Hélène, C. (1992) *Proc. Natl. Acad. Sci. U.S.A. 89*, 504–508.

Ecker, D. J., Vickers, T. A., Bruice, T. W., Freier, S. M., Jenison, R. D., Manoharan, M., & Zounes, M. (1992) *Science 257*, 958–961.

Egholm, M., Buchardt, O., Nielsen, P. E., & Berg, R. H. (1992) *J. Am. Chem. Soc. 114*, 1895–1897.

Ehresmann, C., Baudin, F., Mougel, M., Romby, P., Ebel, J.-P., & Ehresmann, B. (1987) *Nucleic Acids Res. 15* 9109–9128.

Elder, P. X., DeVine, R. J., Dagle, J. M., & Walder, J. A. (1991) *Antisense Res. Dev. 1*, 141–151.

Erickson, R. P., & Izant, J. G., Eds., (1992) *Gene Regulation. Biology of Antisense RNA and DNA*, Raven Press, New York.

Ferguson, D. M., & Kollman, P. A. (1991) *Antisense Res. Dev. 1*, 243–254.

Francois, J.-C., Saison-Behmoaras, T., Thuong, N., & Hélène, C. (1989) *Biochemistry 28*, 9617–9619.

Froehler, B., Ng, P., & Matteucci, M. (1988) *Nucleic Acids Res. 16*, 4831–4839.

Gait, M. J., Jones, A. S., & Walker, R. T. (1974) *J. Chem. Soc. Perkin Trans. 1*, 1684–1686.

Gao, W., Stein, C. A., Cohen, J. S., Dutschman, G. E., & Cheng, Y,-C. (1989) *J. Biol. Chem. 264*, 11521–11526.

Geselowitz, D. A., & Neckers, L. M. (1992) *Antisense Res. Dev. 2*, 17–25.

Ghosh, M., & Cohen, J. S. (1991) in *Progress Nucleic Acid Research and Molecular Biology* (Moldave, K. & Cohn, W., Eds.), pp. 79–126, Academic Press, New York.

Ghosh, M. K., Ghosh, K., & Cohen, J. S. (1992) *Antisense Res. Devel. 2*, 111–118.

Giles, R. V., & Tidd, D. M. (1992) *Nucleic Acids Res. 20*, 763–770.

Giovannangeli, C., Montenay Garestier, T., Rougee, M., Chassignol, M., Thuong, N. T., & Hélène, C. (1991) *J. Am. Chem. Soc. 113*, 7775–7777.

Giovannangeli, C., Thuong, N., & Hélène, C. (1992) *Nucleic Acids Res. 20*, 4275–4281.

Goodchild, J. (1990) *Bioconjugate Chem. 1*, 165–187.

Goodchild, J. (1992) *Nucleic Acids Res. 20*, 4607–4612.

Goodchild, J., Agrawal, S., Civeira, M. P., Sarin, P. S., Sun, D., & Zamecnik, P. C. (1988a) *Proc. Natl. Acad. Sci. U.S.A. 85*, 5507–5511.

Goodchild, J., Carroll, E., & Greenberg, J. R. (1988b) *Arch. Biochem. Biophys. 263*, 401–409.

Gorshkova, I. I., Zenkova, M. A., Karpova, G. G., Levina, A. S., & Solov'ev, V. V. (1986) *Mol. Biol. 20*, 1084–1091.

Griffin, L. C., & Dervan, P. B. (1989) *Science 245*, 967–971.

Griffin, L. C., Kiessling, L. L., Beal, P. A., Gillespie, P., & Dervan, P. B. (1992) *J. Am. Chem. Soc. 114*, 7976–7982.

Grigoriev, M., Praseuth, D., Robin, P., Hemar, A., Saison-Behmoaras, T., Dautryvarsat, A., Thuong, N. T., Hélène, C., & Harel-Bellan, A. (1992) *J. Biol. Chem. 267*, 3389–3395.

Grigoriev, M., Praseuth, D., Guieysse, A. L., Robin, P., Thuong, N. T., Hélène, C., & Harel-Bellan, A. (1993) *Proc. Natl. Acad. Sci. U.S.A. 90*, 3501–3505.

Guerrier-Takada, C., Gardiner, K., Marsh, T., Pace, N., & Altman, S. (1983) *Cell 35*, 849–857.

Han, H. Y., & Dervan, P. B. (1993) *Proc. Natl. Acad. Sci. U.S.A. 90*, 3806–3810.

Han, F., Watt, W., Duchamp, D. J., Callahan, L., Kezdy, F. J., & Agarwal, K. (1990) *Nucleic Acids Res. 18*, 2759–2767.

Haner, R., & Dervan, P. B. (1990) *Biochemistry 29*, 9761–9765.

Hanvey, J. C., Peffer, N. J., Bisi, J. E., Thomson, S. A., Cadilla, R., Josey, J. A., Ricca, D. J., Hassman, C. F., Bonham, M. A., Au, K. G., Carter, S. G., Bruckenstein, D. A., Boyd, A. L., Noble, S. A., & Babiss, L. E. (1992) *Science 258*, 1481–1485.

Hausheer, F. H., Rao, B. G., Saxe, J. D., & Singh, U. C. (1992) *J. Am. Chem. Soc. 114*, 3201–3206.

Hawley, P., & Gibson, I. (1992) *Antisense Res. Dev. 2*, 119–127.

Heidenreich, O., & Eckstein, F. (1992) *J. Biol. Chem. 267*, 1904–1909.

Hélène, C., & Toulme, J.-J. (1989) in *Oligonucleotides, Antisense Inhibitors of Gene Expression* (Cohen, J. S., Ed.) pp 137–172, Macmillan Press, London.

Hélène, C., & Toulme, J.-J. (1990) *Biochim. Biophys. Acta 1049*, 99–125.

Hélène, C., Montenay-Garestier, T., Saison, T., Takasugi, M., Toulme, J. J., Asseline, U., Lancelot, G., Maurizot, J. C., Toulme, F., & Thuong, N. T. (1985) *Biochimie 67*, 777–783.

Hendry, P., McCall, M. J., Santiago, F. S., & Jennings, P. A. (1992) *Nucleic Acids Res. 20*, 5737–5741.

Hjalt, T., & Wagner, E. G. H. (1992) *Nucleic Acids Res. 20*, 6723–6732.

Hogrefe, R. I., Vaghefi, M. M., Reynolds, M. A., Young, K. M., & Arnold, L. J., Jr. (1993) *Nucleic Acids Res. 21*, 2031–2038.

Homann, M., Tzortzakaki, S., Rittner, K., Sczakiel, G., & Tabler, M. (1993) *Nucleic Acids Res. 21*, 2809–2814.

Howard, F. B., Frazier, J., Lipsett, N. M., & Miles, H. T. (1964) *Biochem. Biophys. Res. Commun. 17*, 93–102.

Huang, C.-Y., Cushman, C. D., & Miller, P. S. (1993) *J. Org. Chem. 58*, 5048–5049.

Hubbard, J. M., & Hearst, J. E. (1991) *Biochemistry 30*, 5458–5465.

Iverson, P. L., Zhu, S., Meyer, A., & Zon, G. (1992) *Antisense Res. Dev. 2*, 211–222.

Iyer, R. P., Egan, W., Regan, J. B., & Beaucage, S. L. (1990) *J. Am. Chem. Soc. 112*, 1253–1254.

Izant, J. G. (1992) in *Gene Regulation. Biology of Antisense RNA and DNA* (Erickson, R. P., & Izant, J. G., Eds.), pp. 183–195, Raven Press, New York.

Jaeger, J. A., Turner, D. H., & Zucker, M. (1989) *Proc. Natl. Acad. Sci. U.S.A. 86*, 7706–7710.

Jager, A., & Engles, J. (1984) *Tetrahedron Lett. 25*, 1437–1440.

Jager, A., Levy, M. J., & Hecht, S. M. (1988) *Biochemistry 27*, 7237–7246.

Jaroszewski, J. W., & Cohen, J. S. (1991) *Adv. Drug Delivery Rev. 6*, 235–250.

Jayaraman, K., McParland, K., Miller, P., & Ts'o, P. (1981) *Proc. Natl. Acad. Sci. U.S.A. 78*, 1537–1541.

Jayasena, S. D., & Johnston, B. H. (1992) *Biochemistry 31*, 320–327.

Jetter, M. C., & Hobbs, F. W. (1993) *Biochemistry 32*, 3249–3254.

Jones, D. S., & Tittensor, J. R. (1969) *J. Chem. Soc. Chem. Commun.*, 1240.

Kashni-Sabet, M., Funato, T., Tone, T., Jiao, L., Wang, W., Yoshida, E., Kashfinn, B. I., Shitara, T., Wu, A. M., Moreno, J. G., Traweek, S. T. Ahlering, T. E., & Scanlon, K. J. (1992) *Antisense Res. Dev. 2*, 3–15.

Kean, J. M., Murakami, A., Blake, K. R., Cushman, C. D., & Miller, P. S. (1988) *Biochemistry 27*, 9113–9121.

Kiessling, L. L., Griffin, L. C., & Dervan, P. B. (1992) *Biochemistry 31*, 2829–2834.

Kikuchi, Y., & Sasaki, N. (1991) *Nucleic Acids Res. 19*, 6751–6755.

Knorre, D. G., & Vlassov, V. V. (1985) *Prog. Nucleic Acid Res. Mol. Biol. 32*, 291–320.

Koizumi, M., Kamiya, M. H., & Ohtsuka, E. (1992) *Gene 117*, 179–184.

Kool, E. T. (1991) *J. Am. Chem. Soc. 113*, 6265–6266.

Koshlap, K. M., Gillespie, P., Dervan, P. B., & Feigon, J. (1993) *J. Am. Chem. Soc. 115*, 7908–7909.

Kozak, M. (1989) *J. Cell Biol. 108*, 229–241.

Koziolkiewicz, M., Uznanski, B., & Stec, W. J. (1986) *Chemica Scripta 26*, 251–259.

Krawczyk, S. H., Milligan, J. F., Wadwani, S., Moulds, C., Froehler, B. C., & Matteucci, M.D. (1992) *Proc. Natl. Acad. Sci. U.S.A. 89*, 3761–3764.

Kruger, K., Grabowski, P. J., Zaug, A.J., Sands, J., Gottschling, D. E., & Cech, T. R. (1982) *Cell 31*, 147–157.

Kulka, M., Smith, C. C., Aurelian, L., Fishelevich, R., Meade, K., Miller, P., & Ts'o, P. O. P. (1989) *Proc. Natl. Acad. Sci. U.S.A. 86*, 6868–6872.

Kulka, M., Wachsman, M., Miura, S., Fishelevich, R, Miller, P. S., Ts'o, P. O. P., & Aurelian, L. (1993) *Antiviral Res. 20*, 115–130.

Lancelot, G., Asseline, U., Thuong, N. T., & Hélène, C. (1985) *Biochemistry 24*, 2521–2529.

LaPlanche, L. A., James, T. L., Powell, C., Wilson, W. D., Uznanski, B., Stec, W. J., Summers, M. F., & Zon, G. (1986) *Nucleic Acids Res. 14*, 9081–9093.

Larrouy, B. Blonski, C., Boiziau, C., Stuer, M., Moreau, S., Shire, D., & Toulme, J.-J. (1992) *Gene 121*, 189–194.

Lavignon, M., Tounekti, N., Rayner, B., Imbach, J.-L., Keith, G., Paoletti, J., & Malvy, C. (1992) *Antisense Res. Dev. 2*, 315–424.

Le Doan, T., Perrouault, L., Chassignol, M., Thuong, N. T., & Hélène, C. (1987) *Nucleic Acids Res. 15*, 8643–8659.

Lee, B. L., Murakami, A., Blake, K. R., Lin, S.-B., & Miller, P. S. (1988) *Biochemistry 27*, 3197–3203.

Lee, J. S., Woodsworth, M. L., Latimer, L. J. P., & Morgan, A. (1984) *Nucleic Acids Res. 12*, 6603–6614.

Leonetti, J. P., Rayner, B., Lemaitre, M., Gagnor, C., Milhaud, P. G., Imbach, J.-L., & Lebleu, B. (1988) *Gene 72*, 323–332.

Leonetti, J. P., Mechti, N., Degols, C., Gagnor, C., & Lebleu, B. (1991) *Proc. Natl. Acad. Sci. U.S.A. 88*, 2702–2706.

Lesnikowski, Z. J., Jaworska, M., & Stec, W. J. (1990) *Nucleic Acids Res. 18*, 2109–2115.

Liebhaber, S. A., Cash, F. E., & Shakin, S. H. (1984) *J. Biol. Chem. 259*, 15597–15602.

Lima, W. F., Monia, B. P., Ecker, D. J., & Freier, S. M. (1992) *Biochemistry 31*, 12055–12061.

Lin, S.-B., Blake, K. R., Miller, P. S., & Ts'o, P. O. P. (1989) *Biochemistry 28*, 1054–1061.

Lipsett, M. N. (1963) *Biochem. Biophys. Res. Commun. 11*, 224–228.

Lisziewicz, J., Sun, D., Klotman, M., Agrawal, S., Zamecnik, P., & Gallo, R. (1992) *Proc. Natl. Acad. Sci. U.S.A. 89*, 11209–11213.

Lisziewicz, J., Sun, D., Metelev, V., Zamecnik, P., Gallo, R. C., & Agrawal, S. (1993) *Proc. Natl. Acad. Sci. U.S.A. 90*, 3860–3864.

Loke, S., Stein, C., Zhang, X., Mori, K., Nakanishi, M., Subasinghe,C., Cohen, J., & Neckers, L. (1989) *Proc. Natl. Acad. Sci. U.S.A. 86*, 3474–3478.

Macaya, R. F., Gilbert, D. E., Malek, S., Sinsheimer, J. S., & Feigon, J. (1991) *Science 254*, 270–274.

Maher, L. J., Wold, B., & Dervan, P. B. (1989) *Science 245*, 725–730.

Maher, L. J., Dervan, P. B., & Wold, B. (1992) *Biochemistry 31*, 70–81.

Majumdar, C., Stein, C. A., Cohen, J. S., Broder, S., & Wilson, S. H. (1989) *Biochemistry 28*, 1340–1346.

Malcolm, A. D. B., Coulson, J., Blake, N., & Archard, L. C. (1992) *Biochem. Soc. Trans. 20*, 762–764.

Manzini, G., Xodo, L. E., Gasparotto, D., Quadrifoglio, F., van der Marel, G. A., & van Boom, J. H. (1990) *J. Mol. Biol. 213*, 833–843.

Marcus-Sekura, C. J., Woerner, A. M., Shinozuka, K. (1987) *Nucleic Acids Res. 15*, 5749–5763.

Marti, G., Egan, W., Noguchi, P., Zon, G., Matsukura, M., & Broder, S. (1992) *Antisense Res. Dev. 2*, 27–39.

Matsukura, M., Shinozuka, K., Zon, G., Mitsuya, H., Reitz, M., Cohen, J. S., & Broder, S. (1987) *Proc. Natl. Acad. Sci. U.S.A. 84*, 7706–7710.

Matteucci, M. (1990) *Tetrahedron Lett. 31*, 2385–2388.

Matteucci, M., Lin, K. Y., Butcher, S., & Moulds, C. (1991) *J. Am. Chem. Soc. 113*, 7767–7768.

Mergny, J.-L., Sun, J.-S., Rougee, M., Montenay-Garestier, T., Barcelo, F., Chomilier, J., & Hélène, C. (1991) *Biochemistry 30*, 9791–9798.

Mertes, M. P., & Coates, E. A. (1969) *J. Med. Chem. 12*, 154–157.

Miller, P. S. (1992) in *Antisense RNA and DNA* (Murray, J.A.H., Ed.), pp. 79–95, Wiley–Liss, New York.

Miller, P. S., & Cushman, C. D. (1993) *Biochemistry 32*, 2999–3004.

Miller, P. S., & Ts'o, P. O. P. (1988) *Ann. Rep. Med. Chem. 23*, 295–304.

Miller, P. S., Braiterman, L., & Ts'o, P. O. P. (1977) *Biochemistry 16*, 1988–1996.

Miller, P. S., Yano, J., Yano, E., Carroll, C., Jayaraman, K., & Ts'o, P. O. P. (1979) *Biochemistry 18*, 5134–5142.

Miller, P. S., Dreon N., Pulford, S., & McParland, K. (1980) *J. Biol. Chem. 255*, 9659–9665.

Miller, P. S., Cushman, C. D., & Levis, J. T. (1991) in *Oligonucleotides and Analogues. A Practical Approach* (Eckstein, F., Ed.), þp. 137–154, IRL Press, Oxford.

Miller, P. S., Bhan, P., Cushman, C. D., & Trapane, T. L. (1992) *Biochemistry 31*, 6788–6793.

Milligan, J. F., Matteucci, M. D., & Martin, J. C. (1993) *J. Med. Chem. 36*, 1923–1937.

Minshull, J., & Hunt, T. (1986) *Nucleic Acids Res. 14*, 6433–6451.

Minshull, J., & Hunt, T. (1992) in *Antisense RNA and DNA* (Murray, J. A. H., Ed.), pp. 195–212, Wiley–Liss, New York.

Monia, B. P., Johnston, J. F., Ecker, D. J., Zounes, M. A., Lima, W. F., & Freier, S. M. (1992) *J. Biol. Chem. 267*, 19954–19962.

Morrison, R. (1991) *J. Biol. Chem. 266*, 728–734.

Morvan, F., Rayner, B., Leonetti, J. P., & Imbach, J.-L. (1988) *Nucleic Acids Res. 16*, 833–847.

Morvan, F., Rayner, B., & Imbach, J.-L. (1991) *Anti-Cancer Drug Design 6*, 521–529.

Morvan, F., Proumb, H., Degols, G., Lefebvre, I., Pompon, A., Sproat, B. S., Rayner, B., Malvy, C., Lebleu, B., & Imbach, J. L. (1993) *J. Med. Chem. 36*, 280–287.

Mungall, W. S., & Kaiser, J. K. (1977) *J. Org. Chem. 42*, 703–706.

Murakami, A., Blake, K. R., & Miller, P. S. (1985) *Biochemistry 24*, 4041–4046.

Murray, J. A. H., Ed., (1992) *Antisense RNA and DNA*, Wiley–Liss, New York.

Murray, J. A. H., & Crockett, N. (1992) in *Antisense RNA and DNA* (Murray, J. A. H., Ed.), pp. 1–49, Wiley–Liss, New York.

Nellen, W., Hildebrandt, M., Mahal, B., Mohrle, A., Kroger, P., Maniak, M., Oberhauser, R., & Sadiq, M. (1992) *Biochem. Soc. Trans. 20*, 750–754.

Nielsen, P. E., Egholm, M., Berg, R. H., & Buchardt, O. (1991) *Science 254*, 1497–1500.

Nielsen, P. E., Egholm, M., Berg, R. H., & Buchardt, O. (1993a) in *Antisense Research and Applications* (Crooke, S. T., & Lebleu, B., Eds.), pp. 364–373, CRC Press, Boca Raton, FL.

Nielsen, P. E., Egholm, M., Berg, R. H., & Buchardt, O. (1993b) *Nucleic Acids Res. 21*, 197–200.

Odai, O., Kodama, H., Hiroaki, H., Sakata, T., Tanaka, T., & Uesugi, S. (1990) *Nucleic Acids Res. 18*, 5955–5960.

Ojwang, J. O., Hampel, A., Looney, D. J., Wong Staal, F., & Rappaport, J. (1992) *Proc. Natl. Acad. Sci. U.S.A. 89*, 10802–10806.

Ono, A., Ts'o, P. O. P., & Kan, L.-S. (1991) *J. Am. Chem. Soc. 113*, 4032–4033.

Ono, A., Ts'o, P. O. P., & Kan, L.-S. (1992) *J. Org. Chem. 57*, 3225–3230.

Orson, F. M., Thomas, D. W., McShan, W. M., Kessler, D. J., & Hogan, M. E. (1991) *Nucleic Acids Res. 19*, 3435–3441.

Paolella, G., Sproat, B. S., & Lamond, A. I. (1992) *EMBO J. 11*, 1913–1919.

Parker, R., Muhlrad, D., Deshler, J. O., Taylor, N., & Rossi, J. J. (1992) in *Gene Regulation. Biology of Antisense RNA and DNA* (Erickson, R. P., & Izant, J. G., Eds.), pp. 55–70, Raven Press, New York.

Pascolo, E., Blonski, C., Shire, D., & Toulme, J. J. (1993) *Biochimie 75*, 43–47.

Paterson, B. M., Roberts, B. E., & Kuff, E. L. (1977) *Proc. Natl. Acad. Sci. U.S.A. 74*, 4370–4374.

Pieles, U., & Englisch, U. (1989) *Nucleic Acids Res. 17*, 285–298.

Pilch, D. S., Levenson, C., & Shafer, R. H. (1991) *Biochemistry 30*, 6081–6087.

Piotto, M. E., Granger, J. N., Cho, Y., Farschtschi, N., & Gorenstein, D. G. (1991) *Tetrahedron Lett. 47*, 2449–2453.

Povsic, T. J., & Dervan, P. B. (1989) *J. Am. Chem. Soc. 111*, 3059–3061.

Prakash, G., & Kool, E. T. (1991) *J. Chem. Soc. Chem. Commun.* 1161–1163.

Prakash, G., & Kool, E. T. (1992) *J. Am. Chem. Soc. 114*, 3523–3527.

Praseuth, D., Chassignol, M., Takasugi, M., Le Doan, T., Thuong, N. T., & Hélène, C. (1987) *J. Mol. Biol. 196*, 939–942.

Pyle, A. M. (1993) *Science 261*, 709–714.

Quartin, R. S., & Wetmur, J. G. (1989) *Biochemistry 28*, 1040–1047.

Radhakrishnan, I., & Patel, D. J. (1993) *J. Am. Chem. Soc. 115*, 1615–1617.

Radhakrishnan, I., de los Santos, C., & Patel, D. J. (1991a) *J. Mol. Biol. 221*, 1403–1418.

Radhakrishnan, I., Gao, X., de los Santos, C., Live, D., & Patel, D. J. (1991b) *Biochemistry 30*, 902ᴢ–9030.

Radhakrishnan, I., Patel, D. J., Veal, J. M., & Gao, X. L. (1992) *J. Am. Chem. Soc. 114*, 6913–6915.

Reynolds, M. A., Beck, T. A., Hogrefe, R. I., McCaffrey, A., Arnold, L. J., Jr., & Vaghefi, M. M. (1992) *Bioconjugate Chem. 3*, 366–374.

Roberts, R. W., & Crothers, D. M. (1992) *Science 258*, 1463–1466.

Saison-Behmoaras, T., Tocque, B., Rey, I., Chassignol, M., Thuong, N. T., & Hélène, C. (1991) *EMBO J. 10*, 1111–1118.

Salunkhe, M., Wu, T., & Letsinger, R. L. (1992) *J. Am. Chem. Soc. 114*, 8768–8772.

Sarin, P. S., Agrawal, S., Civeira, M. P., Goodchild, J., Ikeuchi, T., & Zamecnik, P.C. (1988) *Proc. Natl. Acad. Sci. U.S.A. 85*, 7448–7451.

Scanlon, K. J., Jiao, L., Funato, T., Wang, W., Tone, T., Rossi, J. J., & Kashani-Sabet, M. (1991) *Proc. Natl. Acad. Sci. U.S.A. 88*, 10591–10595.

Sexena, S. K., & Ackerman, E. J. (1990) *J. Biol. Chem. 265*, 17106–17109.

Shakin, S. H., & Liebhaber, S. A. (1986) *J. Biol. Chem. 261*, 16018–16025.

Shaw, J. P., Kent, K., Bird, J., Fishback, J., & Froehler, B. (1991a) *Nucleic Acids Res. 19*, 747–750.

Shaw, J. P., Milligan, J. F., Krawczyk, S. H., & Matteucci, M. (1991b) *J. Am. Chem. Soc. 113*, 7765–7766.

Shimizu, M., Konishi, A., Shimada, Y., Inoue, H., & Ohtsuka, E. (1992) *FEBS Lett. 302*, 155–158.

Shoji, Y., Akhtar, S., Periasamy, A., Herman, B., & Juliano, R. L. (1991) *Nucleic Acids Res. 19*, 5543–5550.

Sigman, D. S., Bruice, T. W., Mazumder, A., & Sutton, C. L. (1993) *Acc. Chem. Res. 90*, 4206–4210.

Singleton, S., & Dervan. P. (1992) *J. Am. Chem. Soc. 114*, 6957–6965.

Sioud, M., Natvig, J. B., & Forre, O. (1992) *J. Mol. Biol. 223*, 831–835.

Skoog, J. U., & Maher, L. J. (1993) *Nucleic Acids Res. 21*, 2131–2138.

Smith, C. C., Aurelian, L., Reddy, M. P., Miller, P., & Ts'o, P. O. P. (1986) *Proc. Natl. Acad. Sci. U.S.A. 83*, 2787–2791.

Souza, D., & Kool, E. (1992) *J. Biomol. Struct. Dynamics 10*, 141–152.

Spiller, D. G., & Tidd, D. M. (1992) *Anti-Cancer Drug Design 7*, 115–129.

Stec, W. J., Zon, G., Egan, W., & Stec, B. (1984) *J. Am. Chem. Soc. 108*, 6077–6079.

Stein, C. A. (1992) *Leukemia 6*, 967–974.

Stein, C. A., & Cheng, Y.-C. (1993) *Science 261*, 1004–1012.

Stein, C., Mori, K., Loke, S., Subasinghe, C., Shinozuka, K., Cohen, J., & Neckers. L. (1988a) *Gene 72*, 333–341.

Stein, C. A., Shinozuka, K. Subasinghe, C., & Cohen, J. S. (1988b) *Nucleic Acids Res. 16*, 3209–3221.

Stein, C. A., Tonkinson, J. L., Zhang, L.-M., Yakubov, L., Gervasoni, J., Taub, R., & Rotenberg, S. A. (1993) *Biochemistry 32*, 4855–4861

Stephenson, M. L., & Zamecnik, P. C. (1978) *Proc. Natl. Acad. Sci. U.S.A. 75*, 285–288.

Stilz, H. U., & Dervan, P. B. (1993) *Biochemistry 32*, 2177–2185.

Stirchak, E. P., & Summerton, J. E. (1987) *J. Org. Chem. 52*, 4202–4206.

Strobel, S. A., Doucette-Stamm, L. A., Riba, L., Housman, D. E., & Dervan, P. B. (1991) *Science 254*, 1639–1642.

Stull, R. A., Taylor, L. A., & Szoka, F. C., Jr. (1992) *Nucleic Acids Res. 20*, 3501–3508.

Sun, J. S., Asseline, U., Rouzaud, D., Montenay-Garestier, T., Taylor, N. R., Kaplan, B. E., Swiderski, P., Li, H., & Rossi, J. J. (1992) *Nucleic Acids Res. 20*, 4559–4565.

Teare, J., & Wollenzien, P. (1989) *Nucleic Acids Res. 17*, 3359–3372.

Thaden, J. J., & Miller, P. S. (1993) *Bioconjugate Chem. 4*, 386–394.

Thiele, D., & Guschlbauer, W. (1971) *Biopolymers 10*, 143–157.

Thuong, N. T., & Hélène, C. (1987) *Nucleic Acids Res. 15*, 6149–6158.

Thuong, T., Asseline, U. Roig, V., Takasugi, M., & Hélène, C. (1987) *Proc. Natl. Acad. Sci. U.S.A. 84*, 5129–5133.

Tidd, D. M. (1990) *Anticancer Res. 10*, 1169–1182.

Tidd, D. M., & Varenius, H.M. (1989) *Br. J. Cancer 60*, 343–350.

Toulme, J., Krisch, H. M., Loreau, N., Thuong, N. T., & Hélène, C. (1986) *Proc. Natl. Acad. Sci. U.S.A. 83*, 1227–1231.

Uhlenbeck, O. C. (1993) in *Antisense Research and Applications* (Crooke, S. T., & Lebleu, B., Eds.), pp. 84–96, CRC Press, Boca Raton, FL.

Uhlmann, E., & Peyman, A. (1990) *Chem. Rev., 90*, 543–584.

van Blokland, R., de Lange, P., Mol, J. N. M., & Kooter, J. M. (1993) in *Antisense Research and Applications* (Crooke, S. T., & Lebleu, B., Eds.), pp. 125–148, CRC Press, Boca Raton, FL.

Vasanthakumar, G., & Ahmed, N. A. (1989) *Cancer Commun. 1*, 225–232.

Vlassov, V. V. (1993) in *Antisense Research and Applications* (Crooke, S. T., & Lebleu, B., Eds.), pp. 236–250, CRC Press, Boca Raton, FL.

Wagner, R. W., Matteucci, M. D., Lewis, J. G., Butierrez, A. J., Moulds, C., & Froehler, B. (1993) *Science 260*, 1510–1514.

Walder, J. A. (1988) *Gene Dev. 2*, 502–504.

Walder, R. Y., & Walder, J. A. (1988) *Proc. Natl. Acad. Sci. U.S.A. 85*, 5011–5015.

Wang, E., Malek, S., & Feigon, J. (1992) *Biochemistry 31*, 4838–4846.

Webb, T. R., & Matteucci, M. D. (1986) *Nucleic Acids Res. 14*, 7661–7674.

Wickstrom, E., Ed. (1991) *Prospects for Antisense Nucleic Acid Therapy of Cancer and AIDS*, Wiley–Liss, New York.

Woo, J., & Hopkins, P. B. (1991) *J. Am. Chem. Soc. 113*, 5457–5459.

Xodo, L. E., Manzini, G., & Quadrifoglio, F. (1990) *Nucleic Acids Res. 18*, 3557–3546.

Xodo, L. E., Manzini, G., Quadrifoglio, F., van der Marel, G. A., & van Boom, J. H. (1991) *Nucleic Acids Res. 19*, 5625–5631.

Yakubov, L. A., Deeva, E. A., Zarytova, F., Ivanova, E., Ryte, A., Yurchenko, L., & Vlassov, V. (1989) *Proc. Natl. Acad. Sci. U.S.A. 86*, 6454–6458.

Yoon, K., Hobbs, C. A., Koch, J., Sardaro, M., Kutny, R., & Weis, A. L. (1992) *Proc. Natl. Acad. Sci. U.S.A. 89*, 3840–3844.

Young, S. L., Krawczyk, S. H., Matteucci, M. D., & Toole, J. J. (1991) *Proc. Natl. Acad. Sci. U.S.A. 88*, 10023–10026.

Zamecnik, P. C., & Stephenson, M. C. (1978) *Proc. Natl. Acad. Sci. U.S.A. 75*, 280–284.

Zamecnik, P. C., Goodchild, J., Taguchi, Y., & Sarin, P. S. (1986) *Proc. Natl. Acad. Sci. U.S.A. 83*, 4143–4146.

Zhao, Q., Matson, S., Herrera, C. J., Fisher, E., Yu, H., & Krieg, A. M. (1993) *Antisense Res. Dev. 3*, 53–66.

Zon, G., & Geiser, T. (1991) *Anti-Cancer Drug Design 6*, 539–543.

Chapter 13

Avery, O. T., MacLeod, C. M., & McCarty, M. (1994) *J. Exp. Med. 79*, 137–158.

Baer, M. F., Wesolowski, D., & Altman, S. (1989) *J. Bacteriol. 171*, 6862–6866.

Bass, B. L., & Cech, T. R. (1984) *Nature 308*, 820–826.

Benzal-Herranz, A., Joseph, S., & Burke, J. M. (1993) *EMBO J. 6*, 2567–2574.

Biebricher, C. K., Eigen, M., & McCaskill, J. S. (1993) *J. Mol. Biol. 231*, 175–179.

Brown, J. W., & Pace, N. R. (1992) *Nucleic Acids Res. 20*, 1451–1456.

Burke, J. M., Belfort, M., Cech, T. R., Davies, R. W., Schweyen, R. J., Shub, D. A., Szostak, J. W., & Tabak, H. W. (1987) *Nucleic Acids Res. 15*, 7217–7221.

Calcutt, M. J., & Schmidt, F. J. (1992) *J. Bacteriol. 174*, 3220–3226.

Carter, B. J., Vold, B. S., & Hecht S. M. (1990) *J. Biol. Chem. 265*, 7100–7103.

Cech, T. R. (1990) *Annu. Rev. Biochem. 59*, 543–568.

Cech, T. R., & Bass, B. L. (1986) *Annu. Rev. Biochem. 55*, 599–629.

Cech, T. R., Herschlag, D., Piccirilli, J. A., & Pyle, A. M. (1992) *J. Biol. Chem. 267*, 17479–17482.

Cech, T. R., & Uhlenbeck, O. C. (1994) *Nature 372*, 39–40.

Chastain, M., & Tinoco, I., Jr. (1991) *Prog. Nucl. Acid Res. Mol. Biol. 41*, 131–177.

Chowrira, B. M., Benzal-Herranz, A., Keller, C. F., & Burke, J. M. (1993) *J. Biol. Chem. 268*, 19458–19462.

Dange, V., Van Atta, R. B., & Hecht, S. M. (1990) *Science 248*, 585–588.

Doudna, J. A., Couture, S., & Szostak, J. W. (1991) *Science 251*, 1605–1608.

Eigen, M. (1992) *Steps Toward Life*. Oxford University Press, New York.

Forster, A. C., & Altman, S. (1990) *Science 249*, 783–786.

Gesteland, R. F., & Atkins, J. F., Eds. (1993) *The RNA World*, Cold Spring Harbor Laboratory, Cold Spring Harbor, NY.

Gott, J. M., Shub, D. A., & Belfort, M. (1986) *Cell 47*, 81–87.

Green, C. J., & Vold, B. S. (1988) *J. Biol. Chem. 263*, 652–657.

Green, R., & Szostak, J. W. (1992) *Science 258*, 1910–1915.

Guerrier-Takada, C., Gardiner, K., Marsh, T., Pace, N. R., & Altman, S. (1983) *Cell 35*, 849–857.

Guthrie, C. (1991) *Science 253*, 157–163.

Guthrie, C., & Atchison, R. (1980) in *Transfer RNA: Biological Aspects* (Soll, D., Abelson, J. N., & Schimmel, P. Eds), pp. 83–97, Cold Spring Harbor Laboratory, Cold Spring Harbor, NY.

Haas, E. S., Morse, D. P., Brown, J. W., Schmidt, F. J., & Pace, N. R. (1991) *Science 254*, 853–856.

Hansen, F. G., Hansen, E. B., & Atlung, T. (1985) *Gene 38*, 85–93.

Herschlag, D. (1991) *Proc. Natl. Acad. Sci. U.S.A. 88*, 6921–6925.

Herschlag, D., & Jencks, W. P. (1990) *Biochemistry 29*, 5172–5179.

Jaeger, L., Westhof, E., & Michel, F. (1991) *J. Mol. Biol. 221*, 1153–1164.

James, B. D., Olsen, G. J., & Pace, N. R. (1989) *Methods Enzymol. 180*, 227–239.

Joyce, G. F., & Orgel, L. E. (1993) in *The RNA World* (Atkins, J. F., & Gesteland, R. F., Eds.), pp. 1–25, Cold Spring Harbor Laboratory, Cold Spring Harbor, NY.

Kauffman, S. A. (1993) *The Origins of Order*, Oxford University Press, New York.

Kazakov, S., & Altman, S. (1992) *Proc. Natl. Acad. Sci. U.S.A. 89*, 7939–7943.

Kole, R., & Altman, S. (1979) *Proc. Natl. Acad. Sci. U.S.A. 76*, 3795–3799.

Kruger, K., Grabowski, P. J., Zaug, A. J., Sands, J., Gottschling, D. E., & Cech, T. R. (1982) *Cell 31*, 147–157.

LaGrandeur, T. E., Darr S. C., Haas, E. S., & Pace, N. R. (1993) *J. Bacteriol. 175*, 5043–5048.

Lehman, N., & Joyce, G. F. (1993) *Nature 361*, 182–185.

Lehninger, A. L. (1975) *Biochemistry*, 2nd edition, p. 9, Worth, New York.

Lesser, C. F., & Guthrie, C. (1993) *Science 262*, 1982–1988.

Limmer, S., Hoffman, H.-P., Ott, G., & Sprinzl, M. (1993) *Proc. Natl. Acad. Sci. U.S.A. 90*, 6199–6202.

Long D. M., & Uhlenbeck O. C. (1993) *FASEB J. 7*, 25–30.

McClain, W. H., Guerrier-Takeda, C., & Altman, S. (1987) *Science 238*, 527–530.

Michel, F., & Westhof, E. (1990) *J. Mol. Biol. 216*, 585–610.

Michel, F., Hanna, M., Green, R., Bartel, D. P., & Szostak, J. W. (1989) *Nature 342*, 391–395.

Morse, D. P., & Schmidt, F. J. (1992) *Gene 117*, 61–66.

Morse, D. P., & Schmidt, F. J. (1993) *J. Mol. Biol. 230*, 11–14.

Motamedi, H., Lee, K., Nichols, L., & Schmidt, F. J. (1982) *J. Mol. Biol. 162*, 535–550.

Nichols, L. N., & Schmidt, F. J. (1988) *Nucleic Acids Res. 16*, 2931–2942.

Nolan, J. M., Burke, D. H., & Pace, N. R. (1993) *Science 261*, 762–765.

Noller, H., Hoffarth, V., & Zimniak, L. (1992) *Science 256*, 1416–1419.

Pan, T., & Uhlenbeck, O. C. (1992) *Nature 358*, 560–563.

Perotta, A., & Been, M. D. (1991) *Nature 350*, 434–436.

Piccirilli, J. A., McConnell, T. S., Zaug, A. J., Noller, H. F., & Cech, T. R. (1992) *Science 256*, 1420–1424.

Pleij, C. W. A., & Bosch, L. (1989) *Methods Enzymol. 180*, 289–303.

Pley, H. W., Flaherty, K. M., & McKay, D. B. (1994) *Nature 372*, 68–74.

Pyle, A. M. (1993) *Science 261*, 709–714.

Ramamoorthy, R., & Schmidt, F. J. (1991) *FEBS Lett. 295*, 227–229.

Reich, D., Olsen, G. J., Pace, B., & Pace, N. R. (1988) *Science 239*, 178–181.

Schmidt, F. J., Seidman, J. G., & Bock, R. M. (1976) *J. Biol. Chem. 251*, 2440–2445.

Smith, D., & Pace, N. R. (1993) *Biochemistry 32*, 5273–5281.

Sontheimer, E. J., & Steitz, J. A. (1993) *Science 262*, 1989–1996.

Srivastava, R. A., Srivastava, N., & Apirion, D. (1991) *Biochem. Int. 25*, 57–65.

Steitz, J. A. (1992) *Science 257*, 888–889.

Steitz, T. A., & Steitz, J. A. (1993) *Proc. Natl. Acad. Sci. U.S.A. 90*, 6498–6502.

Svard, S. G., & Kirsebom, L. A. (1993) *Nucleic Acids Res. 21*, 427–434.

Symons, R. H. (1992) *Annu. Rev. Biochem. 61*, 641–671.

Tsuchihashi, Z., Khosla, M., & Herschlag, D. (1993) *Science 262*, 99–102.

Tuschl, T., & Eckstein, F. (1993) *Proc. Natl. Acad. Sci. U.S.A. 90*, 6991–6994.

Vold, B. S., & Green, C. J. (1988) *J. Biol. Chem. 263*, 14390–14396.

Waechterhaueser, G. (1988) *Microbiol. Rev. 52*, 452–488.

Waugh, D. S., & Pace, N. R. (1993) *FASEB J. 7*, 188–195.

Waugh, D. S., Green, C. J., & Pace, N. R. (1989) *Science 244*, 1569–1571.

Westheimer, F. W. (1987) *Science 235*, 1173–1178.

Wise, J. A. (1993) *Science 262*, 1978–1979.

Woese, C. R., & Pace, N. R. (1993) in *The RNA World* (Atkins, J. F., & Gesteland, R. F., Eds.), pp. 91–117, Cold Spring Harbor Laboratory, Cold Spring Harbor, NY.

Yarus, M. (1988) *Science 240*, 1751–1758.

Young, B., Herschlag, D., & Cech, T. R. (1991) *Cell 67*, 1007–1019.

Yu,Y.-T., Maroney, P. A., & Nilsen, T. W. (1993) *Cell 75*, 1049–1059.

Zaug, A. J., Kent, J. R., & Cech, T. R. (1984) *Science 224*, 574–578.

Zuker, M. (1989) *Science 244*, 48–52.

Chapter 14

Arnheim, N., & Erlich, H. (1992) *Annu. Rev. Biochem. 61*, 131–156.

Ballabio, A., Gibbs, R. A., & Caskey, C. T. (1990) *Nature 343*, 220.

Beaudry, A. A., & Joyce, G. F. (1992) *Science 257*, 635–641.

Berzal-Herranz, A., Simpson, J., & Burke, J. M. (1992) *Genes Dev. 6*, 129–134.

Birkenmeyer, L. G., & Mushahwar, I. K. (1991) *J. Virol. Methods 35*, 117–126.

Bock, L. C., Griffin, L. C., Latham, J. A., Vermass, E. H., & Toole, J. J. (1992) *Nature 355*, 564–566.

Chong, S. S., Kristjansson, K., Cota, J., Handyside, A. H., & Hughes, M. R. (1993) *Hum. Mol. Genet. 2*, 1187–1191.

Chou, Q., Russell, M., Birch, D. E., Raymond, J., & Bloch, W. (1992) *Nucleic Acids Res. 20*, 1717–1723.

Chowrira, B. M., Berzal-Herranz, A., & Burke, J. M. (1991) *Nature 354*, 320–321.

Cui, X., Li, H., Goradia, T. M., Lange, K., & Kazazian, H. H., Jr., Galas, D., & Arnhein, N. (1989) *Proc. Natl. Acad. Sci. U.S.A. 86*, 9389–9393.

Cui, X. F., Gerwin, J., Navidi, W., Li, H. H., Kuehn, M., & Arnheim, N. (1992) *Genomics 13*, 713–717.

de Bruijn, M. H. L. (1988) *Parasitol. Today 4*, 293–295.

Donis-Keller, H., Green, P., Helms, C., Cartinhour, S., Weiffenbach, B., Stephens, K., Keith, T. P., Bowden, D. W., Smith, D. R., Lander, E. S., Botstein, D., Akots, G., Rediker, K. S., Gravius, T., Brown, V. A., Rising, M. B., Parker, C., Powers, J. A., Watt, D. E., Kauffman, E. R., Bricker, A., Phipps, P., Muller-Kahle, H., Fulton, T. R., Ng, S., Schumm, J. W., Braman, J. C., Knowlton, R. G., Barker, D. F., Crooks, S. M., Lincoln, S. E., Daly, M. J., & Abrahamson, J. (1987) *Cell 51*, 319–337.

Ecker, D. J., Vickers, T. A., Hanecak, R., Driver, V., & Anderson, K. (1993) *Nucleic Acids Res. 21*, 1853–1856.

Erlich, H. A., & Arnheim, N. (1992) *Annu. Rev. Genet. 26*, 479–506.

Ellington, A. D., & Szostak, J. W. (1990) *Nature 346*, 818–822.

Ellington, A. D., & Szostak, J. W. (1992) *Nature 355*, 850–852.

Famulok, M., & Szostak, J. W. (1992) *J. Am. Chem. Soc. 114*, 3990–3991.

Ferre, F. (1992) *PCR Methods Appl. 2*, 1–9.

Friedman, K. J., Highsmith, W. E., Jr., Prior, T. W., Perry, T. R., & Silverman, L. M. (1990) *Clin. Chem. 36*, 695–696.

Funk, W. D., Pak, D. T., Karas, R. H., Wright, W. E., & Shay, J. W. (1992) *Mol. Cell. Biol. 12*, 2866–2871.

Green, R., & Szostak, J. W. (1992) *Science 258*, 1910–1915.

Handyside, A. H., Kontogianni, E. H., Hardy, K., & Winston, R. M. L. (1990) *Nature 344*, 768–770.

Holodniy, M., Katzenstein, D. A., Sengupta, S., Wang, A. M., Casipit, C., Schwartz, D. H., Konrad, M., Groves, E., & Merigan, T. C. (1991) *J. Infect. Dis. 163*, 862–866.

Holodniy, M., Winters, M. A., & Merigan, T. C. (1992) *Biotechniques 12*, 37–39.

Horwitz, M. S. Z., & Loeb, L. A. (1986) *Proc. Natl. Acad. Sci. U.S.A. 83*, 7405–7409.

Innis, M. A., Gelfand, D. H., Sninsky, J. J., & White, T. J., Eds. (1990) *PCR Protocols: A Guide to Methods and Applications*, pp. 159–166. Academic Press, San Diego.

Lawyer, F. C., Stoffel, S., Saiki, R. K., Chang, S.-Y., Landre, P. A., Abramson, R. D., & Gelfand, D. H. (1993) *PCR Methods Appl. 2*, 275–287.

Li, H., Gyllensten, U. B., Cui, X., Saiki, R. K., Erlich, H. A., & Arnhein, N. (1988) *Nature 335*, 414–417.

Longo, M. C., Berninger, M. S., & Hartley, J. L. (1990) *Gene 93*, 125–128.

Lyons, J. (1992) *Cancer 69*, 1527–1531.

Macaya, R. F., Schultze, P., Smith, F. W., Roe, J. A., & Feigon, J. (1993) *Proc. Natl. Acad. Sci. U.S.A. 90*, 3745–3749.

Mavrothalassitis, G., Beal, G., & Papas, T. S. (1990) *DNA Cell Biol. 9*, 783–788.

Mullis, K. B. (1991) *PCR Methods Appl. 1*, 1–4.

Myers, T. W., & Gelfand, D. H. (1991) *Biochemistry 30*, 7661–7666.

Oliphant, A. R., & Struhl, K. (1987) *Methods Enzymol. 155*, 568–582.

Oliphant, A. R., Brandl, C. J., & Struhl, K. (1989) *Mol. Cell. Biol. 9*, 2944–2949.

Pääbo, S., Gifford, J. A., & Wilson, A. C. (1988) *Nucleic Acids Res. 16*, 9775–9787.

Pan, T., & Uhlenbeck, O. C. (1992) *Biochemistry 31*, 3887–3895.

Rychlik, W., Spencer, W. J., & Rhoads, R. E. (1990) *Nucleic Acids Res. 18*, 6409–6412.

Saiki, R. K., Scharf, S., Faloona, F., Mullis, K. B., Horn, G. T., Erlich, H. A., & Arnheim, N. (1985) *Science 230*, 1350–1354.

Schneider, D., Gold, L., & Platt, T. (1993) *FASEB J. 7*, 201–207.

Thomas, R. H., Schaffner, W., Wilson, A. C., & Pääbo, S. (1989) *Nature 340*, 465–467.

Tsai, D. E., Harper, D. S., & Keene, J. D. (1991) *Nucleic Acids Res. 19*, 4931–4936.

Tsai, D. E., Kenan, D. J., & Keene, J. D. (1992) *Proc. Natl. Acad. Sci. U.S.A. 89*, 8864–8868.

Tuerk, C., & Gold, L. (1990) *Science 249*, 505–510.

Tuerk, C., MacDougal, S., & Gold, L. (1992) *Proc. Natl. Acad. Sci. U.S.A. 89*, 6988–6992.

Tuerk, C., MacDougal, S., & Green, L. (1993) *Clin. Chem. 39*, 722.

Wang, K. Y., McCurdy, S., Shea, R. G., Swaminathan, S., & Bolton, P. H. (1993) *Biochemistry 32*, 1899–1904.

White, B. A., Ed. (1993) *Methods in Molecular Biology, Vol 15; PCR Protocols: Current Methods and Applications*, pp. 205–215. Humana Press, Clifton, NJ.

Zhang, L., Cui, X., Schmitt, K., Hubert, R., Navidi, W., & Arnheim, N. (1992) *Proc. Natl. Acad. Sci. U.S.A. 89*, 5847–5851.

Index